POLYATOMIC MOLECULES

POLYATOMIC MOLECULES

Results of *ab Initio* Calculations

ROBERT S. MULLIKEN

DEPARTMENT OF CHEMISTRY
THE UNIVERSITY OF CHICAGO
CHICAGO, ILLINOIS

WALTER C. ERMLER

DEPARTMENT OF CHEMISTRY AND CHEMICAL ENGINEERING
STEVENS INSTITUTE OF TECHNOLOGY
HOBOKEN, NEW JERSEY

1981

ACADEMIC PRESS

A Subsidiary of Harcourt Brace Jovanovich, Publishers

New York London Toronto Sydney San Francisco

ACADEMIC PRESS, INC.
111 Fifth Avenue, New York, New York 10003

United Kingdom Edition published by
ACADEMIC PRESS, INC. (LONDON) LTD.
24/28 Oval Road, London NW1 7DX

Library of Congress Cataloging in Publication Data
Main entry under title:

Polyatomic molecules.

Includes bibliographical references.
1. Molecular structure. I. Mulliken, Robert
Sanderson. II. Ermler, Walter C.
QD461.P64 541.2'2 80-2764
ISBN 0-12-509860-X AACR2

PRINTED IN THE UNITED STATES OF AMERICA

81 82 83 84 9 8 7 6 5 4 3 2 1

CONTENTS

PREFACE

In recent years a large number of quantum chemical computations on the structure and properties of molecules have been published. W. G. Richards and collaborators have issued an exhaustive bibliography through 1977 of *ab initio* calculations, with information as to computed energies and, in certain cases, computed properties, and shortly are publishing a supplement through 1980. However, aside from some reviews on limited groups of molecules, no comprehensive detailed review of the results of *ab initio* computations has been published.

In 1977 we published such a review for selected diatomic molecules, for papers through 1976. The present volume is a somewhat similar report on results through 1979 for polyatomic molecules. While far from covering all the molecules listed in the Richards bibliography, the coverage in this volume is much more comprehensive than in our diatomic volume. Radicals, known and hypothetical reaction intermediates, and reaction paths and mechanisms are included. Ground and many excited and ionized states are discussed. Emphasis is placed on the most accurate of existing computations. Criteria (methods of computation, basis sets, etc.) for judging the accuracy of the computations are given in nearly every case. The most accurate results are for small molecules. With increasing molecular size, one must be content with increasingly poorer approximations to accurate theory. Semiempirical computations are excluded here as being less reliable and largely only on ground states.

The volume, covering nearly all of the most important polyatomic molecules on which computations have been done, should be helpful especially to experimental chemists wishing to know the current results of *ab initio* theoretical computations. Many comparisons with experiment are made, but also, predictions are made for experimentally unknown or little known systems. For theoretical chemists interested in making new computations, this survey should be valuable as a guide to what has been done and what needs to be done. It should be useful to biochemists, physicists, and other scientists in giving a general idea of the scope and comprehensiveness of *ab initio* computations to date on polyatomic molecules, as well as dealing with particular molecules in which they may be interested. The volume is written on a level intended to be comprehensible to beginning graduate students in chemistry and physics.

In our review, earlier papers, regardless of their quality, have not been cited when later papers (mostly since 1970) cover the main results more accurately. References to these earlier papers may be found in the papers cited. We have omitted reference to numerous computations of small or moderate accuracy on relatively small molecules that are summarized in a book in preparation by Pople and collaborators. Because of the very large number of papers reviewed, our consideration has often had to be superficial.

The distribution of molecules among chapters of the book is for the most part not in terms of chemical relationships, but is based to a large extent on symmetry in ground states. Symmetry is of dominant importance in computations and in molecular orbital and state symbols. Excited and ionized states are rather extensively discussed, but if not otherwise stated reference is to the ground state. Chapter I discusses the various methods of calculation used. Chapters II–V discuss three-atom systems, differing with respect to the number of H atoms in the molecules. Chapter V deals with reaction surfaces. Chapters VI–VIII discuss molecules whose ground states belong to relatively high symmetry groups and Chapters IX and X consider molecules with little or no symmetry. Chapters II–X include organic and inorganic molecules without discrimination. Relatively large organic molecules are treated in Chapter XI and Chapter XII deals with molecular compounds and complexes, including hydrogen bonding and hydration of ions.

The style varies from section to section, but everywhere information is first given to serve as a criterion of reliability or accuracy of the computation. Various computational results are then presented, especially those relevant to spectroscopic transitions (excitation energies and oscillator strengths) and to ionization. Other computed molecular properties are also cited. Comparisons of the theoretical calculations with experimental evidence are made, but details of experimental methods are omitted; for these, the theoretical papers themselves may be consulted. Our discussions often lack specification of the exact molecular dimensions assumed in the computations; reference must be made to the original papers for this information. Sometimes experimental values are used, but often the geometry is that obtained by an energy-minimizing computation, either an approximate preliminary one or a more exact one. Note that computed excitation energies are in many cases those for vertical excitation only.

Since this book was written, a volume by Čársky and Urban has appeared that treats some of the material in Chapter I.

We believe that this survey of current progress through 1979 gives a panorama of what needs to be done, and what can be done, especially with the help of the newly developed CM and MCSCF methods mentioned in Chapter I. In this endeavor, the work of the recently terminated National Resource for Computation in Chemistry should be carried further in a collaboration of chemists, physicists, and others. (See, for example, the report by R. M. Baum in *Chemistry & Engineering News,* Vol. 59, No. 19, p. 28, 1981.)

LIST OF ABBREVIATIONS

AO	Atomic orbital
APSG	Antisymmetrized product of strongly orthogonal geminals
CCA	Coupled cluster approaches
CEPA	Coupled electron pair approximation
CGTF	Contracted Gaussian-type function
CI	Configuration interaction
CIMI	Configuration interaction from molecular integrals
CIPSI	Configuration interaction by perturbation selection by iterative process
CM	Configuration mixing
CPMET	Coupled pair many electron theory
CSF	Configuration state function
CT	Charge transfer
DZ	Double zeta
EOM	Equations of motion
EHP	Electron-hole potential
FSGO	Floating spherical Gaussian orbital
FOCI	First-order configuration interaction
GBT	Generalized Brillouin's Theorem
GLF	Gaussian-lobe function
GTF	Gaussian-type function

GTO	Gaussian-type orbital
GUGA	Graphical unitary group approach
GVB	Generalized valence bond
GVB-PP	Generalized valence bond perfect pairing
HF	Hartree-Fock
HO	Hamiltonian operator
ICSCF	Internally-consistent self-consistent field
IEPA	Independent electron pair approximation
INO	Iterative natural orbital
IVO	Improved virtual orbitals
LCAO	Linear combination of atomic orbitals
LCGLF	Linear combination of Gaussian-lobe functions
LCGTF	Linear combination of Gaussian-type functions
LCSTF	Linear combination of Slater-type functions
LMO	Localized molecular orbital
MAO	Modified atomic orbital
MBPT	Many-body perturbation theory
MCSCF	Multiconfiguration self-consistent field
MECE	Molecular extra correlation energy
MO	Molecular orbital
MRD-CI	Multireference double configuration interaction
NO	Natural orbital
OVC	Optimized valence configurations
PNO	Pair (or pseudo) natural orbitals
POL-CM	Polarization configuration mixing
RHF	Restricted Hartree-Fock
SCEP	Self-consistent electron pairs
SCF	Self-consistent field
STF	Slater-type function
STO	Slater-type orbital
UHF	Unrestricted Hartree-Fock
VB	Valence bond
VCI	Valence configuration interaction

CHAPTER I

INTRODUCTION AND COMPUTATIONAL METHODS

A. GENERAL CONSIDERATIONS

In their discussion of the status of molecular quantum mechanics in 1959 Mulliken and Roothaan referred to "bottlenecks" in the evolution from formalism and qualitative understanding to quantitative applications [1]. Nearly 20 years later in a previous volume on diatomic results [2], and now in this book, we survey the results of the enormous number of applications to molecules that have essentially paralleled the technological developments of the computer age. In 1959, the bottleneck in the "pipeline" from theory to applications was located at the point where the evaluation of electronic repulsion integrals involving two, three, and four nuclear positions was required. It was reported at that time [1] that a minimal basis set SCF calculation for N_2 that took about one year using desk calculators could be accomplished in 35 minutes on the most advanced computer. Today such a calculation requires a few seconds. The situation is now more complicated in that the nature and position of the bottleneck depends on the problem at hand. It may involve the requirement of a large basis set, the need for extensive configuration mixing, integrals transformations, matrix diagonalizations, or combinations of these, or other stages in the calculation. Clearly there are in many cases several bottlenecks in the pipeline. The natures of these are discussed in the following sections.

The first chapter of our earlier volume set forth more or less formally the classes of computational approaches that have been used to obtain the results discussed in the later chapters. Namely, the self-consistent field (SCF), multiconfiguration self-consistent field (MCSCF), and configuration mixing (CM) methods were introduced and briefly illustrated. The recently published series *Modern Theoretical Chemistry,* especially Volume 3, *Methods of Electronic Structure Theory* [3], and Volume 4, *Applications of Electronic Structure Theory* [4], is an excellent reference for more detailed treatises on many of the approaches presently used.

In this chapter we review the methods used for the calculations whose results are discussed in subsequent chapters. Our discussion will be limited to *ab initio* methods, by which we mean that no integrals have been neglected or approximated and no empirical parameters employed. This we do by design, as in our diatomics volume [2], and in no way do we mean to imply that semiempirical methods do not have an important place in the build-up of detailed knowledge of molecular structure and spectra. Many systems are presently accessible only through such approaches. In fact, it is difficult in many cases to justify the preference of, say, minimal basis set (see below) results over those derived from well-conceived semiempirical approaches. Volumes 7 and 8 of *Modern Theoretical Chemistry* [5, 6] contain excellent reviews of semiempirical methods and the range of their applications. The article [5] by Freed on the "Theoretical Basis for Semiempirical Theories" is noteworthy. A recent book on valence electron methods analyzes many empirical techniques [7].

There has been a recent resurgence of qualitative molecular orbital theory, an area that, although pursued by necessity before the computer age, now benefits from the existence of accurate *ab initio* results. A recent review article [8] contains references to the early efforts of workers in this area (also included are references to many of the popular semiempirical methods). Two new books with somewhat modified approaches have been published [9].

Returning to the goals of the present book, we wish to make a fairly comprehensive survey and analysis of *ab initio* calculations on polyatomic molecules. Three volumes of Specialist Periodical Reports that include literature references to 1977 should be mentioned as related efforts [10]. In addition to independent literature searching, the bibliography of Richards *et al.* on *ab initio* wave functions, including two supplements [11], was consulted.

A complete survey of the studies by the group at Carnegie-Mellon University dealing primarily with SCF calculations of molecules composed of light atoms (mostly first and second row) using minimal or moderately larger basis sets is in preparation [12]. Consequently, we have chosen to omit discussions of many noteworthy publications of this group, especially in cases where entire classes of molecules were treated and trends enumerated. The book by Snyder and Basch [13] lists double-zeta SCF results including energies, MO coefficients, population analyses, and one-electron properties for 56 molecules containing H through F and up to 48 electrons.

Schaefer has partially addressed the objectives of the present study in his 1972 book [14] and his 1976 review article [15]. The present volume, while covering many more recent studies, is more comprehensive as to the number of molecules but deals much less with methods and much more with the results of calculations.

The book progresses from the smallest to the largest polyatomic molecules that have been studied by *ab initio* methods. Each chapter begins with some general remarks concerning the class of molecules to be discussed, including group theoretical character tables that define the irreducible representations used to classify electronic states and molecular orbitals. Specific calculations are then discussed. Total energies, given in Hartree units E_h [16], basis set size, and the degree of configuration mixing are usually given as criteria for judging the validity of the conclusions. Other properties are discussed as warranted by the specific interest in the molecule or class.

Polyatomic molecules, although having many characteristics that may be viewed as straightforward extensions of similar properties in diatomic molecules, are inherently more complicated because of the added vibrational and rotational degrees of freedom. Total and orbital energies must be classified according to the point group symmetry of the molecule, with electron configurations similar in form to those of diatomic molecules except that diatomic (and linear molecule) orbital designations σ, π, δ, etc. are replaced by the appropriate (lower case) symbols for irreducible representations.

New features not shared by diatomic molecules involve intramolecular (vibrational) motion in normal coordinates, the complications of many-dimensional potential energy surfaces in the determination of equilibrium geometries and reaction paths, vibronic and rovibronic interactions (Jahn–Teller and Renner–Teller effects), internal rotation, and intermolecular forces. Herzberg's Volume III, *Electronic Spectra and Electronic Structure of Polyatomic Molecules* [17], gives a clear discussion of these phenomena. His book dealing with free radicals [18] also contains useful analyses of many topics relevant to polyatomic molecules. Other more recent works on the general subject of molecular spectroscopy may also be consulted [19] as may references on specific experimental techniques [20] and series and compendia of experimental results [21].

The *ab initio* studies discussed in Chapters II–XII are presented in such a way as to highlight the most accurate calculations and the important or novel characteristics of each molecule. To this end three categories of results are presented: energies, wave functions, and properties.

Energy quantities of interest are the total energy and orbital energies (approximately related to ionization potentials through Koopmans' theorem [2, 3]). Also discussed are excitation energies, both vertical and in many cases adiabatic (i.e., the difference between the energy of a lower electronic state at its equilibrium geometry and that of an upper state at its equilibrium geometry), ionization potentials and electron affinities, bond dissociation energies, barrier heights and activation energies, and intra- and intermolecular interaction energies.

Wave functions are generally more difficult to report because of the large number of parameters involved in their definition. The total wave function from a large CM calculation (see Section D) would require the tabulation of the atomic basis set, the LCAO-MO expansion coefficients (see Section B), and the CM mixing coefficients from the eigensolution of the Hamiltonian matrix for the state in question. Only in simple cases is it usually possible to draw conclusions regarding the bonding, charge distribution, etc. from such detailed data. Often, however, CM coefficients provide useful information about the nature of an electronic state or its dissociation properties. Two- and three-dimensional contour maps of charge densities and corresponding difference maps relative to the constituent atoms are especially useful for displaying information contained in the wave functions [14, 22]. Compton profiles provide another means of analyzing molecular wave functions in that use is made of the density matrix in momentum space [23].

The broad category of properties may be broken down into geometric properties, one- and two-electron properties, spectroscopic constants, and spectroscopic transition probabilities. Geometric properties include equilibrium bond lengths and angles as well as important intermediate positions corresponding to saddle points and barrier peaks in reaction pathways. By "one-electron properties" are meant those that may be calculated from a molecular wave function using a one-electron operator [2–4, 13]. Typical examples are dipole and higher molecular moments, Hellmann–Feynman forces (electric field), electric field gradients, diamagnetic shielding constants, diamagnetic susceptibilities, charge densities at the nucleus, and approximate population analyses [2, 8]. Two-electron properties [24], which generally require a perturbation expansion, include polarizabilities, paramagnetic shielding and susceptibility tensors, hyperfine structure constants, and various spin and orbital interaction properties. Relativistic effects are also considered in a few cases (see Pyykko [24] and Pitzer [24]).

Spectroscopic constants generally refer to the spacings of vibrational and rotational levels in bound electronic states [19, 25]. The constants are defined in terms of a power series expansion in the vibrational quantum numbers corresponding to each normal mode and the rotational quantum numbers for each of the rotational degrees of freedom. The term energy for an electronic state is given by the general formula [25]

$$T = G(v_1, v_2, v_3, \ldots) + F_{[v]}(J) \tag{1}$$

In Eq. (1) the pure vibrational term energy is a sum over the normal modes having degeneracy d_i

$$G(v_1, v_2, v_3, \ldots) = \sum \omega_i(v_i + d_i/2) + \sum_{i,k} x_{ik}(v_i + d_i/2)(v_k + d_k/2) \tag{2}$$

In Eq. (2) ω_i are the harmonic or classical frequencies and the x_{ik} are pure vibrational anharmonicity constants. The term $F_{[v]}(J)$ is a function of the rotational quantum numbers and involves the principal moments of inertia $A_e \geqslant B_e \geqslant C_e$. (The subscript denotes equilibrium values.) For example, for a spherical top molecule $A_e = B_e = C_e$ and

$$F_{[v]}(J) = B_{[v]}J(J + 1) + D_{[v]}J^2(J + 1)^2 \tag{3}$$

where

$$B_{[v]} = B_e - \sum_i \alpha_i(v_i + d_i/2) \tag{4}$$

provides a vibrational–rotational interaction correction and $D_{[v]}$ a centrifugal distortion correction. More complicated formulas arise for symmetric and asymmetric tops [25].

Transition probabilities are the expectation values of the dipole moment or related velocity or acceleration operator with respect to the two electronic and/or vibrational or rotational states in a spectroscopic transition [17]. The simplest application yields selection rules defining electronic-dipole allowed and forbidden transitions. The magnitude of the expectation value is proportional to the transition intensity or the oscillator strength or the inverse of the radiative lifetime.

The remainder of this chapter is devoted to discussions of computational methods now in use for *ab initio* studies of molecular electronic structure, spectra, and reaction surfaces. Fundamental aspects of all such studies, namely basis sets and general formalistic approaches, are reviewed in the next two sections. More specific methods built on these fundamentals are presented in the next sections, with a short summary of some promising new advances in the final section.

B. BASIS SETS

Most of the wave functions to be discussed in later chapters are constructed in some form from molecular orbitals (MOs). The MOs in turn are approximated as linear combinations of basis functions. Thus, the accuracy of a molecular wave function can be first assessed by examination of the quality or flexibility of the basis set. The two types of basis functions generally used are Slater-type functions (STFs) and Gaussian-type functions (GTFs). The normalized STFs in spherical polar coordinates are defined as

$$\chi_{nl}{}^{\mathrm{S}} = N_{\mathrm{S}} r^{n-1} \exp(-\zeta_{nl} r) Y_{lm}(\theta, \phi), \tag{5}$$

where n, l, and m are principal, azimuthal, and magnetic quantum numbers, ζ_{nl} an exponential prefactor or "orbital exponent," and $Y_{lm}(\theta, \phi)$ a spherical harmonic. Linear combinations of STFs are used in approximating atomic orbitals (AOs).

GTFs used in molecular calculations are usually expressed in Cartesian coordinates as

$$\chi_{ijk}^{\mathrm{G}} = N_{\mathrm{G}} x^i y^j z^k \exp(-\zeta_{ijk} r^2) \tag{6}$$

where i, j, and k are positive integers. In Eqs. (5) and (6) the independent variables (r, θ, ϕ) or (x, y, z) refer to the displacement of an electron from chosen (e.g., nuclear position) point of reference and N_{S} and N_{G} are appropriate normalization factors. Gaussian-lobe functions (GLFs) are special cases of Eq. (6) with $i = j = k = 0$, i.e., only the $\exp(-\zeta r^2)$ portion is used and several such lobes are placed at appropriate positions to approximate the angular characteristics of the spherical harmonics.

In the early calculations, MOs were constructed as linear combinations of AOs in which the AO was represented by a single STF, hence the name LCAO-MO. Such a basis set used in approximating an MO is referred to as a *minimal* (AO) *basis set*. For molecular ground states, there are usually as many LCAO-MOs as AOs in the corresponding free atom. In a few cases an additional LCAO-MO is added, corresponding, e.g., to the 2p AO for Li and Be. In MO calculations more accurate than those with a minimal basis set, additional STFs needed for the accurate SCF representation of the atomic-occupied AOs are included (properly termed LCSTF-MOs, although the designation LCAO-MO is still often used). Or instead of LCSTF-MOs one may use LCGTF-MOs in which a set of GTFs have been obtained by expansions of either the relevant minimal set of AOs or of the STFs used to approximate self-consistent field AOs. Thus the basis sets we are discussing here consist of building blocks of approximate AOs, and may be called AO basis sets in contradistinction to the MO basis sets used in CM (see Section D).

We now discuss several details about the construction of basis sets of STFs or GTFs to be used in *ab initio* molecular calculations. Ideally a basis set of adequate size is chosen for a particular state of a molecule and the ζ's are optimized for each geometric configuration at the highest level of approximation to be ultimately employed. This is, of course, extremely time consuming and experience has shown that good results may be obtained using some rather standard procedures [26]. Zeta optimizations carried out on the atoms (i.e., SCF AOs, as discussed above) have generally proven to be adequate for calculations on molecules for ground or valence excited states, but of course the coefficients of the STFs or GTFs are different in the MOs than in the AOs.

However, accurate calculations for molecules require the addition of "polarization functions," i.e., STFs or GTFs having higher l quantum numbers than the occupied atomic orbitals, or alternatively, sometimes functions (usually called "bond functions") centered at locations other than the nuclear positions. Polarization functions are optimized in molecular calculations or are obtained via atomic calculations in which polarizability is made optimum. Further, for computations on negative ions and on Rydberg states of molecules, it is necessary also to include in the basis set functions with relatively small ζ values; these are often called "diffuse functions" [26].

Although STFs are more accurate than GTFs for nonlinear polyatomic molecules, they have been used infrequently because of the present nonavailability of efficient programs for computing the required integrals. However, the POLYCAL program of Stevens [27] based on the Gaussian-transform method [28] for computing integrals over STFs is fairly efficient for basis sets containing only s- and p-type functions or those with d-type functions on one of the nuclei. Minimal basis set calculations using this program have been published for large molecules [29]. Linear polyatomic molecules constitute a special class where integrals involving STFs can be evaluated rapidly. For these, as for diatomic molecules, the ALCHEMY program has been used for highly accurate calculations using large basis sets, including f- and g-type functions [30].

The vast majority of *ab initio* calculations which have been made on polyatomic molecules use Gaussian basis sets. The advantages due to the facilitation of multicenter integral evaluation were described by Boys in 1950 [31] and several efficient integrals programs are in use. These include

IBMOL [32], POLYATOM [33], BIGGMOLI [34], GAUSSIAN 70 [35], MOLECULE [36], and, recently, HONDO [37]; all using the general Cartesian Gaussians of Eq. (6). Several programs that use GLFs have also been written [38]. An approach that uses "subminimal" basis sets of GLFs, termed "floating spherical Gaussian orbitals" (FSGOs), whereby a single FSGO is used for each electron pair, has been employed for large molecules [39]. Several other computer programs using GTFs are also in existence [40].

The use of GTFs in molecular calculations is rendered less time consuming with little loss in accuracy by "contracting" them; i.e., linear combinations of "primitive" GTFs with coefficients fixed at their atom-optimized AO values (CGTFs) are used to reduce the basis set size at the stage in the calculation, e.g., SCF or MCSCF (see Sections D and E), where the MOs are constructed (see [26] and Whitten [38]). In this way an analogy to STFs is made using CGTFs.

The terms "minimal basis set," "double-zeta basis set," containing two STFs for each occupied AO, or a suitable corresponding set of CGTFs; "triple-zeta," etc. are used to refer to either STFs or corresponding CGTFs. As atomic basis sets become larger and larger (extended basis sets) in SCF calculations the so-called Hartree–Fock limit (or, better, SCF limit) is approached; the energy approaches its LCSCF-MO SCF minimum. Several tabulations of basis sets for atoms using STFs and GTFs from minimal to extended are available [13, 26].

Beyond double-zeta, at least, good basis sets must include polarization and in some cases diffuse functions (see above). Several studies have been made, notably those of McLean and Yoshimine using STFs [41] and those of Clementi and Popkie using GTFs [42], in which the basis sets were systematically improved to approach the SCF limit.

The energy of the SCF limit is a number that provides a point of reference for the definition of the "correlation energy" of a system. This is the difference between the exact nonrelativistic energy and the energy computed in the Hartree–Fock approximation [43]. However, the SCF limit itself is of considerable importance in *ab initio* predictions of heats of reaction. It was noted by Snyder in 1967 [44] that for certain reactions, namely those in which the correlation energy is essentially the same for reactants and products, reasonable estimates of heats of reactions can be calculated from SCF energies. The necessary condition is approximately fulfilled if the electron configurations of reactants and products consist of MO closed shells. The general subject has been reviewed by Hurley [45]. Recently George *et al.* have extended this approach by making systematic estimates of SCF limit energies of numerous molecules to determine the effects of electron correlation on reaction heats [46].

The basis sets in general use are quite varied. As mentioned above, probably the most common and simplest are the minimal basis sets. Since only a single STF or contracted GTF is used to represent each atomic subshell, such basis sets lack flexibility and should ideally be ζ-optimized for each geometry and electronic state of the molecule [27]. A systematic study in which SCF-MO calculations on a large number of molecules have been performed indicates that certain properties, such as equilibrium geometries, are generally predicted fairly well using minimal basis sets with exponents that correspond to an average of values optimized for representative molecules [12, 47]. Since calculations using such standardized basis sets are often reported in later chapters, a brief discussion is given here.

Instead of using STFs, one can get nearly the same results by using least squares fits of STFs by linear combinations of GTFs. These have been designated [47] as STO-KG, where $K = 2, 3, \ldots$ is the number of GTFs chosen to represent an STF having a particular exponent ζ. The GAUSSIAN 70 program [35] has been made especially efficient by using simplifying relationships that arise when all ζ values for principal quantum number n are the same, e.g., $\zeta_{2p} = \zeta_{2s}$ etc. However, it should be noted that this practice results in a basis set that should be strictly classified as being of less than minimal quality. There are then *two* approximations inherent in the STO–NG minimal basis as commonly used. The first being the NG expansion error and the second the restriction of the same ζ for all functions having the same n value. The STO-3G minimal basis has been found to yield the most reasonable compromise between accuracy and computational effort for molecules involving atoms H through F [47].

To include more flexibility in the atomic basis sets two extensions were proposed. The first [48], termed the "split-valence" basis, designated as 4-31G, corresponds to representing the inner shells of the heavy atoms (B to F) using a 4-term expansion of the 1s AO and a four-term expansion of the valence 2s and 2p AOs, after which the three GTFs having the largest ζ are contracted together and the GTF having the smallest ζ is left uncontracted. The second extension is more nearly comparable to the double-zeta STF basis set and is termed 6-31G, in which six-term expansions of the 1s AOs of the heavy atom are used together with four-term expansions of the 2s and 2p AOs contracted as in the 4-31G [48]. In each of the extended bases the 1s orbital of the H atom is represented as a four-term expansion contracted so that the three GTFs having the largest ζ's comprise the first CGTF and that having the smallest ζ the second. The inclusion of polarization functions, one set of 3d GTFs on the heavy atoms, is indicated by appending one asterisk (e.g., 4-31G*) and the inclusion of both 3d GTFs on the heavy atoms and one set of 2p GTFs on H are indicated using two asterisks (e.g., 4-31G** [50]). For increased accuracy it has been found that alternative 6-21G and, especially, 3-21G split-valence basis sets are preferable (see Binkley *et al.* [48]).

Studies on a variety of closed-shell molecules of the type AH_n have indicated that the minimal or STO-3G basis sets predict bond lengths and bond angles generally to within about 0.03 Å (too long) and 4° (too small), respectively. Increasing the basis set to the 4-31G level improves the bond lengths to a mean deviation of about 0.01 Å (too short) and angles within about 5° (too large) [47, 48]. The addition of polarization functions is important in the accurate prediction of energies [45, 46, 50] and often for geometries, especially with regard to dihedral angles of many molecules. The inclusion of CM is required for definitive geometry predictions in many cases [51].

Instead of using expansions of STFs in terms of GTFs, it is also a generally accepted procedure to incorporate basis sets of GTFs that have been directly optimized in atomic SCF calculations and contract them using the relative weights in the AOs to form minimal, double-zeta, etc. basis sets. This procedure has been reviewed by Dunning and Hay [26] and by Snyder and Basch [13]. Such procedures lead to what Raffenetti has termed "segmented" CGTF basis sets [52], in which each "primitive" GTF of the atomic basis can occur in only one CGTF of the final basis set. He has proposed a scheme for "general contraction" of primitive GTF atomic basis sets whereby the AO coefficients are used to construct CGTFs, each of which can contain contributions from *all* of the primitive GTFs. This procedure is in some respects a means of generating MOs as linear combinations of modified AOs (MAOs) provided the MAOs are calculated in some "prepared" or hybrid form [2].

Although the notation for CGTF basis sets is not universal, it is usually given in a form whereby the primitive basis for the heavy atoms is listed first, followed by that for the light atoms (usually H), then the respective contracted forms [26]; e.g., for NH_3, (9s 5p 1d/4s 1p)/⟨4s 2p 1d/2s 1p⟩ defines a double-zeta plus polarization CGTF basis set. The sensitivity of molecular properties to the quality of CGTF basis sets continues to be an area where detailed studies are being carried out [53].

An alternative to the standard independently optimized STF and GTF basis sets has been presented by Ruedenberg and co-workers [54]. Termed "even-tempered' basis sets, they are defined as sets of STFs or GTFs with their ζ's in geometric progressions. The ζ's for each atomic subshell or symmetry species use two parameters which are optimized; thus $\zeta_i = \alpha_i\beta_i^k$, with $\alpha_i > 0$, $\beta_i > 1$, and $k = 1, 2, \ldots, m$ for a basis set of m functions for species i. Such a scheme offers the advantage that only two parameters need to be optimized for each species regardless of the size m of the expansion. A recent study indicates a systematic procedure for obtaining SCF-limit results for atoms using even-tempered expansions of GTFs (see Schmidt and Ruedenberg [54] and Feller and Ruedenberg [54]).

C. SELF-CONSISTENT FIELD (SCF) METHODS

As mentioned above, the bottleneck in molecular structure studies using *ab initio* methods depends on what properties are to be determined in a given series of calculations. Presently, for large molecules or for large basis set studies on even relatively small molecules the bottleneck is at the stage where electron repulsion integrals must be evaluated and processed. Consequently SCF-MO studies (for closed-shell systems with only a single Slater determinant or for open-shell systems with a fixed linear combination of a small number of Slater determinants) still constitute the largest class.

The formalism underlying the closed and open shell SCF-MO procedure was outlined in our first volume and extensive references were listed [2]. To establish notation used in later chapters note that each MO may be approximated as a linear combination of m basis functions (STFs, GTFs, CGTFs, etc.)

$$\phi_i = \sum_p^m \chi_p c_{pi} \tag{7}$$

The total closed-shell molecular wave function is an antisymmetrized product (Slater determinant) of molecular spinorbitals: one-electron functions ψ_i that are each a product of an MO and a one-electron spin function. Thus for an N electron molecule

$$\Psi = \hat{A} \prod_i^N \psi_i(i) \tag{8}$$

where $\hat{A}$ is the antisymmetrization operator.

The total energy of a closed-shell molecule represented by $n = N/2$ MOs is obtained using the restricted Hartree–Fock (RHF) procedure

$$E^{\text{SCF}} = 2 \sum_i^n h_i + \sum_{i,j}^n (2J_{ij} - K_{ij}) \tag{9}$$

In Eq. (9) h_i are the one-electron energies and J_{ij} and K_{ij} the Coulomb and exchange energies:

$$h_i = \langle\phi_i|\hat{h}|\phi_i\rangle = \langle\phi_i|\hat{T} + \hat{V}_{\mathrm{na}}|\phi_i\rangle \tag{10a}$$

$$J_{ij} = \langle\phi_i(1)\,\phi_j(2)\left|\frac{1}{r_{12}}\right|\phi_i(1)\,\phi_j(2)\rangle \tag{10b}$$

$$K_{ij} = \langle\phi_i(1)\,\phi_j(2)\left|\frac{1}{r_{12}}\right|\phi_j(1)\,\phi_i(2)\rangle \tag{10c}$$

where $\hat{T}$ is the kinetic energy, $\hat{V}_{\mathrm{na}}$ is the nuclear attraction operator, and $1/r_{12}$ is the electron repulsion operator in atomic units ($e^2 = 1$).

The MOs and orbital energies ε_i are solutions of the Hartree–Fock–Roothaan equations

$$\hat{F}\phi_i = \varepsilon_i\phi_i \tag{11}$$

where the Fock operator is defined in terms of Coulomb and exchange operators

$$\hat{F} = \hat{h} + \sum_i (2\hat{J}_i - \hat{K}_i) \tag{12}$$

with

$$\hat{J}_i(2)\,\phi_j(2) = \langle\phi_i(1)\left|\frac{1}{r_{12}}\right|\phi_i(1)\rangle\phi_j(2) \tag{13}$$

$$\hat{K}_i(2)\,\phi_j(2) = \langle\phi_i(1)\left|\frac{1}{r_{12}}\right|\phi_j(1)\rangle\phi_i(2) \tag{14}$$

The orbital energies are approximations to the ionization potentials (Koopmans' theorem). SCF calculations are presently carried out routinely for quite large polyatomic molecules [12]. Most of the recent advances relate to more efficient formal procedures for open-shell systems [55], increased computational speed through the use of point group symmetry [56], and algorithms for accelerating and/or insuring convergence [57].

The application of SCF procedures to open-shell molecules has also become relatively commonplace. A generalization of Eq. (9), applicable to many open-shell cases is

$$E^{\mathrm{SCF}} = 2\sum_{i=1}^{n} f_i h_i + \sum_{i,j=1}^{n} (a_{ij}J_{ij} - b_{ij}K_{ij}) \tag{15}$$

where n, which is not necessarily $N/2$ in this case, is the number of occupied MOs. The coefficients f_i, a_{ij}, and b_{ij} are characteristic of the specific electronic state of interest. The SCF wave function is a Slater determinant, or a fixed linear combination of Slater determinants, that exhibits the proper spatial and spin symmetry for the open-shell configuration and desired state.

Although the restricted or conventional Hartree–Fock (RHF) method outlined above is used most extensively, there have been several "extended" Hartree–Fock procedures developed. These allow for the removal or relaxation of certain symmetry and equivalence restrictions imposed on the space and/or spin factors of the RHF orbitals while retaining their independent particle forms. Among these methods are the unrestricted Hartree–Fock (UHF) method [58] [also referred to as different orbitals for different spins (DODS)], various spin-projected UHF (PUHF) procedures [59], and spin-optimized procedures (SOHF) [60].

The UHF procedures are the first logical extensions of the RHF wave functions short of using the superposition-of-configurations methods (MCSCF and CM) or many-body perturbation theory or equations-of-motion procedures. Unfortunately, although the UHF procedures are formally more desirable than the RHF and more easily interpretable (in many cases) than the CM procedures, the fact that UHF wave functions are not pure spin states (viz., not eigenfunctions of $\hat{S}^2$) and that the SOHF procedures generally require the use of nonorthogonal orbitals has restricted their application [14]. Certain properties such as the spin density at the nucleus are well defined by UHF wave functions.

D. CONFIGURATION MIXING (CM)

The closed- and open-shell SCF (RHF) functions discussed in the previous section approximate many-electron interactions by the assumption that each electron exists in an average field produced by all the remaining electrons. The energy error resulting from this assumption is called the correlation energy. It is defined as the difference between the exact solution of the nonrelativistic Hamiltonian and its expectation value in the RHF approximation [43]. The latter is the SCF limit (also called the Hartree–Fock limit) energy approached as the basis set approaches completeness (Section B).

The correlation energy, although generally a small fraction of the total molecular energy, is of the same order of magnitude as the strengths of chemical bonds and of electronic excitation energies. In addition, RHF wave functions are usually not flexible enough to remain adequate as one approaches the united atom or molecular dissociation. This is particularly serious when potential energy surfaces and dissociation processes are considered.

The logical and most straightforward improvement of the RHF procedure is through the use of configuration mixing (CM),[1] also commonly termed configuration interaction (CI) [61, 62]. The total molecular wave function is then represented as a linear combination (in principle infinite, but in practice finite, say n in number, where n may be very large) of many-particle functions called configuration state functions (CSFs) or configuration functions (CFs). Each CSF is either a single Slater determinant or a specific linear configuration of Slater determinants possessing the proper symmetry of the desired electronic state. The CSFs are eigenfunctions of all spin and space operators that commute with the Hamiltonian [62]. The CM wave function is defined as

$$\Psi = \sum_{I=1}^{n} C_I \Phi_I \tag{16}$$

where I indexes the (orthonormal) CSFs Φ_I and the coefficients C_I are determined through the use of the linear variation principle as applied to the energy expectation value. This leads to the secular equation

$$(\mathsf{H} - E\mathbf{1})\mathbf{C} = 0 \tag{17}$$

where the symmetric matrix H is made up of elements

$$H_{IJ} = \langle \Phi_I | \hat{H} | \Phi_J \rangle \tag{18}$$

E and $\mathbf{C}$ are the eigenvalue and eigenvector of H for n roots corresponding to successive electronic states, and $\mathbf{1}$ is the unit matrix. The eigenvectors $\mathbf{C}_p$ comprise a linearly independent set that correspond to the eigenvalues E_p, each of which is an upper bound to the corresponding exact eigenvalue of H [61]. In practice, usually only the lowest one or a few roots are considered due, for example, to basis set limitations or the difficulty of selecting a union of CSFs appropriate for several roots.

It is worth noting that in later chapters we refer to both CSFs and configurations. The latter (preferably called *electron* configurations) designate assignments of electrons to MOs without distinguishing their detailed forms but merely indicating the number of electrons occupying an MO of each species (irreducible representation) of the molecular point group. The MOs of any one species are ordered in terms of increasing orbital energy by a number that precedes the lower-case designation of the orbital species. For example, the ground-state electron configuration of HCN (point group $C_{\infty v}$) is

$$1\sigma^2\, 2\sigma^2\, 3\sigma^2\, 4\sigma^2\, 5\sigma^2\, 1\pi^4$$

[1] As in our earlier volume [2] we elect to use Roothaan's term "configuration mixing" as an alternative to Hartree's more appropriate but clumsier suggested term "superposition of configurations."

and that of H_2O (point group C_{2v}) is

$$1a_1^2\ 2a_1^2\ 1b_1^2\ 3a_1^2\ 1b_1^2$$

The "electron configuration" of any electronic state usually means that of the CSF in Eq. (16) where C_I is largest (usually 0.9 or more in ground states). The configuration of this CSF is then commonly called the "reference configuration." The configurations of the other CSFs with smaller coefficients are then obtained by promoting one or more electrons to higher-energy MOs than in the reference configuration. This process is often called "excitation," but a better term which we shall try to use is "substitution" or possibly "replacement." It is better to reserve the term "excitation" for truly excited states in which the reference configuration itself is a promoted one.

Here we note that in many CM calculations, one starts the CM by using more than one reference configuration, i.e., by beginning with more than one Φ_I in Eq. (16), and afterward adding more in a further development to include up to a very large number. Procedures that incorporate CM with a multireference base can solve, to varying degrees, two main problems in calculating correlated wave functions. First, single reference (SCF) CM wave functions are limited, in most cases, to equilibrium geometry because additional (or other) CSFs are needed to assume proper behavior for dissociation, curve crossings, or other phenomena (see Section E.). Here one may resort to multiconfiguration SCF (MCSCF), approximate natural orbital, or other methods, which incidently also provide improved MOs. In the MCSCF method [2] a limited number of Φ_I's are usually used, after which one may (or may not) proceed to add extensive CM (see Section F). Second, with a single reference base it is difficult to compute integrals (Hamiltonian matrix elements) for entire classes of CSFs that correspond to triple, quadruple, and higher substitutions. (However, see below and Section H.) The choice of a multireference base overcomes this difficulty to a considerable extent; i.e., limiting the substitutions to single and double replacements with respect to a multireference base is tantamount to a selection of a subset of the complete set of higher replacements with respect to one reference configuration that should be the most reasonable on physical grounds.

Most of the CM calculations described in the following chapters were carried out in the more or less traditional procedure described by Shavitt [61]. Appropriate one- and two-electron integrals with respect to MOs are computed by means of the usual four index transformation of the basis function integrals. The basis set of MOs used in CM are from an SCF calculation or an MCSCF calculation, or are some form of natural orbitals, etc. (The MO basis set used in CM should be distinguished from the AO basis set discussed in Section B.) Usually some separation of these MOs into subsets such as valence MOs and virtual MOs, or core, internal, and external MOs, is made for purposes of delineating levels of approximation with respect to choice of CSFs [61, 62]. Recall that any subset of the full configuration space (complete CI) results in upper bound energies for the successive electronic states of a chosen symmetry. Usually at the next stage MO integrals are used to construct the Hamiltonian matrix H of Eq. (18) with respect to the chosen set of CSFs and proper electronic total symmetry. The eigenstates of the system are then derived by solving Eq. (17) for the desired number of roots.

It is appropriate at this juncture to mention that there have been several developments of major importance in the past year or so that will impact the computation of correlated molecular wave functions in such a way as to greatly expand the range of applicability of *ab initio* methods. Although relatively few applications of these methods are discussed in the following chapters, we present brief descriptions with appropriate references because of their assured place in future studies. These advances may be classified in two categories as they relate to CM procedures and to MCSCF procedures. The former category will be briefly discussed here and the latter at the end of the next section.

The improvements in CM procedures have also been in two main areas which are presently being reassessed and, in some instances, combined. The most recent derives from the unitary group formulation of the electron correlation problem as set forth by Paldus [63] and extended by Shavitt [64] to what is now known as the graphical unitary group approach (GUGA). A recent article [65] describing how the GUGA has been implemented for computational studies of molecular structure contains the statement: "The power of the graphical unitary group approach, as developed by Paldus and Shavitt, lies with the tremendous amount of insight it provides into the *structure* of the Hamiltonian matrix H. This structure is not only an edifice of mathematical beauty but also a practical tool for the calculus of matrix element evaluation." The GUGA utilizes efficient algorithms for the construction of a distinct row table that defines the desired configuration space. This table is then used to generate all the Hamiltonian matrix elements in a symbolic (rather than explicitly computed) form. The GUGA has been compared to conventional and direct (see below) CM procedures in calculations on methylene, ethylene, AlC_2H_2, and ketene [65]. The increased efficiency and extended range of capability, as many as 23,000 CSFs [66], clearly indicates a promising future for CM procedures. We do not attempt to describe the GUGA in detail but refer the reader to the source quoted above.

The second major advance in CM procedures is based on what Roos and Siegbahn have termed the direct configuration interaction from molecular integrals (CIMI) method or, alternatively, the direct CI method [67]. A major bottleneck in conventional CM procedures is the time-consuming steps of generating Hamiltonian matrix elements by means of either the so-called "formula," "code," or "energy expressions" tape or as an integral step of the CM procedure where the MO integrals are employed [61, 62]. The formula tape procedure, although it may be somewhat slower, has distinct advantages if many nuclear configurations (potential energy surfaces) for a given state symmetry are desired. This is because the formula tape is generated once and can be re-used for all desired geometries. The calculation of the formula tape is quite time consuming for large sets of CSFs. Brooks and Schaefer cite timing data for several large-scale CM calculations which indicate that the construction of the formula tape typically takes ten or more times longer than any other step in the entire procedure (i.e., construction of MOs, integrals transformation, diagonalization of the Hamiltonian matrix).

The CIMI procedure does not require the use of a formula tape, being an MO "integral driven" method. As conventional CM procedures require the eigensolution (diagonalization) of the large Hamiltonian matrix H of Eq. (17), methods have been developed for using iterative procedures for obtaining several lowest energy solutions [61, 68]. Roos and Siegbahn [67] describe this procedure as the multiplication of the Hamiltonian matrix with a trial wave function (e.g., from a related small CM calculation) defined by a vector $\mathbf{C}^{(k-1)}$ that generates an improved vector $\boldsymbol{\sigma}^{(k)}$, i.e.,

$$\sigma^{(k)} = \mathsf{H}\mathsf{C}^{(k-1)} \tag{19}$$

The improved vector is used to construct a new trial vector and the procedure is repeated until convergence is achieved.

The CIMI method makes use of the fact that since H is made up of linear combinations of one- and two-electron integrals, it is possible to derive expressions for the vector $\boldsymbol{\sigma}^{(k)}$ directly in terms of these MO integrals and circumvent completely the explicit construction of H. This procedure is symbolically written in terms of the increment to the element i of the vector $\boldsymbol{\sigma}^{(k)}$ from a particular MO integral:

$$\Delta\sigma_i^{(k)} = A\langle \phi_a(1)\, \phi_b(2) | \phi_c(1)\, \phi_d(2)\rangle C_d^{(k-1)} \tag{20}$$

where A is a coupling coefficient that is appropriate to the spin symmetry of the CSFs involved. The derivation of these coupling coefficients is quite complicated [67] and, although initially the CIMI method was limited to certain special types of CM (e.g., single and double replacements relative to a closed-shell reference configuration), several advances have increased its generality to include triplet states [69] and doublets, quartets, and open-shell singlets [70]. Results using the CIMI method involving 10,000 to 57,000 CSFs have been reported [67, 70].

Important relationships between the CIMI and GUGA methods have been noted [64, 71]; thus the next stage in the development of powerful CM procedures with general applicability appears to be moving toward a combination of the two.

The importance of the development of improved methods for treating the electron correlation problem cannot be overemphasized. It is well known that for a particular basis set of STFs or GTFs in the limit of complete or full CM an unambiguous well-defined limiting wave function would be obtained. Indeed, for a complete AO basis and with full CM an accurate solution for the nonrelativistic Hamiltonian would be attained. In more formal terms [61], the full CM wave function is invariant to any nonsingular transformation of the orbitals, i.e., it depends only on the AO basis set $[\chi_p]$ and not on the forms of the MOs $[\phi_i]$. In addition, if the MOs can be partitioned into subsets whereby the MOs in each subset are mutually orthogonal to those in each other subset, then the CM wave function is invariant to any separate nonsingular linear transformations within each of these subsets. Shavitt designates these as full-class CI wave functions. This distinction is important because it gives rise to certain well-defined levels of CM such as "first order" [72] and full single- and double-replacement wave functions (see Section F).

Unfortunately, because the number of CSFs increases proportionally to m^N for m basis function and N electrons, there are few molecules that can be treated at the level of full CM. Equally unfortunate is the fact that the convergence rate of CM wave functions with respect to successively higher electron configurations is not generally monotonic. Thus, methods such as the GUGA and CIMI that permit full treatments of the CM problem by levels, without introducing truncations of orbitals or perturbation theory approximations, can remove many of the uncertainties in *ab initio* procedures used to date. The proceedings of a workshop on CM procedures contains additional references and brief outlines of the methods mentioned above [73].

E. MULTICONFIGURATION SELF-CONSISTENT FIELD (MCSCF) METHODS

In the brief description of MCSCF wave functions in our previous volume [2] we chose the procedure of Veillard and Clementi [74] to illustrate the formalism. Only a qualitative discussion of the general MCSCF method will be given in this section since it is our intention only to provide a basis for understanding applications to calculations of molecular wave functions. The unique feature of all MCSCF methods is that the total wave function

$$\Psi = \sum_{I=1}^{n} a_I \Phi_I \tag{21}$$

is defined in terms of a linear combination of CSFs Φ_I whereby *both* the coefficients a_I and those defining the CSFs in terms of the MOs ϕ_i of Eq. (7) are simultaneously optimized with respect to the total energy. The MCSCF equations are derived by constructing the expression for the total energy in terms of the MOs and CSFs and applying the variation principle with respect to the appropriate linear coefficients.

The MCSCF methods in current usage may (roughly) be placed in two general categories: Hamiltonian operator (HO) methods and generalized Brillonin theorem (GBT) methods. The HO methods being based essentially on extensions of Hartree–Fock SCF procedures require the construction of appropriate operators using Lagrangian multipliers and iterating between the determination of the coefficients of the MOs and those of the CSFs until this coupled problem attains satisfactory convergence. We categorize the rather specialized procedures such as that of Veillard and Clementi [74], the generalized valence bond (GVB) method [75], and separated pair method [76], as well as the more general MCSCF approaches of Das and Wahl [77], Hinze and Roothaan [78], and Golebiewski *et al.* [79] as HO methods. In each of these approaches coupled equations are solved iteratively for the two kinds of linear coefficients, formally defined through rather complicated algebraic expressions. One of the major problems with the HO methods has been in the rate, and in many cases the assurance, of convergence to the desired wave function.

The GBT method was proposed by Levy and Berthier [80] with subsequent reformulations and applications by Grein and Chang [81] and Banerjee and Grein [82]. Naively, GBT methods exploit the property that matrix elements of the Hamiltonian operator between a proper MCSCF wave function for a chosen electronic state and that due to any well-defined set of excited CSFs are identically zero [80]. Following Grein and Chang [81] this may be written

$$\langle \Psi | H | \Psi(i \rightarrow j) \rangle = 0 \tag{22}$$

where Ψ is defined by Eq. (21) and

$$\Psi(i \rightarrow j) = \sum_{k} a_k [\Phi_k(i \rightarrow j) - \Phi_k(j \rightarrow i)] \tag{23}$$

The Slater determinants $\Phi_k(i \to j)$ are defined as follows. If the MO ϕ_i occurs only once in column λ of Φ_k then Φ_k is replaced by a determinant containing ϕ_j instead of ϕ_i in that column. If ϕ_i occurs twice with electrons having opposite spins in columns λ and μ then Φ_k is replaced by a sum of two Slater determinants with ϕ_i and ϕ_j in the first and ϕ_j and ϕ_i in columns λ and μ, respectively ($\phi_k \equiv \phi_k\alpha$, $\phi_k \equiv \phi_k\beta$). The procedure proposed by Grein and Chang to determine Ψ is as follows:

1. An initial guess of Ψ is determined from the MOs ϕ_i as defined in terms of an orthogonalized set of primitive basis functions $\{\chi_p\}$ [Eqs. (5)–(7)].
2. The coefficients a_k are calculated via the standard CM procedure [Eq. (17)].
3. The single replacement CSFs $\Psi(i \to j)$ and "super" CI wave function $\Psi_{CI} = \Psi + \sum_{i<j} C_i^j\Psi(i \to j)$ are formed with the coefficients c_i^j determined by variation or perturbation techniques.
4. A new set of MOs is formed using $\phi_i = \phi_i + \sum_{j \neq i} c_i^j\phi_j$, a new Ψ is set up, and steps (2)–(4) are repeated until convergence is reached.

Recently, improved MCSCF procedures based on the GBT method have expanded the range of applicability for calculations of accurate wave functions for larger molecules including larger sets of CSFs. The improvements include the incorporation of natural orbital transformations [83], a procedure whereby natural orbitals are used as intermediaries to improve MCSCF wave function [84], the incorporation into the variational problem involving the MOs of an algebra that insures quadratic convergence to the true MCSCF wave function [85, 86], and the development of a procedure to treat the coupling of the CSF and MO coefficients directly [87]. These advances, as well as those in the HO method (see Das and Wahl [77], in Ref. 3; see also [79]), together with parallel strides made in CM procedures (Section D) herald a new era for *ab initio* studies of atoms and molecules.

The importance of MCSCF procedures lies in the fundamental necessity to provide a multireference orbital space to insure sufficient flexibility in the wave function for cases where coefficients of more than one CSF become large. This is most obvious in cases where bonds are being formed or broken or where potential energy curves interact strongly. In general, in those instances where the coefficients of two or more CSFs are ~0.2 or larger, one expects that an MCSCF procedure is best suited as a *first approximation* to the wave functions. The simplest example of this need is for calculating the potential energy curve of the ground state of H_2. The two-configuration MCSCF wave function for H_2 is $\Psi = c_1\Phi_1 +$

$c_2\Phi_2$, with $\Phi_1 \approx 1\sigma_g{}^2$ and $\Phi_2 \approx 1\sigma_u{}^2$. This has $c_1 = c_2$ at $R = \infty$ and c_2 small near R_e. The single-configuration (SCF-MO) wave function is incorrect at $R = \infty$. The method of optimized valence configuration (OVC) devised by Das and Wahl [77], the GVB method of Goddard and co-workers [75], and the method of the full optimized reaction space of Ruedenberg and co-workers [73, 84] were designed to recover this most important portion of the electron correlation that is absent from the SCF-MO wave function. As will be discussed below, such MCSCF procedures also afford a means for defining orbital spaces upon which more extensive CM wave functions may be based. This combined MCSCF-CM procedure has become among the most popular of approaches for constructing accurate *ab initio* wave functions.

F. SELECTION METHODS FOR CSFs AND MOs

The calculation of an accurate *ab initio* wave function involves three fundamental steps: (a) the choice of a "primitive" or *AO basis* $\{\chi_p\}$ [Eqs. (5) and (6)]; (b) the use of the $\{\chi_p\}$ to obtain the *MO basis* $\{\phi\}$, generally using SCF or MCSCF procedures; and (c) the selection of CSFs that make up the *configuration space* (or space of CSFs) that defines the total wave function Ψ [Eq. (16)]. We reemphasize here that the accuracy of a wave function is ultimately grounded in the choice of the AO basis $\{\chi_p\}$. Improvements in the MO basis $\{\phi_i\}$ through more elaborate MCSCF or natural orbital procedures or extensions of the space of CSFs through the addition of higher replacements *cannot* recover inadequacies in the AO basis $\{\chi_p\}$. With this condition firmly in mind the most commonly used procedures for choosing MO bases $\{\phi_i\}$ and selecting configurations are reviewed in this section. These are discussed together because in many procedures the MO and configuration spaces are dealt with in a more or less coupled procedure. (We use the terms MO space or MO basis in preference to the more general term "orbital space" because, except for only a few instances, *ab initio* calculations of molecular wave functions are based on MOs in some form. The terms "one-particle space" and "*n*-particle space" are often used for MO basis and CSF space, respectively.)

The necessity to employ procedures for reducing the size of the space of MOs and the number of CSFs is due to the rapid increase in the latter (roughly m^N for m MOs and N electrons) as complete CM is approached. Clearly the number of CSFs for a given case may be reduced by limiting the choice of CSFs to be included, by reducing the number of MOs that define the MO, space or both.

Historically selection procedures have been largely grounded in perturbation theory approaches [61, 88] whereby the orbital space is partitioned and CSFs are categorized into classes. This may be accomplished by rewriting Eq. (16) as

$$\Psi = \sum_{I=1}^{k} C_I\Phi_I + \sum_{J=I+1}^{n} C_J\Phi_J \tag{24}$$

where the set of CSFs $\{\Phi_I\} \equiv \Phi_0$ define a zeroth-order space or "reference set" and the set $\{\Phi_J\}$ comprises the remainder of the configuration space. The Hamiltonian matrix of Eq. (18) then takes on the form shown in Fig. 1 with the upper triangle H_{aa} due to the reference set of CSFs. The B_k approximation of Gershgorn and Shavitt [88] corresponds to the construction of the Hamiltonian matrix whereby the matrix elements in the portion of Fig. 1 denoted as H_{bb} are set equal to zero with the exception of those along the diagonal (i.e., $H_{IJ} = 0$; $I, J > k$; $I \neq J$). The Hamiltonian matrix in the B_k approximation may be solved to give the first-order eigenfunction $\tilde{\Psi}$ and the second-order eigenvalue $\tilde{E}$ in Brillouin–Wigner perturbation theory [88]. A simple example of the application of perturbation theory to estimate the coefficients of CSFs in the set $\{\Phi_k\}$ is given by Shavitt [61] using the first-order Rayleigh–Schrödinger method

$$C_J = H_{J0}/(H_{00} - H_{JJ}) \tag{25}$$

The corresponding expression for the second-order contribution to the energy is

$$\Delta E_J = |H_{J0}|^2/(H_{00} - H_{JJ}) \tag{26}$$

The more elaborate B_k method [88] includes the interaction of the configurations that contribute to H_{aa} (Fig. 1) with those of the secondary set that contribute to H_{ab} and the diagonal elements H_{JJ}. Thus perturbation theory affords a means for estimating effects of higher substitutions on any calculation involving a relatively small number of reference CSFs.

The reason for introducing the concept of partitioning the Hamiltonian matrix in the manner shown in Fig. 1 is, however, more far-reaching than just for the purpose of defining the perturbation theory procedures. As mentioned in Section E, the necessity to define a set of reference configurations instead of a single SCF configuration as a starting point for improved CM wave functions has become more and more widely accepted. To clarify the various selection procedures it is perhaps best to refer to a specific example.

One for which we have rather extensive experience is the nitrogen molecule. Features discussed are generally found for polyatomic molecules as well. If it is assumed that the AO basis $\{\chi_p\}$ for N_2 is chosen as five $s\sigma$,

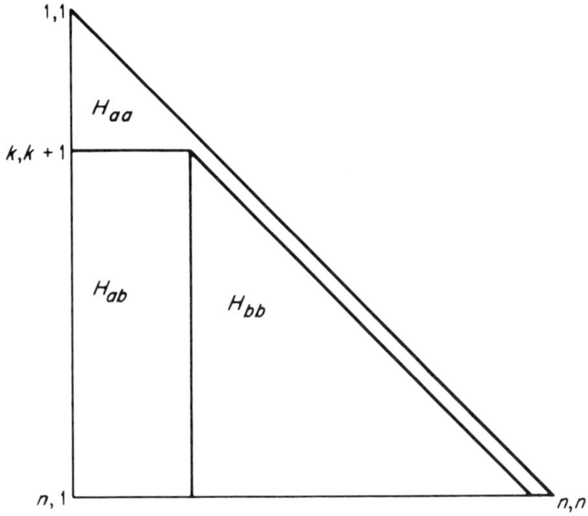

FIG. 1. A partitioned form of the Hamiltonian matrix.

four pσ, and four pπ, and three dσ, dπ, and dδ STFs (a total of 44 basis functions), then the following MOs arise:

$$\begin{array}{l} 1\sigma_g\,2\sigma_g\,3\sigma_g\,4\sigma_g \ldots 12\sigma_g \\ 1\sigma_u\,2\sigma_u\,3\sigma_u\,4\sigma_u \ldots 12\sigma_u \\ 1\pi_u\,2\pi_u \ldots 7\pi_u \\ 1\pi_g\,2\pi_g \ldots 7\pi_g \\ 1\delta_g \ldots 3\delta_g \\ 1\delta_u \ldots 3\delta_u \end{array} \tag{27}$$

The ground-state SCF-MO, which defines the dominant CSF, configuration is

$$1\sigma_g^2\,1\sigma_u^2\,2\sigma_g^2\,2\sigma_u^2\,1\pi_u^4\,3\sigma_g^2 \tag{28}$$

A complete ground-state CM calculation involving all 44 MOs would require about 10^{15} CSFs. On the other hand, a ground-state calculation including all single and double replacements for the ten-electron system, but constraining the 1s electrons to remain "frozen" in the $1\sigma_g$ and $1\sigma_u$ MOs needs only 2670 CSFs. The reduction in the size of the problem from complete CM to "singles and doubles" CM with frozen K-shell electrons is dramatic! If the replacements from the core MOs were allowed, the resulting space of CSFs is increased to about 10,000. This shows that invoking a chemically reasonably assumption, that the 1s core electrons are not expected to influence the bonding and many properties of the molecule, results in a significant reduction in the size of the problem (the inclusion of core correlation does, however, significantly lower the total energy). Most of the calculations discussed in the following chapters make use of the frozen core approximation, some even invoking the assumption that additional shells are unimportant for the properties under investigation.

It is well known [61] that calculations including up to double replacements with respect to a *single* (SCF-MO) reference configuration recover a large portion of the correlation energy. The Hamiltonian matrix element between two CSFs is zero if they differ by more than two orbitals in this special case. Furthermore, Brillouin's theorem states that a closed-shell SCF-MO wave function has zero matrix elements with all single replacement CSFs. Thus this particular case, a closed-shell SCF-MO reference CSF plus all double replacements, defines a CM wave function that is correct to first order and a total energy correct to third order of perturbation theory. The *second-order* wave function then includes single, triple, and quadruple replacements in the set of CSFs having nonvanishing matrix elements with the first-order wave function, etc. However, these higher replacements can make significant contributions to the wave function. The number of such CSFs is enormous and, in general, cannot be included in their entirety.

Davidson and Langhoff [89] have derived a simple formula for approximating the effects of a certain class of quadruple replacements referred to as "unlinked clusters," configurations that correspond to products of double replacements. They give an estimate of the contribution to the correlation energy due to quadruple substitutions as

$$\Delta E_{\mathrm{Q}} = (1 - C_0^2)\Delta E_{\mathrm{D}}, \tag{29}$$

where ΔE_{D} is the correlation energy ($E_{\mathrm{D}} - E_{\mathrm{SCF}}$) due to the double replacement CM calculation and C_0 is the coefficient of the reference SCF-MO configuration. An example of the application of Eq. (29) to the potential energy surface of the water molecule is discussed in Chapter II, Section G. More sophisticated and systematic applications of many-body perturbation theory have been made. These are mentioned briefly in Section H.

The improvement of correlated wave functions through the inclusion of higher replacements is not the most economical and efficient way to proceed because of the rapidly increasing numbers of CSFs and because the convergence to the full CM result is not monotonic. The latter problem may be interpreted as being due to the chemically reasonable fact that for many situations the restriction to a single (SCF-MO) reference configuration is a poor approximation. This was mentioned in Section E with regard to the wave function for H_2 at large distances, where two CSFs are required to properly define the dissociation process. Our example molecule N_2 is an even more dramatic case. A wave function of sufficient flexibility for dissociation to two N atoms in their ground states as represented in the SCF approximation requires a minimum of *ten* CSFs. Lie and Clementi have reported MCSCF calculations for all of the second-row diatomic hydrides and homonuclear molecules, including only those CSFs required for proper dissociation [90]. The nitrogen molecule requires the largest number of CSFs because it is formed from atoms with half-filled p shells. The MOs $3\sigma_g$, $1\pi_u$, $1\pi_g$, and $3\sigma_u$ of (27) are involved in the proper dissociating CSFs, with the six-fold replacement $3\sigma_g{}^2\ 1\pi_u{}^4 \rightarrow 1\pi_g{}^4\ 3\sigma_u{}^2$ being the highest level required. At $R = \infty$ all ten of these CSFs are weighted with large coefficients. Thus, a procedure whereby only the ground SCF-MO configuration is chosen as a reference CSF is not expected to accurately represent the system at all values of the internuclear distance R. This necessity for including several reference configurations by a MCSCF or related procedure such as improved natural orbitals (see below) in the process of designing *ab initio* calculations has given rise to numerous procedures that combine MO space separations and the systematic selection of spaces of CSFs. It can be seen from the MOs of (27) and the preceding discussion of the requirements for proper dissociation that the MO space may be partitioned into four sets. The first (Set I) contains the core MOs and is here considered to be "frozen" in the sense that no replacements from these are allowed in any procedure we may devise. Thus the 14-electron problem reduces to a 10-electron problem at the stage where CM is to be treated. The second (Set II) contains the $2\sigma_g$ and $2\sigma_u$ MOs and may be called the "inner valence" space. This distinction is made because although replacements from these MOs are important for an accurate wave function for the ground and (especially) the excited states over the full range of R values, they are not required for the properly dissociating wave functions discussed above. The third (Set III) is made up of the MOs $3\sigma_g$, $1\pi_u$, $1\pi_g$, and $3\sigma_u$ and may be called the "outer valence" space. Bond formation and valence excitations are described by the MOs in Set III for the low-energy states of N_2. The last (Set IV) contains all the MOs not assigned to I, II, or III and is comprised of the MOs that are required to account for all the rest of the functions in the 44-term STF basis, i.e., $4\sigma_g - 12\sigma_g$, $4\sigma_u - 12\sigma_u$, $2\pi_u - 7\pi_u$, $2\pi_g - 7\pi_g$, $1\delta_g - 3\delta_g$, and $1\delta_u - 3\delta_u$ in (27). Set IV is called the "virtual" or "external" space.

Sets I, II, and III comprise a space of MOs that is precisely the same size as one that results from a minimal STF basis set (1s, $2p_x$, $2p_y$, $2p_z$) centered at each of the N atoms. Functions added to the AO basis, in addition to improving the MOs, also increase the number of virtual orbitals (Set IV) in the entire space of MOs. Indeed if a minimal basis is chosen to represent an atom or a molecule, correlated wave functions can be calculated without the necessity of an SCF or MCSCF preliminary stage. The basis functions may simply be made orthonormal and as long as complete CM is accomplished, any unitary transformation imposed on this orthonormal set will not affect the wave function [61]. This procedure has been carried out for many systems and constitutes a well-defined, albeit usually not very accurate, type of wave function generally called *complete valence CI* (VCI). We adopt the more general (and generally accepted) definition of VCI as the wave function that results when an MO basis is appropriately truncated by symmetry type to the size corresponding to a minimal AO basis set *and* complete CM is accomplished in this space. Thus, VCI defines, for any state of any molecule, a CSF expansion of fixed length. Our earlier volume [2] contains many references to VCI studies. In addition, the VCI procedure is *size consistent* in that the wave function treats the separate and united atom limits properly (See Section H).

As described above the *space of CSF's* in a VCI wave function is well defined. However, due to the constraints that the size of the *space of MOs* be the same as a minimal basis set (regardless of the actual size of the AO basis set) the quality of the VCI wave function varies depending on the nature of the MOs. The VCI wave function, the space of the CSFs being well-defined, naturally finds its place as a starting point for improved CM procedures. Deferring, for the present, discussion of the origin of the MOs comprising the valence space, we discuss such systematic approaches.

The first-order CI (FOCI) procedure of Schaefer *et al.* [14, 91] is perhaps the most well-known, logical first extension of VCI. It is based on a formalism due to Sinanoglu *et al.* [14] and follows the perturbation theory nomenclature. Ideally, a near complete AO basis would be used, in an MCSCF procedure, to define the valence space for the state under investigation; i.e., all CSFs corresponding to the VCI would be used in an MCSCF calculation to determine the MOs that comprise the valence space. In our present example, N_2 as a 10-electron system leads to 96 CSFs if all possible replacements among the orbitals in Sets II and III are included. Such an MCSCF calculation would be rather difficult, although the improved procedures discussed in Section E render it quite feasible. A more conservative, but not unreasonable procedure would be to treat the molecule as a 6-electron system ("frozen" $2\sigma_g$ and $2\sigma_u$ as well as $1\sigma_g$ and $1\sigma_u$). This procedure results in an MCSCF calculation involving 18 CSFs. (This includes 8 CSFs beyond those required for proper dissociation [90] and which therefore do not contribute to the wave function at $R = \infty$). The critical aspect of the MCSCF procedure is that it makes possible a unique definition of all the valence MOs in Sets I–III [90, 92]. In particular, the $1\pi_g$ and $3\sigma_u$ MOs, which are *virtual orbitals* unoccupied in the ground-state SCF-MO wave function [Eq. (28)] obtain their proper form as *valence* orbitals.

A procedure for defining internal or valence orbitals based on the generalized valence bond (GVB) method of Goddard and co-workers [75] has been used in many of the CM studies of polyatomic molecules discussed in subsequent chapters. In our previous volume [2] we showed, very briefly, how the GVB method in correcting for the improper dissociation behavior of H_2 was related to the Coulson–Fischer wave function. It has the form $\hat{A}[\phi_l\phi_r(\alpha\beta - \beta\alpha)]$, where $\langle\phi_l|\phi_r\rangle = S$ and $\hat{A}$ is the antisymmetrization operator. The important feature of this wave function is that the orbitals ϕ_l and ϕ_r describe exactly the electron associated with the left and right nuclei, respectively, at $R = \infty$ (where $S = 0$). At R_e, $S \approx 0.8$, indicating that the SCF-MO form is the predominant (but not the exact) form.

Thus, the closed-shell molecular spin orbitals of Eq. (8) for the special case of two electrons $\psi_1(1)\psi_2(2) = \phi_a\phi_a(\alpha\beta - \beta\alpha)$ are replaced by a "GVB pair" made up of two nonorthogonal singlet-coupled orbitals $\phi_1\phi_2(\alpha\beta - \beta\alpha)$.

In its unrestricted form the GVB wave function consists of an antisymmetrized product of such orbitals where each singlet-coupled electron pair is allowed to correlate. Since the computational problems inherent in the treatment of wave functions consisting of nonorthogonal orbitals are quite difficult [61], a constraint that all orbitals *other* than the two that describe a particular singlet pair be orthogonal is imposed; this is referred to as the GVB *perfect pairing* (GVB-PP) scheme. This so-called "strong orthogonality constraint" is found not to be a serious restriction in most cases, although there are cases, notably where several bonding schemes are nearly equally important, where the GVB-PP wave function is inappropriate (see Bobrowicz and Goddard [75]). However, in the context of the present section, i.e., where the main objectives are the calculation of valence or internal MOs for use in extended CM wave functions, the GVB-PP approach may be used instead of the more general MCSCF procedures. Most of the calculations reported in the later chapters that are based on GVB procedures use the perfect-pairing approximation.

Returning to N_2, the GVB-PP wave function is defined in terms of the AOs on the left and right atoms (see Dunning *et al.* [92])

$$\Psi_{\mathrm{GVB-PP}} = A1s_l^2\, 1s_r^2\, 2s_l^2\, 2s_r^2\, 2p\sigma_l\, 2p\sigma_r\, 2p\pi_{xl}\, 2p\pi_{xr}\, 2p\pi_{yl} \times 2p\pi_{yr}\alpha\beta\alpha\beta\alpha\beta\alpha\beta\,(\alpha\beta - \beta\alpha)(\alpha\beta - \beta\alpha)(\alpha\beta - \beta\alpha)(\alpha\beta - \beta\alpha) \tag{30}$$

The AOs $1s_l$, $2s_l$, $2p\sigma_l$, are, in general, modified atomic orbitals (MAOs): "MO adapted" linear combinations of basis functions (STFs, CGTFs, etc.) centered at the respective nuclei [2]. These MAOS can be combined in minimal basis fashion into MOs that transform according to $D_{\infty h}$ symmetry types; i.e., $1\sigma_{g,u} = 1s_l \pm 1s_r$, $2\sigma_{g,u} = 2s_l \pm 2s_r$, $3\sigma_{g,u} = 2p\sigma_l \pm 2p\sigma_r$, $1\pi_{u,g} = 2p\pi_l \pm 2p\pi_r$. To carry out subsequent CM calculations, each pair of GVB orbitals is combined to form a set of orthonormal MOs as

$$\phi_{g,u} = [2(1 \pm S)]^{-1/2}(\phi_l^{\mathrm{GVB}} \pm \phi_r^{\mathrm{GVB}}); \quad S = \langle\phi_l|\phi_r\rangle \tag{31}$$

The GVB-PP wave function for N_2 can then be expressed as

$$\Psi_{\mathrm{GVP-PP}} = 1\sigma_g^2\, 1\sigma_u^2\, 2\sigma_g^2\, 2\sigma_u^2(3\sigma_g^2 - \lambda_\sigma 3\sigma_u^2)(1\pi_{ux}^2 - \lambda_\pi 1\pi_{gx}^2) \times (1\pi_{uy}^2 - \lambda_\pi 1\pi_{gy}^2)\alpha\beta\alpha\beta\alpha\beta\alpha\beta\alpha\beta\alpha\beta\alpha\beta \tag{32}$$

Note that this form is sufficiently flexible to define all of the valence MOs of N_2. At $R = \infty$ the mixing coefficients λ_σ and λ_π are unity and the separate atom SCF-AOs result, just as for the MCSCF procedure described above. In the actual calculations (see Dunning *et al.* [92]) the orbitals of Eq. (31) are obtained using the GVB-PP scheme for the six (outermost) electron problem. These are used to define the valence orbitals and additional CM is accomplished using conventional methods (Section D). This procedure is generally applicable to any polyatomic molecule provided that an appropriate GVB-PP wave function can be devised such that all the desired valence or internal orbitals are defined.

The MCSCF or GVB-PP procedure then defines the valence or internal space (Sets I–III) uniquely, with the external space (Set IV) remaining simply the set of (orthonormal) complement MOs that account for the remainder of the orbitals derivable from the AO (44-term) basis set. This separation into valence/virtual [61] or internal/external [62] spaces is a quite general procedure that possesses several unique features, perhaps the most important of which is that unitary transformations among orbitals of the internal space or among orbitals of the external space do not change the total wave function for cases of well-defined sets of CSFs.

The first-order wave function is one of these cases. It is defined by adding to the set of CSFs comprising the VCI *all* CSFs that arise when at most *one* electron is substituted into the external orbitals, i.e., when all single replacements are made with respect to the CSFs in the VCI set. Returning to our N_2 example, the FOCI calculation that includes *all* single replacements from Sets II and III to Set IV contains 5441 CSFs for the orbital space defined by Eq. (27). It is noteworthy that this "complete" FOCI has the property that at $R = \infty$ there are included in the wave function some CSFs that result in atomic electron correlation. This is because of the inclusion of single replacements from Set II. This often undesirable feature can however be avoided. To calculate a wave function that recovers only the molecular extra correlation energy (MECE), defined as the difference between the sum of the atomic *SCF* energies and the correlation energy due to CM in the molecule, a "constrained" FOCI wave function may be designed. In our example, if only those single replacements from Set II into Set IV in which the spin coupling is ${}^1\Sigma_g{}^+$ are retained, 1008 CSFs result and the wave function dissociates exactly to two SCF-AO N atoms in their ground states [92].

The general FOCI method, and a related procedure [93] called "polarization CI" (POL-CI), have the advantages of being uniquely defined, of providing improvements upon the VCI procedure, and maintaining, in general, a CSF set of manageable size. The added single replacements introduce what have been termed "semiinternal double" correlation effects; i.e., with respect to some chosen configuration in the valence space a single replacement *within* the valence space *together* with a single replacement to the virtual space is a double replacement. The full set of such CSFs constitutes the set of semiinternal doubles. These, together with the remaining single replacements, which, with respect to the SCF-MO configuration, do *not* contribute to the total wave function by Brillouin's theorem, but *do* contribute for a many-reference CSF internal space, constitute the FOCI set of CSFs. It should also be mentioned that it is a reasonable approximation to use a *common set* of MOs for an entire family of states using the FOCI [94] or POL-CI [93, 95] procedures. The single replacements to the virtual orbitals result in the effective "relaxation" of the wave function to the form appropriate to each respective state [96], i.e., roughly to its SCF-MO form but including some electron correlation.

The VCI and FOCI procedures exemplify one of the most important considerations taken in the design of *ab initio* studies of molecular structure; i.e., a totally accurate wave function being in most cases unattainable, a compromise approach is taken in which the *particular properties under study* are most accurately approached with a clear understanding of the approximations made and the expected cancellation of errors. In the aforementioned procedures relative potential energy surfaces in particular are expected to be reasonably represented in spite of considerable percentages of correlation energy being neglected due to the frozen core approximation and the constraint of dissociation to SCF-MO atomic wave functions (among other known shortcomings).

The next extension to the general procedure based on the internal/external space separation is termed "second-order" CI. This corresponds to adding to the set of CSFs that comprise the FOCI all CSFs that arise when *two* electrons are substituted into the orbitals of the external space. In the N_2 example this increases the total number of CSFs to 14,238 if the restriction is made that double replacements from Set III only to Set IV are allowed. This number is more than 60,000 when the set of double replacements from Set II to Set IV are also included. The space of CSFs here becomes prohibitively large because at each higher order the replacements are referred to the full VCI (96 CSFs), i.e., as high as *eight-fold* substitutions with respect to the ground SCF-MO configuration. Clearly second-order calculations are not feasible in most cases and reasonable alternatives must be sought. The successive "stepping-up" in order is mentioned because it defines a systematic procedure for approaching the complete CM wave function. Several different approaches for making such large CM calculations tractable have been examined.

As mentioned above, the magnitude of a CM calculation is determined by the number of MOs that define the MO space and the number of CSFs that are included in the construction of the Hamiltonian matrix. It is usually necessary both to *truncate* the MO space and to make a *selection* of CSFs if calculations beyond the level of FOCI or single and double replacements with respect to one, or only a few, configurations are required.

To effect a truncation of the orbital space for a given problem the external orbitals must be transformed in some way to allow for their identification as to their degree of importance. This may be accomplished in any of several ways. One commonly used procedure is to identify and exclude the orbitals that are complements to the core or frozen orbitals in the sense that the dominant coefficients in the MO weight the AO basis functions having large (core-like) ζ's. Similarly, the external MOs may be defined by orthonormalizing unit vectors on to the internal MOs. In this way external MOs that correspond, e.g., to the polarization functions of the AO basis may be eliminated [93, 95]. Both of these choices are used routinely by several groups and have proven to result in a significant reduction in the number of possible CSFs with apparently little loss of accuracy. Note that such procedures are independent of the electronic state or nature of the internal MOs.

Two other procedures that are energy and state dependent are also in common use. The first is the straightforward extension of the number of MOs through the inclusion in the MCSCF procedure of several MOs of the same symmetry (see Das and Wahl [77], in Ref. 3; see also [97]); e.g., the $4\sigma_g$, $4\sigma_u$, $2\pi_u$, and $2\pi_g$ MOs of Eq. (28). This is usually done for each state under investigation and generally requires that a choice of CSFs to be included in the MCSCF procedure be made. This choice renders the calculation "orbital dependent" and may seriously affect the relative quality of the wave functions from state to state when excited states are considered.

The safest and most commonly used procedure for defining external MOs that may be arranged in order of decreasing importance, thereby preparing them for truncation, is the transformation to natural orbitals (NOs) [98]. Following the treatment in our first volume [2], we define a first-order density matrix for an N-electron system as

$$\rho(1, 1') = N \int \Psi(\tau_1, \tau_2, \tau_3, \ldots, \tau_N)\Psi^*(\tau_1', \tau_2, \tau_3, \ldots, \tau_N) \, d\tau_2 \, d\tau_3 \ldots d\tau_N \quad (33)$$

where τ may include space coordinates (x_i, y_i, z_i) and spin coordinates η_i. The total wave function Ψ is in general a linear combination of Slater determinants. If Eq. (32) is integrated over spins by first setting $\eta_i' = \eta_i$, the first-order density matrix becomes

$$\rho(1, 1') = \sum_{i,j} \gamma_{ij}\phi_j(1)\phi_i^*(1') \quad (34)$$

In matrix form this becomes

$$\rho(1, 1') = \mathrm{tr}\, \Gamma\mathbf{\Phi}\mathbf{\Phi}' \quad (35)$$

where Γ is the $N \times N$ matrix of coefficients γ_{ij} and $\mathbf{\Phi}(\mathbf{\Phi}')$ is a column (row) vector of MOs. A unitary transformation of the matrix product $\Gamma\mathbf{\Phi}\mathbf{\Phi}'$ is then made:

$$\begin{aligned} \rho(1, 1') &= \mathrm{tr}\ \mathsf{U}^\dagger\Gamma\mathbf{\Phi}\mathbf{\Phi}'\mathsf{U} = \mathrm{tr}\ \mathsf{U}^\dagger\Gamma\mathsf{U}\mathsf{U}^\dagger\mathbf{\Phi}\mathbf{\Phi}'\mathsf{U} \\ &= \mathrm{tr}\ \Lambda(\mathsf{U}^\dagger\mathbf{\Phi})(\mathbf{\Phi}'\mathsf{U}) \end{aligned} \quad (36)$$

This results in a new set of orbitals θ_k such that

$$\mathsf{U}^\dagger\mathbf{\Phi} = \mathbf{\Theta} \quad (37)$$

where Θ is the column vector of NOs. Then from Eq. (34)

$$\rho(1,1') = \lambda_1\theta_1(1)\theta_1^*(1') + \lambda_2\theta_2(1)\theta_2^*(1') + \ldots + \lambda_N\theta_N(1)\theta_N^*(1') \quad (38)$$

Setting $1 = 1'$ results in the charge density being defined in terms of *occupation numbers* λ_k of the NOs θ_k. Thus the eigensolution of the first-order density matrix yields a transformation matrix U that defines the NOs θ_k as a specific linear combination of the MOs θ_i and eigenvalues λ_k having magnitudes that are decreasing as the (energetic) importance of the respective NOs decreases.

Since the NOs are strictly obtainable only *after* the density matrix $\rho(1,1')$ has been calculated, they serve, in their truest form, as a means for analyzing the contribution of the successive members of the space of MOs to the total wave function. Their usefulness for reducing the size of the CSF expansion set rests with their derivation using some subset of CSFs defined in the space of the original ("canonical") virtual orbitals, e.g., via an FOCI or perturbation theory procedure [99], the derivation of the NOs, the truncation of this new orbital space based on occupation number, and the construction of a *new*, more extensive set of CSFs in the truncated space of NOs. Note that unless prior constraints on the MOs being transformed are imposed (Section D) the NO transformation modifies *all* MOs, internal *and* external. This is often undesirable, especially if the internal MOs are MCSCF-MOs.

The iterative natural orbital (INO) procedure of Bender and Davidson is the most well-known method for exploiting the properties of NOs [100]. It may be summarized as follows:

1. Make an initial guess of the NOs in terms of an AO basis set.
2. Construct an approximate wave function using some set of CSFs formed from these NOs selected (usually) by chemical intuition.
3. Compute the NOs for this wave function.
4. Repeat steps 2 and 3 selecting the NOs having the largest occupation numbers until the total energy and NOs are stable to within some threshold.

Clearly the INO procedure is not expected to be monotonically convergent and, depending on the choice of NOs, the total energy may increase or decrease at various stages [101]. Nonetheless the INO procedure has proven to be quite dependable in numerous cases, yielding orbitals that closely approximate corresponding MCSCF orbitals. Shavitt *et al.* and Karlström *et al.* present detailed comparisons of the merits of various orbital spaces in accurate CM studies [102].

Several alternative procedures to the use of NOs for arranging the orbital space to be used in a CM wave function have been suggested. Many of these are based on the redefinition of the virtual MOs that result from SCF-MO or MCSCF calculations. The virtual orbitals that result from solutions of the Hartree–Fock–Roothaan equations (Section C) are known to be related to the $(N + 1)$-electron system and are thus not amenable to the truncation procedures used when NOs are available. Since the occupied and virtual spaces are invariant to linear transformations within the respective spaces, several methods have been proposed for rendering the virtual space more appropriate for CM. Huzinaga and Arnau [103] and Davidson [104] have derived methods for redefining the effective Hamiltonian operator such that the virtual orbitals are modified without changing the occupied orbitals, as distinct from NO procedures that require a predetermined density matrix. Davidson has termed this the "internally consistent" self-consistent field (ICSCF) procedure. Hunt and Goddard introduced the improved virtual orbital (IVO) procedure [105] in which an electron is effectively removed from a virtual MO by the deletion of certain two-electron integrals from the Fock matrix defined for a specific state symmetry. The IVO procedure not only serves to define an energy ordered set of virtual orbitals that may be used in truncated CM procedures in the same way that ordered NOs are used, but also yields approximate electronic excitation energies. Morokuma and Iwata [106] have developed another procedure, which they term the electron-hole potential (EHP) method. The EHP method is somewhat more general than the ICSCF or IVO procedures and has been extended to the multireference case. In this respect the EHP method is a complete procedure leading to results for ground and excited molecular states and as such has not been used extensively to provide truncated orbital spaces for additional CM.

In addition to the reduction of the size of the CM problem resulting from the truncation of the orbital space using methods such as those mentioned above, there are procedures (both systematic and intuitive) by which a selection of CSFs can be made. Systematic methods are generally of the type where entire classes of CSFs are defined (such as in the FOCI procedure) and no attempt is made to examine them individually. Reductions in the CSF space are then made by choosing more modest numbers of reference CSFs and then including *all* replacements of a given type with respect to this set [94]. Alternatively, there are procedures whereby for a given orbital space CSFs are selected based on their actual or expected importance to the total wave function.

Probably the most extensively applied procedure of this genre is the "individualized configuration selection procedure" of Buenker and Peyerimhoff [107]. To paraphrase these authors, the main objective of such a selection scheme is to identify and include those CSFs that make an important contribution to the total wave function and eliminate those that do not. They go on to emphasize that any procedure for selection should also contain a means for estimating the errors that result by the elimination of certain CSFs from the wave function. They have combined the concepts of a reference space of MOs, perturbation theory, and extrapolation into an efficient procedure for calculating accurate correlated wave functions.

In more detail, the method [107] goes somewhat as follows:

1. Choose a fairly extensive AO basis set.
2. Choose a rather small number of *main* CSFs to form a basis function Ψ_0 for an MCSCF or INO calculation; experience shows that it is generally sufficient to include only configurations with a coefficient 0.05 or greater in the final CM.
3. Carry out the MCSCF or INO treatment.
4. From Ψ_0 form all CSFs that can be obtained by single or double excitation into the virtual MOs resulting from the MCSCF or INO calculation (see Fig. 1). Test the importance of each of these one at a time by solving a secular equation including Ψ_0 and the CSF to be tested, say the rth CSF (or use second-order perturbation theory [88] to make the test), and determine the energy increment ΔE_r which results. Choose a restricted number of CSFs including all those which exceed a certain ΔE_r threshold T.
5. Using a relatively large T, carry out a CM calculation and convert the MOs to corresponding NOs by diagonalizing the density matrix for the state of interest.
6. Perhaps revise Ψ_0 to include fewer main configurations, as often becomes possible when NOs are used.
7. Solve the CM secular problem at a sufficiently high level of T, including Ψ_0 and all the singly and doubly excited configurations derived from it using the NOs at that T level. By solving the problem at several levels of T, one can by extrapolation arrive at an estimate of the energy as $T \rightarrow 0$, but this may not be necessary in order to obtain the desired potential curves with sufficient accuracy.

In the procedure outlined above, the effects of triple and higher excitations are probably negligible if enough main CSFs from which excitations are made are included in Ψ_0. In the interest of practicality, excitations from some of the innermost MO shells, and certainly from the K shells, may have to be omitted with some loss of accuracy in the results.

One possible difficulty with this approach is that the wave functions describing successive different geometries on a potential energy surface may be of varying accuracy, leading to a somewhat "bumpy" surface. Raffenetti *et al.* have proposed a related "cumulative selection procedure" that claims to treat potential surfaces more consistently [108].

It is safe to claim as generally accepted that a CM procedure whereby all single and double replacement CSFs are defined with respect to a multireference internal orbital space that contains all important CSFs will lead to an accurate wave function. As exemplified above for N_2, however, a full "second-order" calculation using an adequate AO basis with no MO truncation is prohibitively large. This is the case for nearly all molecules.

Buenker, Peyerimihoff, and co-workers have extended the ideas of the "individual configuration selection procedure" to what they presently term the "multireference double-configuration interaction" (MRDCl) method [109]. Since numerous calculations from these authors are reviewed in later chapters, we outline this method here in further detail. To proceed with an MRDCl calculation one decides on the number *m* of roots (states) of a given symmetry that are desired. Next a set of *n* reference (R) or main (M) CSFs are selected based on their expected importance at all geometries and for all *m* states under study. This choice is denoted "*n*M*m*R" and thus defines the level of calculation. All single and double replacement CSFs with respect to the *n* reference CSFs are then generated. This set of CSFs is generally prohibitively large (e.g., for a 72-term primitive basis and 5 reference CSFs for N_2 a total of 54,267 CSFs result; see Buenker *et al.* [109]). A subset of this total set is chosen using a perturbation theory procedure to estimate the contribution of each CSF from the secondary set. If the contribution exceeds some threshold *T* it is included in the final wave function. The threshold is systematically lowered and extrapolated to zero (see above). At some intermediate stage of the calculation an NO transformation may be invoked to better characterize the reference CSFs. The MRDCI method is thus orbital dependent to the extent that a choice of the nature of the orbital space is required. This is true of several of the other methods mentioned above as well and can possibly bias one state over another. Hopefully with the inclusion of sufficient numbers of CSFs and extrapolation a monotonic convergence to the complete CM result is realized.

One further aspect of the MRDCI method is the employment of a generalization of the Langhoff–Davidson unlinked cluster correction [89] to attain an estimate of the "full CI limit." The correction is given by (see Buenker and Peyerimhoff [109])

$$\Delta E = \left(1 - \sum_{\text{ref}} C_0^2\right)(E_{\text{CI}} - E_{\text{ref}}), \tag{39}$$

where the sum includes coefficients of all of the reference (main) CSFs. This formula was not derived but was written down by analogy to the formula [89] based on a single-reference CSF and is therefore somewhat suspect. However, it has been applied in many calculations and has apparently provided consistent improvements of MRDCI results.

G. ELECTRON PAIR METHODS

Some of the highest quality *ab initio* calculations of molecular structure and properties in recent years have been based on methods that incorporate electron correlation using what may be categorized as being based on the "independent electron pair approximation" (IEPA). Review articles by Kutzelnigg [111] and by Meyer [112] provide excellent discussions of the general topic, including the formal aspects, and contain most of the pertinent references up to 1977. The following discussion, taken mainly from our earlier volume [2], is given here to set forth the notation used in electron pair methods and to demonstrate the forms that the energies and wave functions take on. Many calculations using electron pair methods are discussed in later chapters.

A major bottleneck in the use of configuration mixing techniques is the inherently slow convergence of the expansion of a molecular wave function in terms of CSFs. To illustrate this difficulty the wave function is expressed as a sum of terms, each of which corresponds to a certain level of substitution,

$$\Phi = C_0\Phi_0 + \sum_{I,A} C_I{}^A\Phi_I{}^A + \sum_{\substack{I<J\\A<B}} C_{IJ}{}^{AB}\Phi_{IJ}{}^{AB} + \dots, \tag{40}$$

where the CSFs $\Phi_I{}^A$, $\Phi_{IJ}{}^{AB}$, ..., represent single, double, etc. substitutions of electrons from the occupied $(I, J, \dots)$ to the unoccupied or virtual $(A, B, \dots)$ orbitals, and Φ_0 is either the SCF solution or a small set of dominant reference CSFs obtained from, for example, an MCSCF calculation.

The number of CSFs in each summation in Eq. (40) is nearly proportional to M^i, where M is the number of MOs and i is the level of substitution. It is clear from Eq. (40) that the inclusion of double, triple, and higher substitutions soon results in excessive numbers of CSFs.

The single replacements, represented by the $C_I{}^A$ of Eq. (40) can be made to vanish by Brillouin's theorem if the MO basis is properly chosen (all $\Phi_I{}^A \equiv 0$). They can also be eliminated by an appropriate choice of MOs such that the coefficients $C_I{}^A$ are zero. The MOs are usually termed "Brillouin" or "Brueckner" orbitals, respectively, depending on which of the above choices is made. The total energy may then be expressed [112]

$$E = E_0 + \left|\sum_{I<J}\right| \varepsilon_{IJ} \tag{41}$$

where E_0 is the energy due to $C_0\Phi_0$ and the ε_{IJ} are the so-called pair correlation energies.

The possibility of expressing the total energy as a sum of pair energies has been suggested by several authors [113]. The earliest attempts to compute the ε_{IJ} were based on the use of orbitals termed antisymmetrized products of strongly orthogonal geminals (APSG) [114] $\Phi_k{}^R$ for $R = 1, 3, \dots, n/2$ pairs of electrons and $k = 1, 3, \dots, m$ MOs such that

$$\langle \Phi_k{}^R | \Phi_l{}^S \rangle = \delta_{RS}\delta_{kl} \tag{42}$$

The APSGs are said to account only for intrapair correlation effects. The important consequence of strong orthogonality is that the n-electron system is represented by a sum of localized pairs and the variational procedure carried out on coefficients of the doubly substituted APSGs. Quadruple and higher substitutions may be approximated as products of doubles, termed "unlinked clusters" [96]. Methods that are represented by the form Eq. (41) are usually termed independent electron pair approximations (IEPA) [111]. The IEPA may also include interpair terms, i.e., singlet and triplet coupled substitutions from different geminals. It should also be noted that IEPA energies are, in general, nonvariational.

Having a wave function that is expressed as a sum of pair interaction terms lends itself to the application of natural orbital transformations as outlined in Section F. Edmiston and Krauss [115] showed that in the framework of the IEPA, natural orbitals could be computed for the separate pairs and subsequently employed in a wave function of the form Eq. (40) in a conventional configuration mixing procedure. Since these are not the true natural orbitals that would be derived from a total density matrix for a wave function of the system taken as a whole, they were termed pseudonatural orbitals (PNOs). The designation "pair natural orbitals" is used interchangeably with pseudonatural orbitals and is generally recognized as having the same meaning.

The PNOs derived from an IEPA procedure can be used with standard configuration mixing (Section D) to get new mixing coefficients for the wave function (normalized such that the coefficient of Φ_0 is unity)

$$\Psi_{\text{CI}} = \Phi_0 + \sum_{\substack{A<B \\ B<J}} \Phi_{IJ}{}^{AB} \tag{43}$$

that will insure a variational result for the energy. (We use "CI" instead of "CM" here in deference to the notation of the original authors.) Meyer has formulated a PNO-CI procedure whereby MOs ϕ_a and ϕ_b, into which substitutions are made, are NOs of the IEPA geminals. An important feature is the relaxation of the orthogonality constraint, viz., whereas the $\Phi_{IJ}{}^{AB}$ are mutually orthogonal the PNOs of different geminals are nonorthogonal since for each $\Phi_{IJ}{}^{AB}$ the density matrix for the pair is put into diagonal form and the orbitals suitably rotated into NOs independently of the remaining pairs.

The coefficients in the expansion for Ψ_{CI} of Eq. (43) are obtained as eigenvectors of the Hamiltonian matrix for m pairs:

$$\mathsf{H}^{\mathrm{CI}} = \begin{pmatrix} E_0 & \mathsf{H}_{01} & \dots & \mathsf{H}_{0m} \\ \mathsf{H}_{10} & \mathsf{H}_{11} & \dots & \mathsf{H}_{1m} \\ \cdot & \cdot & & \cdot \\ \cdot & \cdot & & \cdot \\ \cdot & \cdot & & \cdot \\ \mathsf{H}_{m0} & \mathsf{H}_{ml} & \dots & \mathsf{H}_{mm} \end{pmatrix} = \mathsf{H}_{\mu\nu} \tag{44}$$

where E_0 is a constant and the block index μ refers to the number of pairs [116]. The eigenvectors of H are of the same blocked form

$$\mathsf{C}^{\mathrm{CI}} = (1, \mathbf{C}_1{}^{\mathrm{CI}}, \dots, \mathsf{C}_m{}^{\mathrm{CI}}). \tag{45}$$

The m coupled eigenvalue equations for the PNO-CI method are

$$\begin{pmatrix} E_0 & \mathsf{H}_{0\mu} \\ \mathsf{H}_{\mu 0} & \mathsf{H}_{\mu\mu} \end{pmatrix} \begin{pmatrix} 1 \\ \mathsf{C}_\mu{}^{\mathrm{CI}} \end{pmatrix} - E_{\mathrm{CI}} \begin{pmatrix} 1 \\ \mathsf{C}_\mu{}^{\mathrm{CI}} \end{pmatrix} = - \sum_{\nu \neq \mu = 1}^{m} \begin{pmatrix} \mathsf{H}_{0\nu} & \mathsf{C}_\nu{}^{\mathrm{CI}} \\ \mathsf{H}_{\mu\nu} & \mathsf{C}_\nu{}^{\mathrm{CI}} \end{pmatrix} \tag{46}$$

These may be compared with the IEPA equations

$$\begin{pmatrix} E_0 & H_{0\mu} \\ \mathsf{H}_{\mu 0} & H_{\mu\mu} \end{pmatrix} \begin{pmatrix} 1 \\ \mathsf{C}_\mu^{\mathrm{IEPA}} \end{pmatrix} - (E_0 + \varepsilon_\mu^{\mathrm{IEPA}}) \begin{pmatrix} 1 \\ \mathsf{C}_\mu^{\mathrm{IEPA}} \end{pmatrix} = 0 \tag{47}$$

Eqs. (47) may be used in an iterative procedure to obtain the C^{CI} of Eq. (46) with the C^{IEPA} as first approximations [116].

To facilitate the inclusion of unlinked clusters Meyer [112] has proposed the use of what he terms the coupled-electron-pair approximation (CEPA). This extension of the PNO-CI method may be described by writing the energy

$$E_{\mathrm{CI}} = E_0 + \sum_{u} \mathsf{H}_{0\mu} \mathsf{C}_\mu{}^{\mathrm{CI}} \tag{48}$$

and defining the PNO-CI pair correlation energy

$$\varepsilon_\mu{}^{\mathrm{CI}} = \mathsf{H}_{0\mu} \mathsf{C}_\mu{}^{\mathrm{CI}} \tag{49}$$

Eq. (46) then becomes

$$\begin{pmatrix} E_0 & \mathsf{H}_{0\mu} \\ \mathsf{H}_{\mu 0} & \mathsf{H}_{\mu\mu} \end{pmatrix} \begin{pmatrix} 1 \\ \mathsf{C}_\mu{}^{\mathrm{CI}} \end{pmatrix} - (E_0 + \varepsilon_\mu{}^{\mathrm{CI}}) \begin{pmatrix} 1 \\ \mathsf{C}_\mu{}^{\mathrm{CI}} \end{pmatrix} = - \sum_{\nu \neq \mu} \begin{pmatrix} 0 \\ \mathsf{H}_{\mu\nu} \mathsf{C}_\mu{}^{\mathrm{CI}} - \varepsilon_\nu{}^{\mathrm{CI}} \mathsf{C}_\mu{}^{\mathrm{CI}} \end{pmatrix} \tag{50}$$

$$\begin{pmatrix} E_0 & \mathsf{H}_{0\mu} \\ \mathsf{H}_{\mu 0} & \mathsf{H}_{\mu\mu} \end{pmatrix} \begin{pmatrix} 1 \\ \mathsf{C}_\mu^{\mathrm{CEPA}} \end{pmatrix} - (E_0 + \varepsilon_\mu^{\mathrm{CEPA}}) \begin{pmatrix} 1 \\ \mathsf{C}_\mu^{\mathrm{CEPA}} \end{pmatrix} = - \sum_{\nu \neq \mu} \begin{pmatrix} 0 \\ \mathsf{H}_{\mu\nu} \mathsf{C}_\nu^{\mathrm{CEPA}} \end{pmatrix} \tag{51}$$

These are the CEPA equations and lead to a total energy of the form

$$E_{\mathrm{CEPA}} = E_0 + \sum_{\mu} \varepsilon_\mu^{\mathrm{CEPA}} \tag{52}$$

which, in spite of being nonvariational, has been shown to provide significantly improved results for many molecules [111, 112, 116]. Equation (52) has the correct form for noninteracting spaces but contains the interactions between the pairs by virtue of the presence of the $H_{\mu\nu}$ in Eq. (51). The physical distinction between the IEPA and CEPA is that whereas the former treats each electron pair in the *Hartree–Fock* field of the other electrons, the latter treats each pair in the fields of the other *correlated* electrons.

A related method, termed the theory of self-consistent electron pairs (SCEP), has been developed [117]. It combines the features of the PNO-CI approach, first-order perturbation theory, and the direct determination of the correlated wavefunction from the basis function matrix elements (Section B) without the construction of a Hamiltonian matrix. Since no configuration list is required, the use of large basis sets is not prohibitive and, in fact, the dependence on basis set size is nearly proportional to that of a corresponding SCF procedure multiplied by the number of pairs. The SCEP method provides orbitals that are optimum in the space of single and double replacements relative to suitably chosen reference configurations and gives variationally additive pair correlation energies. Dykstra [118] has extended the SCEP method to include certain open-shell cases, the effects of quadruple replacements, and the incorporation of multiconfiguration reference wavefunctions in the context of the GVB-PP theory (Section F).

H. MANY-BODY METHODS

Several methods that provide systematic ways to improve results based on SCF, MCSCF, and CM calculations have been developed. A simple formula for estimating the effects on the total energy of quadruple replacements from a single reference configuration due to Davidson and Langhoff was given in Section F. This formula is the result of a specialization of the general Rayleigh–Schrödinger perturbation expansion of diagrammatic many-body perturbation theory (MBPT).

Applications of MBPTs to the electron correlation problem have been reviewed by Bartlett *et al.* [119], Beck and Nicholaides [120], and Robb *et al.* [121]. Standard CM procedures suffer from the shortcomings that the convergence is slow with respect to levels of substitution, the number of CSFs rapidly becomes prohibitively large, and any CM procedure that invokes truncation of the space of CSFs is not "size consistent." The first two problems were discussed in Section E, namely, the difficulties surrounding the approach to full CM. The third, that of size consistency, is of special concern when dissociation processes are under study. In general, a system of N particles (e.g., atoms or molecules) is size consistent if the energy of a lattice of N noninteracting particles is exactly N times the energy of a single particle [119, 122]. The MBPTs can be made size consistent at various orders and are thus more straightforwardly applicable to such problems as heats of reaction where two calculations (reactants and products) suffice.

As was mentioned in Section F, the most important contribution to the correlation energy is due to double replacements with respect to one dominant reference configuration. The next most important is due to quadruple replacements of which there are two types [113,119]. The first consists of the simultaneous interaction of four electrons and the second of two simultaneous interactions of two electrons. The second type occurs more frequently, is more physically interpretable, and is expected to recover the major portion of the total correlation energy. These "unlinked cluster" contributions form the basis for the several "coupled cluster" approaches to the correlation problem. Cizek [123] has termed the procedure whereby only two-body terms involving the double replacements and the unlinked clusters are included, the "coupled-pair many-electron theory" or CPMET.

If the total wave function is written in the form given in Eq. (40), with Φ_0 the Hartree–Fock (HF) wave function, the total energy is defined by $\langle\Phi_0|\hat{H} - E|\Psi\rangle = 0$, or [119]

$$E = E_{\mathrm{HF}} + \sum_{\substack{I>J\\A>B}} C_{IJ}{}^{AB}\,\langle\Phi_0|\hat{H}|\Phi_{IJ}{}^{AB}\rangle. \tag{53}$$

As long as the coefficients $C_{IJ}{}^{AB}$ are *exact* Eq. (53) is an exact expression for E. This is exploited in the electron pair methods discussed in Section G and in the MBPT and coupled cluster approaches (CCA). In standard CM and IEPA procedures the coefficients corresponding to the double replacements $C_{IJ}{}^{AB}$ are defined through the solution of sets of coupled equations involving the other coefficients, i.e., $\langle\Phi_I{}^A|H - E|\Phi_0\rangle = 0$, $\langle\Phi_{IJ}{}^{AB}|\hat{H} - E|\Phi_0\rangle = 0$, $\langle\Phi_{IJK}^{ABC}|\hat{H} - E|\Phi_0\rangle = 0$, etc. Actual calculation to high accuracy using this approach rapidly becomes impractical. Meyer's CEPA procedure (Section G) is one example of the use of a CCA to improve on less complete CM results. In general, the CCAs treat the expansion of Eq. (40) in terms of cluster operators involving one-body, two-body, etc. interactions. Certain categories of MBPT diagrams are summed to all orders and the resulting nonlinear equations are solved iteratively [119].

Pople and co-workers [122] have developed a procedure for calculating the correlation energy to fourth order in Møller–Plesset perturbation theory. Their approach is based on a single determinant reference configuration derived using a restricted or unrestricted Hartree–Fock calculation. (The MBPT methods discussed above also have this requirement.) The development is analogous to that of MBPT but is based on a purely algebraic, as opposed to a diagrammatic, formalism.

The MBPT, CCA, and Møller–Plesset methods offer alternatives to standard CM procedures. Applied to single determinant SCF-MO calculations, they present several advantages. A single transformation of integrals over the MOs is required, i.e., additional transformations are unnecessary in proceeding to successively higher orders. They are size consistent as long as appropriately selected higher-order terms are included. Two disadvantages are that, at present, the reference space must be comprised of a single determinant and the methods are nonvariational. The single reference determinant constraint manifests itself in slow convergence for cases such as N_2, where 10 CSFs are needed for proper dissociation behavior.

In the limit of infinite order the MBPT and complete CM results are equivalent. However, MBPT has the property that a more predictable (i.e., monotonic) convergence pattern is expected. MBPT procedures are not difficult to incorporate and, because of their low cost in terms of computational effort, they should be appended to most *ab initio* calculations based on standard SCF, MCSCF, and CM approaches.

The equations of motion (EOM)–Green's function methods employ perturbation theory and the Hamiltonian operator in second quantized form in treating the electron correlation problem. The EOM approach has been reviewed by McCurdy *et al.* [124], the Green's function method by Cederbaum and Domcke [125], and the (related) propagator method by Purvis and Ohrn [126].

A common feature among these methods is that energy differences such as ionization potentials, electron affinities, and excitation energies are computed directly. This avoids the undesirable procedure of computing the energy for two states separately and obtaining the desired transition energy as a difference (usually small) between large numbers. These procedures are also amenable to the computation of other transition properties such as oscillator strengths. As in the MBPT methods the EOM–Green's function procedures are guaranteed to yield the full CM result for a given AO basis provided they are taken to infinite order. Although these methods are still under development, it appears that, within the limitation of a single reference determinant, the convergence to reliable results is, in many cases, faster than that of conventional CM procedures. Herman *et al.* [127] have recently analyzed the importance of higher-order terms and compared EOM–Green's function and CM procedures for some prototype systems. To date, the most successful applications have been to ionization and electron attachment problems, where the creation–annihilation operators appear to be accurate for determining the correlation energy difference accompanying such processes. CM procedures are inherently more demanding when two systems with different numbers of electrons must be treated at the same level of approximation [127].

An important distinction between the CM methods discussed in Sections E and G and the many-body methods is that the former yield total wave functions as products of computation. These can be used in forming expectation values of quantum mechanical operators corresponding to one- and two-electron properties. The many-body methods are defined with respect to a reference wave function and yield only energies or energy differences. However, formal developments in parallel fashion can be made to define certain other properties. The EOM procedure for calculating oscillator strengths is one example [124].

I. PSEUDOPOTENTIALS

That electrons in atomic and molecular systems can be decomposed into two characteristic types, "core" and "valence" electrons, is one of the most basic and widely applied concepts of modern chemistry. If it is assumed that all bonding processes involve only the valence electrons, whereby the core electrons remain independent of their chemical environment, then it should be possible to account for the atomic cores in any molecular or solid systems using a quantity derived from, and uniquely characteristic of, the parent atoms.

The first attempts to place this notion on a sound theoretical basis were set forth by Hellmann [128] and by Gombás [129] in the 1930s. They treated the core electrons by statistical methods and included their effects in the Hamiltonian by an empirical repulsive potential. In 1954 Sternheimer [130] demonstrated that a zero-order potential could be formed by dividing the system Hamiltonian by an approximate eigenfunction. However, not until 1959 was it rigorously shown by Phillips and Kleinman [131] that the explicit dependence of the wave function on the core electrons could be removed by the introduction of a nonlocal operator, the pseudopotential, into the Hamiltonian.

Although several extensive analyses and applications of pseudopotential theory are available [132–137] a brief summary of the Phillips–Kleinman formalism is given to develop notation and to put the main idea on a quantitative basis.

In SCF theory for many-electron atoms each electron is assumed to move in an average field due to all the remaining electrons (Section D).

Self-consistency is attained by the iterative solution of a set of integrodifferential equations for the N electrons (in atomic units)

$$\left(-\frac{1}{2}\nabla^2 - \frac{Z}{r} + \hat{V}_i\right)\phi_i = \varepsilon_i\phi_i \tag{54}$$

where Z is the nuclear charge, ε_i the energy eigenvalue of the AO ϕ_i, and $\hat{V}_i$ is a one-electron operator representing the effective potential, which describes the electron repulsions as the interaction of the ith electron with a static field due to $N - 1$ electrons. In Hartree–Fock (HF) theory

$$\hat{V}_i^{\mathrm{HF}} = \sum_{j\neq i}(2\hat{J}_j - \hat{K}_j) \tag{55}$$

where $\hat{J}_j$ and $\hat{K}_j$ are the Coulomb and exchange operators (Section C).

To obtain an expression for the effective one-electron potential due to the core electrons in an atom, it is assumed that Eq. (54) represents the solution for a valence electron ($i = v$) moving in the potential due to N core electrons (e.g., Li, Be^+, B^{2+}). Following Phillips and Kleinman [131], Eq. (54) is transformed to the equivalent form

$$\left(-\frac{1}{2}\nabla^2 - \frac{Z}{r} + \hat{V}_v^{\mathrm{HF}} + \hat{V}_{\mathrm{R}}^{\mathrm{PK}}\right)\chi_v = \varepsilon_v\chi_v \tag{56}$$

where

$$\hat{V}_{\mathrm{R}}^{\mathrm{PK}}\chi_v = \sum_{c=1}^{N}\langle\phi_c|\chi_v\rangle(\varepsilon_v - \varepsilon_c)\phi_c \tag{57}$$

and

$$\chi_v = \phi_v + \sum_{c=1}^{N} a_c\phi_c \tag{58}$$

In Eq. (56) $\hat{V}_v^{\mathrm{HF}}$ is defined in Eq. (55), $\hat{V}_{\mathrm{R}}^{\mathrm{PK}}$ is a repulsive nonlocal potential, the pseudopotential, and χ_v is the pseudoorbital. The function χ_v in Eq. (58) is arbitrary to the extent that any coefficients a_c that effectively remove the HF orthogonality constraint may be employed. This development is perhaps best put in perspective by Weeks *et al.* [134]:

> ... What Phillips and Kleinman have shown is that one can replace orthogonality constraints in an eigenvalue equation by adding a non-local pseudopotential $\hat{V}_{\mathrm{R}}^{\mathrm{PK}}$ to $\hat{H}$ without changing the valence eigenvalues. The main point here is that a conceptual shift has been made, from the study of a function satisfying certain orthogonality constraints and an unmodified Hamiltonian, to the study of an unconstrained function satisfying a modified Hamiltonian ... .

These features notwithstanding, several operational problems have been introduced.

The fact is that the pseudopotential in Eq. (57) is neither unique nor local; the former is troublesome because one must examine a reasonably large number of choices for the a_c in order to obtain a smooth pseudopotential, and the latter because the nonlocal nature of V_R^{PK} makes the evaluation of all two-electron integrals between the valence and core orbitals necessary. Thus, the use of the "exact" pseudopotential results in a problem of comparable difficulty to the initial all-electron problem.

Several methods for deriving pseudopotentials which circumvent the above difficulties have been described [133, 135, 136, 138]. An approach due to Cohen and Heine [135b] rests on the argument that since the pseudopotential is required to be smooth in the core region any pseudowave function which represents a reasonable repulsive potential to the valence region can be employed with some success. The problem of nonuniqueness has been examined in detail by these authors and two of their solutions involve the variational smoothing of the pseudowave function [Eq. (58)]. They showed that the minimization of the kinetic energy operator ($-\frac{1}{2}\nabla^2$) or the operator $V_b^{HF} - V_R^{PK}$ should effectively remove the undesired oscillations in the core region. Thus, a unique and local pseudopotential may be formed by fixing the coefficients a_c at the values which yield a minimum with respect to the chosen operators. This procedure yields a local effective potential

$$U^{HF-PK} = \frac{V_v^{HF}\chi_{min}}{\chi_{min}} + \frac{V_R^{PK}\chi_{min}}{\chi_{min}}, \tag{59}$$

where χ_{min} is the solution of

$$\sum_{c=1}^{N} \frac{\partial}{\partial a_c} [\langle\chi_v|\mathcal{M}|\chi_v\rangle / \langle\chi_v|\chi_v\rangle] = 0 \tag{60}$$

and $\mathcal{M}$ represents the kinetic energy or the nonlocal effective potential. It should be noted that these minimization procedures are based on the reasoning that the principal criterion for a good pseudopotential is that it be smooth in the core region. Thus, although the resultant pseudopotentials are unique, as yet no rigorous justification for the procedure has been shown. Christiansen *et al.* [138] proposed an alternative form for Eq. (58) whereby both terms are multiplied by coefficients such that the pseudoorbital becomes identical to the HF AO beyond a certain radius from the nucleus. The pseudopotential derived from this form of pseudoorbital is generally *not* smooth in the core region, but has been shown to yield excellent results when applied in molecular calculations [138].

The categorization of calculations that employ pseudopotentials as semiempirical or as *ab initio* is based on whether or not experimental atomic data is incorporated. However, the inherent arbitrariness of all pseudopotentials that are made local by procedures such as those mentioned above makes this distinction less clear than for all-electron calculations since this is an additional approximation. Pseudopotentials are also often designated as "model potentials" if a predetermined functional form is assumed. The term "effective core potential" is preferred for those pseudopotentials derived using the pseudoorbital procedure [136].

The important aspect of the use of pseudopotentials is that the *size* of the problem for a given molecule is reduced; i.e., both the AO basis *and* the MO basis may be made smaller since the core electrons are represented by the pseudopotential. The savings are especially significant for molecules containing heavy atoms since the core is comprised of a large number of electrons. The pseudopotentials may be used in the context of any of the methods previously discussed (Sections C–H).

Recently the Dirac–Hartree–Fock and related methods for atomic wave functions have been used to derive pseudopotentials that introduce relativistic effects into molecular wave functions [139]. Thus, in addition to the reduction of the problem size through the replacement of the core electrons, important changes in the description of the nature of bonding in heavy-atom systems have been noted. Calculations using relativistic pseudopotentials for UF_5 [140] and UF_6 [141] have been reported.

J. EQUILIBRIUM GEOMETRIES AND VIBRATIONAL PROPERTIES

One of the most important (and most successful) applications of *ab initio* calculations to polyatomic molecules has been for the confirmation and prediction of molecular geometries. The Born–Oppenheimer approximation and the variation principle allow for the determination of the minimum point on a molecular potential energy surface by a point-by-point mapping through successive calculations of the total energy. Although this procedure must ultimitely lead, for a bound system, to a minimum corresponding to the equilibrium geometry, it becomes increasingly more complicated as the number of atoms is increased. This is due to the $3N - 6$ degrees of freedom ($3N - 5$ for a linear molecule), where N is the number of atoms. Symmetry considerations and a reasonable guess of the equilibrium geometry reduce the complexity to some degree. As discussed in Section B, Pople and co-workers have been successful in the determination of geometries of a large number of molecules using such a point-by-point procedure [47].

It has become apparent over the past several years that a more efficient procedure for determining the characteristics of a potential energy surface is through the calculation of derivatives of the total energy directly from the molecular wave function. Pulay has developed a method by which this can be accomplished through the computation of three types of force integrals: the Hellmann–Feynman, density, and integral forces [142]. These are calculated at the stage where the necessary Hamiltonian matrix elements with respect to the AO basis are constructed. Expressions are derived for these force integrals from the definition of the total energy in terms of the SCF expansion coefficients (Section C) and the nuclear coordinates [142]. Since one of the types, the integral force, is a two-electron operator, the integral stage of such a calculation is made more complex. This is more than compensated for, however, by the savings realized by the necessity to calculate fewer points on the potential energy surface.

In addition to the first derivative or gradient, which is used primarily to determine the equilibrium geometry, higher derivatives can be calculated using this "gradient method." Additional points must be calculated but the number is still fewer than that required in a point-by-point procedure. The optimum increments in geometry for mapping out a surface have been investigated [143]. The gradient method has recently been applied using Møller–Plesset perturbation theory [144] open-shell SCF [145], and MCSCF [146] procedures.

The higher derivatives define the force constants of the potential energy surface. These are related to the spectroscopic constants through the analysis of the Schrödinger equation for nuclear motion [147, 148]. Knowledge of the successive derivatives then allows for the *ab initio* computation of vibration–rotation spectra (within the accuracy of a force constant representation of the potential energy surface). However, the quality of existing *ab initio* potential energy surfaces is generally not sufficient to permit the inclusion of anharmonicity and other refinements in the vibrational analysis; i.e., in most cases only harmonic frequencies can be extracted. Certain triatomic molecules have been investigated in quite thorough analyses of nuclear motion (see Chapters II–IV).

REFERENCES

1. R. S. Mulliken and C. C. J. Roothaan, *Proc. Natl. Acad. Sci. U.S.A.* **45,** 394 (1959).
2. R. S. Mulliken and W. C. Ermler, "Diatomic Molecules: Results of Ab Initio Calculations." Academic Press, New York, 1977.
3. H. F. Schaefer, III, ed., "Modern Theoretical Chemistry, Vol. 3, Methods of Electronic Structure Theory." Plenum, New York, 1977.
4. H. F. Schaefer, III, ed., "Modern Theoretical Chemistry, Vol. 4, Applications of Electronic Structure Theory." Plenum, New York, 1977.
5. G. A. Segal, ed., "Modern Theoretical Chemistry, Vol. 7, Semiempirical Methods of Electronic Structure Calculation, Part A, Techniques." Plenum, New York, 1977.

6. G. A. Segal, ed., "Modern Theoretical Chemistry, Vol. 8, Semiempirical Methods of Electronic Structure Calculation, Part B, Applications." Plenum, New York, 1977.
7. D. B. Cook, "Structures and Approximations for Electrons in Molecules." Wiley, New York, 1978; see also T. H. Halgren, K. A. Klein, J. H. Hall, Jr., L. D. Brown, and W. N. Lipscomb, *J. Am. Chem. Soc.* **100,** 6595 (1978).
8. R. S. Mulliken, *Annu. Rev. Phys. Chem.* **29,** 1 (1978); see also D. A. Ramsay and J. Hinze, eds., "Selected Papers of R. S. Mulliken." Univ. of Chicago Press, Chicago, Illinois, 1975.
9. B. M. Gimarc, "Molecular Structure and Bonding." Academic Press, New York, 1979; J. P. Lowe, "Quantum Chemistry." Academic Press, New York, 1979.
10. "Theoretical Chemistry" (R. N. Dixon and C. Thomson, eds.), Vol. 1. Chem. Soc., London, 1974; Vol. 2, 1975; Vol. 3, 1978.
11. W. G. Richards, T. E. H. Walker, and R. K. Hinkley, "A Bibliography of *Ab Initio* Molecular Wave Functions." Oxford Univ. Press (Clarendon), London and New York, 1971; W. G. Richards, T. E. H. Walker, L. Farnell, and P. R. Scott, "Supplement for 1970–1973." *ibid.*, 1974; W. G. Richards, T. E. H. Walker, and P. R. Scott, "Supplement for 1974–1977." *ibid.*, 1978.
12. R. von R. Schleyer, W. J. Hehre, L. Radom, and J. A. Pople, "A Bibliography of *Ab Initio* Calculations," to be published.
13. L. C. Snyder and H. Basch, "Molecular Wave Functions and Properties: Tabulated from SCF Calculations in a Gaussian Basis Set." Wiley, New York, 1972.
14. H. F. Schaefer, III, "The Electronic Structure of Atoms and Molecules," Addison-Wesley, Reading, Massachusetts, 1972.
15. H. F. Schaefer, III, *Annu. Rev. Phys. Chem.* **27,** 261 (1976).
16. D. H. Whiffen, *Pure Appl. Chem.* **50,** 75 (1978).
17. G. Herzberg, "Molecular Spectra and Molecular Structure, Vol. III, Electronic Spectra and Electronic Structure of Polyatomic Molecules." Van Nostrand, Princeton, New Jersey, 1966.
18. G. Herzberg, "The Spectra and Structure of Simple Free Radicals." Cornell Univ. Press, Ithaca, New York, 1971.
19. See, e.g., I. N. Levine, "Molecular Spectroscopy." Wiley, New York, 1975; D. C. Harris and M. D. Bertolucci, "Symmetry and Spectroscopy." Oxford Univ. Press, London and New York, 1977; P. R. Bunker, "Molecular Symmetry and Spectroscopy." Academic Press, New York, 1979.
20. See, e.g., D. W. Turner, C. Baker, A. D. Baker, and C. R. Brundle, "Molecular Photoelectron Spectroscopy." Wiley, New York, 1970; C. R. Brundle and A. D. Baker, eds., "Electron Spectroscopy: Theory and Techniques," 3 vols. Academic Press, New York, 1979; R. E. Ballard, "Photoelectron Spectroscopy and Molecular Orbital Theory." Halsted, New York, 1978; K. Siegbahn, C. Nordling, G. Johansen, J. Hedman, P. F. Heden, K. Hamrin, U. Gelius, T. Bergmark, L. O. Werme, R. Manne, and Y. Baer, "ESCA Applied to Free Molecules." North-Holland Publ., Amsterdam, 1969; T. A. Carslon, "Photoelectron and Auger Spectroscopy." Plenum, New York, 1975; J. P. Pendry, "Low Energy Electron Diffraction." Academic Press, New York, 1974; L. Azároff, "X-Ray Spectroscopy." McGraw-Hill, New York, 1974; B. S. Rabinovitch, J. M. Schurr, and H. L. Strauss, eds., *Annu. Rev. Phys. Chem.* **30** (1979).
21. See, e.g., K. N. Rao and C. W. Mathews, eds., "Molecular Spectroscopy: Modern Research," 2 vols. Academic Press, New York, 1972, 1976; M. B. Robin, "High Excited States of Polyatomic Molecules," 2 vols. Academic Press, New York, 1974, 1975; J. W. Robinson, ed., "Handbook of Spectroscopy." CRC Press, Cleveland,

Ohio, 1974; R. F. Barrow, D. A. Long, and J. Sheridan, eds., "Molecular Spectroscopy," 5 vols. Chem. Soc., London, 1978.

22. See, e.g., R. F. W. Bader, "An Introduction to the Electronic Structure of Atoms and Molecules." Clarke, Irwin, Vancouver, 1970; J. R. van Wazer and I. Absar, "Electron Densities in Molecules and Molecular Orbitals." Academic Press, New York, 1975.
23. See, e.g., P. E. Cade, W. H. Henneker, and A. Kostyla, *J. Chem. Phys.* **70,** 3715 (1979).
24. See, e.g., S. Fraga and G. Malli, "Many-Electron Systems: Properties and Interactions." Saunders, Philadelphia, Pennsylvania, 1968; S. R. Langhoff and C. W. Kern, in Ref. 4; P. Pyykko, *Adv. Quantum Chem.* **11,** 353 (1978); K. S. Pitzer, *Acc. Chem. Res.* **12,** 271 (1979).
25. G. Herzberg, "Molecular Spectra and Molecular Structure, Vol. II, Infrared and Raman Spectra of Polyatomic Molecules." Van Nostrand, Princeton, New Jersey, 1945.
26. See T. H. Dunning, Jr. and P. J. Hay, in Ref. 3; E. Clementi and C. Roetti, *At. Data Nucl. Data Tables* **14,** 177 (1974).
27. R. M. Stevens, *J. Chem. Phys.* **52,** 1397 (1970); **55,** 1725 (1971).
28. I. Shavitt and M. Karplus, *J. Chem. Phys.* **43,** 415 (1965).
29. See, e.g., J. H. Hall, Jr., D. S. Marynick, and W. N. Lipscomb, *J. Am. Chem. Soc.* **96,** 770 (1974).
30. P. S. Bagus, B. Liu, A. D. McLean, and M. Yoshimine, *in* "Wave Mechanics: The First Fifty Years" (W. C. Price, S. S. Chissick, and T. Ravensdale, eds.). Butterworth, London, 1973.
31. S. F. Boys, *Proc. R. Soc. London Ser. A* **200,** 542 (1950); see also I. Shavitt, *Methods Comput. Phys.* **2,** 1 (1963).
32. D. R. Davis and E. Clementi, IBMOL, Program 92, QCPE, Indiana Univ., Bloomington, Indiana.
33. J. W. Moskowitz and L. C. Snyder, in Ref. 3.
34. R. C. Raffenetti, *J. Chem. Phys.* **58,** 4452 (1973).
35. W. J. Hehre, W. A. Lathan, R. Ditchfield, M. D. Newton, and J. A. Pople, GAUSSIAN 70, Program 236, QCPE, Indiana Univ., Bloomington, Indiana; see also J. S. Binkley, R. A. Whiteside, R. Krishnan, R. Seeger, D. J. DeFrees, H. B. Schlegel, S. Topiol, L. R. Kahn, and J. A. Pople, GAUSSIAN 80, *ibid.*
36. J. Almlöf, USIP Rep. 72-09, Univ. Stockholm, 1972; see also P. S. Bagus, U. Wahlgren, and J. Almlöf, *J. Chem. Phys.* **64,** 2324 (1976).
37. M. Dupuis, J. Rys, and H. F. King, *J. Chem. Phys.* **65,** 111 (1976).
38. See J. L. Whitten, *J. Chem. Phys.* **44,** 359 (1966); R. Ahlrichs, *Theor. Chim. Acta* **33,** 157 (1974); S. Shih, R. J. Buenker, S. D. Peyerimhoff, and B. Wirsam, *Theor. Chim. Acta* **18,** 277 (1970).
39. A. A. Frost, in Ref. 3.
40. See N. C. Handy, J. D. Goddard, and H. F. Schaefer, III, *J. Chem. Phys.* **71,** 426 (1979).
41. A. D. McLean and M. Yoshimine, *IBM J. Res. Dev.* **2,** 190, Suppl. (1967); see also, P. E. Cade, K. D. Sales, and A. C. Wahl, *J. Chem. Phys.* **44,** 1973 (1966).
42. E. Clementi and H. Popkie, *J. Chem. Phys.* **57,** 1077 (1972).
43. P.-O. Löwdin, *Adv. Chem. Phys.* **2,** 207 (1959).
44. L. C. Snyder, *J. Chem. Phys.* **46,** 3602 (1967); L. C. Snyder and H. Basch, *J. Am. Chem. Soc.* **91,** 2189 (1969).
45. A. C. Hurley, *Adv. Quantum Chem.* **7,** 315 (1973).

46. P. George, M. Trachtman, A. M. Brett, and C. W. Bock, *Int. J. Quantum Chem.* **12,** 61 (1977); P. George, C. W. Bock, and M. Trachtman, *ibid.* **15,** 423 (1979).
47. J. A. Pople, in Ref. 4.
48. R. Ditchfield, W. J. Hehre, and J. A. Pople, *J. Chem. Phys.* **54,** 724 (1971); J. S. Binkley, J. A. Pople, and W. J. Hehre, *J. Am. Chem. Soc.* **102,** 939 (1980).
49. W. J. Hehre, R. Ditchfield, and J. A. Pople, *J. Chem. Phys.* **56,** 2257 (1972).
50. P. C. Hariharan and J. A. Pople, *Chem. Phys. Lett.* **16,** 217 (1972); *Theor. Chim. Acta* **28,** 213 (1973).
51. See D. J. De Frees, B. A. Levi, S. K. Pollack, W. J. Hehre, J. S. Binkley, and J. A. Pople, *J. Am. Chem. Soc.* **101,** 4085 (1979).
52. R. C. Raffenetti, *J. Chem. Phys.* **58,** 4452 (1973).
53. See, e.g., J. D. Goddard and I. G. Csizmadia, *J. Chem. Phys.* **68,** 2172 (1978); R. Daudel, R. A. Poirier, J. D. Goddard, and I. G. Csizmadia, *Int. J. Quantum Chem.* **15,** 261 (1979).
54. K. Ruedenberg, R. C. Raffenetti, and R. D. Bardo, *in* "Energy, Structure, and Reactivity" (D. W. Smith, ed.). Wiley, New York, 1973; R. C. Raffenetti, *J. Chem. Phys.* **59,** 5936 (1973); R. D. Bardo and K. Ruedenberg, *ibid.* **59,** 5956, 5966 (1973); R. C. Raffenetti and K. Ruedenberg, *ibid.* **59,** 5978 (1973); M. W. Schmidt and K. Ruedenberg, *J. Chem. Phys.* **71,** 3591 (1979); D. F. Feller and K. Ruedenberg, *Theor. Chim. Acta* **52,** 231 (1979).
55. R. Carbo and J. M. Riera, *in "Lecture Notes in Chemistry"* (G. Berthier *et al.,* eds.). Springer-Verlag, New York, 1978; see also H.-L. Hsu, E. R. Davidson, and R. M. Pitzer, *J. Chem. Phys.* **65,** 609 (1976); F. W. Bobrowicz and W. A. Goddard, III, in Ref. 3.
56. See, e.g., M. Dupuis and H. F. King, *Int. J. Quantum Chem.* **11,** 613 (1977); E. R. Davidson, *J. Chem. Phys.* **62,** 400 (1975); R. M. Pitzer, *ibid.* **59,** 1973.
57. See, e.g., D. Firsht and B. T. Pickup, *Int. J. Quantum Chem.* **12,** 765 (1976).
58. J. A. Pople and P. K. Nesbet, *J. Chem. Phys.* **22,** 571 (1954).
59. P.-O. Löwdin, *Rev. Mod. Phys.* **36,** 966 (1964).
60. R. C. Ladner and W. A. Goddard, III, *J. Chem. Phys.* **51,** 1073 (1969).
61. I. Shavitt, in Ref. 3.
62. A. D. McLean and B. Liu, *J. Chem. Phys.* **58,** 1066 (1973).
63. J. Paldus, *J. Chem. Phys.* **61,** 5321 (1974); *Int. J. Quantum Chem., Symp.* **9,** 165 (1975); *Phys. Rev. A* **14,** 1620 (1976).
64. I. Shavitt, *Int. J. Quantum Chem., Symp.* **11,** 131 (1977); **12,** 5 (1978); *Chem. Phys. Lett.* **63,** 421 (1979).
65. B. R. Brooks and H. F. Schaefer, III, *J. Chem. Phys.* **70,** 5092 (1979).
66. B. R. Brooks, W. D. Laidig, P. Saxe, N. C. Handy, and H. F. Schaefer, III, *Phys. Scr.* **21,** 312 (1980).
67. B. O. Roos and P. E. M. Siegbahn, in Ref. 3; *ibid. Int. J. Quantum Chem.* **17,** 485 (1980).
68. E. R. Davidson, *J. Comput. Phys.* **17,** 87 (1975).
69. R. R. Lucchese and H. F. Schaefer, III, *J. Chem. Phys.* **68,** 769 (1978).
70. N. C. Handy, J. D. Goddard, and H. F. Schaefer, III, *J. Chem. Phys.* **71,** 426 (1979).
71. P. E. M. Siegbahn, *J. Chem. Phys.* **70,** 5391 (1979).
72. H. F. Schaefer, III, R. A. Klemm, and F. E. Harris, *Phys. Rev.* **181,** 137 (1969).
73. "Post Hartree-Fock Configuration Interaction," Natl. Resour. Comput. Chem. Rep. LBL-8233, Berkeley, California, 1978.
74. A. Veillard and E. Clementi, *Theor. Chim. Acta* **7,** 133 (1967); see also M. B. Hall, *Int. J. Quantum Chem.* **14,** 613 (1978).

75. W. A. Goddard, III, T. H. Dunning, Jr., W. J. Hunt, and P. J. Hay, *Acc. Chem. Res.* **6,** 368(1973); F. W. Bobrowicz and W. A. Goddard, III, in Ref. 3.
76. E. L. Mehler, D. M. Silver, and K. Ruedenberg, *J. Chem. Phys.* **52,** 1206 (1970).
77. G. Das and A. C. Wahl, *J. Chem. Phys.* **44,** 87 (1967); *ibid.*, in Ref. 3.
78. J. Hinze and C. C. J. Roothaan, *Suppl. Prog. Theor. Phys.* **40,** 37 (1967); J. Hinze, *J. Chem. Phys.* **59,** 6424 (1974).
79. A. Golebiewski, J. Hinze, and E. Yurtsever, *J. Chem. Phys.* **70,** 1101 (1979).
80. B. Levy and G. Berthier, *Int. J. Quantum Chem.* **2,** 307 (1968); **3,** 247 (1969).
81. F. Grein and T. C. Chang, *Chem. Phys. Lett.* **12,** 44 (1971).
82. A. Banerjee and F. Grein, *J. Chem. Phys.* **66,** 1054 (1977).
83. D. R. Yarkony, *J. Chem. Phys.* **66,** 2045 (1977); R. R. Lucchese and D. R. Yarkony, *ibid.* **68,** 2697 (1978).
84. K. Ruedenberg, L. M. Cheung, and S. T. Elbert, *Int. J. Quantum Chem.* **16,** 1069 (1979).
85. D. Yeager and P. Jørgensen, *J. Chem. Phys.* **71,** 755 (1979).
86. C. C. J. Roothaan, J. Detrich, and D. G. Hopper, *Int. J. Quantum Chem., Quant. Chem. Symp.* **13,** 93 (1979).
87. T. C. Chang and W. H. E. Schwartz, *Theor. Chim. Acta* **44,** 45 (1977).
88. Z. Gershgorn and I. Shavitt, *Int. J. Quantum Chem.* **2,** 751 (1968); see also E. R. Davidson and C. F. Bender, *Chem. Phys. Lett.* **59,** 369 (1978).
89. E. R. Davidson, *in* "The World of Quantum Chemistry" (R. Daudel and B. Pullman, eds.). Reidel Publ., Dodrecht, Netherlands, 1974; S. R. Langhoff and E. R. Davidson, *Int. J. Quantum Chem.* **8,** 61 (1974).
90. G. C. Lie and E. Clementi, *J. Chem. Phys.* **60,** 1275, 1288 (1974).
91. H. F. Schaefer, III, *J. Chem. Phys.* **67,** 2290 (1977).
92. W. C. Ermler, A. D. McLean, and R. S. Mulliken, unpublished results; also T. H. Dunning, Jr., D. C. Cartwright, W. J. Hunt, P. J. Hay, and F. W. Bobrowicz, *J. Chem. Phys.* **64,** 4755 (1978).
93. P. J. Hay and T. H. Dunning, Jr., *J. Chem. Phys.* **64,** 5077 (1976).
94. For O_2, see R. P. Saxon and B. Liu, *J. Chem. Phys.* **67,** 5432 (1977).
95. For O_3, see P. J. Hay and T. H. Dunning, Jr., *J. Chem. Phys.* **67,** 2290 (1977).
96. H. J. Silverstone and O. Sinanoğlu, *J. Chem. Phys.* **44,** 1899 (1966); **44,** 3608 (1966).
97. For CO_2, see W. B. England and W. C. Ermler, *J. Chem. Phys.* **70,** 1711 (1979).
98. E. R. Davidson, "Reduced Density Matrices in Quantum Chemistry." Academic Press, New York, 1976, and reference therein.
99. See, e.g., G. C. Lie, B. Liu, and J. Hinze, *J. Chem. Phys.* **59,** 1872 (1973); P. J. Hay, *J. Chem. Phys.* **59,** 2468 (1973).
100. C. F. Bender and E. R. Davidson, *J. Phys. Chem.* **70,** 2675 (1966).
101. K. H. Thuneman, J. Römelt, S. D. Peyerimhoff, and R. J. Buenker, *Int. J. Quantum Chem.* **11,** 743 (1977).
102. I. Shavitt, B. J. Rosenberg, and S. Palalikit, *Int. J. Quantum Chem. Symp.* **10,** 33 (1976); G. Karlström, B. Jonsson, B. O. Roos, and P. E. M. Siegbahn, *Theor. Chim. Acta* **48,** 59 (1978).
103. S. Huzinaga and C. Arnau, *J. Chem. Phys.* **54,** 1948 (1971).
104. E. R. Davidson, *J. Chem. Phys.* **57,** 1999 (1972).
105. W. J. Hunt and W. A. Goddard, III, *Chem. Phys. Lett.* **3,** 414 (1969).
106. K. Morokuma and S. Iwata, *Chem. Phys. Lett.* **16,** 192 (1972); S. Iwata and K. Morokuma, *Theor. Chim. Acta* **33,** 285 (1974).
107. R. J. Buenker and S. D. Peyerimhoff, *Theor. Chim. Acta* **35,** 33 (1974); **39,** 217 (1975).
108. R. C. Raffenetti, K. Hsu, and I. Shavitt, *Theor. Chim. Acta* **45,** 33 (1977).

109. R. J. Buenker, S. D. Peyerimhoff, and W. Butscher, *Mol. Phys.* **35,** 771 (1978); R. J. Buenker and S. D. Peyerimhoff, *in* "Excited States in Quantum Chemistry" (C. A. Nicholaides and D. R. Beck, eds.). Reidel Publ., Dodrecht, Netherlands, 1978; P. J. Bruna, *Gaz. Chim. Ital.* **108,** 395 (1978).
110. P. Prendergast, G. O. Morrell, Jr., and E. F. Hayes, *J. Chem. Phys.* **71,** 124 (1979).
111. W. Kutzelnigg, in Ref. 3.
112. W. Meyer, in Ref. 3.
113. R. K. Nesbet, *Adv. Chem. Phys.* **9,** 321 (1965); **14,** 1 (1969); O. Sinanoğlu, *ibid.* **6,** 315 (1964); **14,** 337 (1969).
114. A. C. Hurley, J. Lennard-Jones, and J. A. Pople, *Proc. R. Soc. London Ser. A* **220,** 446 (1953); W. Kutzelnigg, *J. Chem. Phys.* **40,** 3640 (1964).
115. C. Edmiston and M. Krauss, *J. Chem. Phys.* **45,** 1833 (1966).
116. R. Ahlrichs, H. Lischka, V. Staemmler, and W. Kutzelnigg, *J. Chem. Phys.* **62,** 1225 (1975).
117. W. Meyer, *J. Chem. Phys.* **64,** 2901 (1976); C. E. Dykstra, H. F. Schaefer, III, and W. Meyer, *ibid.* **65,** 2740 (1976); **65,** 5141 (1976).
118. C. E. Dykstra, *J. Chem. Phys.* **67,** 4716 (1978); **68,** 1829 (1978); **72,** 2928 (1980).
119. R. J. Bartlett and I. Shavitt, *Int. J. Quantum Chem., Quantum Chem. Symp.* **11,** 165 (1977); R. J. Bartlett and G. D. Purvis, III, *ibid.* **14,** 561 (1978); R. J. Bartlett, I. Shavitt, and G. D. Purvis, III, *J. Chem. Phys.* **71,** 281 (1979).
120. D. R. Beck and C. A. Nicholaides, *in* "Excited States in Quantum Chemistry" (C. A. Nicholaides and D. R. Bech, eds.), p. 105. Reidel Publ., Dodrecht, Netherlands, 1978; C. A. Nicholaides and D. R. Beck, *ibid.*, p. 143.
121. M. A. Robb, D. Hegarty, and S. Prime, *in* "Excited States in Quantum Chemistry" (C. A. Nicholaides and D. R. Beck, eds.). Reidel Publ., Dodrecht, Holland, 1978; M. A. Robb, *in* "Computational Techniques in Quantum Chemistry" (G. H. F. Diercksen, ed.). *ibid.*, 1975.
122. J. A. Pople, J. S. Binkley, and R. Seeger, *Int. J. Quantum Chem. Symp.* **10,** 1 (1976); J. A. Pople, R. Seeger, and R. Krishnan, *ibid.* **11,** 149 (1977); R. Krishnan and J. A. Pople, *Int. J. Quantum Chem.* **14,** 91 (1978).
123. J. Cizek, *Adv. Chem. Phys.* **14,** 35 (1969).
124. C. W. McCurdy, Jr., T. N. Rescigno, D. L. Yeager, and V. McKoy, in Ref. 3.
125. L. S. Cederbaum and W. Domcke, *Adv. Chem. Phys.* **36,** 205 (1977).
126. G. D. Purvis and Y. Ohrn, *J. Chem. Phys.* **65,** 917 (1976); *Int. J. Quantum Chem., Quantum Chem. Symp.* **11,** 359 (1977).
127. M. F. Herman, K. F. Freed, D. L. Yeager, *J. Chem. Phys.* **72,** 602 (1980); M. F. Herman, K. F. Freed, D. L. Yeager, and B. Liu, *ibid.* 611 (1980).
128. H. Hellmann, *Acta Physiochim. URSS* **1,** 6 (1935); *ibid.* **4,** 225, 324 (1936).
129. P. Gombás, *Z. Phys.* **94,** 473 (1935).
130. R. M. Sternheimer, *Phys. Rev.* **96,** 951 (1954); see also J. O. Hirschfelder, W. Byers Brown, and S. T. Epstein, *Adv. Quantum Chem.* **1,** 255 (1964).
131. J. C. Phillips and L. Kleinman, *Phys. Rev.* **116,** 287 (1959).
132. W. A. Harrison, "Pseudopotentials in the Theory of Metals." Benjamin, New York, 1966.
133. L. Szasz and G. McGinn, *J. Chem. Phys.* **42,** 2363 (1965); *ibid.* **45,** 2898 (1966); *ibid.* **47,** 3495 (1967); *ibid.* **48,** 2997 (1968); *ibid.* **50,** 1404 (1969); *ibid.* **51,** 1681, 3135 (1969); *ibid.* **53,** 3635 (1970).
134. J. D. Weeks, A. Hazi, and S. A. Rice, *Adv. Chem. Phys.* **16,** 283 (1969).
135. V. Heine, *Solid State Phys.* **24,** 1 (1970); M. L. Cohen and V. Heine, *ibid.* 38; V. Heine and D. Weaire, *ibid.*, 250.

136. L. R. Kahn, P. Baybutt, and D. G. Truhlar, *J. Chem. Phys.* **65,** 3827 (1976).
137. R. N. Dixon, in Ref. 10c.
138. P. C. Christiansen, Y. S. Lee, and K. S. Pitzer, *J. Chem. Phys.* **71,** 4445 (1979).
139. G. Das and A. C. Wahl, *J. Chem. Phys.* **64,** 4672 (1976); Y. S. Lee, W. C. Ermler, and K. S. Pitzer, *J. Chem. Phys.* **67,** 5861 (1977); L. R. Kahn, P. J. Hay, and R. D. Cowan, *ibid.* **68,** 2386 (1978); S. N. Dutta, C. S. Ewig, and J. R. Van Wazer, *Chem. Phys. Lett.* **57,** 83 (1978); P. Hafner and W. H. E. Schwarz, *ibid.* **65,** 537 (1979).
140. W. R. Wadt and P. Jeffrey Hay, *J. Am. Chem. Soc.* **101,** 5198 (1979).
141. P. J. Hay, W. R. Wadt, L. R. Kahn, R. C. Raffenetti, and D. H. Phillips, *J. Chem. Phys.* **71,** 1767 (1979).
142. P. Pulay, in Ref. 4; P. Pulay, G. Fogarasi, F. Pang, and J. E. Boggs, *J. Am. Chem. Soc.* **101,** 2550 (1979).
143. J. G. C. M. Van Duijneveldt-Van De Rijdt and F. B. Van Duijneveldt, *J. Mol. Struct.* **35,** 263 (1976).
144. J. A. Pople, R. Krishnan, H. B. Schlegel, and J. S. Binkley, *Int. J. Quantum Chem., Quantum Chem. Symp.* **13,** 225 (1979).
145. J. D. Goddard, N. C. Handy, and H. F. Schaefer, III, *J. Chem. Phys.* **71,** 1525 (1979).
146. S. Kato and K. Morokuma, *Chem. Phys. Lett.* **65,** 19 (1979).
147. I. M. Mills, in Ref. 10, Vol. 1; W. G. Richards *et al., ibid.*
148. G. D. Carney, L. L. Sprandel, and C. W. Kern, *Adv. Chem. Phys.* **37,** 305 (1978).

CHAPTER II

TRIATOMIC MOLECULES AH_2

A. GENERAL CONSIDERATIONS

In Chapters II–V, we deal with molecules or systems containing just three atoms. Many triatomic molecules in their ground states are linear. Some are symmetric (symmetry $D_{\infty h}$), others unsymmetric (symmetry $C_{\infty v}$). Their MOs and overall states are of the same species as for diatomic molecules. The MO species are σ, π, δ, ϕ, ... for $C_{\infty v}$ and σ_g, σ_u, π_u, π_g, ... for $D_{\infty h}$. The state species are Σ^+, Σ^-, Π, Δ, Φ, ... for $C_{\infty v}$ and Σ_g^+, Σ_g^-, Σ_u^+, Σ_u^-, Π_u, Π_g for $D_{\infty h}$, each prefixed by a superscript spin index (1, 2, 3, 4, 5, ... for spin 0, $\frac{1}{2}$, 1, $1\frac{1}{2}$, 2, ...).

Other triatomic molecules in their ground states are nonlinear (bent), either symmetrical (symmetry C_{2v}) or unsymmetrical (symmetry C_s). The group theoretical character tables for these symmetries may be written as follows:

C_{2v}	E	$C_2(z)$	σ_{xz}	σ_{yz}
A_1	1	1	1	1
A_2	1	1	−1	−1
B_1	1	−1	1	−1
B_2	1	−1	−1	1

C_s	E	σ_{yz}
A′	1	1
A″	1	−1

To avoid confusion in the use of the C_{2v} table, it is necessary to agree on a definite choice of the orientations of the axes x, y, z in the molecule [1]. First, the symmetry axis for the twofold rotation C_2 is called the z axis. There is general agreement about this choice. The x axis may next be taken perpendicular to the plane of the molecule, the position of the y axis then follows. The two symmetry planes are then σ_{xz} and σ_{yz}.

Various authors at times have reversed the choices of the x and y axes and thereby the identifications of σ_{xz} and σ_{yz} which are *defined* as, respectively, perpendicular to the y and x axes. As a result, the meanings of the species symbols B_1 and B_2 are interchanged. To avoid this confusion, the axes will here be consistently chosen in the manner indicated above.

In unsymmetrical bent molecules the one remaining symmetry element is the plane of the molecule σ_{yz}; choices of the y and z axes within the symmetry plane may be arbitrary.

Molecules that are linear in their ground states commonly become nonlinear in their excited states, whereas nonlinear ground-state molecules have some linear excited states. Because of these facts, it is often needful to know how states of a given species for one symmetry correlate with those for a lower symmetry. The following tables show these correlations.

Linear	Bent	Linear	Bent	Linear	Bent
Σ_g^+	A_1	Π_g	$A_2 + B_2$	Σ^+	A′
Σ_u^+	B_2	Δ_g	$A_1 + B_1$	Σ^-	A″
Σ_g^-	B_1	Δ_u	$A_2 + B_2$	Π	A′ + A″
Σ_u^-	A_2	Φ_u	$A_1 + B_1$	Δ	A′ + A″
Π_u	$A_1 + B_1$	Φ_g	$A_2 + B_2$	Φ	A′ + A″

In this and Chapters III and IV, *ab initio* computations on ground and excited states of triatomic systems will be reviewed with emphasis on spectroscopic constants, dipole moments and other properties, and, in Chapter V, reaction surfaces.

In Chapters II–V the various molecules and radicals are reviewed for the most part individually or in closely related groups. However, a paper by McCain and Palke [1a] on the calculation of hyperfine coupling constants in a long list of inorganic radicals (NO_2, CO_2^-, BF_2, FCO, FNN, HCO, HCN^-, HNC^-, NF_2, CF_2^-, OF_2^+, NO_2^{2-}, O_3, O_3^+, O_3^-, FOO, FNO^-, HOO, ClO_2, SO_2^-, PF_2, N_3, CNO, and CF_3) is best considered here. The calculations, of the SCF type, were made using an efficient STF program by R. M. Stevens (see Chapter I). but with a minimal basis set, hence they are not of the highest accuracy. Nevertheless, agreements with experimental values where available are for the most part relatively good. Computed total energies and LCAO coefficients for unpaired electrons are listed. Molecular geometry is explored for NF_2, O_3^+, O_3, O_3^-, and CF_3; the respective results are angle FNF = 101°, R(NF) = 1.35 Å; angle OOO = 132°, R(OO) = 1.28 Å; angle OOO = 117°, R(OO) = 1.29 Å; angle OOO = 114°, R(OO) = 1.37 Å; angle FCF = 111°, R(CF) = 1.38 Å. Other properties reported are isotropic coupling constants of the hyperfine coupling tensor for π radicals, based on the Fermi contact expression; unpaired electron densities at nuclei for σ and π radicals; and anisotropic coupling constants of the hyperfine coupling tensor for σ and π radicals. Here π radicals are planar species with an open-shell MO composed of p atomic orbitals oriented perpendicular to the nuclear plane; σ radicals are defined as those species which have an open-shell MO exhibiting nonzero spin density at one or more of the nuclei.

B. SURVEY

Of the ground and low excited states of the first-row hydrides AH_2, some are linear ($D_{\infty h}$), most are bent (C_{2v}). The electron configurations are

$$\text{Linear: } 1\sigma_g^2\, 2\sigma_g^2\, 1\sigma_u^m\, 1\pi_u^{n+p}$$

$$\text{Bent: } 1a_1^2\, 2a_1^2\, 1b_2^m\, 3a_1^n\, 1b_1^p$$

The core MO $1a_1$ and the $1b_1$ MO are nearly nonbonding, and $2a_1$ is somewhat bonding. The principal bonding MOs are $1b_2$ and $3a_1$. When the orbital energies are plotted as a function of apex angle in a Walsh diagram, the $1b_2$ MO strongly favors a maximum angle (180°), while $3a_1$ strongly favors a small angle. The other MOs are not very angle-sensitive. The resulting geometries of the ground and low excited states of several AH_2 molecules are shown in Table 1. The states are listed in order of increasing energy. In BH_2, CH_2^+, and NH_2 the two lowest states correspond to a Renner–Teller splitting of a $^2\Pi_u$ linear-molecule state.

Of general interest are papers by Bell [2] on the effects of basis set and CM on the predicted geometries of AH_2 molecules. These effects are examined for a large number of computations on H_2O, NH_2, CH_2, H_2O^+, and BH_2.

TABLE 1

Electronic States and Geometries of AH_2 Molecules

Molecule	Population[a] 3a$_1$	Population[a] 1b$_1$	State	Apex angle[b]	R_e(AH)[c]
BeH_2			$^1\Sigma_g^+$	180	1.34
BH_2	1		$^2A_1$	129 (131)	1.20 (1.18)
		1	$^2B_1$ ($^2\Pi_u$)	180	1.18 (1.17)
CH_2^+	1		$^2A_1$	141	1.12
		1	$^2B_1$ ($^2\Pi_u$)	180	1.10
CH_2	1	1	$^3B_1$	132 (134)	1.08 (1.08)
	2		$^1A_1$	101 (102)	1.11 (1.11)
	1	1	$^1B_1$	138 (140 ± 15)	1.07 (1.05)
	0	2	$^1\Sigma_g^+$	180	
CH_2^-	2	1	$^2B_1$	104	1.12
	1	2	$^2A_1$	131	1.08
SiH_2			$^1A_1$	94 (92)	1.54 (1.52)
			$^3B_1$	118	1.47
			$^1B_1$	123 (122)	1.47 (1.48)
NH_2^+	1	1	$^3B_1$	140	
	2		$^1A_1$	120	
	1	1	$^1B_1$	180	
NH_2	2	1	$^2B_1$	103.1 (103.3)	1.029 (1.024)
	1	2	$^2A_1$	143.4 (144 ± 5)	1.000 (1.004 ± 0.03)
	2	2	$^2B_2$[d]	47.5	1.162
NH_2^-	2	2	$^1A_1$	103	1.002
H_2O^+	2	1	$^2B_1$	111.4 (112.4)	0.987 (0.983)
			$^2A_1$	180?	
			$^2B_2$[d]	55	1.140
H_2O	2	2	$^1A_1$	104.93 (104.45)	0.953 (0.956)
H_2S				(92.2)	(1.328)

[a] The electron configuration begins with $1a_1^2\ 2a_1^2\ 1b_2^2$ in all cases.

[b] Degrees; experimental value in parentheses.

[c] Angstroms; experimental value in parentheses.

[d] One electron missing from $1b_2$.

C. BeH_2 AND MgH_2

The molecule BeH_2 is not known experimentally but the linearity of its ground state is assured by the Walsh rules and by several theoretical calculations [3]. Hosteny and Hagstrom [4] made SCF plus CM (up to 80 terms) calculations using basis sets with up to four s, three $2p\sigma$, and two $2p\pi$ STFs on Be and two s, one $2p\sigma$, and one $2p\pi$ on H. They included K-shell correlation estimated to be 78% of the actual, and valence-shell correlation estimated at 55% of the actual. For the best calculation, their E_h was -15.8478 (or -15.85034, see the Table IV in [4]).

A more recent computation by Ahlrichs *et al.* [5] using Gaussian functions and the CEPA–PNO method obtained $E_h = -15.8458$. Their basis set included d functions on Be and H, which they showed to be of appreciable importance. Although their wave functions should thereby be better than those in [4], their omission of K-shell correlation and perhaps the loss of accuracy due to the use of a Gaussian instead of an STF basis would seem to account for their slightly worse E_h.

For a computed Be—H equilibrium distance, [4] gave 2.52 a.u., [5], 2.525 a.u. For the vibration frequencies, [4] yielded 2083 cm^{-1} for ν_1, 2497 for ν_3, and an estimate of 780 ± 50 for ν_2, while [5] gave 2060 for ν_1, 2240 for ν_3, and 745 for ν_2. The values from [5] are to be preferred.

Gallup and Norbeck [5a] made less accurate VB calculations on the ground and several excited states of BeH_2, and give curves of the energy for symmetrical dissociation of the $^1\Sigma$ states into atoms.

Calculations on BeH_2^{2+} have been made [3].

Ahlrichs *et al.* [6] have reported a calculation for MgH_2 similar to that for BeH_2. They predict a binding energy for Mg + H_2 that is 3 kcal/mole less than for H_2, which implies that MgH_2, if linear (this is not quite certain), is not a stable species. In contrast, in earlier papers they predicted BeH_2 to be bound by about 40 kcal/mole with respect to Be + H_2.

D. BH_2 AND CH_2^+

Bender and Schaefer [7] carried out SCF and SCF plus natural orbital CM calculations on the $^2A_1$ ground and $^2B_1(^2\Pi_u)$ first excited states of BH_2 and of CH_2^+, using a double-zeta-size contracted Gaussian basis set. The calculated geometries for BH_2 were $R_e = 1.211$ Å, apex angle $\theta = 129°$ for $^2A_1$, versus the experimental values 1.18 Å and 131°; and for $^2B_1(^2\Pi)$, $R_e = 1.188$ Å versus experimental 1.17 Å. For CH_2^+ the computed results for $^2A_1$ were $R_e = 1.12$ Å, $\theta = 141°$, and for the $^2B_1(^2\Pi)$ state R_e was about 1.08 Å. The $^2A_1$ state of CH_2^+ is predicted to lie about 1060 cm^{-1} below the $^2B_1(^2\Pi)$ state. For BH_2 the predicted interval is 2260 cm^{-1} as compared with the experimental 4300 cm^{-1}. Calculated force constants and dominant CM configurations are listed for BH_2 and for CH_2^+.

A much more accurate calculation on the two states of BH_2 (65 Gaussian lobe functions contracted to 38) has been made by Staemmler and Jungen [8]. They obtain $R_e = 1.204$ Å, $\theta = 128.6°$ for $^2A_1$ and $R_e = 1.176$ for $^2B_1$. They compute $\nu_1 = 2487$, $\nu_2 = 1017$, $\nu_3 = 2725$ cm^{-1} for $^2A_1$ versus the experimental value $\nu_2 = 1030$; and for $^2B_1$, $\nu_1 = 2774$, $\nu_2 =$

992, $\nu_3 = 3054$ versus the experimental value 954 for ν_2. For the $^2A_1 \rightarrow {}^2B_1$ excitation energy they compute 2740, 9400, and 2035 cm^{-1} for adiabatic, vertical, and T_{00} excitation. They consider their calculated T_{00} more accurate than the experimental 4194 cm^{-1}. For the ionization potential they compute 7.85 (adiabatic) and 8.80 eV (vertical), as compared with experimental values in the range 8–10 eV (vertical).

Calculations on BH_2^+ and BH_2^- are available [3]. For calculations on SiH_2^+, and for ionization energies of CH_2, see Section E. CM computations on a third state $1a_1^2\ 2a_1^2\ 2b_2\ 3a_1^2$, $^2B_2$ of CH_2^+ have been made by Liskow *et al.* [9] and by Pearson and Roueff [10]. They obtain C_{2v} potential surfaces as functions of C—H_2 and H—H distances R and r, also for a $C_{\infty v}$ approach (see Chapter V for details).

E. CH_2, SiH_2, NH_2^+, AND PH_2^+

Experimentally, in addition to Rydberg series of higher excited states, four low-energy states of CH_2 have been reported [11]. The ground state and the next two states are certainly bent (C_{2v}) as shown conclusively by *ab initio* computations. The second excited state $^1B_1$ is similar in structure and geometry to the $^3B_1$ ground state. All these states are stable relative to dissociation into C plus H_2. The third excited state, a singlet state, which is known experimentally, is almost certainly also stable, at least with respect to dissociation to ^1D carbon plus H_2. Theoretically, it is linear ($^1\Sigma_g^+$).

The nature of the four low-energy CH_2 states can be better understood by the following survey of forms of their wave functions for bent (C_{2v}) and for linear ($D_{\infty h}$) CH_2,

C_{2v}	$D_{\infty h}$	
$\ldots 1b_1^2 + \lambda 3a_1^2,\ {}^1A_1$	$2^{-1/2}(\ldots 1b_1^2 + \ldots 3a_1^2) = \ldots \pi^+\pi^-,\ {}^1\Sigma_g^+$	
$\ldots 3a_1\, 1b_1,\ {}^1B_1$	$\ldots 3a_1\, 1b_1 = 2^{-1/2}(\pi^+\pi^+ - \pi^-\pi^-)$	$^1\Delta_g$
$\ldots 3a_1^2 - \lambda 1b_1^2,\ {}^1A_1$	$\ldots 2^{-1/2}(\ldots 3a_1^2 - \ldots 1b_1^2) = 2^{-1/2}(\pi^+\pi^+ + \pi^-\pi^-)$	$^1\Delta_g$
$\ldots 3a_1\, 1b_1,\ {}^3B_1$	$\ldots 3a_1\, 1b_1 = \ldots \pi^+\pi^-,\ {}^3\Sigma_g^-$	

On a Walsh diagram (as is also qualitatively obvious), the lower energy $3a_1$ strongly favors an acute angle, while the higher energy $1b_1$ is angle-indifferent. The *ab initio* calculations on the $^3B_1$ and $^1B_1$ states decide conclusively for an apex angle very considerably smaller than 180°. In the lower $^1A_1$ state, an acute angle is observed as expected, because of the influence of the $3a_1$ MO. Conversely, assuming $1b_1$ to be angle-indifferent, one can predict that the energy of the higher $^1A_1$ state should be pushed up as the angle increases, so that the energy minimum occurs for the linear form (π^2, $^1\Sigma_g^+$).

Numerous *ab initio* calculations have been made on the $^3B_1$ and $^1A_1$ states, especially on their energy separation, which is of great interest to organic chemists. Some similar calculations have also been made on the $^1B_1$ state. Less attention has been given to the second $^1A_1$ state. Through 1973, the *ab initio* calculations have been well reviewed by Harrison [12].

O'Neil *et al.* [13] carried out computations on the lowest $^3B_1$, $^1A_1$, $^1B_1$, $^1A_2$, $^3A_2$, $^1B_2$, and $^3B_2$ states, of which the last four are unstable with respect to dissociation to C + H_2. Keeping the K-shell frozen, they used a Gaussian basis set without polarization functions, which later computations have shown to be of some importance. In all calculations except for $^1A_1$ the SCF configuration plus all singly and doubly excited configurations were included, and the iterative NO procedure was used. For the $^1A_1$ state, a two-configuration SCF calculation was used as the starting point. This is needed because CM of ... $1b_1^2$ into ... $3a_1^2$ is rather extensive in $^1A_1$. The work showed conclusively that the $^3B_1$ and $^1B_1$ states are nonlinear by an *energetically* rather small margin, although experimentally it was first thought [9] that the $^3B_1$ is probably a linear $^3\Sigma_g^-$. Later, ESR evidence showed it to be bent.

Most later calculations follow a pattern similar to that of O'Neil *et al.* but with larger basis sets, including polarization functions [14]. Staemmler has made IEPA-PNO calculations on the $^1A_1$ and $^3B_1$ states [15]. There are some good GVB calculations [16,17].

Dykstra *et al.* [18] used the method of self-consistent electron pairs (SCEP) to compute the energy of the $^1A_1$ first excited state of CH_2. They used a double-zeta basis set plus polarization functions (p on H and d on C) to obtain a correlated energy of -39.0396, perhaps the lowest so far.

In recent calculations [17, 19, 22–24], a consensus has been reached on an (adiabatic) separation of about 10.6 ± 1 kcal/mole between the $^3B_1$ and $^1A_1$ states.

Shih *et al.* [19] obtained 11 kcal/mole. They, like others [22], conclude that the interpretation of Zittel *et al.* [20] of electron detachment measurements on CH_2^-, which leads to a much higher value, must be incorrect. They give a reinterpretation in accord with the 11 kcal/mole value. In their work, they used full CM including the K electrons. From their calculations ($E_h = -39.0780$), extrapolation for full CM gave $E_h = -39.0849$ for $^3B_1$. Their reinterpretation of the CH_2^- electron detachment spectrum is in agreement with one by Harding and Goddard [21].

Shih *et al.* made new computations on the $^3B_1$, $^1A_1$, and $^1B_1$ states of CH_2 and the $^2B_1$ and $^2A_1$ states of CH_2^-, and gave a figure showing the energies of these states as a function of apex angles. They also calculated frequencies and energy levels for the bonding vibration. Harding and Goddard obtained in part similar results. The adiabatic excitation energy of the $^1B_1$ level is calculated as 1.10 eV (1.08 by Harding and Goddard).

Using very large basis sets, Meadows and Schaefer [25] have made SCF (and two-configuration calculations—see above—in the case of $^1A_1$) on the $^3B_1$ and lower $^1A_1$ states of CH_2 and on the analogous states of silylene (SiH_2). For CH_2 the adiabatic energy difference is 10.9 kcal/mole, almost exactly the same as in the full CM treatment. For SiH_2 the $^1A_1$ state is predicted to lie *below* the $^3B_1$ state by 10.4 kcal/mole, with a computed bond distance and angle of 1.536 Å and 93.8°; the experimental values are 1.521 Å and 92.1°. Computed SCF data for the lowest $^1B_1$ states of CH_2 and SiH_2 are also given [25]. For CH_2 $R_e = 1.065$ Å, $\theta = 141°$, $E_h = -38.871$ are given, for SiH_2 1.468 Å, 123.5°, -289.964. In their CM calculation, O'Neil *et al.* [13] obtained $E_h = -38.911$ for $^1B_1$ of CH_2. This could be improved. Harding and Goddard [17] discuss this state, but do not give an E_h. On methylsilylene, see Chapter IX, Section H,1.

Ball and Thomson [25a] using a (12,9/6) Gaussian basis set contracted to (6,4/2), plus polarization functions (p on H and d on Si and P) have made SCF calculations on SiH_2^+, SiH_2, SiH_2^-, PH_2, PH_2^+, and PH_2^- (for PH_2 and its ions see also Section F). For SiH_2, they obtained, for $^1A_1$, $E_h = -290.0088$, not as good as Meadows and Schaefer's $E_h = -290.0489$. They tabulate orbital energies and E_h values for the ground states of SiH_2, PH_2, and their ions, each at computed optimum geometries: R(SiH) = 1.474, 1.516, and 1.551 Å and θ = 120.0°, 93.44°, and 93.7° for SiH_2^+, SiH_2, and SiH_2^-, respectively. They discuss the results of a population analysis. They give computed dipole moments (0.335 *D* for SiH_2), stretching and bending force constants, net charge on H and on heavy atom, for SiH_2, PH_2, and their ions, and a large number of other computed properties for SiH_2. The SCF ionization energies (better than Koopmans' theorem values, but not yet allowing for electron correlation) are 765 and 829 kJ/mole for SiH_2 and PH_2, respectively. The computed electron affinities are too unreliable.

For calculations on the core-ion states of CH_2, see Section D on CH_2^+.

Hase *et al.* [25b] have made SCF plus extensive perturbation CM calculations on all the vertical ionization potentials of the $1a_1^2\,2a_1^2\,1b_2^2\,3a_1\,1b_1$, $^3B_1$ and $1a_1^2\,2a_1^2\,1b_2^2\,3a_1^2$, $^1A_1$ states of CH_2, using a 6-31G** (double-zeta plus polarization) Gaussian basis set. The results in electron volts, with the ionized MO in parentheses, are as follows: from $^3B_1$, $^2A_1(1b_1)$, 10.28; $^2B_1(3a_1)$, 10.66; $^4A_2(1b_2)$, 14.15; $^2A_2(1b_2)$ 15.90; $^2A_2(1b_2)$, 16.55; $^4B_1(2a_1)$, 19.31; $^2B_1(2a_1)$, 23.09; $^2B_1(2a_1)$, 25.64; $^4B_1(1a_1)$, 292.95; $^2B_1(1a_1)$, 295.59; $^2B_1(1a_1)$, 295.90; from $^1A_1$, $^2A_1(3a_1)$, 10.02; $^2B_2(1b_2)$, 14.97; $^2A_1(2a_1)$, 22.87; $^2A_1(1a_1)$, 295.99.

Clark *et al.* [25c] have made SCF calculations on core-ion ($1s^{-1}$ CH_2^+) states of CH_2, CHF, and CF_2, using a (9s 5p 1d/4s) basis contracted to (4s 2p 1d/2s). They compute 293.07 eV for core ionization of $^3B_1$ to $^4B_1$ in CH_2 (they do not compute the two $^2B_1$ core-ion states) and 295.30 eV for ionization of $^1A_1$ to $^2A_1$. These values allow for changes in geometry (R from 2.03 to 1.99 a.u. and θ from 129° to 136° for $^3B_1$, R from 2.08 to 2.02 a.u. and θ from 104° to 112° for $^1A_1$). Force constants and barrier heights (for rotation through a linear conformation) are given. Comparison with 1s binding energies for CH_4 and C_2H_4 are given. Changes in gross population on core ionization are given.

Several *ab initio* calculations of the zero-field spin fine structure of the $^3B_1$ state of CH_2 have been made. The fine structure consists of three components with energies $D/3 + E$, $D/3 - E$, and $-2D/3$, where D and E are two parameters which can be computed. Experimentally, D is 0.76 ± 0.02 cm^{-1}, while E is much smaller. The parameters D and E each contain a spin–spin and a spin–orbit contribution. Using good CM wave functions, Langhoff and Davidson [26] compute the spin–spin contributions to D and E at the angle 135° as 0.781 cm^{-1} and 0.050 cm^{-1}, respectively. Langhoff [27] computes spin–orbit contributions to D of 0.023 and 0.0245 cm^{-1} at 135° and 180°, respectively; to E the contribution is 0.000 cm^{-1}. For the total computed splittings at 135°, $D = 0.807$ cm^{-1}, $E = 0.049$ cm^{-1}.

Several other computed properties of $^3B_1$ of CH_2 are listed by Langhoff and Davidson.

Rather good CM calculations on the three lowest states of CH_2 and NH_2^+ have been made by Harrison and Eakers [28]. For NH_2^+ they report charge distributions and show a contour map of the $3a_1$ MO. Somewhat more accurate CM calculations on CH_2 and NH_2^+ are reported by Chu *et al.* [29]. Their calculation indicates that this state at equilibrium is linear ($^1\Sigma_g^+$) for both CH_2 and NH_2^+. The state is known experimentally, but not well enough to establish linearity.

There is one earlier paper by Harrison in 1971 [30]. The most accurate calculation for the second $^1A_1$ state of CH_2 is apparently one by Bauschlicher and Yarkony [30a] using a polarized basis set and starting with MCSCF calculations on the two $^1A_1$ states, and continuing with CM using all single and double excitations except from 1s. The 2^1A_1 state is predicted to be almost but not quite linear.

Using a double-zeta polarization basis, Bender *et al.* [31] have performed an *ab initio* study of potential energy surfaces for the ... $1b_2^2\,3a_1\,1b_1$, $^3B_1$ ground state and the ... $3a_1^2\,1b_2\,1b_1$, $^3A_2$ excited state of NH_2^+, with a view to elucidating the path of the reaction $N^+ + H_2 \rightarrow NH_2^+$. The work is similar to that reported above under CH_2^+ for the

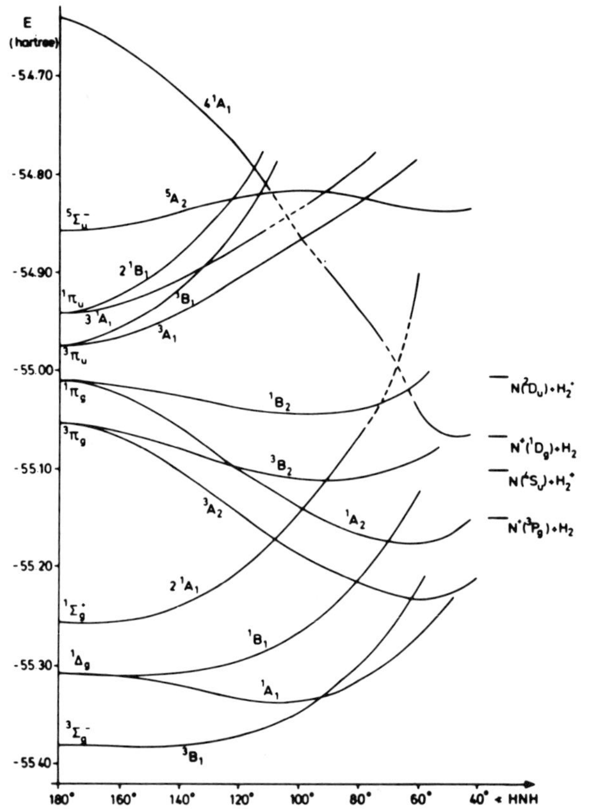

FIG. 1. Potential curves showing energy (hartrees) as a function of apex angle for 14 states of NH_2^+, for R_{NH} = 1.03 Å. From [31a].

reaction $C^+ + H_2 \rightarrow CH_2^+$. The $^3A_2$ surface has a surprisingly deep minimum, explainable by the presence of $3a_1^2$, about 60 kcal/mole below $N^+ + H_2$, at R_e(NH) about 1.26 Å and θ about 43°. After approach of H_2 to N^+ on $^3A_2$, deformation to C_s symmetry permits transfer to $^3B_1$ ground-state NH_2^+. Energy contours for R (N–mid H_2) versus r (HH distance) are given for the $^3A_2$ and $^3B_1$ states.

Peyerimhoff and Buenker [31a], using their MRD–CI SCF + CM extrapolation method (see Chapter I) with a (9s 5p 2d/5s) basis contracted to (5s 3p 2d/2s) plus an s and a p bond function in the middle of each NH bond, computed angular and NH symmetric stretch potential curves for the lowest 14 states of NH_2^+, with some consideration also of asymmetric conformations. Figure 1 from their article shows angular potential curves (note: all for $R_{NH} = 1.03$ Å) at the estimated full-CM level of treatment. Note the small angle $\theta = 60°$ which occurs at R(NH) = 1.15 Å for equilibrium for $^3A_2$, with dissociation energy 64 kcal/mole (2.79 eV) to $N^+ + H_2$, in rough agreement with Bender *et al.* The corresponding data for the $^1A_2$ state, $\theta = 63°$, R(NH) = 1.16 Å, and dissociation energy 31 kcal/mole are also of interest. Similar conclusions may probably be expected for the $^3A_2$ and $^1A_2$ states of CH_2. The 0 to 0 energy between the $^3B_1$ ground state and the $^1A_1$ first excited state is calculated as 30 kcal/mole. Oscillator strengths for allowed transitions from these two to higher states are given. The dissociation energy D_e for the $^3B_1$ state is computed as 6.45 eV. The second $^1A_1$ state is found to be linear as in CH_2.

Hirst [31b] using a moderately good polarized basis plus CM has also discussed the $^3B_1$ and $^3A_2$ surfaces of NH_2^+, noting that they intersect and must then become $1^3A''$ and $2^3A''$ of C_s symmetry. Surfaces of $C_{\infty v}$ and C_{2v} symmetry correlate as follows: $^3\Sigma^- \rightarrow 1^3A'' \rightarrow {}^3A_2$, $^3\Pi \rightarrow 2^3A'' \rightarrow {}^3B_1$. For collinear and near collinear collisions of $N^+ + H_2$ the reaction $N^+ + H_2$ is expected to occur by a direct mechanism on the $^3\Sigma^- \rightarrow 1^3A''$ surface. For C_{2v} geometries motion will occur on the $^3A_2$ surface. A low-energy route to the $^3B_1$ ground state is possible for C_s collisions of near C_{2v} geometry. Triplet surfaces correlating with $N + H_2^+$ are repulsive and involve higher energies.

Some derivatives of CH_2 are discussed in Sections III,E, IV,G, and IX,E,3.

F. CH_2^-, NH_2, H_2O^+, SiH_2^-, AND PH_2

The calculations of Shih *et al.* on CH_2^- have been mentioned in Section E (see Table 1 for the data). Calculations by Ball and Thomson are also reported in Section E.

Relatively few *ab initio* calculations on NH_2 have been made. One of the best on the three lowest states is that of Bell and Schaefer [32]. They used a double-zeta plus polarization basis and extensive CM for the $\ldots 1b_2^2\, 3a_1^2\, 1b_1$, $^2B_1$ ground state and the $\ldots 1b_2^2\, 3a_1\, 1b_1^2$, $^2A_1$ and $\ldots 3a_1^2\, 1b_2\, 1b_1^2$, $^2B_2$ excited states. The $^2B_1$ and $^2A_1$ states represent the Renner–Teller components of what in the linear molecule would be a $\ldots 1\sigma_u^2\, 1\pi_u^3$, $^2\Pi_u$ state. Using a somewhat improved polarized basis set and their MRD-CI method for electron correlation, Peyerimhoff and Buenker [32a] have calculated potential curves for the lowest three states of NH_2 as a function of apex angle. For the $^2A_1$ state, a barrier to linearity (to become part of $^2\Pi_u$) of 1178 cm^{-1} is found. Calculated apex angles for the $^2B_1$ and $^2A_1$ states are 103° and 114°, respectively, in good agreement with experiment; for the $^2B_2$ state, 50° with $R_e = 1.13$ Å. The oscillator strength of the $^2A_1 - {}^2B_1$ transition is 0.00196, which appears consistent with the experimental evidence. The $^2B_2 - {}^2B_1$ transition is predicted to be very weak.

The ground state of NH_2^+ is predicted [31a] to be nearly linear (150° apex angle); if linear, it would be ... $1\pi_u^2$, $^3\Sigma_g^-$, with the excited states $^1\Delta_g$ and $^1\Sigma_g^+$. Peyerimhoff and Buenker [32a] have made extensive MRD-CI calculations on the related NH_2 Rydberg states ... $1\pi_u^2\,3s$, $^4\Sigma_g^-$, $^2\Sigma_g^-$, $^2\Delta_g$, and $^2\Sigma_g^+$ and found for $^4\Sigma_g^-$, apex angle $\theta = 180°$; $^2\Sigma_g^-$ (2^2B_1), 154°; $^2\Delta_g$, 2^2A_1 component, $\theta = 180°$, 3^2B_1 component, $\theta \approx 100°$; $^2\Sigma_g^+(3^2A_1)$, 180°. Energies are given. Intensities of various transitions are computed; a rather strong $^2\Sigma_g^-(2^2B_1) - \tilde{A}\,{}^2A_1$ transition is predicted.

For H_2O^+ several calculations have been made, relevant especially to ionization of H_2O. The most accurate are PNO–CI calculations by Meyer [33]. The ground state of H_2O is $1a_1^2\,2a_1^2\,1b_2^2\,3a_1^2\,1b_1^2$, $^1A_1$. For the removal of $1a_1$, $2a_1$, $1b_2$, $3a_1$, and $1b_1$, respectively, Meyer computes 19.8308 (19.821), 1.1887 (1.183), 0.6929 (0.688 ± 0.008), 0.5395 (0.545 ± 0.004), 0.4585 (0.4698) in a.u., where the numbers in parentheses are the experimental values corrected to verticality. For the several states of H_2O^+, Meyer computes R_e values and apex angles as follows: $^2B_1$, 0.978 Å, 112.5°; $^2A_1(3a_1)$, 0.976, 180°; $^2B_2$, 1.101 Å, 57.7°. For some direct computations of the ionization potentials of H_2O and H_2S, see the end of Section G.

Smith *et al.* [34] made limited basis-set SCF calculations on the above five states of H_2O^+, and gave computed force constants for vibrations. With a small basis set (STO-3G), Sakai *et al.* [35] made unrestricted SCF calculations on the three lowest states of H_2O^+ and H_2S^+. They give electron and spin populations, and spin density contour maps. Fortune *et al.* [36] made MCSCF/CM computations on the $^2B_1$ and $^2B_2$ states, and give computed geometries, force constants, vibrational frequencies, and a number of computed vibrational state energies for the $^2B_1$ state, also a number of points on the $^2B_1$ and $^2B_2$ potential surfaces.

In a VB computation of relatively low accuracy, Balint-Kurti and Yardley [37] pointed out that the $^2B_2$ state of H_2O^+, plotted as a function of apex angle, is crossed (see their figure) near its minimum by the rapidly rising $^2A_1$ curve. The situation is exactly analogous to that of the interactions of the ground state $^2A_1$ and excited state $^2B_2$ curves of CH_2^+ and of the $^3B_1$ and $^3A_2$ curves of NH_2^+, leading via a relaxation to C_s symmetry to a reaction $A^+ + H_2 \rightarrow AH_2^+$. The $^2A_1$ and $^2B_2$ states of NH_2 again show the same relation.

Using an ample basis set, Leclerc *et al.* [37a] made SCF calculations on potential energy surfaces of several doublet and quartet states of H_2O^+, and presented contour maps showing cross sections for some with one OH bond length fixed. Various dissociation processes are discussed.

The SCF calculations of Ball and Thomson [25a] on PH_2^+, PH_2, and PH_2^- have in part been discussed in Section E in connection with SiH_2 and its ions. Also obtained were the following results on the optimum geometries of PH_2^+, PH_2, and PH_2^-, respectively: $R(PH) = 1.40$, 1.43, and 1.48 Å, $\theta = 118°$, 95°, and 94°. Orbital energies and E_h values are also given. Populations are discussed and some molecular properties listed (e.g., dipole moment 0.82*D* for PH_2).

Peric *et al.* (37b) have made an MRD-CI calculation on the potential surfaces of the $^2B_1$ ground state and the $^2A_1$ first excited state of PH_2. From the results they have computed vibrational levels for each state in PH_2 and PD_2, and also Franck–Condon factors for the transition between the states, with results in very good agreement with experiment where available. Computed apex angles (91° for $^2B_1$, and 122° for $^2A_1$) agree with experiment.

G. H_2O, H_2F^+, H_2S, NH_2^-, PH_2^-, AND H_2S^-

Very many *ab initio* computations have been made on the ground state of water. A comprehensive review of theoretical work on H_2O to 1972 is given by Kern and Karplus [38]. We mention first a nonvariational calculation by Miller and Kelly [39] with a single-center basis set using diagrammatic perturbation theory through third order. They obtained $E_h = -76.233$ a.u., leading with estimated higher-order corrections to a final estimate of -76.48 ± 0.07. This compares with an experimental value of -76.481, or after removing relativistic contributions of -0.049 a.u., a value of -76.432.

In a related calculation by a united-atom many-body perturbation procedure, Lee and Das [40] have calculated a *correlation energy* of -0.3375 ± 0.01 a.u. for H_2O. This, added to the estimated SCF E_h limit for H_2O of -76.0675 [41], gives a total of -76.405. (Lee and Das obtained a computed SCF energy only of -75.998 ± 0.15 a.u. Earlier, Moccia [42], using a one-center basis set, obtained -75.922.)

The best SCF calculation to date was carried out by Popkie *et al.* [43], yielding $E_h = -76.06598$, which is estimated to be 0.002 a.u. above the SCF limit. The lowest energy so far computed by strictly variational methods is Meyer's [33] PNO–CI value -76.36834; by his not-quite-variational CEPA method, he obtained -76.38289. Aside from Meyer's calculations, the best available, and giving many more details, appear to be the conventional SCF plus CM calculations of Rosenberg and Shavitt [41]. They tried several Gaussian and STF basis sets, with and without frozen $1a_1^2$ core, as well as including an interesting table comparing independent electron pair results with those of [33]. The best computed energy E_h, obtained with the largest STF set, was -76.33980.

Using extensive polarized basis sets and PNO-CI and CEPA for correlation, Werner and Meyer [43a] have computed static dipole polarizability components for H_2O, with corrections for zero-point vibrations. Agreement with experiment for the mean polarizability is excellent.

Rosenberg and Shavitt report calculations on the dipole and quadrupole moments, $\langle r^2 \rangle$, and diamagnetic susceptibility and shielding, and quadrupole coupling tensors. For the dipole moment, they obtain 1.9193 D (experimental 1.8546 ± 0.0006). For the quadrupole moment tensor H in buckinghams (10^{-26} esu cm^2), relative to the center of mass for $H_2\,^{16}O$, they calculate $\mathsf{H}_{xx} = -2.517$, $\mathsf{H}_{yy} = 2.610$, $\mathsf{H}_{zz} = -0.093$ (the respective experimental values are -2.50 ± 0.02, 2.63 ± 0.02, and -0.13 ± 0.03). For the expectation value $\langle r^2 \rangle$ for the electrons in Å^2 relative to the center of mass, they compute 5.479 (experimental 5.1 ± 0.7; they conclude that the true value is 5.5 Å^2).

For the diamagnetic susceptibility tensor (in 10^{-6} cm^3/mole) they compute $\chi_{xx}{}^d - \chi_{av}{}^d = -0.959$, $\chi_{yy}{}^d - \chi_{av}{}^d = 0.891$, $\chi_{zz}{}^d - \chi_{av}{}^d = 0.067$, $\chi_{av}{}^d = -15.497$ as compared with respective experimental values -1.06 ± 0.01, 0.95 ± 0.01, 0.11 ± 0.02, and -14.6 ± 2.0 [44]. Computed average and component values of the diamagnetic shielding σ^d of the nuclei are reported. For the H atom, the computed $\sigma_{av}{}^d$ is 102.33, as compared with an experimental of 102.4.

Computed values for the ^{17}O quadrupole coupling tensor are 10.92 MH$_z$ (*xx* component) and -9.71 (*yy* component) as compared to experimental values of 10.17 ± 0.07 and -8.89 ± 0.03. Values for the D quadrupole coupling tensor are -349.55 KH$_z$ (*aa* component in the principal axis system) and 151.10 (*bb* component) with corresponding experimental values of -307.95 ± 0.14 and 133.13 ± 0.14.

In a companion paper, Rosenberg *et al.* [45] report a potential energy surface for the ground state of H_2O built from 36 molecular geometries in the vicinity of equilibrium. Hennig *et al.* [46] have recently generated a potential energy surface of comparable accuracy including 55 geometries. In addition to obtaining the R_e and θ values given, these authors report harmonic, cubic, and quartic force constants [45, 46], Taylor series expansions for a series of one-electron properties as a function of changes in geometry (ΔR_1, ΔR_2, $\Delta\theta$) [45], and vibrational and rotational constants based on a perturbation theory approach [46]. Comparisons of the geometry dependence of the energy and various properties are made for surfaces derived from SCF, all singles plus doubles CM (SD), and an energy surface including the Langhoff–Davidson approximate correction (see Chapter I) for the effect of quadrupole excitations (SDQ). The shapes of surfaces for various one-electron properties are generally not especially sensitive to improvements in the wave functions by CM, as is exhibited in comparisons [46] of the respective expansion coefficients, even though the equilibrium values of the properties are, in many cases, significantly affected.

Bartlett *et al.* [46a] have applied many-body perturbation theory (MBPT) and coupled cluster (CC) methods to the SD wave functions of [45] to examine the effects of such improvements on the geometric parameters R_e and θ_e and the force constants through fourth degree. Presented in Table 2 are the results due to the SCF, SD, and SDQ calculations of [45] together with two selected sets of constants resulting from the application of MBPT and CC methods by Bartlett *et al.* Also shown are constants derived from spectroscopic data. The results labeled D-MBPT (∞) correspond to the inclusion of double excitation diagrams to all orders and those labeled CCD to the coupled cluster doubles model whereby an interactive procedure is used to sum double excitation diagrams to all orders [46a]. The improvements in the molecular constants are of the largest magnitude as one progresses from the SCF to the SD levels, with noticeable improvements especially in the stretching force constants f_{RR}, $f_{RR'}$, and f_{RRR}.

Figure 2, taken from [46a], summarizes the results of Table 2 in an interesting fashion. It clearly shows the necessity for going beyond the SCF and SD levels of approximation to accurately represent certain molecular distortions. Included in the figure are results due to MBPT contributions summed to third order. It is worthwhile noting that the simple unlinked cluster approximation (SDQ) shows a consistent improvement of the SD, although it is apparent through the increase in standard deviations that there is some introduction of errors.

The water molecule has been used as a test case for numerous methods intended to be applied to the detailed understanding of many properties of polyatomic molecules. For example, Lamanna *et al.* [47] have formulated an approach using an equations of motion method for computing a series of second-order electric and magnetic properties of H_2O, including static electric polyarizabilities, magnetic susceptibility and magnetic shielding tensors, and dynamic electric dipole polarizabilities. They also report results for the molecular moments through the hexadecapole moment and provide useful comparisons with other computed results. An encouraging feature is that there is generally good agreement among values due to

TABLE 2

Potential Energy Surface Constants for H_2O[a]

Constant	SCF[b]	SD[b]	SDQ[b]	D-MBPT (∞)[c]	CCD[c]	Experiment[d]
R_e (Å)	0.9398	0.9527	0.9573	0.9576	0.9558	0.95781 (9)*
θ_e (°)	106.08	104.93	104.58	104.59	104.74	104.4776 (57)
f_{RR}	9.7934 (9)	8.8757 (4)	8.539 (5)	8.5120 (16)	8.6739 (16)	8.454
$f_{\theta\theta}$	0.87515 (6)	0.81351 (8)	0.8011 (9)	0.7987 (3)	0.8033 (3)	0.761
$f_{RR'}$	−0.0662 (7)	−0.0829 (3)	−0.085 (3)	−0.0967 (11)	−0.0965 (11)	−0.101
$f_{R\theta}$	0.25764 (6)	0.27076 (7)	0.2741 (8)	0.2732 (3)	0.2713 (3)	0.2475 (34)*
f_{RRR}	−10.356 (16)	−9.703 (6)	−9.50 (3)	−9.497 (10)	−9.555 (14)	−9.47 (48)
$f_{\theta\theta\theta}$	−0.1492 (4)	−0.1302 (2)	−0.1261 (16)	−0.1268 (5)	−0.1282 (6)	−0.12332 (20)*
$f_{RRR'}$	0.001 (13)	−0.021 (5)	−0.004 (26)	−0.031 (8)	−0.025 (12)	0.12 (72)
$f_{RR\theta}$	−0.033 (14)	−0.039 (5)	−0.016 (29)	−0.034 (9)	−0.041 (12)	−0.001 (15)*
$f_{RR'\theta}$	−0.535 (25)	−0.510 (10)	−0.55 (5)	−0.512 (15)	−0.502 (21)	−0.40 (28)
$f_{R\theta\theta}$	−0.1584 (9)	−0.1560 (4)	−0.159 (3)	−0.1565 (10)	−0.1564 (11)	−0.12 (5)
f_{RRRR}	13.58 (17)	13.97 (21)	15.3 (22)	14.0 (7)	14.0 (7)	14.7 (24)
$f_{\theta\theta\theta\theta}$	−0.047 (3)	−0.033 (3)	−0.052 (33)	−0.032 (10)	−0.032 (11)	−0.0353 (18)*
$f_{RRRR'}$	−0.09 (12)	−0.03 (15)	−1.2 (17)	−0.05 (53)	−0.04 (55)	−0.8 (27)
$f_{RRR'R'}$	0.13 (16)	0.12 (20)	−1.4 (21)	0.06 (68)	0.11 (70)	0.1 (17)
$f_{RRR\theta}$	−0.21 (32)	−0.26 (39)	−2.3 (42)	−0.2 (13)	−0.3 (14)	—
$f_{RRR'\theta}$	0.17 (30)	0.17 (36)	2.3 (39)	0.1 (12)	0.2 (13)	—
$f_{RR\theta\theta}$	−0.080 (53)	−0.075 (63)	0.00 (68)	−0.08 (22)	−0.08 (23)	−0.374 (43)*
$f_{RR'\theta\theta}$	0.29 (10)	0.32 (11)	0.2 (12)	0.35 (39)	0.34 (41)	0.31 (25)
$f_{R\theta\theta\theta}$	0.126 (5)	0.109 (6)	0.095 (65)	0.104 (21)	0.106 (22)	—

[a] The force constants are defined in Eq. (3) of [45] and are given in 10^5 dyn/cm, with angles defined in radians. For the theoretical results the values in parentheses are statistical uncertainties of the fitting procedure, given as *three* times the least squares standard deviations relative to the last figure quoted in each case. For the experimental uncertainties see the corresponding references.

[b] [45]. (See the text for explanations of the column headings.)

[c] See Bartlett *et al.* [46a]. (See the text for explanations of the column headings.)

[d] Experimentally derived values of Hoy *et al.* [46b]. The values followed by asterisks are refinements due to Hoy and Bunker [46c].

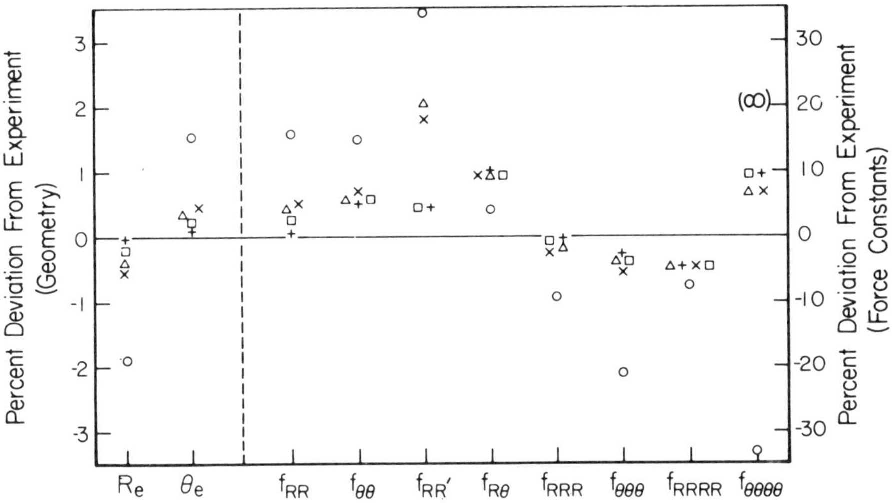

FIG. 2. Percent difference of theoretical predictions relative to experimentally derived values for the geometry selected force constants (right scale) for the H_2O molecule. Experimental values are from Hoy, Mills, and Strey and where available, from the revised fit by Hoy and Bunker (see Table 2). (□) SCF, (△) MBPT (3), (×) SD–CI, (○) D-MBPT(∞), (⊗) CCD.

wave functions derived from a variety of reasonably good basis sets. Compton profiles have also been computed using rather accurate wave functions for H_2O [48].

The availability of accurate *ab initio* potential energy surfaces has made it convenient for several groups to use the water molecule for applications of methods for obtaining vibrational-rotational properties of polyatomic molecules (see Chapter I). There is a series of papers [49] detailing a perturbation theory approach for obtaining *ab initio* vibrational properties with applications to the water molecule: included are such quantities as fundamental frequencies, vibrational excitation energies, expectation values of one-electron properties with respect to vibrational wave functions, and transition moments. The more accurate variational methods discussed in Chapter I have also been applied by Whitehead and Handy [50] using a potential energy surface derived from a somewhat less accurate CM procedure than that of [45, 46]. Vibrational excitation energies are computed to an accuracy of about 3% using spectroscopic constants derived using the formalism of [49] and the potential constants from [45]. Table 3 shows the results for H_2O and HDO [50a]. Similar results are obtained by Henning *et al.* [46], who also report rotational constants.

Since all the valence-shell MOs are filled in H_2O, the excited states are Rydberg states. The lowest-energy Rydberg MOs must, however, have some valence-shell character. For example, the lowest-energy Rydberg MO $4a_1$ should have the form $a3s + b2s_0 + c2p\sigma_0 - \alpha(1s_a + 1s_b)$ (in linear conformations, $c = 0$) with some O—H antibonding character [51]. Several *ab initio* calculations have been reported on these states, of two kinds: (1) calculations of vertical excitation energies, (2) exploration of potential surfaces for dissociation into H + OH.

The most detailed and accurate of the former calculations is perhaps the moderately accurate SCF + CM work of Winter *et al.* [52] on the 16 singlet and triplet excited states arising from excitation out of the $1b_1$ or $3a_1$ MO into one of the Rydberg MOs 3s [$4a_1$], $3p_x$, $3p_y$, $3p_z$. For five experimentally well-established states $^1B_1(1b_1 \rightarrow 3sa_1)$ at 7.49 eV,

TABLE 3
Spectroscopic Constants for H_2O and HDO (cm^{-1})[a]

Constant	H_2O				HDO			
	SCF	SD	SDQ	Experiment[b]	SCF	SD	SDQ	Experiment[b]
ω_1	4132	3931	3856	3832	3040	2894	2839	2824
ω_2	1772	1703	1688	1648	1553	1492	1479	1440
ω_3	4238	4036	3959	3942	4184	3986	3909	3890
x_{11}	−43.7	−41.8	−45.0	−42.6	−49.6	−43.8	−42.5	−43.4
x_{22}	−23.3	−18.6	−20.5	−16.8	−18.4	−12.5	−13.7	−11.8
x_{33}	−51.1	−48.7	−44.3	−47.6	−85.1	−83.9	−73.7	−82.9
x_{12}	−3.3	−9.3	−13.5	−15.9	−19.2	−14.6	−7.3	−8.6
x_{13}	−175.8	−166.1	−149.6	−165.8	−7.0	−9.5	−15.0	−13.1
x_{23}	−5.3	−16.7	−11.7	−20.3	6.4	−10.4	−19.2	−20.1
G	−3.9	1.1	2.6	—	−17.7	−12.0	−8.4	—
E (0, 0, 0)	4991	4761	4683	4634	4328	4130	4062	4032
ΔE (1, 0, 0)	3955	3760	3684	3657	2928	2794	2742	2782
ΔE (0, 1, 0)	1721	1652	1634	1595	1510	1455	1439	1403
ΔE (0, 0, 1)	4045	3847	3790	3756	4014	3808	3744	3707

[a] Values derived from potential energy surfaces of [45]. The spectroscopic constants are defined according to the formulas [49]

$$E(v_1, v_2, v_3) = G + \sum_{i=1}^{3} \omega_i(v_i + \tfrac{1}{2}) + \sum_{i \le 1}^{3} x_{ij}(v_i + \tfrac{1}{2})(v_j + \tfrac{1}{2})$$

and

$$\Delta E(v_1, v_2, v_3) = E(v_1, v_2, v_3) - E(0, 0, 0).$$

[b] Benedict *et al.* [50b].

$^3A_1(1b_1 \rightarrow 3pb_1)$ at 9.81 eV, $^1A_1(1b_1 \rightarrow 3pb_1)$ at 10.172 eV, $^1B_1(1b_1 \rightarrow 3pa_1)$ at 9.996 eV, and $^1A_1(3a_1 \rightarrow 3sa_1)$ at 9.73 eV, the calculated and observed values agree within 0.07 eV on the average. The remaining predicted values should then be significant. In a similar but seemingly more accurate paper by Buenker and Peyerimhoff [53]—they included d functions in the basis, which Winter *et al.* omit, and they get lower computed energies for the ground state, by 0.020 a.u. for SCF and 0.050 a.u. for CM—the authors get similar agreement but excitation energies about 0.3 eV lower. They give few details, but an additional feature is a computation of oscillator strengths where agreement with observation for the transitions to 2^1A_1 and 1^1B_1 is good.

Liquid water shows a very weak absorption at 4.5 eV, and low-energy electron impact shows a feature at the same position. At first this feature was attributed to excitation by the $1b_1 \rightarrow 3s$ transition to $^3B_1$, but the theoretical calculations conclusively place this near 7 eV, not far below the $^1B_1$. (Singlet-triplet intervals are never large in Rydberg states.) An equation-of-motion calculation [54] again confirms this verdict. The nature of the 4.5-eV transition has remained a mystery, but a very recent study [54a] with extremely pure water finds no 4.5 eV peak.

Svensson *et al.* [55] have made SCF and limited CM calculations on the shake-up spectrum of H_2O. The primary K-shell ionization ESCA peak $(1a_1 \rightarrow \infty)$ is followed by several shake-up peaks in which additionally an excitation of a valence-shell MO has occurred. The computations (including intensities for transitions to the two doublet states of H_2O^+ associated with each shake-up) permit satisfactory identification of the shake-up spectrum observed by Siegbahn. The following shake-ups were computed: $1b_1 \rightarrow 2b_1$ and $1b_1 \rightarrow 3b_1$, $3a_1 \rightarrow 4a_1$ and $3a_1 \rightarrow 5a_1$, $1b_2 \rightarrow 2b_2$ and $1b_2 \rightarrow 3b_2$, $2a_1 \rightarrow 4a_1$.

Further calculations on the shake-up spectrum have been made by Wahlgren [55a], using the IVO and equivalent core concepts. CM calculations show considerable interactions among computed shake-up states.

With the help of *ab initio* calculations, potential surfaces for dissociation of the lowest two excitations $1b_1 \rightarrow 4a_1$ and $3a_1 \rightarrow 4a_1$ into OH + H have been discussed by several authors. For the $1b_1 \rightarrow 4a_1$ excitations the SCF approximation is adequate [56–58] to give correct dissociation relations. This can be seen by writing the electron configurations in the symbolism of each of the three symmetries $C_{\infty v}$, C_s, and C_{2v} involved in the reaction $OH(^2\Pi) + H(^2S) \leftrightarrow H_2O(^{3,1}B_1)$.

$C_{\infty v}$: OH, $1\sigma^2\, 2\sigma\, 3\sigma^2\, 1\pi^3$ plus H, 1s $\leftrightarrow$ H_2O, $1\sigma^2\, 2\sigma^2\, 3\sigma^2\, 1\pi^3\, 4\sigma$

C_s: OH, $(1a')^2\,(2a')^2\,(3a')^2\,(4a')^2\,1a''$ plus H(a') $\leftrightarrow$

H_2O, $(1a')^2\,(2a')^2\,(3a')^2\,(4a')^2\,1a''\,5a'$

C_{2v}: OH, $1a_1^2\,2a_1^2\,3a_1^2\,4a_1^2\,1b_1$ plus $H(a_1) \leftrightarrow$

H_2O, $1a_1^2\,2a_1^2\,1b_2^2\,3a_1^2\,1b_1\,4a_1$

Here π of OH splits into 4a′ and 1a″ under C_s, which become $3a_1$ and $1b_1$ in C_{2v}. The absorption spectrum to the $^1B_1$ state is very diffuse, indicating that dissociation is occurring (arrows to the left, starting with C_{2v}, in the above formulation). In fact, the SCF calculations show the excited H_2O flying apart into $OH(^2\Pi)$ + H. The calculations of Miller *et al.* [56] and of Gangi and Bader [57] give portions of the potential surface for the dissociation process.

For the second pair of excited states $(3a_1 \rightarrow 4a_1)$, $^{1,3}A_1$ one can write configurations as above for C_s and C_{2v} symmetries; for $C_{\infty v}$ the formulation is the same as for $(1b_1 \rightarrow 4a_1)$, $^{1,3}B_1$. For $(3a_1 \rightarrow 4a_1)$, $^1A_1$ there is a complication (see below); for $^3A_1$ we have

C_s: OH, $(1a')^2\,(2a')^2\,(3a')^2\,4a'\,(1a'')^2$ plus H $\leftrightarrow$

H_2O, $(1a')^2\,(2a')^2\,(3a')^2\,4a'\,(1a'')^2\,5a'$

C_{2v}: OH, $1a_1^2\,2a_1^2\,3a_1^2\,4a_1\,1b_1^2 \leftrightarrow H_2O$, $1a_1^2\,2a_1^2\,1b_2^2\,3a_1\,(1b_1)^2\,4a_1$

Horsley and Fink [51] have made Gaussian-lobe SCF computations on the potential surfaces of the $(3a_1 \rightarrow 4a_1)$, $^1A_1$ and $^3A_1$ states. They show that their energies are probably minimized, with respect to bending, in a linear conformation, but that they are unstable with respect to dissociation into OH + H. They also showed that the $4a_1$ MO, although mostly 3s at the ground-state equilibrium conformation, rapidly become O—H antibonding $a2s_0 - b(1s_H + 1s_H)$ as dissociation into OH + H occurs.

Matters are straightforward for the $^3A_1$ state, but for the $^1A_1$ state there is a complication because it has the same symmetry as the ground state. The matter has been discussed by Gangi and Bader [57] and by Flouquet and Horsley [58]. The latter use a limited CM treatment, in which for all angles except 180° there is a strong interaction between the $^1A_1$ configuration $(3a'^2\,4a'\,1a''^2\,5a'$ and the ground-state configuration $(3a')^2\,(4a')^2\,(1a'')^2$ at all distances of OH from H greater than 2.5 a.u. (R_e = 1.81 a.u.), so that the two states repel each other, the ground state dissociating to $OH(^2\Pi)$ + H and the excited state to OH, $\sigma\pi^4$, $^2\Sigma^+$ plus H. At increasing distances, the configuration $(1a')^2\,(2a')^2\,3a_1\,(4a')^2\,(1a'')^2\,(5a')$ becomes predominant and leads to $OH(^2\Sigma^+)$ + H, the 3a′ orbital becoming the 3σ orbital of OH and the 4a′ and 1a″ becoming 1π. Photochemically, dissociation to $O(^1D)$ plus H_2 also occurs, and this is discussed in [4].

For H_2S the outer shell is ... $4a_1^2\,2b_2^2\,5a_1^2\,2b_1^2$ analogous to that in H_2O.

Shih *et al.* [59] have made SCF + CM calculations using a moderately large AO basis set including various polarization and Rydberg-type functions to investigate the ground and excited states. For the ground state their E_h is −398.8282 when keeping five core MOs doubly occupied, but (with a somewhat smaller basis) −398.8884 with only the K-shell "frozen." Earlier, in an SCF computation Rothenberg *et al.* [60] obtained −398.6862 a.u. Roos and Siegbahn obtained similar results [60a]. Both these sets of authors computed dipole and multipole moments, magnetic susceptibilities, diamagnetic shielding quantities, and electric field gradients at the nuclei. They also made a population analysis. This indicated a charge of about −0.1 *e* on the S atom, an overlap population of 0.675 *e*, and a sulfur d-orbital population of 0.06 *e*.

Ahlrichs *et al.* [6], using a good polarized basis set including two d and one f orbital, and excluding the core MOs, made PNO-CI calculations on the ground state of H_2S, obtaining $E_h = -398.87126$, very considerably below that of the corresponding calculation of Shih *et al.* They give a table of SCF orbital energies.

Shih *et al.* [59] calculated vertical excitation energies and transition oscillator strengths for a number of excited states. The absorption spectrum and its interpretation resemble those of H_2O. The lowest-energy computed states are ($2b_1 \rightarrow 4sa_1$ or $6a_1$), $^3B_1$ at 5.74 eV and $^1B_1$ at 6.22 eV, then ($2b_1 \rightarrow 3db_2$), $^3A_2$ at 6.17 eV, and $^1A_2$ at 6.80 eV. The longest-wavelength observed absorption is a very broad rather weak (oscillator strength $f \approx 0.04$) continuum with maximum at 6.4 eV. This no doubt corresponds to the $^1A_1 \rightarrow {}^1B_1$ transition, although Shih *et al.* suggest that $^1A_1 \rightarrow {}^1A_2$, although forbidden under C_{2v} symmetry, may to some extent be mixed in, as is allowed for C_s symmetry. In analogy to the case of $4a_1$ in H_2O (see above), $6a_1$ here should not be pure 4s, but should have the form $a4s + b3s - c(1s_H + 1s_H)$, with 4s predominant for the vertical transition, but decreasing greatly as dissociation into H + SH occurs. (A mistaken but widely held idea, apparently accepted by Shih *et al.*, is that there are two separate transitions in such cases, to 4s and to $3s - (1s_H + 1s_H)$ in the present case [61].)

Flouquet [61a] in a minimal basis SCF calculation plus limited CM has discussed the potential energy surface of the $\ldots 5a_1\, 2b_1^2\, 6a_1$, $^1A_1$ state driving it to decomposition to SH + H. The case is similar to that of the lowest excited $^1A_1$ state of H_2O (see above). Contour diagrams for the potential energy surfaces for these two cases (H_2O and H_2S) are shown.

Shih *et al.*'s computed MO $3db_2$ is an interesting novelty. The orbital exponent for 3d here is relatively large. They report that it has considerable H atom character in such a way as to give bonding. This would account for the fact that its computed energy is considerably (2 eV) lower than for the $3s^2\,3p^3\,3d$ states of the S atom. Apparently we have here an excited MO with predominant d character which is strongly bonding, although in the ground state d participation is small.

Shih *et al.* deal with a number of higher excited states and make several identifications with observed Rydberg levels.

Guest and Rodwell [61b] made SCF plus CM computations apparently very similar to those of Shih *et al.* They give ground-state SCF orbital energies but not oscillator strengths of transitions. For the best basis set they report for the ground state $E_h = -398.6875$ in their SCF calculation, -398.8387 using CM, considerably lower than Shih *et al.* Their calculations on the excited states may be the best to date.

A many-body Green's function method using up to d and f sets of polarization functions in the basis has been used by von Niessen *et al.* [62] to compute directly the three lowest ionization potentials of H_2O and of H_2S. For H_2O ($1b_1$, $3a_1$ $1b_2$) the results are as follows: calculated, 12.90, 14.85, and 18.91 eV; experimental, 12.78, 14.83, and 18.72 eV. Comparisons are given with accurate calculations by other authors. For H_2S ($2b_1$, $5a_1$, and $2b_2$) the results are as follows: calculated, 10.38, 13.36 and 13.58 eV; experimental, 10.48; 13.4 and 15.5 eV.

For $4a_1$ of H_2S, 22.0 and 23.3 are observed; Domcke *et al.* [63], using a slightly less accurate Green's function method, show that for $4a_1$ two main lines at 20.84 (relative intensity 0.19) and at 22.65 (relative intensity 0.40) are expected theoretically, together with numerous satellite lines of relatively small intensity extending up to at least 35.7 eV. They also compute a weak satellite for each of the $2b_1$ and $5a_1$ ionizations.

Diercksen *et al.* [64] have made an SCF computation on FH_2^+, yielding $R_e(\mathrm{FH}) = 0.95$ Å, a bond angle of 115°, and a proton binding energy of 120 kcal/mole.

Bell [65] has made improved SCF calculations on NH_2^- with a double zeta plus polarization basis with the addition of some diffuse functions. The diffuse functions considerably alter the predicted geometry, giving $R(\mathrm{NH}) = 1.019$ Å, $\theta = 104°$, with $E_h = -55.5372$.

Calculations on PH_2^- have been mentioned in Section F.

Nagata *et al.* [66] discuss calculations on H_2S^- using the unrestricted SCF method with spin annihilation, for STO-3G and 4-31G basis sets. They consider two states, a ground state $^2B_2$ with the added electron in a $3b_2$ antibonding MO, and $^2A_1$ with it in a bonding $6a_1$ ($3d\sigma$) MO. With $R(\mathrm{SH})$ fixed at 1.335 Å as in H_2S, the energy is minimized for a 160° angle with either basis set for the $^2B_2$ state and (for the STO-3G basis) at an angle of about 80° for the $^2A_1$ state. Variation of $R(\mathrm{SH})$ then gave an energy minimum at 1.55 Å by STO-3G and 1.65 Å by 4-31G. Calculations on hyperfine coupling constants at $\theta = 160°$ are reported; correlation with experiment is poor. Spectral excitation energies ($2b_1 3b_2^2 \leftarrow 2b_1^2 3b_2$, $4a_1 3b_2^2 \leftarrow 4a_1^2 3b_2$, and $6a_1 \leftarrow 3b_2$) are calculated; two experimental transitions have been reported.

REFERENCES

1. R. S. Mulliken, *J. Chem. Phys.* **23,** 1977 (1955).
1a. D. C. McCain and W. E. Palke, *J. Chem. Phys.* **56,** 4957 (1972).
2. S. Bell, *J. Chem. Phys.* **68,** 3014 (1978); **69,** 3879 (1978).
3. See W. G. Richards, T. E. H. Walker, and R. K. Hinkley, "A Bibliography of *Ab Initio* Molecular Wave Functions." Oxford Univ. Press (Clarendon), London and New York, 1971; W. G. Richards, T. E. H. Walker, L. Farnell, and P. R. Scott, "Supplement for 1970–1973." *ibid.,* 1974; W. G. Richards, P. R. Scott, E. A. Colbourn, and A. F. Muchington, "Supplement for 1974–1977." *ibid.,* 1978.
4. R. P. Hosteny and S. A. Hagstrom, *J. Chem. Phys.* **58,** 4396 (1973).
5. R. Ahlrichs, F. Driessler, H. Lischka, and V. Staemmler, *J. Chem. Phys.* **62,** 1235 (1975).
5a. G. A. Gallup and J. M. Norbeck, *Chem. Phys.* **2,** 19 (1973).
6. R. Ahlrichs, F. Keil, H. Lischka, W. Kutzelnigg, and V. Staemmler, *J. Chem. Phys.* **63,** 455 (1975).
7. C. F. Bender and H. F. Schaefer, III, *J. Mol. Spectrosc.* **37,** 423 (1971).
8. V. Staemmler and M. Jungen, *Chem. Phys. Lett.* **16,** 187 (1972).
9. D. H. Liskow, C. F. Bender, and H. F. Schaefer, III, *J. Chem. Phys.* **61,** 2507 (1974).
10. P. K. Pearson and E. Roueff, *J. Chem. Phys.* **64,** 1240 (1976).
11. G. Herzberg, "Molecular Spectra and Molecular Structure, Vol. III, Electronic Spectra and Structure of Polyatomic Molecules." Van Nostrand, Princeton, New Jersey, 1966.
12. J. F. Harrison, *Acc. Chem. Res.* **7,** 378 (1974).
13. S. V. O'Neill, H. F. Schaefer, III, and C. F. Bender, *J. Chem. Phys.* **55,** 162 (1971); cf. also D. R. McLaughlin, C. F. Bender, and H. F. Schaefer, III, *Theor. Chim. Acta* **25,** 352 (1972); G. F. Tantardini, M. Raimondi, and M. Simonetta, *Int. J. Quantum Chem.* **7,** 893 (1973).
14. C. F. Bender, H. F. Schaeffer, III, D. R. Franceschetti, and L. C. Allen, *J. Am. Chem. Soc.* **94,** 6888 (1972).
15. V. Staemmler, *Theor. Chim. Acta* **31,** 49 (1973); **35,** 309 (1974).
16. P. J. Hay, W. J. Hunt, and W. A. Goddard, III, *Chem. Phys. Lett.* **13,** 30 (1972); *J. Am. Chem. Soc.* **94,** 8293 (1972).
17. L. B. Harding and W. A. Goddard, III, *J. Chem. Phys.* **67,** 1777 (1977).
18. C. E. Dykstra, H. F. Schaefer, III, and W. Meyer, *J. Chem. Phys.* **65,** 2740 (1976).
19. S.-K. Shih, S. D. Peyerimhoff, R. J. Buenker, and M. Perić, *Chem. Phys. Lett.* **55,** 206 (1978).
20. P. F. Zittel, G. B. Ellison, S. V. O'Neil, E. Herbst, W. C. Lineberger, and W. P. Reinhardt, *J. Am. Chem. Soc.* **98,** 3731 (1976).
21. L. B. Harding and W. A. Goddard, III, *Chem. Phys. Lett.* **55,** 217 (1978).
22. R. R. Lucchese and H. F. Schaefer, III, *J. Am. Chem. Soc.* **99,** 6765 (1977).
23. B. O. Roos and P. M. Siegbahn, *J. Am. Chem. Soc.* **99,** 7716 (1977).
24. C. W. Bauschlicher, Jr. and I. Shavitt, *J. Am. Chem. Soc.* **100,** 739 (1978).
25. J. H. Meadows and H. F. Schaefer, III, *J. Am. Chem. Soc.* **98,** 4383 (1976).

25a. J. R. Ball and C. Thomson, *Int. J. Quantum Chem.* **14,** 39 (1978).
25b. H. L. Hase, G. Laner, K. W. Schute, A. Schweig, and W. Thiel, *Chem. Phys. Lett.* **54,** 494 (1978).
25c. D. T. Clark, B. J. Crowarty, and A. Sgamellotti, *J. Electron Spectrosc.* **17,** 237 (1979).
26. S. R. Langhoff and E. R. Davidson, *Int. J. Quantum Chem.* **7,** 759 (1973); see also S. R. Langhoff, S. T. Elbert, and E. R. Davidson, *Int. J. Quantum Chem.* **7,** 999 (1973).
27. S. R. Langhoff, *J. Chem. Phys.* **61,** 3881 (1974).
28. J. F. Harrison and C. W. Eakers, *J. Am. Chem. Soc.* **95,** 3467 (1973).
29. S. Y. Chu, A. K. Q. Siu, and E. F. Hayes, *J. Am. Chem. Soc.* **94,** 2969 (1972).
30. J. F. Harrison, *J. Am. Chem. Soc.* **93,** 4112 (1971).
30a. C. W. Bauschlicher, Jr. and D. R. Yarkony, *J. Chem. Phys.* **69,** 3875 (1978).
31. C. F. Bender, J. H. Meadows, and H. F. Schaefer, III, *Discuss. Faraday Soc.* **62,** 59 (1976).
31a. S. D. Peyerimhoff and R. J. Buenker, *Chem. Phys.* **42,** 167 (1979).
31b. D. M. Hirst, *Mol. Phys.* **35,** 1559 (1978).
32. S. Bell and H. F. Schaefer, III, *J. Chem. Phys.* **67,** 5173 (1977).
32a. S. D. Peyerimhoff and R. J. Buenker, *Can. J. Chem.* **57,** 3182 (1979).
33. W. Meyer, *Int. J. Quantum Chem., Symp.* **5,** 341 (1971).
34. J. A. Smith, P. Jørgensen, and Y. Öhrn, *J. Chem. Phys.* **62,** 1285 (1975).
35. H. Sakai, S. Yamabe, K. Fukui, and H. Kato, *Chem. Phys. Lett.* **25,** 541 (1974).
36. P. J. Fortune, B. J. Rosenberg, and A. C. Wahl, *J. Chem. Phys.* **65,** 2201 (1976).
37. C. G. Balint-Kurti and R. N. Yardley, *Chem. Phys. Lett.* **36,** 342 (1975).
37a. J. D. Leclerc, J. A. Horsley, and J. C. Lorquet, *Chem. Phys.* **4,** 337 (1974).
37b. M. Perić, R. J. Buenker, and S. D. Peyerimhoff, *Can. J. Chem.* **57,** 2491 (1979).
38. C. W. Kern and M. Karplus, *in* "Water—A Comprehensive Treatise" (F. Franks, ed.), Vol. 1. Plenum, New York, 1972.
39. J. H. Miller and H. P. Kelly, *Phys. Rev. A* **4,** 480 (1971).
40. T. Lee and T. P. Das, *Phys. Rev. A* **6,** 968 (1972).
41. B. J. Rosenberg and I. Shavitt, *J. Chem. Phys.* **63,** 2162 (1975).
42. R. Moccia, *J. Chem. Phys.* **40,** 2186 (1964).
43. H. Popkie, H. Kistenmacher, and E. Clementi, *J. Chem. Phys.* **59,** 1325 (1973).
43a. H. J. Werner and W. Meyer, *Mol. Phys.* **31,** 855 (1976).
44. See also P. Lazzeretti, R. Zanasi, and B. Cadioli, *J. Chem. Phys.* **67,** 382 (1977).
45. B. J. Rosenberg, W. C. Ermler, and I. Shavitt, *J. Chem. Phys.* **65,** 4072 (1976).
46. P. Hennig, W. P. Kraemer, and G. H. F. Diercksen, *Theor. Chim. Acta* **47,** 233 (1978).
46a. R. J. Bartlett, I. Shavitt, and G. D. Purvis, *J. Chem. Phys.* **71,** 281 (1979).
46b. A. R. Hoy, I. M. Mills, and G. Strey, *Mol. Phys.* **24,** 1265 (1972).
46c. A. R. Hoy and P. R. Bunker, *J. Mol. Spectrosc.* **74,** 1 (1979).
47. U. T. Lamanna, C. Guidotti, and G. P. Arrighire, *J. Chem. Phys.* **67,** 604 (1977).
48. L. C. Snyder and T. A. Weber, *J. Chem. Phys.* **63,** 113 (1977); S. Palalikit and I. Shavitt, unpublished results; S. Palalikit, Ph.D. Thesis, Ohio State Univ., Columbus, 1976.
49. See W. C. Ermler and B. J. Krohn, *J. Chem. Phys.* **67,** 1360 (1977); B. J. Krohn and C. W. Kern, *J. Chem. Phys.* **69,** 5310 (1979).
50. R. J. Whitehead and N. C. Handy, *J. Mol. Spectrosc.* **55,** 356 (1975); *ibid.* **59,** 459 (1976).
50a. W. C. Ermler, B. J. Rosenberg, and I. Shavitt, unpublished results.
50b. W. S. Benedict, N. Gailar, and E. K. Plyler, *J. Chem. Phys.* **24,** 1139 (1956).
51. J. A. Horsley and W. H. Fink, *J. Chem. Phys.* **50,** 750 (1969).

52. N. W. Winter, W. A. Goddard, III, and F. W. Bobrowicz, *J. Chem. Phys.* **62,** 4325 (1975).
53. R. J. Buenker and S. Peyerimhoff, *Chem. Phys. Lett.* **29,** 253 (1974).
54. D. Yeager, V. McCoy, and G. A. Segal, *J. Chem. Phys.* **61,** 755 (1974).
54a. T. L. Quickenden and J. A. Irvin, *J. Chem. Phys.* **72,** 4416 (1980).
55. S. Svensson, H. Agren, and U. Wahlgren, *Chem. Phys. Lett.* **38,** 1, (1976).
55a. U. Wahlgren, *Mol. Phys.* **33,** 1109 (1977).
56. K. J. Miller, S. R. Mielczarek, and M. Krauss, *J. Chem. Phys.* **51,** 26 (1969).
57. R. A. Gangi and R. F. W. Bader, *J. Chem. Phys.* **55,** 5369 (1971).
58. F. Flouquet and L. A. Horsley, *J. Chem. Phys.* **60,** 3767 (1974).
59. S. Shih, S. D. Peyerimhoff, and R. J. Buenker, *Chem. Phys.* **17,** 391 (1976).
60. S. Rothenberg, R. H. Young, and H. F. Schaefer, III, *J. Am. Chem. Soc.* **92,** 3243 (1970).
60a. B. Roos and P. Siegbahn, *Theor. Chim. Acta* **21,** 386 (1971).
61. See last paragraph in R. S. Mulliken, *Acc. Chem. Res.* **9,** 7 (1976).
61a. F. Flouquet, *Chem. Phys.* **13,** 257 (1976).
61b. M. F. Guest and W. R. Rodwell, *Mol. Phys.* **32,** 1075 (1976).
62. W. von Niessen, G. H. F. Diercksen, and L. S. Cederbaum, *J. Chem. Phys.* **67,** 4124 (1977); W. von Niessen, L. S. Cederbaum, W. Domcke, and G. H. F. Diercksen, *J. Chem. Phys.* **66,** 4893 (1977). On the method, see L. S. Cederbaum, *Mol. Phys.* **28,** 479 (1974).
63. W. Domcke, L. S. Cederbaum, J. Schirmer, W. von Niessen, and J. P. Maier, *J. Electron Spectrosc.* **14,** 59 (1978).
64. G. H. F. Diercksen, W. von Niessen, and W. P. Kraemer, *Theor. Chim. Acta* **31,** 205 (1973).
65. S. Bell, *J. Chem. Phys.* **69,** 3879 (1978).
66. S. Nagata, T. Yamabe, and K. Fukui, *Bull. Chem. Soc. Jpn.* **51,** 2795 (1978).

CHAPTER III

TRIATOMIC MOLECULES HAB AND HA_2

A. SURVEY

Many of the entities discussed in this section are of astrophysical importance, as inhabitants either of interstellar clouds or of stellar atmospheres. Of the ground and low-energy excited states, some are linear ($C_{\infty v}$), most are bent (C_s). The electron configurations are as follows

$$\text{Linear: } 1\sigma^2\, 2\sigma^2\, 3\sigma^2\, 4\sigma^2\, 5\sigma^m\, 1\pi^{n+p}\, 2\pi^{r+s}$$

$$\text{Bent: } (1a')^2\, (2a')^2\, (3a')^2\, (4a')^2\, (5a')^m\, (6a')^n\, (1a'')^p\, (7a')^r\, (2a'')^s$$

The computed (and in some cases observed) geometries of the ground and low-energy excited states are set out in Table 1. The method used in the bent *ab initio* computations is indicated briefly. In some cases additional computations on higher excited (including Rydberg) states have been made.

In a paper reporting on MRD-CI calculations on isomerization energies of HOS, HNP, and HCSi and their positive ions, Buenker *et al.* [1] note that in HAB molecules where A is a first-row and B a second-row atom, there is a strong preference for structures with the second atom at the ter-

TABLE 1
Low-Energy Electron States of HAB and HA_2 Molecules[a]

Molecule	Population $C_{\infty v}$[b] 5σ	1π	2π	C_s[c] 5a′	6a′	1a″	7a′	2a″	State	Geometry <HAB[d]	R_e (AH)[e]	R_e (AB)[e]
HCN^+	2	3							$^2\Pi$	180	1.08*	1.18*
	1	4							$^2\Sigma^+$	180		
HC_2	1	4							$^2\Sigma^+$	180	1.06	1.242
	2	3							$^2\Pi$	180	1.06	1.316
HC_2^-	2	4							$^1\Sigma^+$	180	1.06	1.373
HCN	2	4							$^1\Sigma^+$	180	1.066	1.149
											(1.066)	(1.153)
	(2	3	1)	2	2	1	1		$^1A''$	123	1.08	1.33
										(125)	(1.14)	(1.297)
	(2	3	1)	2	1	2	1		$^1A'$	120	1.10	1.35
										(114.5)	(1.14)	(1.334)
	(2	3	1)	2	1	2		1	$^1\Delta$ ($^1A''$)	≈180	1.077	1.337
HNC	2	4							$^1\Sigma^+$	180	0.995	1.169
	(2	3	1)	2	2	1	1		$^1A''$	113	1.022	1.461
	(2	3	1)	2	1	2	1		$^1A'$	120	1.025	1.417
	(2	3	1)	2	1	2	1		$^1\Delta$ ($^1A''$)	157	0.996	1.418
HCO^+	2	4							$^1\Sigma^+$	180	1.091	1.102
HOC^+	2	4							$^1\Sigma^+$	180	0.988	1.153
HN_2^+	2	4							$^1\Sigma^+$	180	1.032	1.095
HCO	(2	4	1)	2	2	2	1		$^2A'$	130	1.098	1.188
	2	4	1	(2	2	2		1)	$^2\Pi$ ($^2A''$)	180		
	2	3	2[f]	2	1	2	2[f]		$^2A'$	~100		
HN_2	(2	4	1)	2	2	2	1		$^2A'$	124[g]	1.04	
	2	4	1	(2	2	2		1)	$^2\Pi$ ($^2A''$)	180	0.98	
HNO^+	(2	4	1)	2	2	2	1		$^2A'$	125	1.068	1.158
	2	4	1	(2	2	2		1)	$^2\Pi$ ($^2A''$)	180		
HON^+	(2	4	1)	2	2	2	1		$^2A'$	118	1.019	1.204
	2	4	1	(2	2	2		1)	$^2\Pi$ ($^2A''$)	180		
HNO	(2	4	2)	2	2	2	2		$^1A'$	109	(1.062)	(1.211)
										(108.5)		
	(2	4	2)	2	2	2	1	1	$^3A''$	120		
	(2	4	2)	2	2	2	1	1	$^1A''$	117	(1.035)	(1.241)
										(116.3)		
	(2	4	2)	2	2	2		2	$^1A'$	180		
HCF	(2	4	2)	2	2	2	2		$^1A'$	102.2	1.111	1.325
										(101.6)	(1.121)	(1.314)
	(2	4	2)	2	2	2	1	1	$^3A''$	120.4	1.077	1.321
HO_2	(2	4	3)	2	2	2	2	1	$^2A''$	(106)	(0.968)	(1.34)
				2	2	2	1	2	$^2A'$			

(continued)

TABLE 1 (*Continued*)

	Population		Geometry			
	$C_{\infty v}$ [b]	C_s [c]				
Molecule	5σ 1π 2π	5a′ 6a′ 1a″ 7a′ 2a″	State	<HAB[d]	R_e (AH)[e]	R_e (AB)[e]
HOF	(2 4 4)	2 2 2 2 2	$^1A'$	101 (97)	1.08* (0.94)	1.45* (1.44)
HF_2, HF_2^- (see text)						

[a] From CM calculations except SCF where marked by *.
[b] Electron configuration begins with $1\sigma^2 2\sigma^2 3\sigma^2 4\sigma^2$.
[c] Electron configuration begins with $(1a')^2 (2a')^2 (3a')^2 (4a')^2$.
[d] Degrees; experimental value in parentheses.
[e] Angstroms; experimental value in parentheses.
[f] Mixed with 7a′ → 3s.
[g] Or possibly unstable.

minal position. In cases studied to date the largest and smallest distinctions in relative stabilities of HAB and HBA are found for systems with 10 and 12 valence electrons, respectively (e.g., 110 kcal/mole for HCS^+—HSC^+ and 2 kcal/mole for $^1A'$ HOS^+—HSO^+). Isomerization energies tend to increase with the electronegativity of the second-row atoms.

B. LiCH, LiN^+H, LiOH, $LiOH^-$; HC_2^+, HCN^+, HC_2, HC_2^-; HCSi, HSiC, $HCSi^+$, $HSiC^+$

Using a (9s, 5p) basis set contracted to (3s, 2p) for C, 8s contracted to 3s for Li, and 4s contracted to 2s for H, Harrison *et al.* [1a] in an SCF calculation found a linear nearly Li^+CH^- arrangement for LiCH, with a $^3\Sigma^-$ ground state and two fairly low $^3\Pi$ excited states. A similar calculation on $(LiNH)^+$ shows a linear $^3\Sigma^-$ ground state and a $^1\Delta$ state about 42 kcal/mole higher up.

Blickensderfer and Jordan [2] have made SCF calculations on LiOH and $LiOH^-$. They use an (11s, 4p) basis set contracted to (4s,2p) for Li, (9s,5p) contracted to (4s,2p) for O, and (5s,1p) contracted to (3s,1p) for H. They quote an unpublished calculation of McLean on LiOH, giving a dipole moment of 4.87 D; they get 4.58 D. LiOH is believed linear; they report that $LiOH^-$ is linear but with a small force constant for bending. Like other LiX^- compounds, the added electron in $LiOH^-$ is found loosely bound at the Li in the dipole field of the LiOH.

The molecules HNB, HCN^+, N_2^+, and HC_2 are isoelectronic. HCN^+ is known experimentally from the photoelectron spectrum of HCN. So and Richards [3] have made an SCF calculation on HCN^+ using a large basis of STFs on the $^2\Pi$ ground state and $A^2\Sigma^+$ first excited state. Note that the order of these two states is reversed as compared with that for N_2^+ or HC_2. Population analyses relative to HCN are given. Hansoul *et al.* [4] have made an SCF calculation plus single-substitution CM from the reference configurations of the potential surface of the $B^2\Sigma^+$ second excited state of HCN^+. They concluded that it is a mixed state with two potential minima separated by a saddle point; the two minima correspond mainly to the respective configurations $\ldots 4\sigma\, 5\sigma^2\, 1\pi^4$ and $\ldots 4\sigma^2\, 5\sigma\, 1\pi^3\, 2\pi$. Anomalous features of the photoelectron spectrum for this state are elucidated. Higher $^2\Pi$ and $^4\Pi$ states are very briefly discussed.

Hirst [4a], using much more extensive CM than Hansoul *et al.* (1953 configurations versus 246), has recalculated the $A^2\Sigma$ and $B^2\Sigma^+$ states. For $A^2\Sigma^+$ he reports minima at $R_{CH} = 2.03a_0$ and $R_{CN} = 2.25a_0$, for $B^2\Sigma^+$ one minimum at $R_{CH} = 2.75a_0$, $R_{CN} = 2.26a_0$ and a maximum of height 0.27 eV relative to $R_{CH} = 10.0a_0$. It appears that this maximum is due to an avoided crossing in which $\ldots 1\pi^4\, 5\sigma^2\, 4\sigma$ is dominant (among 11 configurations used for reference) for R_{CH} below $3.5a_0$ but a mixture of $\ldots 1\pi^4\, (4\sigma\, 5\sigma)\, 6\sigma$ and $\ldots 1\pi^4\, 4\sigma^2\, 6\sigma$ for R_{CH} greater than $4a_0$.

Botschwina [5], using the CEPA method, has calculated the energies and force constants of the lowest two states of HNB and HBN. HNB has a $^2\Sigma^+$ ground state, and an excited $^2\Pi$ state about 18,000 cm^{-1} higher. HBN is calculated to have a $^2\Pi$ ground state, 21 kcal/mole above the ground state of HNB, and an excited $^2\Sigma^+$ state.

Experimentally, the ethynyl radical HC_2 is known from ESR spectra and fragmentary vibrational and electronic spectra of the radical trapped in an organic matrix, HC_2^- is well known from negative ion mass spectra. Values of the electron affinity in the literature are mostly in the range 1.8–2.5 ev. Vasudevan and Grein [6], using the accurate MRD-CI procedure of Peyerimhoff and Buenker and a basis set like that used by Shih *et al.* on HC_2 (see below), have made very accurate CM calculations of the ground states of HC_2 and HC_2^-. Thereby, using optimized R_{CH} and R_{CC} distances, they computed the adiabatic electron affinity (EA) of HC_2 as 2.14 eV, the vertical EA as 2.09 eV, and the vertical detachment energy of HC_2^- as 2.17 eV. From these values, the heat of formation of HC_2 and the C—H bond dissociation energy of C_2H_2 were estimated to be 121 ± 3 and 119 ± 3 kcal/mole, respectively. The geometries were determined as follows: R_{CH} = 1.062 Å for HC_2 and 1.063 Å for HC_2^-, R_{CC} = 1.242 Å for HC_2 and 1.273 Å for HC_2^-. Earlier, Pacansky and Orr [7] had made SCF, hence less accurate, calculations on HC_2 and HC_2^-. They give orbital energies and a population analysis.

In SCF calculations using a double-zeta plus 3d and 4f STF basis, So and Richards [8] have dealt with the ground ... $5\sigma\ 1\pi^4$, $^2\Sigma^+$ and the first excited ... $5\sigma^2\ 1\pi^3$, $^2\Pi$ excited states of HC_2. They computed stretching frequencies ν_{CH} = 3512, ν_{CC} = 2311 cm^{-1} for the $^2\Sigma$, and ν_{CH} = 3582, ν_{CC} = 1928 cm^{-1} for the $^2\Pi$ state. They give orbital energies and the following population summary: $^2\Sigma^+$, $H^{0.75}C^{6.35} \equiv C^{5.90}$; $^2\Pi$, $H^{0.75}C^{6.24} \equiv C^{6.04}$. They identify the observed absorptions near 10,000 Å with the $^2\Pi \leftarrow {}^2\Sigma^+$ transition. Their SCF energy is $E_h = -76.1623$.

Hillier *et al.* [8a], using a polarized double-zeta basis followed by large-scale CM, made calculations on the three lowest states $^2\Sigma$, $^2\Pi$, and $^2\Sigma$ of HC_2 optimizing bond lengths. They also calculated the isotropic hyperfine coupling constants, dipole moment, and electric fields at the nuclei.

Shih *et al.* [9], with a large Gaussian basis set including midbond functions for polarization, have made an MRD-CI calculation of the vertical excitation energies for a considerable number of excited states. These include the $^2\Pi$ states already mentioned, five other valence-shell doublet and quartet excitations, mostly $\pi^* \leftarrow \pi$, and several Rydberg states.

The CM treatment places the $^2\Pi$ state vertically 0.96 eV up, much higher than by SCF. Their SCF energy for the ground state is -76.1299 a.u. They compute oscillator strengths for transitions to excited states and also give potential curves for the $^2\Sigma^+$ and the $^2\Pi$ states; these cross when R(CC) is increased to about 1.36 Å. They compare the states of HC_2 with those of the isoelectronic molecule CN.

They also give SCF energies for five states of HC_2^+: ... $4\sigma^2\ 1\pi^3\ 5\sigma$, $^3\Pi$ (ground state) and $^1\Pi$, $E_h \approx -75.7633$ and -75.7086 (9.97 eV and 11.46 eV up); ... $4\sigma^2\ 1\pi^4$, $^1\Sigma^+$, $E_h \approx -75.6078$ (14.20 eV up), and ... $4\sigma^2\ 1\pi^2\ 5\sigma^2$, $^3\Sigma^-$ and $^1\Delta$ with $E_h = -75.7270$ and -75.6813 (10.96 and 12.20 eV up).

In a later paper [10], Shih *et al.* have extended their calculations on C_2H, with the same basis and method as before, to investigate the behavior of the various states on C—C stretching and C—C—H bending, in order to obtain a better understanding of the emission spectrum. They give angular potential curves for the excited states, many of which in equilibrium are bent (C_s symmetry) and have increased R(C–C) values. From these they obtain an interpretation of the emission spectrum. Acetylene photolysis to form C_2H is discussed. It is suggested that the calculated crossing of the ground state $^2\Sigma^+$ curve by the low $^2\Pi$ curve when the

C—C bond is stretched may account for the irregular vibrational spacings observed in the infrared in the 3600 cm^{-1} region.

Montgomery and Dykstra [11] have made very accurate calculations on the equilibrium structures of the $^1\Sigma^+ \cdots 4\sigma^2\ 1\pi^4$ and $4\sigma^2\ 1\pi^4\ 5\sigma^2$ states of HC_2^+ and HN_2^+ using the triple-zeta plus two d basis set, followed by CM using single and double and some quadruple substitutions in the SCEP method. Equilibrium geometry is determined as follows $R(CC) =$ 1.203 Å, $R(CH) =$ 1.083 Å in HC_2^+, $R(NN) =$ 1.095 Å, $R(NH) =$ 1.029 Å. The rotational constants B_e and the frequencies ν of the $1 \leftarrow 0$ rotational transition are calculated for use in predictions of interstellar spectra: B_e = 44.3 GHz and ν_{10} = 88.2 GHz for HC_2^+. Values for HN_2^+, which is known to have a $^1\Sigma^+$ ground state, are well established. Unfortunately, the SCF results of Shih *et al.*, even with all due allowance for CM, indicate that the ground state of HC_2^+ is $^3\Pi$.

Buenker *et al.* [1] have made MRD-CI calculations on the ground and low-energy excited states of HSiC and HCSi and their positive ions, using a good polarized basis set. Potential curves as a function of the C—Si distance are given. For the neutral systems, $\sigma^2\pi^3$, $^2\Pi$ of HCSi is lowest, $\sigma\pi^4$, $^2\Sigma^+$ of HCSi is 29 kcal/mole higher, $^2\Pi$ of HSiC is 61 and $^2\Sigma^+$ of HSiC 67 kcal/mole higher. The states are linear except for a small bending of one component ($^2A''$) of the $^2\Pi$ of HSiC. Note that the relative energies of $^2\Pi$ and $^2\Sigma^+$ are reversed as compared with HC_2. In the positive ions, for $HCSi^+$, the relative energies of the lower states in kilocalories per mole, are $\sigma^2\pi^2$, $^3\Sigma^-$, ground; $^1\Delta$, 22 higher, $^1\Sigma^+$, 35 higher; then $\sigma\pi^3$, $^3\Pi$, 43 higher, $^1\Pi$, 56 higher. For $HCSi^+$, the heights (above $\sigma^2\pi^2$, $^3\Sigma^-$ of $HSiC^+$) are 69, 84, 90, 82, and 112 kcal/mole. Equilibrium SiC and HSi or HC values for all the aforementioned states are given; the Si—C distances are considerably shorter, and the Si—H or C—H distances usually slightly shorter, for the states with more π electrons. In $HSiC^+$, some of the states are slightly bent, hence deviated toward the stabler isomer.

C. HCN, HNC, HCO^+, HOC^+, HN_2^+, HBO, HCS^+, HCP, HBS, ETC.

The molecules HC_2^-, HCN, HNC, HCO^+, HOC^+, HN_2^+ are isoelectronic, and may be expected to be spectroscopically similar. In their ground states, all are linear.

There are a number of SCF computations on the ground state of HCN. The most accurate, by McLean and Yoshimine [12], gives $E_h = -92.9147$. There are also some CM calculations, mostly devoted to excited states. One of these, by Wahlgren *et al.* [13], using a double-zeta basis set plus polarization functions, (11s 6p 1d) contracted to (6s 3p 1d) for C and N, (6s 1p) contracted to (3s 1p) for H, and for CM including all single and double MO replacements except for the 1s shells, gave $E_h = -93.1924$. This paper is devoted primarily to the computations of force constants; these agree reasonably well with experiment. R_e values are also computed. SCF population analysis is obtained, giving a charge distribution $H^{0.81}C^{6.05}N^{7.14}$.

Using a less accurate SCF plus CM calculation ($E_h = -93.0033$ by CM for HCN), Haese and Woods [14] have calculated dipole moments for HCN (3.08D, experimental 2.975 D), HNC (3.14 D, experimental 3.05 D), HN_2^+ (3.50 D), HCO^+ (4.07 D), HBF^+ (3.98 D), and $HCNH^+$ (0.23 D). They also computed bond distances and E_h values.

Pearson *et al.* [15] briefly report accurate CM calculations on the ground states of HCN and HNC. Their wave function for HCN is the same as that of [13]. Their R_e values are slightly different; those of [13] are perhaps slightly preferable. They are listed in Table 1 along with those computed in [22] for HNC. Pearson *et al.* find an energy 0.63 eV higher for HNC than for HCN, the two being separated by a substantial barrier. This work gives support to the identification of a galactic microwave absorption line at 90.665 GHz to HNC; a value of 90.48 GHz is computed.

Kraemer *et al.* [16] have performed large-scale SCF + CM calculations on the ground states of HCN, HNC, HCO^+, HOC^+, and HNN^+. The Gaussian basis set was of improved double-zeta quality plus polarization functions. For CM all single and double excitations were included (presumably omitting 1s excitation of C, N, O). The energy hypersurface was determined for 28 configurations near the minimum, and geometry and force constants were determined.

From spectroscopic constants, rotational transition frequencies $\nu(J = 1 - 0)_{v=0}$ were calculated in gigahertz for comparison with astrophysical observations. These frequencies are computed for a number of isotopic molecules, and a small empirical correction (based on HCN) is added. There is thus good agreement with experiment for HCN (all isotopes), HN_2^+ (principal isotopes), HCO^+ (three isotopes), and moderate agreement for HNC (eight isotopes). The astrophysical X-ogen line has been identified with HCO^+. Wahlgren *et al.* have made similar calculations [17].

Taylor *et al.* [18], using a good AO polarized basis set, have made a CM calculation using their coupled-pair approximation (CPA) with nonorthogonal independent-pair natural orbitals and have made accurate calculations on geometry, rotational constants, and force constants for the vibrational stretching modes, obtaining somewhat better agreements with experiment than variational CM calculations. Taylor *et al.* [19] have analyzed valence-shell correlation energies in HCN and HNC into pair contributions at the independent-pair, CM, and coupled-pair levels. They also show contour maps for localized bonding and lone-pair MOs. They used a good polarized basis set and conventional CM as well as CEPA and other coupled-pair methods.

Details on the potential energy surface and the reaction path between HCN and HNC are given in another paper by Pearson *et al.* [20]. They find the saddle point for the motion of the H atom around the CN to be at 74° if 0° represents HCN and 180° HNC, with an energy barrier 35 kcal/mole higher than HCN and with $R_{CH} = 1.18$ Å, $R_{MH} = 1.18$ Å (M is center of mass of CN). They used the same basis set and CM as in [7]. Pople *et al.* [21] report more or less similar results in calculations using a 6-31G* basis and Møller–Plesset third-order corrections.

Schwenzer *et al.* [22] have made computations, by a method exploiting Brillouin's theorem, on the six lowest singlet and six lowest triplet excited states of HCN. Their basis set was of double-zeta quality without polarization functions. Most of the excitation is of the $\pi \rightarrow \pi^*$ type, resulting in bent excited states (C_s symmetry): 6a′ or 1a″ corresponding to π is excited to 7a′ or 2a″ corresponding to π^*. The energies and geometries of the states are computed. The lowest excited state ... 6a′ 2a″ 7a′, ${}^1A''$ can be identified with the spectroscopically observed A ${}^1A''$ state, with angle 125.0° (computed 127.2°). The second observed excited state was designated B ${}^1A''$ by Herzberg and Innes, with a 114.5° angle [23], and configuration ... 6a′ $1a''^2$ 2a″. Schwenzer *et al.* [22] point out that the observed angle is inconsistent with Walsh's rules (see Fig. 1), which predict a linear geometry for this configuration. They suggest that the B state is (with considerable CM) primarily a ... 6a′ $(1a'')^2$ 7a′, ${}^1A'$ state, for which Walsh's rules predict a small angle. In a later paper [24], using a basis set including polarization functions, Schwenzer *et al.* confirm the conclusions just stated.

The matter has been further clarified by Perić *et al.* [25], including polarization functions and a normal CM treatment. They agree with the identification of A ${}^1A''$, and point out that this would correspond to $\pi^3 \pi^*$, ${}^1\Sigma^-$ in a linear molecule. They also agree that the B state is probably ${}^1A'$; the B state together with the unidentified ... 6a′ $(1a'')^2$ 2a″, ${}^1A''$ are then two components of the linear molecule $\pi^3 \pi$, ${}^1\Delta$ state, with the ${}^1A'$ component strongly bent and the other linear, i.e., ${}^1A''$ (${}^1\Delta$) (see Fig. 2). Perić *et al.* have further examined the vibrational structure in the A ← X and B ← X transitions; their calculations for HCN and DCN are in rather good agreement with experiment. They also discuss the observed predissociation in

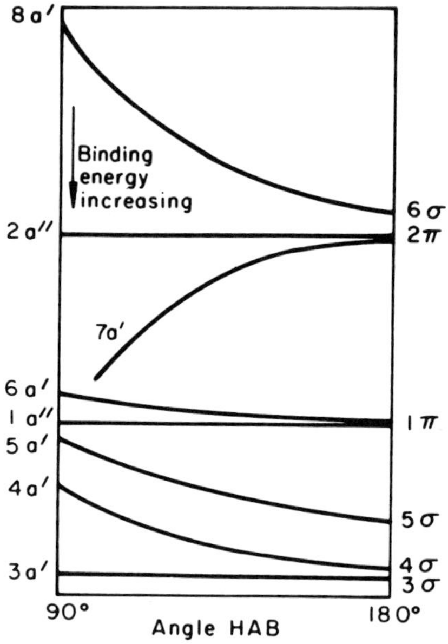

FIG. 1. Original Walsh diagram for HAB molecules, except for current notation of MOs. From Schwenzer *et al.* [27].

the A state, and express it in terms of the gradual conversion of the π^* MO of bent HCN into a hydrogen atom 1s A0 as the molecule dissociates to H + CN($^2\Pi$).

Valquez and Gouyet [26], with the help of 4-31G SCF plus CM calculations, discuss H—CN dissociation curves for excited states.

Schwenzer *et al.* [27] have done moderately accurate computations using the same method as in [22], on the six lowest singlet and also six lowest triplet excited states of HNC. They find these to be similar to those of HCN, but mostly (not all) lying higher above the ground state than for corresponding states of HCN. Their results on geometry are listed in Table 1. They find that Walsh's rules hold fairly well for HNC (see above concerning HCN).

Using a many-body Green's function method with a good polarized basis set, von Niessen *et al.* [28] have computed vertical valence ionization potentials for HCN and HNC. SCF orbital energies are listed. For HCN, the calculated values (with experimental values in parentheses) are 1π, 13.50 eV (13.61); 5σ, 13.92 (14.01); 4σ, 20.49 (~20.1). For HNC, the calculated values are 5σ, 11.77 eV; 1π, 13.91; 4σ, 22.43. Although HCN is 0.44 eV more stable than HNC, HNC^+ in its lowest state is 1.3 eV more stable than HCN^+.

Clark and Colling [29], using an STO-31G basis set, have compared calculated bond lengths and stretching force constants for HCN with those of (1s) core-ionized HCN^+ and computed overall band profiles for the core-ionized levels.

Using moderate basis sets, Pulay *et al.* [29a] have computed force constants for HCN, in fair agreement with experiment.

Using a 4-31G basis set in SCF computations, Summers and Tyrrell [30] have obtained minimum-energy reaction paths for the reactions

$$H^+ + CO \rightarrow HCO^+,\ CO + H^+ \rightarrow COH^+,\ H^+ + N_2 \rightarrow N_2H^+,\ H^+ + NO^+ \rightarrow HNO^{2+},\ NO^+ + H^+ \rightarrow NOH^{2+}$$

together with energy barriers for these reactions. The minimal-energy approach paths are all linear.

In another paper [31], using the same basis set, they have given interatomic distances and rotational constants, E_h values and valence-electron orbital energies, and total electronic populations for HCF^{2+}, HFC^{2+}, HBC^{2-}, HCB^{2-}, HBN^-, HNB^-, HNO^{2+}, HN_2^+, HCO^+, HOC^+, HCN, HNC, HBO, HOB, HCC^-, HBF^+, and HFB^+.

Good SCF plus CM computations on HCO^+ and HOC^+ by Bruna *et al.* [32] and by Vasudevan *et al.* [33] on HN_2^+ are also available. These include calculations on a number of excited states. Contour diagrams of the MOs are given, and interestingly detailed comparisons with the MOs and excited states of CO and N_2 are made. Potential surfaces for ground and

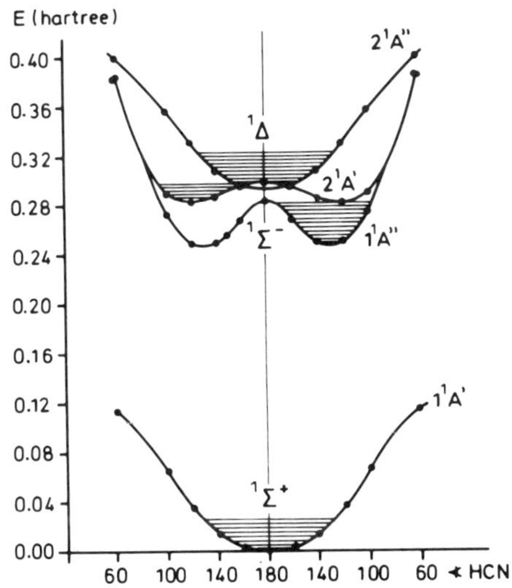

FIG. 2. Angular potential curves for low-lying states of HCN from CM calculations; also showing calculated vibrational levels. From Peric *et al.* [25].

excited states are discussed and mapped out. The HCO^+ species at equilibrium is found to be more stable than HOC^+ by 18.4 kcal/mole. The protonation energy ($H^+ + CO \rightarrow HCO^+$) of CO to HCO^+ is computed as 144.3 kcal/mole, becoming 139.3 kcal/mole after correction for zero-point energy, in good agreement with the experimental value of 137. There is no potential barrier for protonation at either end of the CO molecule. In some of the excited states, dissociation is to CO + H (the π^* MO becomes $1s_H$). Similar relations hold for HN_2^+.

Thomson and Wishart [34] made an SCF calculation on HBO and HBO^+ using a good polarized basis set. They report valence-orbital energies, dipole and quadrupole moments, force constants, and field gradients at the nuclei for HBO, and give a population analysis of the MOs. The computed charge distribution is $H^{-0.07}B^{0.83}O^{-0.76}$. The valence-shell MOs are 3σ, a B—O bound MO; 4σ, an H—B bound MO; 5σ, mainly a lone-pair MO; and 1π, B—O bound MOs. Optimum geometry was found to be H—A, 1.16 Å, B—O, 1.18 Å.

Botschwina [5] has made SCF, PNO—CI, and CEPA calculations on geometry, energies, and force constants of HBO and HOB, and SCF calculations for six states of HBO^+ (ground state $^2\Pi$). He finds $R(\text{HB}) =$ 1.16 Å, $R(\text{BO}) = 1.21$ Å for HBO and $R(\text{HO}) = 0.95$ Å, $R(\text{OB}) = 1.30$ Å for HBO.

Bruna *et al.* [35] have made an SCF plus CM study of potential surfaces of ground and excited states for HCS^+ and HSC^+ similar to their study of HCO^+ and HOC^+. HCS^+ in the ground state is linear like HCO^+, but it dissociates to $CS^+ + H$ in contrast to HCO^+, which goes to $CO + H^+$. HSC^+ is not stable; if formed, the H atom would rotate continuously until HCS^+ is reached; however, linear HCS^+ is represented by a saddle point, somewhat below the asymptote of $CS^+ + H$, but 110 kcal/mole above the most stable HCS^+ structure. The lowest excited states are π,π^* species as in HCO^+. Comparisons with the CS states are made.

Preuss *et al.* [36] have made MRD-CI calculations on HNSi and HSiN using good polarized basis sets. Both are linear ($^1\Sigma^+$). HNSi is calculated as 68 kcal/mole more stable than HSiN, in contrast to the pair HNC, HCN. The barrier between them is found as 13 kcal/mole. The N—Si and H—N distances in HNSi are calculated as 1.53 and 0.99 Å, respectively, in good agreement with experiment, and the N—Si and H—Si distances in HSiN as 1.59 and 1.46 Å, respectively. The binding energy for H + SiN ($^2\Sigma^+$) is calculated as 122 kcal/mole. Orbital energies and some CM coefficients are listed. Comparisons with isoelectronic and related molecules are made.

In a later paper [37] the same authors calculate bending and Si—N stretching potential curves for the ground and 11 excited singlet and triplet states of HNSi and HSiN. Most of the (π,π^*) excited states are bent but the (σ,π^*) states are linear.

Thomson [38] has made an SCF calculation on the ground state of HCP, using a polarized double-zeta STF basis. Various one-electron properties are computed. A population analysis gives gross population $H^{0.8}C^{6.4}P^{14.8}$. Botschwina *et al.* [39] have made an *ab initio* calculation of force constants and vibration frequencies.

Thioborine (HBS), although not stable in the gas phase, is the simplest molecule to exhibit a B=S double bond. Accurate SCF calculations [40] predict that it is linear with R(HB) = 1.17 Å and R(BS) = 1.60 Å and yields gross populations $H^{0.86}B^{4.98}S^{16.16}$. Fundamental frequencies and the dipole moment are also computed.

Roelandt *et al.* [41] have made SCF calculations on the $^1\Sigma^+$ ground state of HNSi using basis sets including one of polarized double-zeta quality. With this, R(NSi) = 1.525 Å and R(NH) = 0.985 Å are found (experiment, 1.54 and 1.00 Å). Force constants and vibration frequencies are computed in moderate agreement with experiment. Orbital energies are listed, and a population analysis is given. The dipole moment is computed. The isomer HSiN is not mentioned; it must be less stable (cf. Section A).

D. HCN^-, HCO, HN_2, HNO^+, HON^+, HBF

Pacansky *et al.* [42] have made SCF calculations on the ground-state potential surface of HCN^-, using basis sets contracted to (5s,4p) for C and N and to (3s,1p) for H. This anion, although known experimentally in matrix environments, is found unstable by 1.95 eV to autoionization. R(CH) = 1.12 Å, R(CN) = 1.23 Å, and θ = 122° are found.

The formyl radical HCO is known experimentally from its spectrum and other evidence. Bruna *et al.* [43] have made good SCF plus CM calculations on HCO and HOC. The calculated dissociation energy of HCO is 18.8 kcal/mole, in good agreement with experiment. HOC is calculated to be 23 kcal/mole less stable, hence not bound relative to H + CO, but with a definite local minimum protected by a small barrier. The lowest-energy optical absorption (known experimentally) is from the bent ($^2A'$) ground state to the $^2A''$ component of a linear ($^2\Pi$) upper state (see Table 1). The $^2A'$ and $^2\Pi$ ($^2A''$) states are related in the Renner–Teller manner. The higher-frequency emission bands are tentatively identified with an upper state which is a mixture of 6a′ → 7a′ and 7a′ → 3r (Rydberg) transitions. The 7a′ unpaired electron in the ground state is primarily π^* at small C—H distances, but becomes $1s_H$ on dissociation. As in HCO^+, the orbital energy is greater for 5σ (5a′) than for 1π (6a′, 1π). In COH, the order is reversed, as in CO. Vertical excitation energies and oscillator strengths are computed, including excitation energies for two quartet states.

Tanaka and Davidson [44] have further examined the ground and first two excited states of HCO, and also two related bent states $^4A'$ and $^4A''$. They used 4-31G calculations for a general picture, and a polarized double-zeta basis with extensive CM near the equilibrium geometries of the ground and some excited states. They point out that the interaction of the $1^2A'$ ground and first excited $2^2A'$ (part of $^2\Pi$) states corresponds to a conical intersection. Identification of the hydrocarbon flame-bands with in part the $2^2A' \rightarrow 1^2A'$ transition and with a bent $2^2A'' \rightarrow 1^2A'$ transition ($1^2A''$ is one component of the linear $^2\Pi$ state) is indicated.

Adams *et al.* [45] have made many-body perturbation and couple-cluster doubles calculations of the ground state of HCO and its dissociation to H plus CO. They predict the following: dissociation energy 16.6 kcal/mole, R(CH) = 1.1 Å, R(CO) = 1.19 Å, θ = 124°, in good agreement with experiment. They predict a barrier of 18.5 kcal/mole on the way to dissociation, with R(CH) = 1.77 Å, R(CO) = 1.14 Å, and θ = 115°.

Carter *et al.* [46], using spectroscopic data and in part the calculations of Bruna *et al.* [43], have derived an analytic formula for the potential surface of the HCO—HOC system, and give an instructive contour map displaying this.

With a rather small basis set, Botschwina [47] has calculated force constants and vibration frequencies for HCO.

Isoelectronic with HCO is HN_2, on which Vasudevan *et al.* [48] have made good SCF plus CM calculations. These give a ground-state dissociation energy of 3 kcal/mole, not enough to make it certain that the molecule is actually stable. The vertical spectrum of HN_2 is computed, and the spectrum and dissociation relations are discussed.

Also isoelectronic are HNO^+ and HON^+, which are of interest especially for the possibility of their occurrence in interstellar clouds. Marian *et al.* [49] made double-zeta SCF plus CM calculations for the ground states and for vertical excitation energies and transition oscillator strengths for excited states up to 10 eV. The electronic spectrum should resemble that of HCO. Orbital energies for the ground state are listed and relations to NO states are discussed, as are dissociation energies to H + NO^+ and to H^+ + NO. The ground states $^2A'$ of both conformers are bent (C_s symmetry), but the first excited state is linear ($^2A''$ component of a $^2\Pi$ state, the other component being the ground state). The barrier toward interconversion of the conformers is higher for the ground state than for dissociation into H plus NO^+, but in the first excited state interconversion is easier than dissociation (here to H^+ plus NO).

McLean *et al.* [50] have made improved SCF plus CM calculations of the ground and first excited states, using a good polarized basis set. They obtained optimized geometries and revised dissociation energies for the two states. They show an instructive figure giving energy and geometry curves as a function of ⟨HON from 0° (HON^+) to 180° (HNO^+). This is reproduced here as Fig. 3. The vertical excitation energy from the ground to the first excited state is calculated as 1.78 eV, in good agreement with Herzberg's 7200 Å (1.73 eV) absorption spectrum. In NOH^+ the equivalent excitation is at 1.39 eV. McLean *et al.* list computed rotational constants for HNO^+ and HON^+.

Bruna and Marian [51] have improved on Marian's calculations by

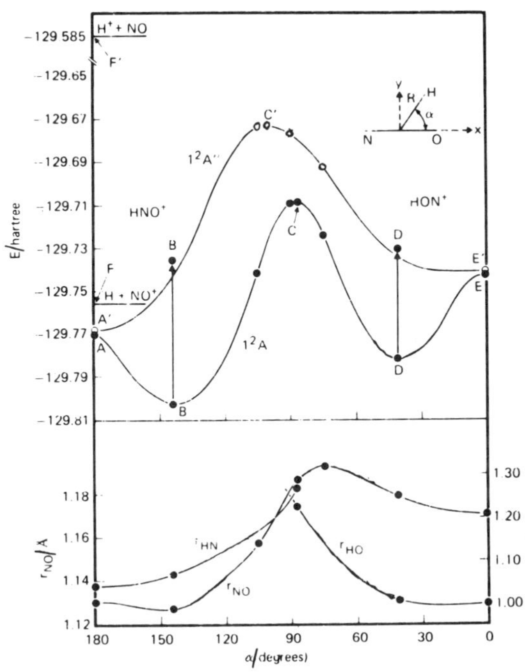

FIG. 3. The upper panel shows ground, $1^2A'$, and first excited, $1^2A''$, potential energy curves for HNO^+ and NOH^+. The abscissa is the variable α illustrated in the upper right of the top panel. The dissociation limit of the lowest surface is point F (H + NO^+), for the excited surface it is F′ (H^+ + NO), shown as energies on the vertical axis. Points B and D, above the excited surface, are the excited-state energies after vertical excitation. The lower panel shows bond-length variation with α for points on the $1^2A'$ potential curve. Bond-length variation along the excited curve shows the same behavior. (In more accurate calculations points A, A′ should be identical, as should E, E′.) From McLean *et al.* [50].

using a polarized basis set and by using the MRD–CI method for CM. Their results on the two lowest states agree fairly well with those of McLean *et al.* Their results for the ground-state geometry are given in Table 1. For the dissociation energies to H plus NO^+ they obtain for HNO^+ $D_e = 31.5$ kcal/mole (McLean *et al.* give 35 ± 2) and $D_0 = 26.3$; for HON^+, $D_e = 14.9$ (McLean *et al.* 21 ± 2) and $D_0 = 8.6$ kcal/mole.

Gaussian SCF computations on the ground and three excited states of another isoelectronic molecule HBF are reported by Brotchie and Thomson [52]. A population analysis gives $H^{0.99}B^{4.53}F^{9.46}$. Table 2 shows the results of their other computations.

Buenker *et al.* [1] have made MRD-CI calculations on the ground states of HNP, HPN and their positive ions, using a good polarized basis set. The $^2A'$ ground state of HNP is found to be 25 kcal/mole more stable than that of HPN, and the $^1\Sigma^+$ ground state of HNP^+ 87 kcal/mole more stable than that of HPN^+. Geometries of the states are given.

TABLE 2
Geometries and Energies for Electronic States of HBF

State	MO configuration	R(HB) (a.u.)	θ (deg)	R(BF) (a.u.)	E (a.u.)
X $^2A'$	... $(6a')^2\,(1a'')^2\,(7a')$	2.60	122	2.45	−123.67877
A $^2A''$ (π)	... $(6a')^2\,(1a'')^2\,(2a'')^2$	2.46	180	2.51	−123.61876
B $^2A''$	... $(6a')^2\,(1a'')\,(7a')^2$	2.70	89	3.80	−123.50842
C $^2A'$	... $(6a')\,(1a'')^2\,(7a')^2$	3.65	91	2.52	−123.47693

E. HNO, HPO, HCF, HNF^+, HO_2^+

Wu *et al.* [53] have made a rather good SCF plus CM calculation on the ground state and vertical excitation energies for the lowest four triplet and four singlet excited states of HNO. The ground state is a singlet, but a triplet comes only 1.7 eV higher. The geometries of the excited states were studied by means of SCF calculations. The NH dissociation curves (to H + NO) for the lowest excited, $^3A''$ and $^1A''$, states are found to contain potential maxima near 1.6 Å. This can account for the observed predissociation in the $^1A'' \leftarrow {}^1A'$ absorption spectrum.

Using spectroscopic and earlier *ab initio* data of Peslak *et al.* [54], Carter *et al.* [46] have derived an analytic formula for the potential surface of the HNO—HON system, and give an instructive contour map displaying this.

Using a rather small unpolarized basis set, Botschwina [54a] has calculated the general quartic force field for the ground state of HNO.

In SCF calculations including d functions, Collins and Duke [55] find HNS more stable than HSN by 0.04 a.u. Similarly HNO is more stable than HON by 0.04 a.u. according to Peslak *et al.* [54, 56]. The calculated geometry for HSN is $R(NH) = 1.01$ Å, $R(SN) = 1.55$ Å, H—N—S angle 100°.

Tapia *et al.* [57] report SCF computations on HPO, using d functions on P and f functions on H but no d function on O. By a partial optimization they calculate $R(PO) = 1.536$ Å, $R(PH) = 1.475$ Å, bond angle near 102° (experimental 104°). Comparisons with HNO are made. The computed charge distribution is $H^{-0.03}P^{0.55}O^{-0.52}$. Dipole and quadrupole moments are computed.

HCF is of interest principally as one of the carbenes (see CH_2 and CF_2). The first *ab initio* calculation on HCF was by Harrison [58]. The best SCF plus CM study is that of Bauschlicher *et al.* [59]. They found the $^3A''$ state 9 kcal/mole above the $^1A'$ ground state. In addition to those on HCF, they report calculations on CHCl and CHBr.

Clark *et al.* [60] have made SCF calculations on the lowest singlet and triplet states of HCF (also CH_2 and CF_2), and also of the corresponding core hole states.

Another isoelectronic molecule is the nitrenium ion HNF^+, on which Harrison and Eakers [61] have made SCF plus limited CM computations (Gaussian lobe functions without polarization). They found the $^3A''$ state (135°) 4 kcal/mole below the $^1A'$ state (107°), but one cannot be certain of the order without a more accurate computation. Comparisons are made with the carbenes. Charge distributions are calculated.

Van Lenthe and Ruttink [62], using a polarized basis set, have made SCF or MCSCF computations on the three lowest states of HO_2^+. The ground state is found to be ... $(1a'')^2\,(7a')\,2a''$, $^3A''$ with $R(OH) = 0.995$ Å, $R(OO) = 1.22$ Å, bond angle 111°, $E_h = -149.8243$. The corresponding singlet state has $R(OH) = 1.005$ Å, $R(OO) = 1.28$ Å, bond angle 109°, $E_h = -149.7862$. For these a single configuration is used. For the third state, an MCSCF calculation is used, resulting in $\Psi(^1A') = 0.97(\ldots 1a''^2\,7a'^2) - 0.24(\ldots 1a''^2\,2a''^2)$ at equilibrium, with $R(OH) = 1.00$ Å, $R(OO) = 1.16$ Å, bond angle 112°, $E_h = -149.7875$, thus just slightly below the $^1A''$ state. A higher-energy $^1A'$ state (mostly ... $1a''^2\,2a''^2$) is not computed. Energy barriers for transfer of the H atom from one O to the other were computed to be 1.35 to 1.8 eV via a linear arrangement, 2.6 to 2.9 eV via a C_{2v} conformation, for the three computed states. Computations were made on three modes of formation of HO_2^+ (from $H_2O_2^+$, from $H_2^+ + O_2$, and from $H_2 + O_2^+$), and it was concluded that HO_2^+ is formed in its ground state in all these. (For HOS^+, see Section III.F.)

F. HO_2, HNF, AND HOS

In calculations on ground and excited states of HO_2 and HO_2^+, Shih *et al.* [63] used the MRD-CI method with a basis set including (9s,5p) contracted to (4s,2p) plus two d polarization functions, for oxygen, with functions contracted to (2s,1p) for hydrogen, plus additional short-range and also long-range s and p functions in the O—O bond. They calculated vertical transition energies and oscillator strengths. For the ground state $\ldots 6a'^2\, 1a''^2\, 7a'^2\, 2a''$, $1^2A''$, $E_h = -150.5848$. Computed excited states and transitions are as follows: $7a' \rightarrow 2a''$ to $^2A'$ with $\Delta E = 0.93$ eV (experimental absorption bands, 0.88 eV), $f = 4 \times 10^{-6}$; $1a'' \rightarrow 2a''$ to $2^2A''$, $\Delta E = 5.90$ eV, $f = 6.5 \times 10^{-2}$, identifiable with an observed broad, strong ultraviolet absorption continuum peaking near 2100 or 2050 Å; $6a' \rightarrow 2a''$ to $2^2A'$, $\Delta E = 6.49$ eV, $f = 1.2 \times 10^{-3}$; $7a' \rightarrow 8a'$ (approximately a Rydberg 3s) to $3^2A''$, $\Delta E = 8.02$ *eV*, $f = 1.5 \times 10^{-3}$; $2a'' \rightarrow 8a'$ to $3^2A'$, $\Delta E = 8.04$ eV, $f = 4.5 \times 10^{-4}$; also another $^2A'$, $7a'^2 \rightarrow 2a''\, 8a'$, $\Delta E = 11.37$, $f = 2.5 \times 10^{-6}$. The three main coefficients in the CM for each state are listed; in all, the main configuration predominates very strongly.

Three vertical ionization energies are computed: $^3A''$ from $7a' \rightarrow \infty$, 11.65 eV; $^1A''$ from $7a' \rightarrow \infty$, 12.57 eV; $^1A'$ from $2a'' \rightarrow \infty$, 12.34 eV. Experimental values appear uncertain.

Langhoff and Jaffe [64] in similar calculations on the four lowest states of HO_2 get similar results. They give adiabatic potential energy curves as a function of H—O—O bond angle, O—H bond distance, and O—O bond distance, and determine theoretical equilibrium geometries for the four lowest states. They also show computed dipole moments and transition moments to the excited states as functions of the same three variables. In the transition moment diagram, they note that some avoided crossings occur among the excited states. They also compute vibration frequencies for the two lowest states; there is moderate agreement with experiment. It is concluded that the ultraviolet absorption continuum $2^2A'' \leftarrow 1^2A''$ (also the predicted weaker $2^2A' \leftarrow 1^2A''$) correspond to dissociation into $O(^3P)$ plus $OH(^2\Pi)$. A diagram showing observed and computed absorption intensities in this continuum shows good agreement between the two. For the much weaker infrared bands, Langhoff and Jaffe compute $f = 1.6 \times 10^{-5}$ while Shih *et al.* report 3.9×10^{-6}; but both values are so small that the difference is hardly significant.

In an SCF calculation using up to large polarized basis sets, Komornicki and Jaffe [65] calculated infrared frequencies and intensities for the ground states of HO_2 and DO_2, and made comparisons with experiment.

The formation of HO_2 from $H + O_2$ via an activated $^2A''$ HO_2 intermediate has been discussed in MCSCF plus CM calculations by Melius and Blint [66] (see Chapter V).

Buenker *et al.* [1] have made MRD-CI calculations on the ground states of HOS and HSO and their positive ions, using a good polarized basis set. HOS is found to be more stable than HSO by 9.7 kcal/mole (both $^2A''$). For the positive ions, the lowest state is $^3A''$ (10a′ 3a″) of HOS^+, with $^1A'$ ($10a'^2$) and $^1A''$ (10a′ 3a″), respectively, 13 and 25 kcal/mole higher. For HSO^+, $^1A'$ ($10a'^2$), $^3A''$ (10a′ 3a″), and $^1A''$ (10a′ 3a″) are, respectively, 15, 41, and 53 kcal/mole above $^3A''$ of HOS^+. Geometries of all the states are given.

Earlier, Sannigrahi *et al.* had made calculations on the ground and $^2A'$ first excited states of HOS [67] and HSO [68]. They also calculated dipole and transition moments.

Using a contracted basis set of six s and four p groups on S, two s groups and one p group on H, plus s and pπ bond functions at the S—S midpoint, Sannigrahi *et al.* [69] have made SCF plus CM calculations to determine geometries of the $3a''^2\ 13a'^2\ 4a''$, $^2A''$ ground state and the $3a''^2\ 13a'\ 4a''^2$, $^2A'$ (n,π*) first excited state of HS_2, and also calculated vibration frequencies for HS_2 and DS_2. Comparisons with the corresponding observed spectrum are made. The calculated vertical excitation energy is 0.86 eV. Calculations on five higher excited states are also reported, and a transition to $3a''\ 13a'^2\ 4a''^2$, $^2A''$ is identified with the observed absorption peaking at 3.6 eV. Dipole moments and their components for states, and oscillator strengths of transition, are given. The most important CM components for each state are given. Comparisons with HO_2 and HOS are made.

Brotchie and Thomson [52] reported a double-zeta quality SCF calculation on the ... $(7a')^2\ (1a'')^2$, $^2A''$ ground state and ... 7a′ (1a″), $^2A'$ first excited states of HNF. A population analysis gives $H^{0.70}N^{6.89}F^{9.31}$ for the ground state. The computed geometry was R(NH) = 1.16 Å, R(NF) = 1.41 Å, θ = 101° for the ground, and 1.11 Å, 1.38 Å, 125° for the excited states.

G. HOF AND HOCl

In an SCF geometry search, Kim and Sabin [70] have calculated the force constants and dipole moment of HOF. A population analysis gives $H^{0.67}O^{8.21}F^{9.12}$, and an H—O overlap population of 0.49*e*. They also obtain orbital energies of 13.97, 14.94, and 17.16 eV for the three highest MOs, which compare reasonably with the photoelectron values of 12.71, 14.50, and 17.16 eV.

Hirsch *et al.* [71] have made good SCF plus CM calculations on HOCl and HClO using the MRD-CI method. Orbital energies of the MOs are listed. The dipole moment is computed to be 1.633 D (experimental 1.3 ± 0.3); it is roughly in the O–N direction with the O atom negative. Computations are reported on vertical excitation and ionization processes and on dissociation processes in some detail. It is concluded that all low-lying excited states are unstable relative to dissociation into $OH(^2\Pi)$ + $(Cl(^2P)$. More recently, Jaffe and Langhoff [72], using double-zeta polarization plus CM calculations have discussed the excited states of HOCl in much detail, and their bearing on the photodissociation of HOCl.

In an SCF calculation using up to large polarized basis sets, Komornicki and Jaffe [65] calculated improved frequencies and intensities for the ground states of HOCl and DOCl.

H. $HFCl^+$, $HClF^+$, HF_2, HF_2^-, HCl_2^-; AR_2H, Ar_2H^+, He_2H^+, Li_2H^+, Be_2H^+, Li_2H^-

Joshi and Morokuma [73] have made 4-31G SCF calculations on the ground states of $HFCl^+$ and $HClF^+$, and find the former (R_{HF} = 0.96 Å, R_{FCl} = 1.87 Å) more stable than the latter (R_{HCl} = 1.34 Å, R_{ClF} = 1.71 Å) by 36 kcal/mole; the dissociation energies to H^+ + ClF are, respectively, 127 and 91 kcal/mole. They give an energy decomposition analysis.

According to good SCF plus CM calculations by Preuss *et al.* [74], the radical HF_2 exists only in the form of an extremely weakly bound complex of stable HF with an F atom. Either the F end or the H end of the HF may be oriented toward the second F atom (HF---F or FH---F). In the HF---F case, the calculations indicate a bent structure with HFF angle roughly 110° and R_e(HF) = 0.903 Å, R_e(F---F) = 2.72 Å. A binding energy of 100 cm^{-1} and a bending frequency of about 70 cm^{-1} are computed. The electron configuration and state are ... $(7a')^2\,(2a'')^2\,8a'$, $^2A'$, which would correspond to ... $2\pi^4\,6\sigma$, $^2\Sigma^+$ if the molecule were linear. A very low-energy excited state ... $2\pi^3\,6\sigma^2$, $^2\Pi$ of linear configuration is predicted.

In the FH---F case, the molecule is linear in both the corresponding states, with the ground state $^2\Pi$ bound only by 0.75 kcal/mole, and the $^2\Sigma^+$ excited state repulsive. To reach a symmetrical ($D_{\infty h}$) conformation FHF, it is necessary in the $^2\Pi$ state to raise the energy 45 kcal/mole according to SCF calculations; at this level, the energy has a minimum with respect to symmetrical distortion at an H—F distance of about 1.11 Å.

In HF_2^-, of course, the linear symmetrical structure is stable, with an HF distance of 1.13 Å. For a very good SCF computation, see McLean and Yoshimine [75]. Thomson *et al.* [76] have made a good SCF computation on linear symmetrical HCl_2^-, with an HCl distance of 1.57 Å.

Lozes and Sabin [77], using an ample polarized basis, briefly report results of SCF + CM calculations on a number of singlet and triplet excited states of HF_2^-. The $D_{\infty h}$ excited states, with minima, all lie above the energy of HF_2 and should be autoionizing. The $C_{\infty v}$ states all appear to be vibrationally dissociative.

In connection with the stability of the "hydrogen bridge" in HF_2^- and HCl_2^-, some early Gaussian SCF calculations on $(HeHHe)^+$ [78], $(LiHLi)^+$ [79], and $(BeHBe)^+$ [80] are of interest, although perhaps not entirely conclusive without CM. In all cases, the stable form is found to be linear and symmetrical (linearity not *proved* for Li_2H^+?). The respective computed A—H distances were 0.94, 1.66, and 1.61 Å. He_2H^+ was found to be stable by 8 kcal/mole for dissociation to HeH^+ plus He, Li_2H^+ by 60 kcal/mole relative to LiH plus Li^+. Calculations on $(LiHLi)^-$ [81] indicated that it possessed a stable linear symmetrical state ($R_{LiH} = 1.80$ Å) lower in energy than LiH plus Li^- by 7.61 kcal/mole.

More recently, Millem *et al.* [82] have reached the same conclusions on $HeHHe^+$, finding 11.15 kcal/mole for the dissociation energy to He + HeH^+. Similarly, Matcha *et al.* [83] have reached corresponding conclusions for $NeHNe^+$, finding 11.42 kcal/mole for dissociation to Ne + NeH^+ and 61.06 kcal/mole for dissociation to 2Ne + H^+. Qualitatively, Ar_2H^+ and Kr_2H^+ must be similar, but Xe_2H^+ somewhat different since the dissociation products of XeH^+ are Xe^+ + H.

Matcha and Millem [84] have made SCF calculations on ArHAr and $ArHAr^+$, using an unpolarized but otherwise good basis for Ar, and a polarized basis for H. $ArHAr^+$ ($^1\Sigma_g^+$, isoelectronic with HCl_2^-), together with the $^2\Pi_u$ and $^2\Sigma_u^+$ excited states of ArHAr, are found stably linear with R_{ArH} about 1.6 Å; the ground state of the ion is moderately stable with respect to ArH^+ plus H. (ArH^+ is very stable.) The ground state of ArHAr is unstable to dissociation, but the Ar—H repulsive energy is slightly decreased on bending (C_{2v}) to 120°. Vibration frequencies of Ar_2H^+ are computed. The limited basis set and lack of CM precluded obtaining accurate results.

REFERENCES

1. R. J. Buenker, P. J. Bruna, and S. D. Peyerimhoff, to be published.

1a. J. F. Harrison, R. C. Liedtke, and J. F. Liebman, *J. Am. Chem. Soc.* **101,** 7162 (1979).

2. R. P. Blickensderfer and K. D. Jordan, *Chem. Phys.* **41,** 193 (1979).
3. S. P. So and W. G. Richards, *J.C.S. Faraday Trans. II* **71,** 62 (1975).
4. P. J. Hansoul, C. Galloy, and J. C. Larghet, *J. Chem. Phys.* **68,** 4105 (1978).
4a. D. M. Hirst, *Mol. Phys.* **38,** 2017 (1979).
5. P. Botschwina, *Chem. Phys.* **28,** 231 (1978).
6. K. Vasudevan and F. Grein, *J. Chem. Phys.* **68,** 1418 (1978).
7. J. Pacansky and G. Orr, *J. Chem. Phys.* **67,** 5952 (1977).
8. S. P. So and W. G. Richards, *J.C.S. Faraday Trans. II* **71,** 660 (1975).
8a. I. H. Hillier, J. Kendrick, and M. F. Guest, *Mol. Phys.* **30,** 1133 (1975).
9. S. Shih, S. D. Peyerimhoff, and R. J. Buenker, *J. Mol. Spectrosc.* **64,** 167 (1977).
10. S.-K. Shih, S. D. Peyerimhoff, and R. J. Buenker, *J. Mol. Spectrosc.* **74,** 124 (1979).
11. J. A. Montgomery, Jr. and C. E. Dykstra, *J. Chem. Phys.* **71,** 1380 (1979).
12. A. D. McLean and M. Yoshimine, *IBM J. Res. Dev.* **2,** Suppl. (1967).
13. U. Wahlgren, J. Pacansky, and P. S. Bagus, *J. Chem. Phys.* **63,** 2874 (1975).
14. N. H. Haese and R. C. Woods, *Chem. Phys. Lett.* **61,** 396 (1979).
15. P. K. Pearson, G. L. Blackman, H. F. Schaefer, III, B. Ross, and U. Wahlgren, *Astrophys. J.* **184,** L19 (1973).
16. W. P. Kraemer, P. Hennig, and G. H. F. Diercksen, *Proc. Int. Astrophys. Symp., 21st, Liege, 1977.*
17. U. Wahlgren, B. Liu, P. K. Pearson, and H. F. Schaefer, III, *Nature* (*London*), Phys. *Sci.* **246,** 4 (1973).
18. P. R. Taylor, G. B. Backskay, N. S. Hush, and A. C. Hurley, *J. Chem. Phys.* **69,** 1971 (1978).
19. P. R. Taylor, G. B. Bacskay, N. S. Hush, and A. C. Hurley, *J. Chem. Phys.* **69,** 4669 (1978).
20. P. K. Pearson, H. F. Schaefer, III, and U. Wahlgren, *J. Chem. Phys.* **62,** 350 (1975).
21. J. A. Pople, R. Krishnan, H. B. Schlegel, and J. S. Binkley, *Int. J. Quantum Chem.* **14,** 545 (1978).
22. G. M. Schwenzer, S. V. O'Neil, H. F. Schaefer, III, C. P. Baskin, and C. F. Bender, *J. Chem. Phys.* **60,** 2787 (1974).
23. G. Herzberg and K. K. Innes, *Can. J. Phys.* **35,** 842 (1957).
24. G. M. Schwenzer, C. F. Bender, and H. F. Schaefer, III, *Chem. Phys. Lett.* **36,** 179 (1975).
25. M. Perić, S. D. Peyerimhoff, and R. J. Buenker, *Can. J. Chem.* **55,** 3664 (1977).
26. G. J. Valquez and J.-F. Gouyet, *Chem. Phys. Lett.* **65,** 515 (1979).
27. G. H. Schwenzer, H. F. Schaefer, III, and C. F. Bender, *J. Chem. Phys.* **63,** 569 (1975).
28. W. von Niessen, L. S. Cederbaum, W. Domcke, and G. H. F. Diercksen, *Mol. Phys.* **32,** 1057 (1976).
29. D. T. Clark and L. Colling, *J. Electron Spectrosc.* **13,** 317 (1978).
29a. P. Pulay, A. Ruoff, and W. Sawodny, *Mol. Phys.* **30,** 1123 (1975).
30. N. L. Summers and J. Tyrrell, *Theor. Chim. Acta* **47,** 223 (1978).
31. N. L. Summers and J. Tyrrell, *J. Am. Chem. Soc.* **99,** 3960 (1977).
32. P. J. Bruna, S. D. Peyerimhoff, and R. J. Buenker, *Chem. Phys.* **10,** 323 (1975); P. J. Bruna, *Astrophys. Lett.* **16,** 107 (1975).
33. K. Vasudevan, S. D. Peyerimhoff, and R. J. Buenker, *Chem. Phys.* **5,** 159 (1974).
34. C. Thomson and B. J. Wishart, *Theor. Chim. Acta* **35,** 267 (1974).
35. P. J. Bruna, S. D. Peyerimhoff, and R. J. Buenker, *Chem. Phys.* **27,** 33 (1978).
36. R. Preuss, R. J. Buenker, and S. D. Peyerimhoff, *J. Mol. Struct.* **49,** 171 (1978).
37. R. Preuss, R. J. Buenker, and S. D. Peyerimhoff, *Chem. Phys. Lett.* **62,** 21 (1979).

38. C. Thomson. *Theor. Chim. Acta.* **35,** 237 (1974).
39. P. Botschwina, K. Pecul, and H. Preuss, *Z. Naturforsch., Teil A* **30,** 1015 (1975).
40. O. Gropen and E. Wisloff-Nilssen, *J. Mol. Struct.* **32,** 21 (1976).
41. F. F. Roelandt, D. F. Van de Vondel, and G. P. Van Der Kelen, *J. Mol. Struct.* **54,** 221 (1979).
42. J. Pacansky, N. S. Dalal, and P. S. Bagus, *Chem. Phys.* **32,** 183 (1978).
43. P. J. Bruna, R. J. Buenker, and S. D. Peyerimhoff, *J. Mol. Struct.* **32,** 217 (1976).
44. K. Tanaka and E. R. Davidson, *J. Chem. Phys.* **70,** 2904 (1979).
45. G. F. Adams, G. D. Bent, G. D. Purvis, and R. J. Bartlett, *J. Chem. Phys.* **71,** 3697 (1979).
46. S. Carter, I. M. Miller, and J. N. Murrell, *J.C.S. Faraday Trans. II* p. 148 (1979).
47. F. Botschwina, *Chem. Phys. Lett.* **29,** 98 (1974).
48. K. Vasudevan, S. D. Peyerimhoff and R. J. Buenker, *J. Mol. Struct.* **29,** 285 (1975).
49. C. Marian, P. J. Bruna, R. J. Buenker, and S. D. Peyerimhoff, *Mol. Phys.* **33,** 63 (1977).
50. A. D. McLean, G. H. Loew, and D. S. Berkowitz, *Mol. Phys.* **36,** 1359 (1978).
51. P. J. Bruna and C. M. Marian, *Chem. Phys.* **37,** 425 (1979).
52. D. A. Brotchie and C. Thomson, *Chem. Phys. Lett.* **22,** 338 (1973).
53. A. A. Wu, S. D. Peyerimhoff, and R. J. Buenker, *Chem. Phys. Lett.* **35,** 316 (1975); see also A. A. Wu, *Chem. Phys.* **21,** 173 (1977).
54. J. Peslak, Jr., D. S. Klett, and C. W. David, *J. Am. Chem. Soc.* **93,** 5001 (1971).
54a. P. Botschwina, *Mol. Phys.* **32,** 729 (1976).
55. M. P. S. Collins and B. J. Duke, *J. C. S. Dalton Trans.* p. 277 (1978).
56. J. Peslak, D. S. Klett, and C. W. David, *J. Chem. Phys.* **55,** 1993 (1971).
57. O. Tapia, M. Allavena, and M. Larzilliere, *Chem. Phys. Lett.* **56,** 25 (1978).
58. J. F. Harrison, *J. Am. Chem. Soc.* **93,** 4112 (1971).
59. C. W. Bauschlicher, Jr., H. F. Schaefer, III, and P. S. Bagus, *J. Am. Chem. Soc.* **99,** 7106 (1977).
60. D. T. Clark, B. J. Cromarty, and A. Sgamellatti, *J. Electron Spectrosc.* **17,** 237 (1979).
61. J. F. Harrison and C. W. Eakers, *J. Am. Chem. Soc.* **95,** 3467 (1973).
62. J. H. Van Lenthe and P. J. A. Ruttink, *Chem. Phys. Lett.* **56,** 20 (1978).
63. S.-K. Shih, S. D. Peyerimhoff, and R. J. Buenker, *Chem. Phys.* **28,** 299 (1978).
64. S. R. Langhoff and R. L. Jaffe, *J. Chem. Phys.* **71,** 1475 (1979).
65. A. Komornicki and R. L. Jaffe, *J. Chem. Phys.* **71,** 2150 (1979).
66. C. F. Melius and R. J. Blint, *Chem. Phys. Lett.* **64,** 183 (1979).
67. A. B. Sannigrahi, S. D. Peyerimhoff, and R. J. Buenker, *Chem. Phys.* **20,** 381 (1977).
68. A. B. Sannigrahi, K. H. Thunemann, S. D. Peyerimhoff, and R. J. Buenker, *Chem. Phys.* **20,** 25 (1977).
69. A. B. Sannigrahi, S. D. Peyerimhoff, and R. J. Buenker, *Chem. Phys. Lett.* **46,** 415 (1977).
70. H. Kim and J. R. Sabin, *Chem. Phys. Lett.* **20,** 215 (1973).
71. G. Hirsch, P. J. Bruna, S. D. Peyerimhoff, and R. J. Buenker, *Chem. Phys. Lett.* **52,** 442 (1977).
72. R. L. Jaffe and S. R. Langhoff, *J. Chem. Phys.* **68,** 1638 (1978).
73. B. D. Joshi and K. Morokuma, *J. Am. Chem. Soc.* **101,** 1714 (1979).
74. R. Preuss, S. D. Peyerimhoff, and R. J. Buenker, *J. Mol. Struct.* **40,** 117 (1977).
75. A. D. McLean and M. Yoshimine, *IBM J. Res. Dev.* **2,** Suppl. (1967); M. Yoshimine and A. D. McLean, *Int. J. Quantum Chem., Symp.* **1,** 313 (1967).
76. C. Thomson, D. Clark, T. C. Waddington, and H. D. B. Jenkens, *J.C.S. Faraday Trans. II* **71,** 1942 (1975).

77. R. Lozes and J. R. Sabin, *Int. J. Quantum Chem.* **16,** 273 (1979).
78. P. Swanstrøm, R. Janoschek, and H. Preuss, *Int. J. Quantum Chem.* **3,** 115 (1969).
79. G. Diercksen and H. Preuss, *Int. J. Quantum Chem.* **1,** 637 (1967).
80. R. Janoschek, G. Diercksen, and H. Preuss, *Int. J. Quantum Chem.* **2,** 159 (1968).
81. H. Preuss and G. Diercksen, *Int. J. Quantum Chem.* **1,** 641 (1967).
82. M. B. Milleur, R. L. Matcha, and E. F. Hayes, *J. Chem. Phys.* **60,** 674 (1974).
83. R. L. Matcha, M. B. Milleur, and P. F. Meier, *J. Chem. Phys.* **68,** 4748 (1978).
84. R. L. Matcha and M. B. Milleur, *J. Chem. Phys.* **69,** 3016 (1978).

CHAPTER IV

TRIATOMIC MOLECULES AB_2 AND ABC

A. SURVEY

Optical spectra of many molecules of the types AB_2 and ABC are known, including [1] those of C_3, SiCC, CCN, CNC, NCN, NCO, NCS, N_2O^+, N_3, BO_2, CO_2^+, CO_2, CS_2^+, CS_2, N_2O, OCS^+, OCS, $HgCl_2$, NO_2, CF_2, SiF_2, O_3, SO_2, S_2O, ONF, NF_2, ClO_2, $TeCl_2$, and XeF_2. For many of these and also for other experimentally known or unknown molecules, *ab initio* calculations of varying degrees of accuracy have been made.

Many of these molecules are linear ($D_{\infty h}$ or $C_{\infty v}$) in their ground states, but often are bent in excited states. Many others are bent (C_{2v} or C_s) in their ground states but may be linear in some excited states. Best known are molecules composed of first-row atoms. Ground-state electron configurations can be given for these as follows, with the MOs listed in the order of decreasing binding energy for typical cases; note, however, that the relative order for some MOs is altered in many other examples (more about this later):

$D_{\infty h}$ (e.g., CO_2): $1\sigma_g^2\ 1\sigma_u^2\ 2\sigma_g^2\ 3\sigma_g^2\ 2\sigma_u^2\ 4\sigma_g^2\ 3\sigma_u^2\ 1\pi_u^4\ 1\pi_g^4\ 2\pi_u^0\ 5\sigma_g^0\ 4\sigma_u^0,\ {}^1\Sigma_g^+$

C_{2v} (e.g., NO_2): $1a_1^2\ 1b_2^2\ 2a_1^2\ 3a_1^2\ 2b_2^2\ 4a_1^2\ 3b_2^2\ 1b_1^2\ 5a_1^2\ 1a_2^2\ 4b_2^2\ 6a_1\ 2b_1^0\ 7a_1^0\ 5b_2^0,\ {}^2A_1$

$C_{\infty v}$ (e.g., NNO): $1\sigma^2\ 2\sigma^2\ 3\sigma^2\ 4\sigma^2\ 5\sigma^2\ 6\sigma^2\ 7\sigma^2\ 1\pi^4\ 2\pi^4\ 3\pi^0\ 8\sigma^0\ 9\sigma^0,\ {}^1\Sigma^+$

C_s (e.g., ONF): $(1a')^2\ (2a')^2\ (3a')^2\ (4a')^2\ (5a')^2\ (6a')^2\ (7a')^2\ (1a'')^2\ (8a')^2\ (2a'')^2\ (9a')^2\ (10a')^2\ (3a'')^0\ (11a')^0\ (12a')^0,\ {}^1A'$

In molecules lighter than N_3 or NCO, $1\pi_u$ comes below $3\sigma_u$, or 1π below 7σ. In a few cases, the relative order of $1\sigma_u$ and $2\sigma_g$ (K-shell electrons as is $1\sigma_g$), or of $1b_2$ and $2a_1$, may be reversed. In molecules with fewer electrons than the examples cited, more MOs are vacant. An extreme case is Li_2O, a $D_{\infty h}$ molecule with configuration $1\sigma_g^2\ 2\sigma_g^2\ 1\sigma_u^2\ 3\sigma_g^2\ 2\sigma_u^2\ 1\pi_u^4,\ {}^1\Sigma^+$.

In 1942, Mulliken noted that AB_2 molecules with 16 or less valence-shell electrons are linear and those with 17–20 are bent [1]. He explained this in terms of the ionization energies I of MOs as a function of apex angle. (In view of Koopmans' theorem, for ionization energies one can read SCF MO orbital energies ϵ.) The apex angle was seen to be a result of a competition between occupied MOs with I rising or falling with variation in it. For example, CO_2 is linear, but with addition of another electron in NO_2 the molecule is bent; this is explained by supposing that the I of the added electron in NO_2 decreases rapidly in energy as the molecule becomes bent. A diagram was given showing roughly how the I of each MO varies with the apex angle.

In 1953 Walsh presented a somewhat revised and very extensive generalization of these ideas that covered the geometries of various types of polyatomic molecules [2]. The resulting specifications are generally known as Walsh's rules. Recently, Buenker and Peyerimhoff have given a comprehensive review of the interpretation of Walsh's rules in the light of *ab initio* theory and computations [3]. As they point out, these rules seem to imply that (e.g., in the case of AB_2) energy as a function of apex angle can be measured by ΣI or $\Sigma\epsilon$ taken over the occupied MOs. Instead, the variation of the nuclear repulsion with angle should be included; also, $\Sigma\epsilon$ includes interelectronic repulsions twice, whereas they should be included only once. However, these difficulties prove not to be very significant for the qualitative comparisons for which Walsh's rules are used.

Figure 1 illustrates the Walsh–Mulliken rules as applied to a representative AB_2 molecule. It is adapted with a few modifications from one of Walsh's works and from Mulliken's 1942 article. It has only qualitative validity and, of course, even then cannot be expected to be strictly correct for all AB_2 molecules.

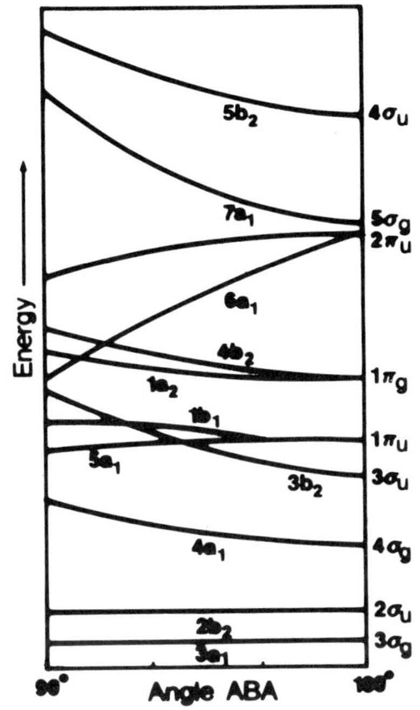

FIG. 1. Correlation diagram for the general AB_2 molecule.

B. 3-VALENCE-ELECTRON (VE) MOLECULES Li_3 AND Na_3 AND 4-VE Li_2C AND Li_2N^+

Kendrick and Hillier [4] have made SCF plus CM calculations on the two lowest states of Li_3, using a fairly extended (double-zeta s, triple-zeta p) unpolarized basis set. They find the two lowest states, of C_{2v} symmetry, to deviate by rather small amounts from an equilateral (D_{3h}) configuration. These are an obtuse-angled $^2B_2$ and an acute-angled $^2A_1$ state, with R_{12} and R_{23} (base of triangle), respectively, 3.2 and 2.8 Å for the former, 2.7 and 3.1 Å for the latter, and an energy of dissociation to Li_2 + Li of 9.2 kcal/mole for both. A $^2\Sigma_u^+$ linear configuration is 4.9 kcal/mole above the $^2B_2$ and $^2A_1$ states.

Martin and Davidson [5] have made SCF plus CM calculations on the analogous states of Na_3, using a polarized basis set: (11s, 7p, 1d) contracted to (6s,4p,1d). They find as ground state the obtuse-angled $^2B_2$, with the acute-angled $^2A_1$ and the linear $^2\Sigma_u^+$ forms as saddle points, respectively, 0.6 and 3.2 kcal/mole higher. The dissociation energy to Na_2 plus Na is found as 8.5 kcal/mole. Correlation involving the low-lying p and d orbitals is important in making the molecule stable, in contrast to the instability of H_3 for dissociation into H_2 plus H. Spin densities in the $^2B_2$ and $^2A_1$ states are discussed in connection with the ESR spectrum. The vertical ionization potential is calculated as 3.8 eV (experimental, 3.9 ± 0.1 eV). Comparison with those of Na (5.1 eV) and Na_2 (4.9 eV) and with the work function of the metal (2.3 eV) is of interest. Twelve approximate excitation energies and corresponding oscillator strengths were computed. These correspond mostly to 3p ← 3s excitation in the atom.

Comparisons of the mean interatomic distances $\overline{R}$ and the dissociation energies D per atom for the diatomic and triatomic molecules and the metal are discussed for Li and Na. For $\overline{R}$ the respective values are 1.43, 1.53, and 1.59 Å for lithium and 1.6, 1.85, and 1.95 Å for sodium, and for D 24, 11, and 9.5 kcal/mole for lithium and 17, 8.5, and 6.5 kcal/mole for sodium. These results show a rapid approach toward metallic binding, much more rapid than in the case of Be or Fe clusters. Be_3 (probably) and Mg_3 are not stable (see Section VII,C).

Gerbar and Schumacher [6] have reconsidered the Li_3 problem, using the not quite variational CEPA approximation, which accounts for at least 80% of the valence-electron correlation energy (the K shells were kept frozen). They point out that the $^2B_2$ and $^2A_1$ states of Li_3 (see above) are so close together and so shaped that in effect they essentially constitute an E′ state of D_{3h} symmetry. However, this is subject to a dynamic Jahn–Teller effect by interaction with an e′ vibration. They made calculations of the resulting vibronic levels, and also of rotational (J, K) levels of the lowest A_2' vibronic state. They show a number of contour diagrams of the adiabatic potential and of the nuclear probability densities for certain vibronic states. The authors are making further computations.

Using a (9s, 5p) basis set contracted to (3s, 2p) for C, 8s contracted to 3s for Li, and 4s contracted to 2s for H, Harrison *et al.* [6a] find a linear highly ionic structure for LiCLi, with a $^3\Sigma_g^-$] ground state and a $^1\Delta_g$ excited state about 23 kcal/mole higher. A similar calculation on LiN^+Li shows a $^3\Sigma_g^-$ ground state; three excited $^3\Pi$ states are computed. The article further discusses the effect of substituent electronegativity on substituted carbenes and related molecules. High substituent electronegativity favors a singlet ground state, electropositive behavior favors a triplet ground state.

C. 8-VE MOLECULES (Li_2O, Li_2S, AND Na_2S)

For the molecule Li_2O, the Walsh rules are scarcely applicable. As a derivative of H_2O one might expect it to be bent, but experimentally it is linear. This fact can be explained in terms of the strong positive charges on the Li atoms.

SCF calculations on Li_2O using Gaussians have been made by Wagner [7]. They indicate Li_2O to be linear. Hinchliffe and Dobson [8] have also made SCF calculations on Li_2O, Li_2S, and Na_2S. All are calculated linear, but the potential surface is not much higher for a 90° MSM angle. They give population analyses and density difference maps (versus ions). The structures have much ionic character, falling as the molecule is bent. Comparisons are made with H_2S.

D. 10- TO 12-VE MOLECULES (LiNC, BCB, BCC, BNC, LiNO NaNO, Al_2O, C_3, Li^+N_2, Li^+CO)

We begin with the experimentally well-known 12-VE molecule C_3. Here the close-lying MOs $4\sigma_g$, $3\sigma_u$, and $1\pi_u$ (see Fig. 1 or 2) have just been filled. The C_3 is just barely linear, as is indicated by its very small bending frequency (63 cm^{-1}). As compared with the stably linear 16-VE molecule CO_2, it lacks the $1\pi_g^4$ shell, which evidently favors linearity. The situation is discussed by Liskow *et al.* [9], who have made SCF and CM calculations. They conclude that a reasonably large basis set including d functions is necessary to account for the small bending frequency, but that CM is not needed for this purpose. They computed orbital energies ϵ as a function of angle. They plotted those for the occupied valence MOs in a Walsh-type diagram, which is reproduced here in Fig. 2. From a population analysis they deduce a charge distribution $C^{0.098}C^{-0.197}C^{0.098}$.

Perić-Radić *et al.* [10] have made an MSCF plus extensive CM calculation on the ground and first two ($3\sigma_u \rightarrow 1\pi_g$, $^3\Pi_u$ and $^1\Pi_u$) excited states of C_3, which are experimentally known. The excitation energies for the $^1\Pi_u \leftarrow {}^1\Sigma_g^+$] transition agree very well with experiment. They also compute 0.03 for the oscillator strength of this transition, in rough agreement with experiment. The higher bending frequency (experimental and theoretical) for the $^1\Pi_u$ than for the $^1\Sigma_g^+$ ground state shows that the $1\pi_g$ MO

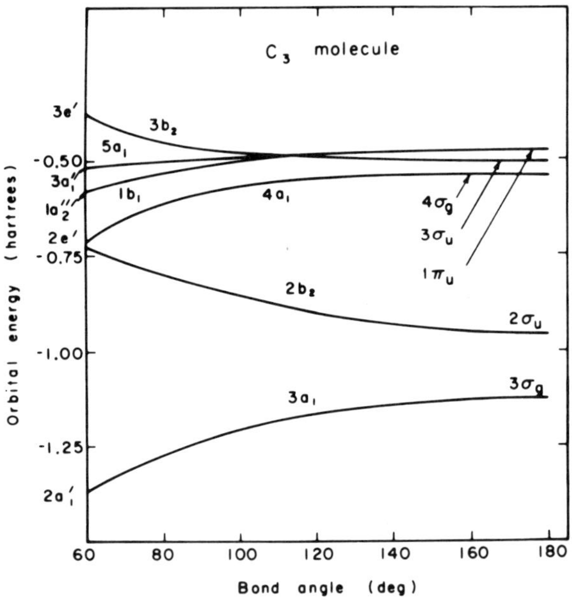

FIG. 2. Orbital energies of C_3 as a function of bond angle. The C—C bond distance was 2.41 bohr in all calculations. From Liskow *et al.* [9].

favors a linear geometry more strongly than the $3\sigma_u$; this relation would not be clear from Fig. 1. The computed and observed vibration frequencies for the various states agree well with the experimental values. A Renner–Teller splitting into A_1 and B_1 components on bending is computed for the $^3\Pi_u$ and $^1\Pi_u$ states; this is observed for the $^1\Pi_u \leftarrow {}^1\Sigma_g{}^+$ transition. The slightly greater bond length (experimental and theoretical) for the $^1\Pi_u$ than for the ground state indicates that the $3\sigma_u$ is slightly more bonding (or less antibonding) than the $1\pi_g$ MO, which should be nearly nonbonding.

Römelt *et al.* [11], using the MRD-CI method, have extended the calculations of Perić-Radić *et al.* to include further excited states and $C_3{}^+$. They used somewhat varied basis sets. The 1s cores are kept doubly occupied. They calculate vertical excitation energies for all the singlet and triplet states arising from excitation from the $3\sigma_u$, $4\sigma_g$, and $1\pi_u$ MOs to $1\pi_g$, also the Rydberg excitation $1\pi_u$ to 3s, and $1\pi_u \rightarrow \infty$ ($C_3{}^+$ production). They agree rather poorly with some equations-of-motion calculations. The calculated vibration frequencies for the ground state are slightly different from those of Perić-Radić. The $\pi\pi^*$ $^1\Sigma_u{}^+$ state is calculated at 8.0 eV, essentially unmixed with a Rydberg state. Ionization to the ground state $^2\Pi_u$ of the ion is calculated at 12.95 eV.

Al_2O is a 12-VE molecule like C_3. Gaussian SCF calculations indicate that it is linear, but experimental evidence is conflicting [7].

Other 12-VE molecules are LiNO and LiON. According to Gaussian SCF calculations by Peslak *et al.* [12] these molecules are strongly bent (C_s symmetry), unlike C_3. For LiNO, R(LiN) = 1.82 Å and bond angle LiNO 68° is found, with R(N–O) = 1.32 Å, and for LiON, R(LiO) = 1.68 Å and angle LiON 81°, with R(NO) = 1.33 Å. Contour maps of the electron density in the molecular plane are given. Roughly speaking, the Li atom is on a line perpendicular to the N—O axis. Charge distributions $Li^{0.82}N^{-0.30}O^{-0.51}$ and $Li^{0.81}O^{-0.56}N^{-0.24}$ are found, and overlap populations are given for each MO.

Similar results are reported [13] for NaNO and NaON. For NaNO, R(NaN) = 2.14 Å, R(NO) = 1.31 Å; for NaON, R(NaO) = 2.03 Å, R(NO) = 1.32 Å. Charge distributions were $Na^{0.57}N^{-0.22}O^{-0.34}$ and $Na^{0.61}O^{-0.25}N^{-0.36}$. Contour maps of electron density are given.

BNC and BCN are also 12-VE molecules, while BCC and CBC are 11-VE, and BCB and BBC are 10-VE molecules. Thomson [14] has made SCF calculations on the ground states of all these molecules, assuming them to be linear, using a double-zeta or best-atom, plus polarization, STF basis. The order of binding of the MOs is found to be appreciably different from that indicated in Fig. 1, namely, ... $2\sigma_u^2\, 3\sigma_g^2\, 1\pi_u^4\, 4\sigma_g^2\, 3\sigma_u^n$, where n = 0 for BCB ($^1\Sigma_g^+$), 1 for BCC ($^2\Sigma_g^+$) and 2 for BNC ($^1\Sigma_g^+$), with the more stable isomer given here in each case. Bond distances were found as follows: in BCB, R(BC) = 1.38 Å; in BBC, R(BB) = 1.43, R(BC) = 1.42; in BCC, R(BC) = 1.36, R(CC) = 1.26; in CBC, R(BC) = 1.38; in BNC, R(BN) = 1.43, R(NC) = 1.17; in BCN, R(BC) = 1.55, R(CN) = 1.13 Å. Dipole moments were computed, as were gross populations from which atomic charges can be obtained. In view of what was said above about C_3, it appears not quite certain whether these molecules, especially BNC, are actually linear rather than bent.

Additional 10-electron molecules are LiCN and LiNC, on which Bak and Clementic *et al.* [15, 16] have made SCF calculations with an extensive STF basis for the linear cases, supplemented by a Gaussian basis for bent conformations. For the linear conformation, LiNC is found to be 0.01 a.u. more stable than LiCN [16]; a population analysis [15] gives charge distributions $Li^{0.82}N^{-0.81}C^{-0.00}$ for LiNC and $Li^{0.47}C^{-0.29}N^{-0.18}$ for LiCN. On bending, the energy rises steadily as the Li (or Li^+) rotates about the CN (or CN^-) to convert LiNC into LiCN (see Fig. 3). Since the energy difference between LiNC and LiCN is so small, thermal energy at moderate temperatures is sufficient to activate this rotation, so that typically a molecule should be in an intermediate state of rotation between the two linear extremes. Clementi *et al.* refer to this situation as an example of a

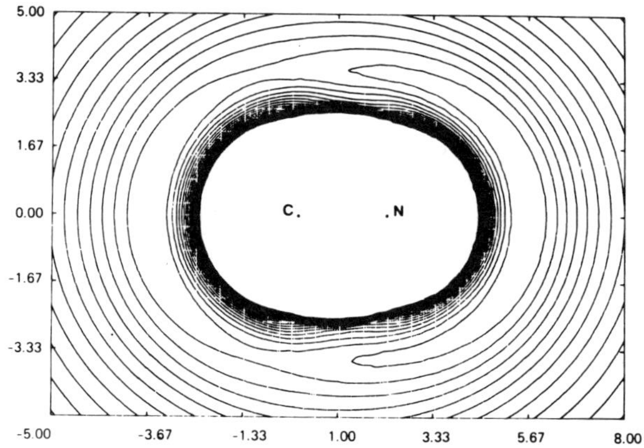

FIG. 3. Energy surface for the motion of a lithium atom around the CN group. The C—N bond length is kept constant at 1.16 Å. The origin of the axis system is at the carbon nucleus (x = 0.0 a.u., y = 0.0 a.u.). The nitrogen nucleus is on the x axis, at x = 2.19 a.u. and y = 0.0 a.u. The interval between each contour corresponds to an energy difference of 0.01 a.u. The most repulsive contour is the most inner one. From Clementi *et al.* [16].

"polytopic bond." It seems not unlikely that the molecules BCB, BCC, and BNC discussed above may also be examples. For LiCN and LiNC, Clementi *et al.* [16] obtain the following equilibrium distances R(LiC) = 1.93, R(CN) = 1.13; R(LiN) = 1.77; R(NC) = 1.15 Å.

As a contribution to an understanding of inelastic scattering of Li^+ ions by N_2 and CO, Staemmler has made SCF calculations on the potential energy surfaces of LiN_2 [17] and of Li with CO [17a]. He used good polarized Gaussian lobe basis sets. For Li^+CO he considered linear approaches Li^+CO and Li^+OC, for which he found respective stabilization energies of 0.53 eV at R(LiC) = $5.5a_0$ and 0.63 eV at R(LiO) = $4.5a_0$; approach perpendicular to the CO unit gave only slight stability. Electron correlation has only minor effects on these results.

E. 13- TO 15-VE MOLECULES (CNC, LiO_2, NCN, OCC, NCO^+, N_3^+, CO_2^+, N_2O^+, N_3, NCO, CSO^+)

In 13- to 16-VE molecules, $1\pi_g$ or 2π electrons stabilize a linear structure, culminating in the 16-VE CO_2, N_2O, NO_2^+, and N_3^- with a complete $1\pi_g$ or 2π shell in the ground state. As compared with 63 cm^{-1} in C_3, the bending frequency is 667 cm^{-1} in CO_2 and 589 cm^{-1} in N_2O. Intermediate values are known experimentally for CCN, CNC, NCN, and NCO.

However, a rather special 13-VE molecule is LiO_2, for which O'Neil *et al.* [18] have made Gaussian SCF calculations, using a moderate basis set without polarization functions. The computed energy is lowest for the C_{2v} state ... $1a_2\,3b_2^2$, $^2A_2$, with the $1a_2^2\,3b_2$, $^2B_2$ state 14 kcal/mole higher. However, the computed energy is only 0.00002 a.u. higher for the ... $2a''\,8a'^2$, $^2A''$ C_s form with θ(LI–O–O) = 68° than for the corresponding $^2A_2$ C_{2v} form. Further, the $^2\Pi$ $C_{\infty v}$ form is only 0.00192 a.u. higher than the $^2A_2$ C_{2v} form. All this suggests a largely ionic polytopic bond like that in LiCN (see Section D). For C_{2v} symmetry, the computed R(LiO) = 1.82 Å with θ(O–Li–O) = 44.5°; for the $^2B_2$ state, 1.76 Å and 46.5°. For $C_{\infty v}$ geometry, R(LiO) = 1.62 Å and R(OO) = 1.35 Å is predicted.

Several 14-VE and especially 15-VE molecules, all of which are linear, have been computed. Thomson [19] has made SCF calculations, using a double-zeta plus polarization STF basis set, and calculated bond lengths, on the ground states of the 14-VE linear molecules NNC and NCN ($2\pi^2$, $^3\Sigma^-$ and $1\pi_g^2$, $^3\Sigma_g^-$), of which he finds NCN the more stable by 0.014 a.u., and on the 13-VE CNC and NCC ($1\pi_g$, $^2\Pi_g$ and 2π, $^2\Pi$), with CNC the more stable by 0.0028 a.u. The calculated bond lengths agree reasonably well with experiment for NNC, NCC, and CNC. Atomic populations are computed, corresponding to charge distributions $N^{0.04}N^{0.08}C^{-0.12}$, $N^{-0.06}C^{0.13}N^{-0.06}$, $N^{0.15}C^{-0.11}C^{-0.04}$, and $C^{-0.06}N^{0.12}C^{-0.06}$.

Thomson and Wishart [20] have reported on SCF calculations using a double-zeta plus polarization STF basis set on the 14-VE $^3\Sigma^-$ radical OCC and on its positive and negative ions. They give tables of computed orbital energies, atomic populations, force constants, and several other properties, and find R(OC) = 1.12 and R(CC) = 1.37 Å. Relevant to OCC^+ and OCC^- they give vertical ionization energy (0.3976 a.u.) and electron affinity (−0.0183 a.u.) of OCC, also atomic populations for OCC^+ and OCC^-.

Wu and Schlier [21] have made Gaussian-lobe SCF computations of moderate accuracy on the 14-VE ... $7\sigma^2\,2\pi^2$, $^3\Sigma^-$, ... $7\sigma\,2\pi^3$, $^3\Pi$, and ... $7\sigma^2\,2\pi\,8\sigma$, $^3\Pi$ states of NCO^+, CNO^+, and CON^+, of which NCO^+ is found to be the most stable. They have also made limited CM computations on the ground and excited states (excitation mostly to 3π) of NCO^+. They show possible potential surfaces of the three lowest states, of interest in connection with recent scattering experiments of N^+ on CO.

Thomson and Wishart [22] have made SCF calculations on the ground state of 15-VE NCO, using a good STF basis set, and computed a number of properties of the molecule (e.g., dipole moment 0.50 D, quadrupole moment −9.48 barn). They predict $R(NC) = 1.23$ Å, $R(CO) = 1.13$ Å. A table of total populations on N, C, and O for each MO is given. This corresponds to $N^{-0.07}C^{0.27}O^{-0.20}$.

Spin–orbit coupling constants for $^2\Pi$ states of the 13-VE radical NCC and the 15-VE CO_2^+, COS^+, N_3, N_2O^+, and NCO have been computed by Horsley and Hall [23], making use of SCF wave functions of double-zeta accuracy which they have computed with the Stevens polyatomic program using STFs. Agreements with experiment are mostly within about 10%. Total energies at specified geometries, and spin–orbit constants, but none of the other usual details, are given.

Archibald and Sabin made Gaussian SCF computations on the ground states of the 14-, 15-, and 16-VE molecules N_3^+, N_3, and N_3^- [24]. They report orbital energies, force constants, and a population analysis for each. However, there must be something wrong with the population analyses, since they give totals for less than the number of electrons (e.g., 21.30 for N_3^-). Of special interest are the computed equilibrium bond lengths: $N_3^-(^1\Sigma_g^+)$, $R_{12} = R_{23} = 1.175$ Å; $N_3(^2\Pi)$, $R_{12} = 1.24$, $R_{23} = 1.14$; $N_3^+(^3\Sigma^-)$, $R_{12} = 1.64$, $R_{23} = 1.095$ Å. It is thus indicated that N_3 and N_3^+ are asymmetrical. In N_3^+, the shorter bond distance is found almost equal to that of N_2. The N—N bond distances found in N_3 are very similar to those in HN_3.

Claxton *et al.* [24a] made minimal basis unrestricted SCF calculations with spin annihilation on N_3 (symmetric and asymmetric) and on N_3^{2-}.

Venanzi and Schulman [24b], using a double-zeta basis, made calculations on N_3^+ and found a local minimum of D_{3h} symmetry with bond length 1.33 Å. Force constants were computed. However, the $^3\Sigma^-$ ground state of N_3^+ is much lower in energy [24].

In an effort to elucidate the course of the ionospheric reaction $O^+(^4S)$ + $N_2(X^1\Sigma_g^+) \rightarrow NO^+(X^1\Sigma^+) + N(^4S)$, with 1.10 eV reaction energy, Hopper [25] has made SCF plus CM computations on the lowest quartet state of N_2O^+, of C_{2v} configuration ... $1a_2\,4b_2\,6a_1$, $^4B_1$, or C_s configuration ... $2a''\,9a'\,10a'$, $^4A''$. This corresponds to a linear-molecule state ... $1\pi_g^2$, $^3\Sigma_g^-\ 2\pi_u$, $^4\Pi_u$, but the state is found to be bent (apex angle 126°) for the R(NN) and R(NO) values of N_2O at equilibrium, and bent for somewhat stretched bonds, but linear if R(NO) is increased 40% or more. It is concluded that the reaction mentioned above can probably go adiabatically via the transient $^4A''$. Although the $^4A''$ in the interaction region correlates to the $^4\Pi_u$ state in either asymptotic region of bond lengths it correlates to ... $1\pi_g^2$, $^3\Sigma_g^-\ 5\sigma_g$, $^4\Sigma_g^-$.

Regarding N_2O^+ and CO_2^+, see discussions of the ionization potentials of N_2O and CO_2 in Section F.

F. 16-VE MOLECULES (CO_2, SiO_2, SCO, NO_2^+, N_2O, N_2S, N_3^-, OCN^-, SCN^-, FCN, ClCN, BeF_2, MgF_2, CaF_2, ZnF_2); ALSO UO_2^{2+}

For the ground states of the 16-VE molecules CO_2, N_2O, OCN^-, FCN, SCO, SCN^-, and ClCN, the very accurate SCF calculations by McLean and Yoshimine [26] are available. These give the basis set coefficients, orbital energies, and total energy. For ClCN, Støgård [27] gives less accurate calculations, also for $ClCN^+$; the energy differences agree well with the four observed ionization potentials of ClCN.

Gilberg *et al.* [28], using SCF calculations with results close to those of McLean and Yoshimine, show excellent agreements for the same energy differences.

The electron configuration of CO_2 in its $^1\Sigma_g^+$ ground state, and the corresponding MO configuration in terms of C_{2v} symmetry as in its bent excited states, are as follows,

$$1\sigma_g^2\ 1\sigma_u^2\ 2\sigma_g^2\ 3\sigma_g^2\ 2\sigma_u^2\ 4\sigma_g^2\ 3\sigma_u^2\ 1\pi_u^4\ 1\pi_g^4\ 2\pi_u^0\ 5\sigma_g^0,\ {}^1\Sigma_g^+$$

$$1a_1^2\ 1b_2^2\ 2a_1^2\ 3a_1^2\ 2b_2^2\ 4a_1^2\ 3b_2^2\ 5a_1^2\ 1b_1^2\ 1a_2^2\ 4b_2^2\ 6a_1^0\ 2b_1^0\ 7a_1^0,\ {}^1A_1.$$

Similarly for N_2O in $C_{\infty v}$ and C_s symmetries, we have

$$1\sigma\ 2\sigma^2\ 3\sigma^2\ 4\sigma^2\ 5\sigma^2\ 6\sigma^2\ 1\pi^4\ 7\sigma^2\ 2\pi^4\ 3\pi^0\ 8\sigma^0,\ {}^1\Sigma^+$$

$$1a'^2\ 2a'^2\ 3a'^2\ 4a'^2\ 5a'^2\ 6a'^2\ 7a'^2\ 1a''^2\ 8a'^2\ 9a'^2\ 2a''^2\ 10a'^0\ 3a''^0\ 11a'^0,\ {}^1A'$$

For N_2O, the calculations of McLean and Yoshimine appear to be the best available. SCF studies of the energies of NNO and NON as a function of apex angle, also limited SCF plus limited CM vertical excitation energies of a number of excited states of NNO, were made by Peyerimhoff and Buenker [29]. Winter has made similar slightly more accurate calculations on N_2O, and given identifications of some of the computed with experimentally observed levels [30].

With a fairly good polarized basis set Domcke *et al.* [31] computed the ionization potential spectrum for the valence electrons by a Green's function method. For the outer valence electrons (6σ, 1π, 7σ, 2π) there is one main component with a few satellites, but for the inner valence electrons (4σ, 5σ) the spectrum is broken up, as in other cases, into numerous individual lines. A detailed discussion and experimental data are given.

Using a moderate unpolarized basis set, Darko *et al.* [31a] computed SCF wave functions for the three N_2O^+ states with K shell (1σ, 2σ, 3σ)

vacancies in N_2O. The resulting MOs were then used as a basis for CM calculations to describe more highly excited states of the core hole ions which give rise to satellite peaks in the ESCA spectrum. The resulting information permitted them to interpret numerous satellite peaks. Especially conspicuous are $\pi \rightarrow \pi$ transitions.

Collins and Duke [32] in an SCF calculation on N_2S with a double-zeta (with d?) basis set found $R(NN) = 1.13$ Å and $R(NS) = 1.63$ Å in a linear structure, with charge distribution $N^{-0.09}N^{0.03}S^{0.07}$ and a dipole moment 0.18 D.

For the 16-VE NO_2^+, Pfeiffer and Allen [33] report moderately accurate Gaussian-lobe calculations as a function of apex angle, and give charge-density contours for many of the MOs at 110° and 180°. They also give orbital energies for 180° and graphs of these as a function of apex angle and MO coefficients, and population analyses (gross and overlap population) for each MO at 110° and 180°. Energies as a function of angle for these excited states are given. Comparisons with NO_2^- are made. For later calculations on excited states of NO_2^+, see at the end of the discussion on NO_2 in Section G.

Cremaschi and Simonetta [34], using an STF basis set with 1s, 2s, 2p, and 3d functions on each atom, have also made SCF computations on (linear) NO_2^+. They list orbital energies, also total energies as a function of apex angle. Bending constants have been calculated. They are nearly the same as those for CO_2. Concerning excited states of NO_2^+, see Shih *et al.* in Section G.

For the ground state of CO_2, the SCF energy surface was constructed for linear geometry and, from this, force constants were computed by Pacansky *et al.* [35]. A moderately good Gaussian basis including polarization functions was used.

Morrison and Hay [36] used a (9s 5p 1d) basis set contracted to (3s 2p 1d) plus diffuse s, p, and d functions in an SCF computation of static polarizabilities, quadrupole moments, and dipole moment derivatives for bending and asymmetric stretching.

A number of calculations on states of CO_2 have been made [37–46]. On the basis of Fig. 1, excitations from $1\pi_g$ or $1\pi_u$ or $3\sigma_u$ to the $6a_1$ component of $2\pi_u$ should give a bent excited state, but perhaps excitation to the $2b_1$ component and certainly further excitation to $5\sigma_g$ or $4\sigma_u$ or any Rydberg state, or ionization to the 15-VE CO_2^+, should give a linear molecule.

England *et al.* [39] have made MCSCF plus CM computations to obtain vertical excitation energies to the $1\pi_g \rightarrow 2\pi_u$ ${}^3\Sigma_u^+$, ${}^3\Delta_u$, ${}^3\Sigma_u^-$, ${}^1\Sigma_u^-$, and ${}^1\Delta_u$ states. They used Gaussian functions including 3d with exponent 1, and valence-shell-only CM. They also computed vertical ionization potentials for the four experimentally known states of CO_2^+. Using an unpolarized basis set, they also made SCF computations of CO_2, CO_2^+, and CO_2^- bent to 135° with R = 1.24 Å. They list the resulting energies and give figures showing the conclusions. For excitation to the $2b_1$ component of $2\pi_u$ stability of the linear geometry is indicated, as expected [40]. Later England *et al.* [41] made EOM (equations of motion) calculations for the $(1\pi_g) \rightarrow 2\pi_u$ vertical excitation energies, using a Gaussian plus 3d basis. However, these seem to be less accurate.

England *et al.* [42] made accurate vertical SCF calculations on all the states that arise from excitation from the occupied MOs $4\sigma_g$, $3\sigma_u$, $1\pi_u$, and $1\pi_g$ to the valence-shell MOs $2\pi_u$, $5\sigma_g$, and $4\sigma_u$ and the Rydberg MOs $1\delta_g$ and $2\pi_g$. All computations were made at the equilibrium bond length 1.161 Å of normal CO_2. They used a very complete STF basis, like that of McLean and Yoshimine, plus even-tempered sets of 1s, 2pσ, 2pπ, 3dπ, and 3dδ diffuse carbon atom orbitals to take care of semi-Rydberg and Rydberg excited MOs. They computed orbital sizes ($\bar{r}^2$ values, also $\bar{x}^2$ and $\bar{z}^2$) and n^* values, and also quadrupole moments, for the excited MOs. They showed that the MOs $5\sigma_g$ and $4\sigma_u$, although nominally valence-shell MOs (e.g., cf. Fig. 1), have Rydberg character ($5\sigma_g$ partially and $4\sigma_u$ fairly fully). The term values (ionization minus excitation energy) were found to be roughly independent of the MO out of which excitation occurred. Their n^* values agree well with those which have been determined experimentally for a number of Rydberg states.

In England *et al.* [42], the SCF energies of $(1\pi_g)^{-1}\,5\sigma_g$, $^1\Pi_g$ and $(1\pi_g)^{-1}\,2\pi_u$, $^1\Delta_u$ are almost degenerate as is also true of the corresponding triplet states. Since $^1\Pi_g$ and $^1\Delta_u$ both give rise to $^1A_2$ and $^1B_2$ states on bending (and correspondingly for the triplet states), the question of the exact positions of the states becomes relevant. The Δ_u states tend to bend, the Π_g states to be linear (cf. Fig. 1).

England and Ermler [45] have addressed this question by making careful MCSCF–CM calculations for the $^1\Sigma_g^+$, $^1\Pi_g$, and $^1\Delta_u$ states in $D_{\infty h}$ geometry, also for the $^2\Pi_g$ state of CO_2^+. They used the McLean–Yoshimine basis plus five small even-tempered 1s STFs centered on carbon. The result is the best available ground state wave function and the conclusion that the $^1\Delta_u$ state is about 0.4 eV higher than the $^1\Pi_g$ state. Consequently upon decreasing the apex angle from 180° an avoided crossing of the downward-tending $^1\Delta_u$ and the upward-tending $^1\Pi_g$ must occur, with the result that the former gives linear and the latter bent $^1A_2$ and $^1B_2$ states. The $^1B_2$ state can now be identified with the experimentally known $^1B_2$ state [46]. For the $^3\Delta_u$ and $^3\Pi_g$ states, very likely similar relations occur, but no computation was made. With the vertical positions of the $^1\Delta_u$ and $^1\Pi_g$ states established, those of other excited states can be estimated fairly well using previous papers [39, 42]. Correlation with experimental work on vertical energies (electron impact and spectroscopic) is discussed, but conclusions are rather uncertain. Quadrupole moments are computed for $^1\Sigma_g^+$ (computed: -4.31×10^{-26} esu cm^2, experimental: -4.1), $^1\Pi_g$ (2.55), and $^1\Delta_u$ (-3.57).

Using a Green's function method with a fairly modest polarized Gaussian basis, von Niessen *et al.* [43] have computed the $4\sigma_g$, $3\sigma_u$, $1\pi_u$, and $1\pi_g$ ionization potentials of CO_2 and obtained agreements with experiment within 0.1–0.3 eV for all four. These agreements are better than those obtained by England *et al.* [39] by MCSCF-CM and, of course, than those obtained by SCF [42]. For the $3\sigma_g$ and $2\sigma_u$ ionizations, Domcke *et al.* [31] show that the ionization potentials split up into a number of components. They give a detailed discussion.

Schirmer *et al.* [47] have made Green's function calculations on the ionization potentials of CS_2. More accurate calculations were used for the main lines of the outer valence electrons ($6\sigma_g$, $5\sigma_u$, $2\pi_u$, and $2\pi_g$), and less accurate calculations for the satellites and the inner valence electrons ($5\sigma_g$, $4\sigma_u$). For the latter, as in other similar cases, the ionization is split up into a number of components. Comparisons with experiment are made for the outer valence electrons.

In contrast to CO_2, the molecule SiO_2 is scarcely known experimentally. Theoretically, the first SCF computation on SiO_2 was published in 1978 by Pacansky and Hermann [48], using a moderately accurate basis set including d functions for all atoms. Not surprisingly, the molecule is computed to be linear and symmetrical; R(Si–O) is computed as 1.498 Å. The orbital energies and a gross atomic population analysis are given. The population analysis gives a charge distribution $O^{-0.7}Si^{1.4}O^{-0.7}$, as compared with $O^{-0.3}C^{0.6}O^{-0.3}$ in CO_2. There is considerable d bonding from Si; the dσ Si population is 0.147 e in $4\sigma_g$ and the silicon dπ population is 0.473 e in the last MO $1\pi_g$. Force constants and vibration frequencies ($\nu_1 = 1100$, $\nu_2 = 300$, $\nu_3 = 1570$ cm^{-1}) are computed.

Rothenberg and Schaefer [49] have made double-zeta plus polarization (d function on each atom) SCF computations on the ground states of the triatomic difluorides BeF_2, BF_2, CF_2, NF_2, and OF_2. For each of these they give total and component computed energies, geometries, orbital energies, and a series of computed properties: dipole, quadrupole, and octupole moments, second and third moments of the electronic charge distribution, diamagnetic susceptibility and shielding, and electric field gradients. Comparisons are made with experiment where available. Population analyses correspond to $Be^{1.05}(F^{-0.53})_2$, $B^{0.40}$, $C^{0.36}$, $N^{0.44}$, $O^{0.23}$; d populations range from 0.1 on O and 0.03 on each F to 0.2 on B and 0.07 on each F in BF_2.

Gole *et al.* [50], using a [basis including d orbitals, have made SCF calculations as a function of apex angle on the ground states of the 16-VE molecules BeF_2, MgF_2, and CaF_2. The effect of the d orbitals in MgF_2 and CaF_2 is to favor a bent molecule. However, MgF_2, like BeF_2, is computed to be linear. CM studies also support this conclusion, while experimental evidence is conflicting. CaF_2 is definitely bent (predicted: 145°, experiment: 141°).

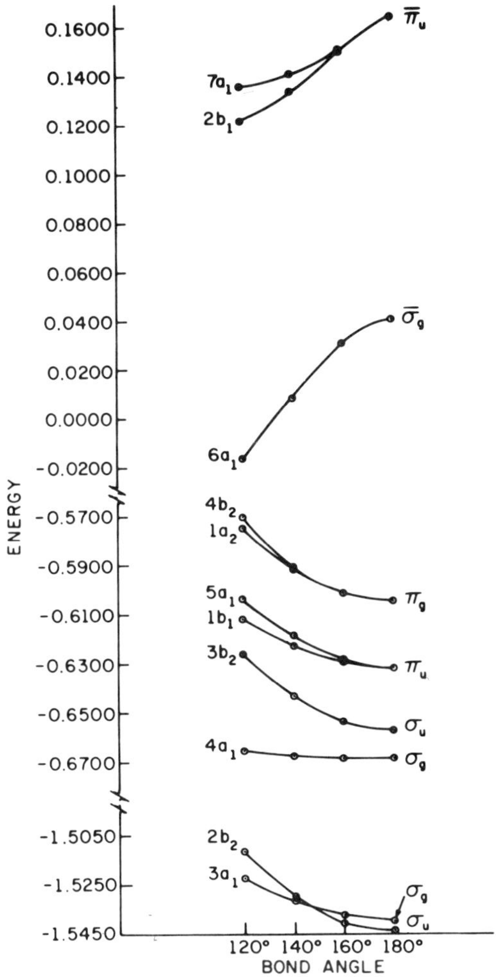

FIG. 4. Correlation diagram for valence orbitals and lowest lying molecular orbitals in BeF_2 (using the s–p–d basis set (energy in hartrees). From Gole [51].

Gole [51] has made computations also on the excited MOs $5\sigma_g$ and $2\pi_u$, and finds that $5\sigma_g$ is much lower than $2\pi_u$ (see Fig. 4), in contrast to CO_2 (cf. Fig. 1). This is explained, like the anomalous linearity of the 8-VE Li_2O, by the very strong polarity of the bonds. For the occupied MOs, the angular correlations are roughly as in Fig. 1. Gole gives energy curves as a function of angle for the $^3A_2$ and $^3B_2$ states which correspond to the excitation $1\pi_g \rightarrow 5\sigma_g$ ($1a_2$ or $4b_2$ to $6a_1$ for the bent molecule); minima occur at about 80° for the $^3B_2$ (lower) and about 100° for $^3A_2$ (higher). Populations of atomic orbitals are given, but do not add up to correct totals.

Yarkony *et al.* [51a], with a somewhat larger basis set than Gole *et al.*, made an SCF calculation on the $^1A_1$ ground state of CaF_2 which surprisingly predicted a linear molecule. However, they considered other matters more important and went on to compute the energies of the $^1A_1$ and the excited $^3B_2$ and $^1B_2$ states as a function of apex angle, and also several properties of all these states (geometry: 54° apex angle in the B states; dipole and quadrupole moments; electric field gradients; etc.). Their objective was to understand chemiluminescence following collision of Ba and Cl_2.

Pulay *et al.* [51b], using 7s 3p and 5s 2p basis sets augmented by bond functions, have calculated force constants for FCN and FN_2^+

Pendergast and Hayes [52] made SCF calculations similar to those of Gole *et al.* on the ground state of MgF_2. The results indicate a linear molecule with $R(\text{MgF}) = 1.77$ Å. Force constants and vibration frequencies are computed, and agree rather well with experiment. Several properties are computed (diamagnetic susceptibility tensor, potential, field gradient, and quadrupole tensor at Mg nucleus).

Seiders *et al.* [53] have made SCF calculations on BeF_2^- and MgF_2^- using rather small basis sets. When bent, the stably linear BeF_2 and MgF_2 obtain a dipole moment which, if it exceeds 1.625 D, can attract an electron. Adiabatic electron affinities of -0.3 and $+0.5$ eV are obtained, with apex angles 135° and 117°, for BeF_2 and MgF_2, respectively. Inclusion of electron correlation would be expected to increase the computed stability of the anions, but BeF_2^- is probably only metastable.

Yarkony and Schaefer [54], using an approximately double-zeta basis set, have made an SCF calculation on ZnF_2, which experimentally and theoretically is linear and symmetrical. They tabulate orbital energies as a function of bond angle, also percentage s, p, and d characters of Zn and s and p characters of F MOs. The computed charge distribution is $Zn^{1.39}(F^{-0.70})_2$. There is one strongly bonding MO, the $7\sigma_g$, which is 17% Zn 4s, 8% Zn $3d\sigma$, 1% F 2s, and 74% F $2p\sigma$. The other MOs are mostly localized on the Zn or the F atom. These results lead to a qualitative discussion of the AF_2 molecules from ZnF_2 to ScF_2.

Using a relativistic effective core potential for Hg, Hay *et al.* [55] made two-configuration SCF valence-electron calculations on $HgCl_2$ using an unpolarized basis set for the Cl atoms. *R*(HgCl) agrees well (2.30 versus 2.29 Å) with experiment, while the calculated dissociation energy *D* (60 versus 106 kcal/mole) is too small, but the omission of d orbitals on Cl and of CM can account for the latter discrepancy. Several ionization potentials are computed and agree well with experiment where observations are available.

With a differently obtained but nearly identical (in the valence region) relativistic core potential for Hg, Basch and Topiol [56] have also made calculations on $HgCl_2$, obtaining R(HgCl) = 2.31 Å and $D = 59$ kcal/mole.

Although there is no real similarity of UO_2^{2+} to the 16-VE molecules just discussed, there is a superficial resemblance (cf. Fig. 3) in that the electron configuration is ... $12\sigma_g^2\, 9\sigma_u^2\, 7\pi_u^4\, 4\pi_g^4$, giving a $^1\Sigma_g^+$ state. This is shown in an X-α method calculation by Boring *et al.* [57], and a discrete variational method by Ellis *et al.* [58] using SCF Dirac–Slater relativistic models. The orbital energies of the four last-occupied MOs are close together, and their order differs in the two calculations. For the spectrum, the lowest energy levels involve excitations from $9\sigma_u$, $2\pi_u$, and $4\pi_g$ to $f\phi_u$ and $f\delta_u$ orbitals.

G. 17-VE MOLECULES (CO_2^-, BF_2, FCO, NO_2, BeF_2^-, MgF_2^-)

The 17-VE CO_2^- ion is unstable: the electron affinity of CO_2 is -0.36 eV according to an unpolarized Gaussian SCF calculation by Pacansky *et al.* [35]. However, a metastable state is observed experimentally. This is explained by the situation depicted in Fig. 5, from a paper on an SCF calculation by Krauss and Neumann [59]. The two upper curves correspond to addition of an electron in the $6a_1$ (to give $^2A_1$) and the $1b_1$ (to give $^2B_1$) component of the $2\pi_u$ MO, the former giving a bent and the latter a linear state. For the minimum of the $^2A_1$ state, Pacansky *et al.* give a computed bending angle of 135° and *R*(CO) = 1.20 Å; an experimental bending angle is about 134°. The absorption transition $^2B_1 \leftarrow {}^2A_1$ is observed experimentally at about 3.5 eV, in agreement with the theoretical Fig. 5. England *et al.* [39] make approximate SCF calculations on CO_2^- bent to 135° with $R = 1.24$ Å, list computed energies for ten states, and give a figure showing correlations with the linear geometry. Almlöf *et al.* [60] used unrestricted SCF calculations with $S = \frac{3}{2}$ annihilation to

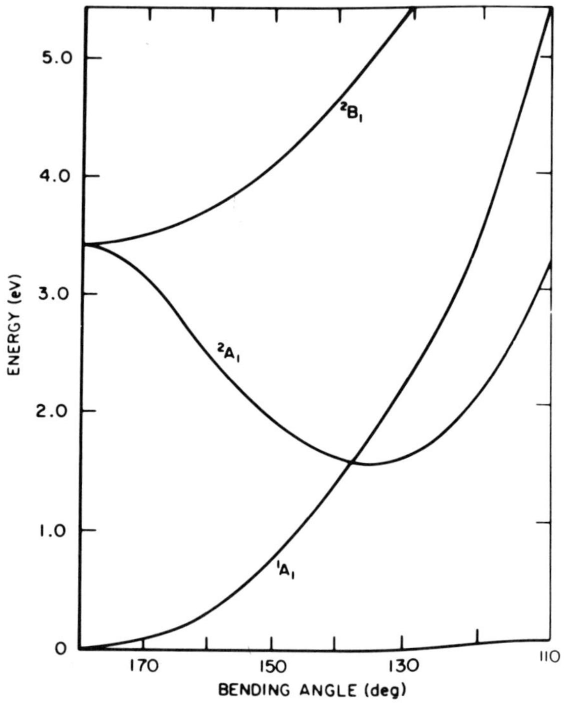

FIG. 5. Energy as a function of bending angles for $CO_2(^1A_1)$, $CO_2^-(^2A_1)$, and $CO_2^-(^2B_1)$. R(CO) held fixed at 2.19 a.u. From Krauss and Neumann [59].

compute magnetic hyperfine coupling constants, and also electric quadrupole coupling constants, for the $^2A_1$ state of CO_2^-.

Unpolarized SCF computations have been made by Thomson and Brotchie [37, 38] on the $^2A_1$ ground state and the $^2B_1$ and two other excited states of BF_2. The computed geometries are $R(\text{BF}) = 1.32$ Å, $\theta = 120°$ for the ground state, while the $^2B_1$ state is linear. Orbital energies are listed. Spin densities at nuclei and hyperfine coupling constants are computed. On BF_2, see also Rothenberg and Schaefer in Section C.

Thompson and Brotchie [60a], using a fairly large unpolarized basis set, have made SCF calculations on FCO. They tabulate orbital energies and compute isotropic hyperfine coupling constants. They obtain $E_h = -212.0693$. Geometrical optimization gives $R(\text{FC}) = 1.32$ Å, $R(\text{CO}) = 1.16$ Å, and FCO angle 126°. They compare FCO with results on HCO and HBF.

Bleicher and Botschwina [60b] used the unrestricted SCF method in computations on FCO, obtaining $E_h = -211.7983$, $R(\text{FC}) = 1.35$ Å, $R(\text{CO}) = 1.17$ Å, and FCO angle 129°. They computed the complete quadratic force field and the diagonal cubic constants. They also investigated FCO^+ and found it linear with R(FC) = 1.23 Å and R(CO) = 1.11 Å. They calculated its stretching force constants.

The strong influence of the $6a_1$ MO (cf. Fig. 1) causes the ground state of the 17-VE molecule NO_2 to be bent (R_{NO} = 1.19 Å, apex angle 134° by experiment). Rothenberg and Schaefer [61], using a double-zeta plus polarization (d) basis set, computed an SCF wave function for the ground state. The computed energy, −204.0679 a.u., is estimated to lie about 0.06 a.u. above the SCF limit. With this wave function multipole moments, diamagnetic susceptibility, and shielding tensors, and electric field gradient tensors, were computed. Orbital energies and components of the total energy are listed. Atomic properties correspond to $N^{0.45}(O^{-0.23})_2$; d populations are 0.19 e on N, 0.05 e on each O.

In an MCSCF plus CM calculation on the ground state (the most accurate until recently) Jackels and Davidson [62] obtained an energy of −204.37346 a.u. They used a Gaussian double-zeta plus incomplete polarization basis. In CM they held the K shells filled and included virtual orbitals except the three highest-energy ones. (For some critical comments, see below concerning Benioff.) They computed several molecular properties (dipole moment; potentials; field and field gradients at the nuclei; quadrupole coupling constants; magnetic hyperfine splitting parameters). They show contour diagrams of the $4b_2$, $5a_1$, $1b_1$, $1a_2$, $6a_1$, and $2b_1$ MOs. Those are reproduced here as Figs. 6–11.

Experimentally, NO_2 presents a very complicated spectrum in the visible and near infrared, followed by a beautiful system near λ 2500 whose structure shows its upper state to be $^2B_2$, assuming $^2A_1$ for the ground state. Many *ab initio* calculations have been made that sought to elucidate

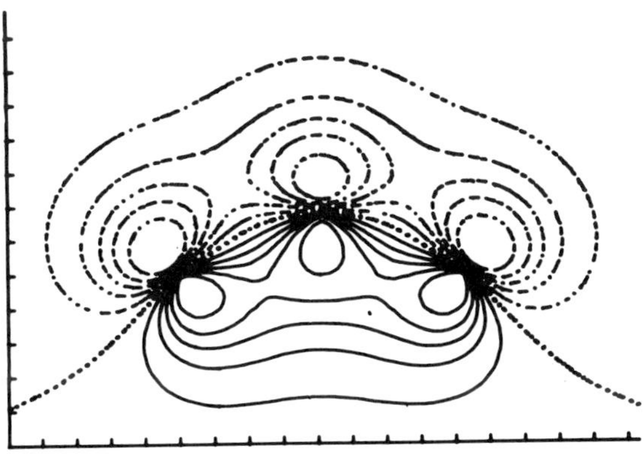

FIG. 6. Electron density contour plot of the $5a_1$ ICSCF molecular orbital of NO_2 in the molecular plane. From Jackels and Davidson [62].

the spectrum. Some of these will be discussed below. In the meantime, Table 1 surveys some salient results of computations on what are probably the lowest-energy excited states. The visible spectrum is now seen to involve the 1^2B_2-1^2A_1, 1^2B_1-1^2A_1, and probably 2^2B_1-1^2A_1 transitions. As has been suggested previously, the λ 2490 system should be ascribed to 2^2B_2-1^2A_1. The ground state ... $6a_1$, $^2A_1$ and the 1^2B_1 state form a Renner–Teller pair corresponding to a $^2\Pi_u$ linear molecule, with (as observed) the $^2A_1$ component bent and the $^2B_1$ component linear.

Attention was first called to the 2^2B_1 state of Table 1 in calculations by Hay [63]. He gives a figure showing a relatively flat 2^2B_1 curve with minimum near 75°, rising to cross the 1^2B_1 curve near 134°. Evidently the crossing should be avoided and two new curves produced, but the matter was not discussed by Hay and has not been discussed by others.

In an SCF plus small CM study using an unpolarized basis set, Jackels and Davidson [64] point out that at small angles the $4b_2$ and $6a_1$ MO curves cross (see Fig. 1), and likewise the ... $4b_2^2\,6a_1$, 1^2A_1 and ... $4b_2\,6a_1^2$, 1^2B_2

TABLE 1
Computed Excited States of NO_2 (C_{2v} Symmetry)

	Populations[a]				Geometry[b]		Excitation Energy (ev)[d]		
State	$1a_2$	$4b_2$	$6a_1$	$2b_1$	R (Å)	Apex angle ϕ (deg)	Adiabatic[b,c]	Vertical[b,c]	Vertical
$1\,^2A_1$	2	2	1	—	1.23	133	0.00	0.00	0.00
$1\,^2A_2$	1	2	2	—	1.32	102	0.95	2.15	3.25
$1\,^2B_2$	2	1	2	—	1.30[e]	104[e]	0.96[e]	2.43[e]	3.22[e]
$1\,^2B_1$	2	2	—	1	1.23	180	1.48	2.46	2.79
$2\,^2B_1$	2	—	2	1	~1.40	~75	2.98	—	—
$1\,^4B_2$	1	2	1	1	—	—	—	2.88	4.04
$1\,^4A_2$	2	2	1	1	—	—	—	3.10	4.09
$2\,^2A_2$[f]	2	1	1	1	1.34	113	3.40	4.14	—
$2\,^2B_2$[f]	1	2	1	1	1.33[g]	122	3.73	4.24	—
$1\,^2\Sigma^+$	2	2	$7a_1$		1.34[h]	180	3.90	—	—

[a] Inner electrons $1a_1^2$, $1b_2^2$, $2a_1^2$, $3a_1^2$, $2b_2^2$, $3b_2^2$, $5a_1^2$, $1b_1^2$. Note that $1a_2$ and $4b_2$ correspond to $1\pi_g$ of the linear conformation, and $6a_1$ and $2b_1$ to $2\pi_u$.

[b] Jackels and Davidson [64].

[c] Assuming R(N–O) = 1.24 Å, ϕ = 133° in the ground state (experimental, 1.19 Å and 134°).

[d] Shih *et al.* [70].

[e] Interacts with $1\,^2A_1$ with C_s symmetry (see [63] and text).

[f] This refers presumably to the lower of the *two* predicted states of the configuration.

[h] Two unequal bond lengths (1.41 and 1.19 Å) according to Gillispie *et al.* [66].

[g] There is experimental evidence for two unequal bond lengths [41, 67].

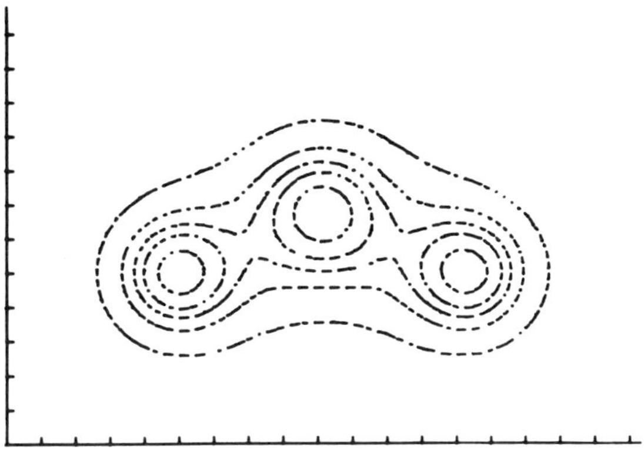

FIG. 7. Electron density contour plot of the $1b_1$ ICSCF molecular orbital of NO_2 0.5 bohr above the molecular plane. From Jackels and Davidson [62].

state curves. They conclude that these two states at small enough angles (far below the equilibrium angle of the ground state) undergo a strong Jahn–Teller reaction which results in two $^2A'$ states (symmetry C_s) in which the two NO bond lengths become unequal. The lower $^2A_1$ state still keeps its observed 134° $^2A_1$ minimum, but develops a second higher ($^2B_2$) minimum at smaller angles (about 103°). Contour diagrams of various MOs under C_s and C_{2v} symmetry are given. The C_s symmetry corresponds to a bond picture

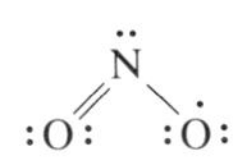

The upper $^2A'$, or $2^2A'$, has a conical Jahn–Teller surface with minimum at 112°, where the $^2A_1$ and $^2B_2$ curves in C_{2v} symmetry would intersect ("conical intersection")—the bond structure is an equal mixture of that just shown and one with the location of the double bond (but not the bond lengths) reversed. These results help to explain some of the complexities in the visible spectrum.

Jackels and Davidson [65], in a moderately accurate (no polarization in the basis) SCF + CM calculation, have made a potential energy surface study of a large number of excited states of NO_2, and given computed geometries for 14 doublet and 4 quartet states and vertical excitation energies for 20 doublet and 12 quartet states. CM configurations are chosen to give correct dissociation relations; SCF MOs corresponding to the lowest $^4B_2$ state were used, and the 6 lowest-energy MOs are kept doubly occupied. They note that in C_{2v} symmetry, crossings of potential curves at small angles result in Jahn–Teller interaction to C_s symmetry, as discussed above [64] for 1^2A_1 and 1^2B_2, and occur also between 2^2A_1 and

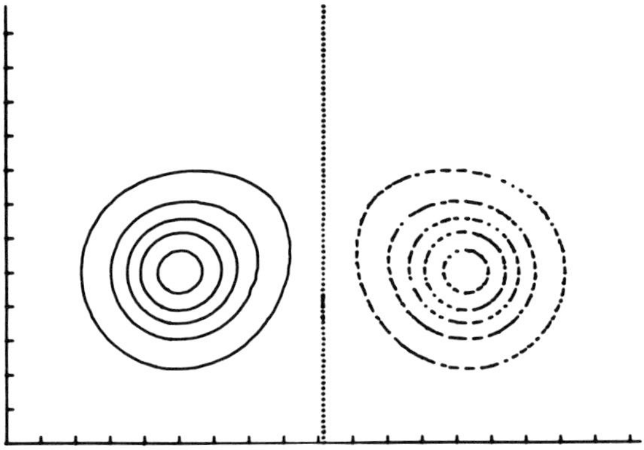

FIG. 8. Electron density contour plot of the $1a_2$ ICSCF molecular orbital of NO_2 0.5 bohr above the molecular plane. From Jackels and Davidson [62].

2^2B_2 and between 1^4A_1 and 1^4B_2. Also, the four lowest states are Renner–Teller coupled. Table 1 is based largely on tables in [65].

Gillispie *et al.* [66], using a double-zeta plus polarization basis, have made a MCSCF plus limited CM (valence and $3\pi_u$ virtual MOs) calculation on the 1^2A_1, 1^2A_2, 1^2B_2, 1^2B_1, 1^4B_2, 1^4A_2, and $1^2\Sigma^+$ states. No excitations out of the three K shells and the two following (mainly O_{2s}) MOs were allowed. For each state they survey the potential surface for some distance around its minimum, using the OVC method for each. Coefficients of OVC configurations are listed. The application of the OVC approach to triatomic MCSCF calculations is discussed. Previous computations on NO_2 are reviewed. Predictions of adiabatic and vertical excitation energies are reported for the several states mentioned, as also dipole

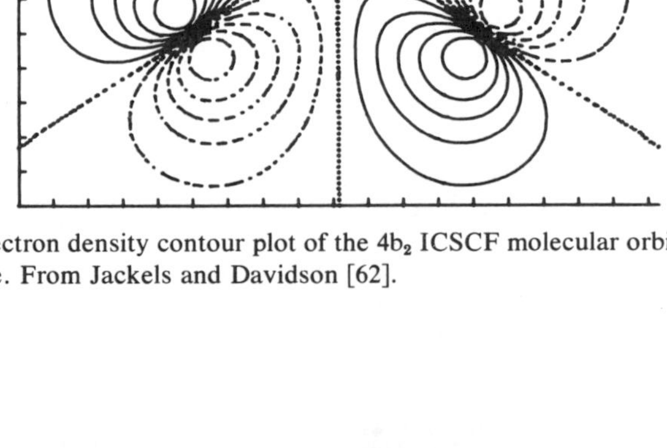

FIG. 9. Electron density contour plot of the $4b_2$ ICSCF molecular orbital of NO_2 in the molecular plane. From Jackels and Davidson [62].

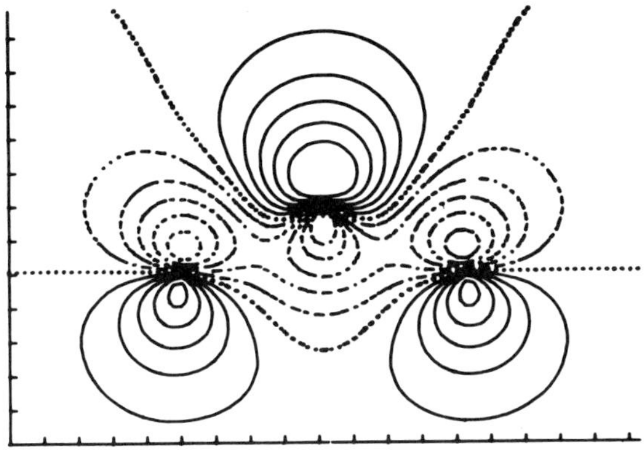

FIG. 10. Electron density contour plot of the $6a_1$ ICSCF molecular orbital of NO_2 in the molecular plane. From Jackels and Davidson [62].

moments, geometries, force constants, and vibration frequencies ν_1 and ν_2. Figure 12, taken from Gillispie *et al.*, shows how the energies of the states they calculated vary as a function of apex angle.

Gillispie *et al.* assume C_{2v} symmetry except for 1^2B_1 and $1^2\Sigma_g^+$, whereas, as noted previously, one needs for 1^2B_2 to consider C_s symmetry, with two unequal bond lengths. The calculations of Gillispie *et al.* on the $1^2\Sigma^+$ state show a linear geometry but with unequal bond lengths (1.41 and 1.19 Å) and a barrier to inversion of 800 cm^{-1}. As first suggested by Mulliken [40] on experimental grounds, and predicted in [65], unequal bond lengths also occur for the 2^2B_2 state; Coon *et al.* [67] have found evidence for this situation also in some states of SO_2 and ClO_2. Gillispie *et al.* suggest that the asymmetry of the $1^2\Sigma^+$ state may be related to its potential dissociation to NO + O.

Benioff [68] improved the calculations of Gillispie *et al.* using more configurations (also somewhat revised d orbital exponents), and extended them to NO_2^+ and NO_2^- (see Section H). Like Gillispie *et al.*, he excluded excitations from the five lowest MOs, but in a trial calculation, showed that the computed energy is then only 0.0045 a.u. higher than if only the K shells are frozen. He presents a critical discussion of various factors which affect the computed energy of the ground state. Major effects are produced by inclusion of polarization in the basis set. In spite of this inclusion of polarization functions, Benioff's best energy is 0.1115 a.u. above that of Jackels and Davidson [64], who did not use these. Fortune *et al.* [69] made similar calculations, but without d functions in the basis set.

Shih *et al.* [70], using the MRD-CI method with an unpolarized Gaussian basis, have calculated vertical excitation energies for most of the

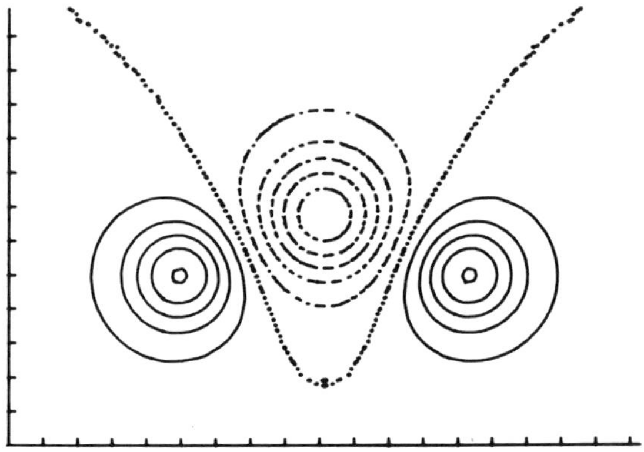

FIG. 11. Electron density contour plot of the $2b_1$ ICSCF molecular orbital of NO_2 0.5 bohr above the molecular plane. From Jackels and Davidson [62].

states in Table 1 and also for several Rydberg states, and for the $6a_1^{-1}$, ${}^1A_1$; $1a_2^{-1}$, ${}^3A_2$ and ${}^1A_2$; and $4b_2^{-1}$, ${}^3B_2$ and ${}^1B_2$ states of NO_2^+. Agreement with experiment to within 0.1–0.2 eV is found where experimental data are available. Jackels [71], using a polarized double-zeta basis, has used an essentially MRD-CI method to recalculate the vertical excitation energies of the ${}^4A_2$ and ${}^4B_2$ states (see Table 2). He also gives vertical deexcitative energies 3.3 eV for ${}^4A_2$ and 3.6 eV for ${}^4B_2$.

The predicted excitation energies in the various works cited above differ considerably, as might be expected from the varied simplifying assumptions made in the calculations. Benioff gives a critical survey of the effects of these, and a comparative survey of the results of various authors for vertical excitation energies. This table is reproduced in modified form as Table 2. In general, inclusion of polarization (d functions) in the basis set lowers the computed energy of a state very considerably, and has a similar but lesser effect on excitation energies. Most authors have used only very limited CM outside the valence shell, but Shih *et al.*, although they used an unpolarized basis, have comprehensively covered the virtual MOs in their CM.

Handy *et al.* [72] have extended the "direct CI" method of Roos to doublet and quartet states, and applied it to the ground and five low-lying states of NO_2. Usually they exclude the first five MOs (or just the first three, K-shell MOs) from correlation. With a polarized basis, but five MOs excluded, they are able to obtain $E_h = -204.4246$ for the ground state, lower than Jackels and Davidson's [64] best value of -204.3743 with an unpolarized basis. Without polarization, but with only three MOs excluded from correlation, they obtain $E_h = -204.2764$. With a polarized basis but five MOs excluded, Benioff obtained $E_h = -204.2520$. Handy *et

TABLE 2

Vertical Electronic Excitation Energies of NO_2(eV)

	This work[a]				Other calculations						
	SCF		MCSCF–CM								
State	[4s 3p]	[4s 3p 1d]	[4s 3p]	[4s 3p 1d]	[63][b]	[66][c]	[65][d]	[13][e]	[14][f]	[71]	Experiment
$X\,^2A_1$	0	0	0	0	0	0	0	0	0	—	0
$1\,^2B_1$	2.47	2.91	2.60	2.93	2.45	2.8	2.46	2.79	2.52	—	2.81–2.85
$1\,^2B_2$	3.82	3.91	3.36	3.40	3.03	3.4	2.43	3.22	3.33	—	3.1
$1\,^2A_2$	2.81	3.06	3.54	3.61	3.09	3.4	2.15	3.25	3.52	—	—
$1\,^4A_2$	2.09	2.66	4.40	4.72	3.81	4.7	3.10	4.09	4.41	4.3	—
$1\,^4B_2$	2.10	2.81	4.43	4.85	3.80	4.6	2.88	4.04	4.48	4.5	—

[a] Benioff [68].

[b] These are minimal basis SCF + CM calculations.

[c] These are (4s 3p 1d) MCSCF–CM calculations.

[d] The calculations were done at $d_{N-O} = d_{N-O'} = 2.35$ bohr, $\theta_{ONO'} = 132.99°$ instead of at the NO_2 equilibrium geometry parameters $d_{N-O} = 2.25$ bohr. An unpolarized basis was used.

[e] These are multireference CM calculations, with very extensive virtual-space CM, but unpolarized basis.

[f] These are (4s 3p) MCSCF–CM calculations, no polarization.

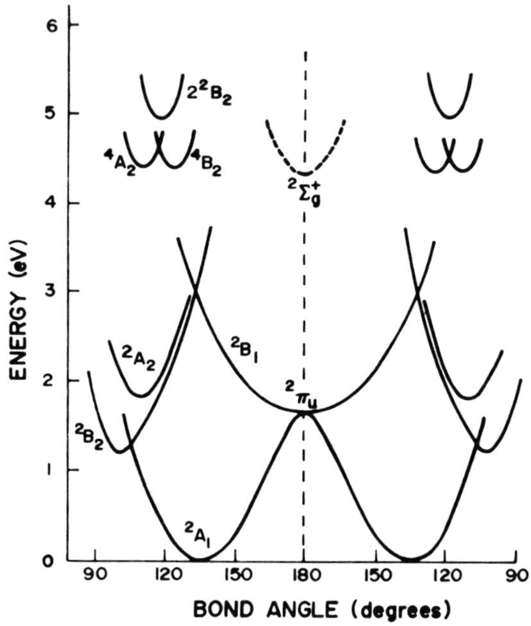

FIG. 12. Electronic energy as a function of bond angle for the low-lying states of NO_2. From Gillispie *et al.* [66].

al. computed vertical CM energies also for five excited states, polarized basis with five MOs frozen, and unpolarized basis with three frozen. They also made computations on NO_2^- (see Section H).

Almlöf *et al.* [60] used unrestricted SCF calculations with $S = \frac{3}{2}$ annihilation to compute magnetic hyperfine coupling constants, and also nuclear quadrupole coupling constants, for the ground state of NO_2. There is good agreement with experiment.

Several authors have made *ab initio* computations of the vertical ionization energies of NO_2 to various states of NO_2^+ (see above for Shih *et al.*). Benioff gives a table of computed excitation energies of NO_2^+ relative to the ground state for the calculations of Shih *et al.*, Fortune *et al.*, and his own, and for experiment. Agreements are not outstanding.

H. 18-VE MOLECULES (NO_2^-, O_3, SO_2, CF_2, CCl_2, SiF_2, NOCl, NSF, NF_2^+)

For the 18-VE NO_2^-, most extensive are the MCSCF plus moderate CM calculations of Benioff [68], giving an energy of −204.1942 a.u. for a

Gaussian basis without polarization but augmented by diffuse functions, as is desirable for a negative ion. This energy was calculated at the equilibrium geometry of NO_2^-. At the geometry of NO_2, inclusion of polarization (d functions) in the basis lowered the energy of NO_2^- by 0.0935 a.u. (inclusion only of diffuse functions lowered it by 0.024 a.u.). Assuming about the same effect at the geometry of NO_2^-, one obtains an estimate of -204.288 a.u. for the combined polarized and diffuse basis set.

With the help of calculations at the two geometries mentioned, computed values of the vertical and adiabatic electron affinity (VEA and AEA) of NO_2 and the vertical electron detachment energy (VDA) of NO_2^- were obtained. Comparisons with experiment are given for AEA; agreements are not very good.

Andersen and Simons [73] used the equations-of-motion (EOM) method to calculate the VDA for NO_2^-. They used a typical Gaussian basis plus diffuse functions, and first calculated an SCF potential surface for NO_2^-, with energy -204.0427 a.u., R(NO) = 1.265 Å (experimental: 1.23 Å), and bond angle 118°. (For comparison, Benioff at the SCF level for equilibrium geometry obtained -204.0487 a.u.) From this they obtained the NO_2 surface by adding the calculated EOM VDA at each of varied geometries. The calculated VDA for NO_2^- at equilibrium is 2.66 eV. They also calculated 1.43 eV for the VEA of NO_2 and 2.23 eV for the AEA of NO_2 [experimental: (2.28–2.36) ± 0.1 eV].

Further, Fortune *et al.* [69] computed the SCF energies and excitation energies for seven excited states of NO_2^-, and Benioff [68] computed the same quantities not only by SCF but also by MCSCF plus CM using a polarized basis. Polarization here increases the excitation energies by about 0.3 eV (not more, since the effects of polarization on ground and excited states largely cancel). Benioff notes that inclusion of polarization functions for the ground states of NO_2^-, NO_2, and NO_2^+ lower the computed energy by 0.093, 0.099, and 0.107 a.u., respectively.

Benioff gives the electronic excitation energies of NO_2^- in two tables, the first computed at the equilibrium conformation of NO_2, the second at that of NO_2^-. The latter shows good agreement with experiment.

Handy *et al.* [72], with a double-zeta unpolarized basis and a frozen core of the first five MOs, used the "direct CI" method to obtain E_h = -204.3039 for the ground state of NO_2^-. This may be compared with Benioff's E_h = -204.1942. Handy *et al.* also computed vertical excitation energies for five higher states. The results do not differ greatly from Benioff's, or from experiment where available.

A short article by Pearson *et al.* [74] reports unpolarized SCF computations on the potential surface of the ground state of NO_2^- near three isomeric configurations. These are the normal configuration R(NO) =

1.264 Å, apex angle θ = 117°, atomic populations N, 6.83; O, 8.58; the peroxy isomer

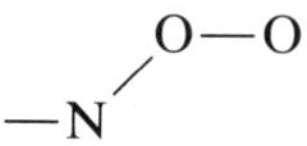

R(NO) = 1.245 Å, R(OO) = 1.493 Å, θ = 119°, energy up 3.20 eV, population N, 7.21; O_1, 8.22; O_2, 8.57; and the "ring" isomer, analogous to the D_{3h} form of O_3, R(NO) = 1.502 Å, θ = 59°, energy up 4.35 eV, populations N, 7.24; O, 8.38.

As compared with NO_2, the presence of two $6a_1$ MOs in the ground state of the 18-VE molecule O_3 causes a further decrease in apex angle, to 110.8°; R(OO) = 1.278 Å. Rothenberg and Schaefer [61] using a double-zeta plus polarization (d) basis set, computed an SCF wave function for the ground state. The computed energy, −224.3093 a.u., is estimated to be about 0.06 a.u. above the SCF limit. With this wave function, multipole moments, diamagnetic susceptibility and shielding tensors, and electric field gradient tensors were computed. Orbital energies and components of the total energy are listed. The lowest orbital energies are as follows: $4b_2$, −0.563; $6a_1$, −0.552; $1a_2$, −0.486; $2b_1$, −0.046; $7a_1$, 0.282; the order is not quite the same as in Fig. 1. Atomic populations correspond to charges +0.30 on the central O atom, −0.15 on the outer atoms; d populations are 0.15 e on the central and 0.05 e on the outer atoms.

Curtiss *et al.* [74a], using a variety of approximate wave functions up to one with a double-zeta plus polarization basis followed by fairly extensive CM, computed dipole moment, dipole moment derivations, and infrared intensities for the ground state of O_3. Agreements with experiment are not outstanding.

Most of the papers on O_3 deal with its excited states and spectrum, with something included on O_3^+. The ground state is then discussed incidentally. For it, the lowest reported energy is −224.893 a.u. for a polarized basis set with d functions (set C) in an MCSCF plus CM computation of Thunemann *et al.* [75]. Two main configurations ... $4b_2^2\ 6a_1^2\ 1a_2^2$, 79% and ... $4b_2^2\ 6a_1^2\ 2b_1^2$, 11% are used, and all 14,197 double excitations from these are considered, but the actual computation is limited to 3093 configurations using natural orbitals; a perturbation theory extrapolation to the expected result for 14,197 configurations is then made: the typical Peyerimhoff and Buenker MRD-CI procedure.

Computed results of Hay and Dunning in a POL-CM calculation [76], and of Thunemann *et al.* on the lowest 9 states of O_3 are given in Table 3. Thunemann *et al.* also give vertical excitation energies for 22 additional excited states involving excitation to $2b_1$, $7a_1$, and Rydberg MOs.

Thunemann *et al.* also compute four vertical ionization potentials ($6a_1$,

TABLE 3
Computed Excited States of O_3 (C_{2v} Symmetry)

	Population				Geometry[a]		Excitation energy (eV)		
State	$1a_2$	$4b_2$	$6a_1$	$2b_1$	R (Å)	θ (deg)	Adiabatic[a]	Vertical[a]	Vertical[b]
$1\,^1A_1$	2	2	2	—	1.299	116	0	0	0
$1\,^3B_2$	1	2	2	1	1.382	108	0.92	1.60	1.20 } 6 electron impact peaks (1.29–1.92)
$1\,^3A_2$	2	1	2	1	1.366	100	1.35	2.09	1.44 }
$1\,^3B_1$	2	2	1	1	1.347	124	1.74	2.01	1.59 }
$1\,^1A_2$	2	1	2	1	1.374	101	1.66	2.34	1.72 }
$1\,^1B_1$	2	2	1	1	1.370	118	2.06	2.41	1.95 (2.13 Chappuis)
$2\,^3B_2$	2	1	1	2	1.463	104	2.92	4.17	3.27
$2\,^1A_1$	2	—	2	2[c]	1.449	60	1.20[e]	4.58	3.7[c] (3.7 Huggins)
$1\,^1B_2$	1	2	2	1	1.405[d]	108	5.54	6.12	4.97 (4.88 Hartley)

[a] From Hay and Dunning [76].
[b] From Thumemann *et al.* [75]; experimental identification in parentheses.
[c] [75] gives, for vertical excitation, population 36% 2 0 2 2, 45% 2 2 0 2; for adiabatic excitation (cyclic configuration), mostly 2 0 2 2.
[d] Subject to asymmetrical distortion (see text).
[e] A thorough study by Burton [82] gives a probable value of 0.5 ± 0.1 eV.

$4b_2$, $1a_2$, and $1b_1$). Hay and Dunning also give computed ionization potentials for the $6a_1$, $4b_2$, and $1a_2$ MOs. Thunemann *et al.* give a table comparing these and other computed values with experiment. Subject to a question of identification, agreement is moderately good, perhaps especially for Hay *et al.* [77]. Possibly as good are agreements with MCSCF plus CM computations of Basch [78] using only a double-zeta basis without polarization. Basch also discusses shake-up transitions.

The geometry and adiabatic excitation potentials in Table 3 are from the calculations of Hay and Dunning, who used a double-zeta basis with d polarization. On the other hand, Thunemann *et al.* use extensive CM but less polarization (s-type *bond* functions only) in their calculations on excited states. These are much less effective than d polarization functions. For the ground and the ionized states, however, they made three calculations: A, s-type bond functions only; B, s- and p-type bond functions; C (best), d polarization functions. For the interesting 2^1A_1 state, they give for *vertical* excitation a composition 36% ... $6a_1^2\,1a_2^2\,2b_1^2$ and 45% ... $4b_2^2\,1a_2^2\,2b_1^2$; however, at equilibrium (see below) the equilateral form is almost entirely ... $6a_1^2\,1a_2^2\,2b_1^2$.

Thunemann *et al.* identify the various computed states with various experimental energies as shown in Table 3. In this connection, Hay and

Dunning and Thunemann *et al.* make calculations of oscillator strengths which agree with experiment in making the Hartley bands about 10,000 times as strong as the Chappuis and Huggins bands. Other authors have hesitated to identify the Huggins bands.

According to SCF calculations, 1^3B_2 should be the ground state of O_3, but extensive CM in 1^1A_1 reverses the order of the two states. Much attention has been given to the 2^1A_1 state, which although rather high for vertical excitation proves to have a minimum for an equilateral configuration (cyclic state, D_{3h} symmetry, at least approximately), which makes it the second lowest excited state and probably metastable. Several papers have been devoted to calculating the (adiabatic) height of the cyclic state above the ground state. Nearly the most recent and thorough are those of Hay and Dunning [76] as reported in Table 3, of Harding and Goddard [79], and of Lucchese and Schaefer [80], all of whom used polarized basis sets and CM. The results are 1.20 eV or 28 ± 3 kcal/mole (HD and HG), and (after an estimated correction) 0.95 eV or 22 kcal/mole. These values may be compared with the dissociation energy for $O_3 \rightarrow O_2 + O$ of 1.05 ± 0.2 eV or 24 ± 4 kcal/mole. In one of the less accurate calculations, Shih *et al.* [81] discuss the question of the barrier to dissociation. That the cyclic state is stable toward dissociation now seems to be fairly well assured through an exhaustive study by Burton [82] of the effects of various basis sets on its computed adiabatic energy. He concludes that the correct value is within the range 0.4–0.9 eV, his most probable estimate being 0.5 ± 0.1 eV.

Of the states 1^3B_2 and 1^1B_2 of the same configuration, Devaquet and Ryan [83] showed, by minimal-basis SCF and then SCF plus CM computations, that the $^3B_2$ state is destabilized but the $^1B_2$ stabilized, by asymmetric distortions to C_s symmetry. The $^1B_2$ situation is analogous to that indicated by experimental evidence in the analogous $^1B_2$ state of SO_2. Further, there is some interaction with the 1^1A_1 and 2^1A_1 states, both of which, along with $^1B_2$, are $^1A'$ states under C_s symmetry. Hay and Dunning in more accurate calculations confirmed these results and found the energy of $^1B_2$ decreasing for $R_1 - R_2$ increasing (R_1 and R_2 refer to the two legs of the triangle), at least out to 0.8 Å, for the average R fixed at 1.406 Å with $\theta = 108.4°$. Here there seems to be a minimum, but they point out that since the distorted geometry was not fully optimized, it is possible that the distortion continues out to dissociation. However, account must be taken of a weakly avoided crossing with the 2^1A_1 state.

SCF calculations on the ground state of SO_2 have been made by Rothenberg and Schaefer [84] (computed energy −547.2089 a.u.) and by Huzinaga and Yoshimine [85] (computed energy −547.2246 a.u. or, with the use of a model potential of the S and O atom cores, −547.26796 a.u.).

With their wave function, Rothenberg and Schaefer computed multipole moments, diamagnetic shielding tensors, and electric field gradients. Both sets of authors list orbital energies, in near agreement; the lowest ones are $4b_2$, -0.697; $7a_1$, -0.695; $2b_1$, -0.666; $5b_2$, -0.539; $1a_2$, -0.514; $8a_1$, -0.492. Atomic populations correspond to $S^{1.12}(O^{-0.56})_2$ with S—O overlap population 0.69 *e*; d orbital populations, 0.425 *e* on S, 0.063 *e* on each O. The SCF calculation of HY is apparently the same as that of RS except that HY computed the geometry $R(SO) = 1.421$ Å, $\theta = 118.4°$, in good agreement with experiment (1.432 Å, 119.5°); and computed the force constants, in rather good agreement with experiment. With a less accurate SCF calculation, Roos and Siegbahn [86] give contour diagrams of the nine outer MOs.

In SCF calculations Gelius *et al.* [87] compare sulfur atom ESCA chemical shifts (changes in K-shell ionization potential) with gross electron populations on the S atom in a number of sulfur compounds. They find a linear relation. Their calculation for SO_2 gives a population $S^{+0.925}(O^{-0.46})_2$.

Cederbaum *et al.* [87a], using a Green's function method with a moderate basis set (d functions on S only), have computed electron attachment spectra for O_3 and SO_2. For the electron affinities of O_3 and SO_2, they calculate 1.76 eV vertical and 2.26 eV adiabatic for O_3, and 0.84 eV vertical and 1.08 eV adiabatic for SO_2, both in good agreement with experiment. They go further and predict the vibrational structures of these spectra. That for SO_2 corresponds very well to Celotta's experimental photodetachment spectrum of SO_2^-. For O_3 no comparison is available. While a photoattachment spectrum should be expected, the O_3^- ion is unstable toward dissociation into O^- plus O_2. In SO_2, the SO_2^- ion is stable toward similar dissociation.

Among 18-VE molecules, CF_2 is of considerable interest chemically, along with CHF (see Section III.E), as a derivative of CH_2. However, in structure and spectrum it is rather different. Its ground state is a singlet, whereas that of CH_2 is a triplet; in CHF, singlet and triplet compete more closely. CCl_2 and SiF_2 are 16-VE molecules related to CF_2.

Rothenberg and Schaefer [49] made a double-zeta plus d polarization (on both atoms) SCF calculation for the ground state, and computed numerous molecular properties (see Section F). Bauschlicher *et al.* [88] made an improved calculation ($E_h = -236.7472$ as compared with -236.7207) and also dealt with the low-energy triplet analogous to the ground state of CH_2. The calculation was of SCF type except that two configurations ($\ldots 6a_1^2$ and $\ldots 2b_1^2$) were used for the ground state. They made calculations also on CCl_2, and again found $^1A_1$ to be the most stable, although by a much smaller margin than in CF_2. Staemmler [89] also made calculations on CF_2 using CM (independent electron-pair approach) but with a poorer basis set. Clark *et al.* [90] also made SCF calculations on the lowest singlet and triplet states of CF_2 (also CH_2 and CHF), and of the corresponding core hole states.

Ha *et al.* [91] made SCF + CM (well-selected single and double substitutions on the partially polarized double-zeta basis set) calculations on the ground state of CCl_2 ($^1A_1$) and of seven low-lying singlet and triplet excited states, with oscillator strengths of transition to them. The lowest excited states are calculated (presumably vertically) as $^3B_1$ at 0.83 eV and $^1B_1$ at 3.90 eV (f = 0.012 for the transition to the latter, which can be identified with an observed absorption band at 3.76 eV). Dipole moments of 1.03 D for the ground and 0.57 D for the $^1B_1$ state were computed.

So [92], using a minimal STO-3G basis set plus a Cl 3d orbital, has made SCF computations on ClCF. The ground state is $^1A'$. The first excited state, $^3A''$, is computed and the second, $^1A''$, estimated. The observed 4000-Å emission spectrum (T_0 = 24,983 cm^{-1}) can be identified with the predicted $^1A'' - {}^1A'$ transition (computed adiabatic transition energy 26,235 cm^{-1}).

Thomson [93] has made an SCF calculation on the $^1A_1$ ground state of SiF_2, using a basis containing d orbitals only on the Si. He gives a list of orbital energies and finds a charge distribution $Si^{0.92}(F^{-0.46})_2$.

Støgård [27] gives SCF calculations with 3d on Cl, on the 18-VE NOCl. He lists orbital energies and six computed ionization potentials. The computed geometry is R(NCl) = 1.86 Å (experimental 1.95 Å) and θ = 114° (experimental: 116°).

Harrison and Eakers [94] have made unpolarized SCF plus limited CM calculations on the 18-VE ion NF_2^+. Like CF_2 and SiF_2, the ground state is $^1A_1$. Predicted apex angles are 120° for the $^1A_1$ and 122° for the $^3B_1$ first excited state. Predicted charge distributions are $N^{0.96}(F^{0.02})_2$ in the $^1A_1$ and $N^{0.65}(F^{-0.17})_2$ in the $^3B_1$ state. A graph showing the CM energies of the $^1A_1$, $^3B_1$, and $^1B_1$ states as a function of apex angle is shown.

Numerous *ab initio* computations on ONF and also one or two on its less stable isomer NOF have been made. Most recently, Lawlor *et al.* [95] have made SCF plus CM computations on these, using a good polarized basis set and rather good CM (K-shell frozen) extrapolated to zero threshold in the MRD-CI manner. For ONF, SCF alone gives R_{NF} = 1.39 Å; CM gives 1.52 Å in good agreement with the experimental value 1.512 Å. Even with a good basis set SCF gives only 58 kcal/mole for dissociation to NO plus F, as compared with an estimated experimental value of 200 kcal/mole; CM gives 121 kcal/mole. NOF is found to be 44 kcal/mole less stable, with R_{OF} = 1.76 Å: E_h = −228.7954 (−228.7768 with another basis set) for ONF, −228.7073 by the second basis set for NOF. Computed dipole moments (SCF) are 0.93 D for ONF and 5.64 D for NOF, indicating considerable amounts of structure $(O^+\equiv N)F^-$ or $(N\equiv O^+)F^-$ in the wave functions. Population analyses are given. Natural orbitals are listed and shown along with SCF MOs in contour diagrams. The nature of the bonding in ONF and NOF is discussed.

With polarized basis sets of double-zeta quality, Rosmus *et al.* [96] have made SCF calculations on NSF and SSO at experimental geometries. Calculated orbital energies, dissociation energies, dipole moments, and population analyses are given. Seven vertical ionization energies, also some vibrational spacings, are computed for NSF and SSO, and compared with those for SO_2, and with photoelectron spectra. A diagram is given comparing orbital energies of the 18-VE SSO, NSF, SO_2, O_3, ONF, and CF_2 with Walsh–Mulliken predictions.

I. 19- TO 23-VE MOLECULES (NF_2, PF_2^+, OF_2, SF_2, ClF_2^+, Cl_2F^+, ClF_2, ClF_2^-, KrF_2, XeF_2, $Ar_2^+F^-$, $Kr_2^+F^-$); ALSO NiCO AND TiO

In the 19- and 20-VE molecules NF_2 and OF_2, one and two electrons are added in the $2b_1$ MO. According to the SCF calculations of Rothenberg and Schaefer [49], the apex angles are here reduced to 104° (NF_2) and 103° (OF_2). Many other properties are computed (see Section F for details).

In SCF plus CM computations using *all* single excitations, Kendrick *et al.* [97] have calculated ESR coupling constants for NF_2 and PF_2 in rather good agreement with experiment.

In a double-zeta plus phosphorus d basis SCF calculation, Cobb and Hinchliffe [98] compute ESR coupling constants for the 19-VE PF_2 which agree well with experiment. For the 20-VE SF_2, Hay [99] has made GVB plus CM calculations at the experimental geometry (R = 1.589 Å, θ = 98.3°). SCF orbital energies are listed. The nature of the bonding is compared with that in SF_4 and SF_6. From SCF functions, a charge distribution $S^{0.766}(F^{-0.383})_2$ is found. The dipole moment, quadrupole moment, and electric field gradient at the S nucleus are computed.

Solouki *et al.* [100], using d functions on each atom, made SCF computations on SCl_2. They found total populations corresponding to $S^{0.25}(Cl^{-0.13})_2$, a dipole moment of 0.61 D, and d populations of 0.13 for sulfur and 0.03 for chlorine. They give orbital energies and a Koopmans' interpretation of the photoelectron spectrum, and comparisons with those of OCl_2 and OF_2.

According to Fig. 1, addition of electrons in the $7a_1$ MO tends strongly to increase the angle, so that the 22-VE F_3^- might be linear. In fact all the higher analogs of F_3^-, e.g., I_3^-, are known to be linear (though easily bent; in some crystals, I_3^- *is* bent). Riedel and Willett [101] have reported an SCF calculation on CL_3^-, using a Gaussian s and p basis set (no d orbitals). They found the molecule to be linear and symmetrical, and listed MO orbital energies.

Ungemach and Schaefer [102] have made unpolarized double-zeta quality SCF calculations on ClF_2^+ (predicted bent, 97°), ClF_2 (predicted bent, 145°), and ClF_2^- (predicted linear), with respective R(ClF) values 1.70, 1.83, and 1.93 Å. On adding a d function on Cl to the basis set, R(ClF) and θ for the 21-VE ClF_2 are changed to 1.72 Å and 149°. Orbital energies for ClF_2^+, ClF_2, and ClF_2^- are listed. Here the $9a_1$ MO corresponds to the $7a_1$ MO in F_3^-, and evidently has similar characteristics. The first excited state of ClF_2, a $^2B_1$ state, is predicted to be linear (part of $^2\Pi_u$). A population analysis of ClF_2 gives a charge distribution $Cl^{0.66}(F^{-0.33})_2$. Guest *et al.* [103] have made similar calculations on ClF_2^+ and ClF_2^-, leading to similar bond angles (ClF_2^+, 98°, experiment: 99.5 ± 3.5; ClF_2^-, 180°, experiment: 180°).

Using a 4-31G basis, Joshi and Morokuma [104] have made SCF computations and a Morokuma energy decomposition analysis for $(ClFCl)^+$ and $(FClCl)^+$, and found the former more stable (E_h = − 1016.7358 versus − 1016.7509), with R(ClF) = 1.80 Å and Cl—F—Cl angle 140°, stabilization energy versus Cl^+ plus ClF, 127 kcal/mole. In $(FClCl)^+$ the F—Cl—Cl angle found is 99°, with R(ClF) = 1.70 Å and R(ClCl) = 2.27 Å. A population analysis for $(ClFCl)^+$ gives a charge distribution $CL^{0.71}F^{-0.42}Cl^{0.71}$.

According to Coulson [105] in valence-bond theory, the experimentally linear 22-VE molecules KrF_2 and XeF_2 are stabilized mainly by resonance between $F{-}A^+F^-$, $F^-\ A^+{-}F^-$, and $\overline{FAF}$. Good SCF plus fairly extensive CM nonrelativistic computations by Bagus *et al.* [106] for KrF_2, [107] for XeF_2) confirm this idea. They used somewhat better than double-zeta plus good polarization STF basis sets and somewhat better than first-order CM. They list orbital energies (... $5\sigma_u^2\, 3\pi_u^4\, 2\pi_g^4\, 8\sigma_g^2\, 4\pi_u^4$, $^1\Sigma_g^+$ for KrF_2 and ... $7\sigma_u^2\, 4\pi_u^4\, 3\pi_g^4\, 10\sigma_g^2\, 5\pi_u^2$, $^1\Sigma_g^+$ for XeF_2). The two bonding MOs ($5\sigma_u$ and $8\sigma_g$ for KrF_2, $7\sigma_u$ and $10\sigma_g$ for XeF_2) at equilibrium are both strongly polar in the direction Rg^+F^-, but lose their polarity if one goes toward dissociation into F + Rg + F. The potential curve for this process shows a small maximum 10.22 eV above the dissociation limit for KrF_2, but not for XeF_2. This maximum may be interpreted as resulting from an avoided crossing between a large-R nonpolar valence bond function and a small-R ion-pair structure. For KrF_2 the forms of the MOs near equilibrium are given in detail.

Calculated bond distances agree well with experiment: R(KrF) = 1.907 Å computed, 1.889 Å observed; R(XeF) = 1.999 Å computed, 1.977 Å observed. For XeF_2 a detailed population analysis (s, p, d, f population in each MO) is given. In the bonding MO $7\sigma_u$ the distribution is 0.62 in Xe 5p, 1.37 on F; in $10\sigma_g$, it is 0.30 on Xe 5s and 5d, 1.64 on F. The overall population at equilibrium is $Xe^{1.04}(F^{-0.52})_2$. For KrF_2, quadrupole moments and electric-field gradients at the nuclei are reported. The quadrupole moment at equilibrium (~ -12.7 a.u.) agrees well with that for $F^{-0.5}Kr^{1}F^{-0.5}$. At larger R it soon falls off in magnitude as the structure becomes less ionic. The electric field gradient at Kr is not far from 6.77, the SCF value for Kr^+ at R_e, but near the potential maximum in the Kr + 2F dissociation curve, drops precipitously, to zero as for the Kr atom, as dissociation is completed. The gradient at fluorine increases monotonically with R toward its value (3.02 a.u.) for the F atom.

Rosén and Ellis [108] have given an approximate discussion of a relativistic wave function for XeF_2.

The diatomic rare gas (Rg) molecules RgF are unstable or barely stable in their ground states, but have stable excited states, the lowest of which have the ionic structure Rg^+F^-, $^2\Sigma^+$. The latter are important in laser action. The ions Rg_2^+ are stable in their ground states (and somewhat in a low excited state). Recently *ab initio* computations have shown that the 23-VE Rg_2F, although unstable in their ground states, are stable in their lowest, $Rg_2^+F^-$, excited states relative to dissociation to Rg plus the excited Rg^+F^- structure, though not relative to dissociation to 2Rg + F. On the other hand, observed emission spectra can be identified with the transition from $Rg_2^+F^-$, $^2B_2$ to Rg_2F (→ 2Rg + F), $^2B_2$ and $^2A_1$. All this is of possible relevance to laser work with the rare-gas fluorides.

The foregoing conclusions on $Rg_2^+F^-$ are based on the work of Wadt and Hay [109] on electronic states of Ar_2F and Kr_2F and of Bender and Schaefer [110] on ionic excited states of Ne_2F. They made SCF plus first-order CM calculations using fairly good basis sets (without d orbitals for Ne_2F, but of course with them for Ar_2F and Kr_2F). Wadt and Hay explored the geometry for Ar_2F and concluded that, although the linear ArArF is distinctly stable, the most stable conformation is that of an isosceles triangle with apex angle about 48° (Ar—Ar distance 2.48 Å, Ar—F 2.52 Å; dissociation energy to Ar^+F^-, 0.60 eV). Oscillator strengths for the two transitions mentioned above are computed (they are not very large), along with lifetimes for the upper states. The situation here does not conform to Fig. 1, which would predict a linear structure; moreover,

$5b_2$ here is found more tightly bound than $7a_1$ in Ne_2F, and similarly in the others. In the second paper of Wadt and Hay there is extensive discussion of spin–orbit coupling, which in a complete treatment must be considered. Resemblances of the spectra to those of Rg_2^+ are discussed.

GVB plus CM calculations have been made by Walch and Goddard [111] on the molecule NiCO, which, however, falls outside the group of molecules we have been discussing. It is relevant to the problem of the adsorption of CO on metallic nickel. Here NiCO is assumed linear. The three lowest states are found to be $^3\Delta$, $^3\Sigma^+$, and $^3\Pi$, in that order, at 0, 0.24, and 0.29 eV, all derived from Ni atom states of configuration d^9 s. Higher states (3–5 eV vertically), derived from Ni d^8 s^2, are found to be repulsive. For the $^3\Delta$ ground state, R(Ni–C) is computed as 1.90 Å. Atomic orbital populations for the three states are listed. The overall charge distribution is reported as $Ni^{0.00}C^{0.11}O^{-0.11}$.

Bagus and Hermann [112] made SCF calculations on the $^3\Delta$ ground state of linear NiCO, using "extended Gaussian basis sets," also computed ionization potentials for those MOs which correspond to CO MOs. In NiCO, the MO corresponding to 5σ in CO acquires 13% Ni $3d\sigma$ character, and furnishes Ni—C bonding, while in the Ni 12σ $1\delta^3$ contribution, 4s (12σ) acquires 27% admixture of $4p\sigma$ character directed away from the C. The computed ionization potentials are compared with those from the photoelectron spectrum of CO acsorbed on nickel. The CO ionization potentials are analyzed as depressed about 1.1 eV, except 5σ by 3.5 eV, then raised about 0–1 eV.

Clark *et al.* [113] made SCF calculations on linear and bent NiCO, having noted empirical evidence that adsorbed CO is oriented corresponding to the latter. They computed linear and bent $^3\Delta$ NiCO from Ni 4s $3d^9$, and found the bent form to be 1 kcal/mole more stable. They also computed $^1\Sigma^+$ NiCO from Ni $3d^{10}$; here they found the linear form to be more stable by 68 kcal/mole. The results are discussed in terms of binding and relaxation energies; here bending has a large influence. Bending is found to increase the binding energy of 1π by 1 eV. A population analysis for linear $^3\Delta$ NiCO yields $Ni^{+0.11}C^{+0.46}O^{-0.57}$.

Using polarized basis sets (double-zeta for the valence electrons), Clark *et al.* [114] have made SCF calculations on ground and core-hole states (Ni_{2s}, N_{1s}, O_{1s}) of linear and bent configurations of NiNO ($^2\Pi$), using $3d^{10}$ nickel. The ground state is bent by perhaps 20°. Core-ionization relaxation energies of the outer electrons are computed and discussed. Population analyses are given for the ground and core-ionized states.

With a view toward understanding the carbonyls (see Section XII.C.3), Mortola and Goddard [115] made GVB calculations on a number of electronic states of TiCO and $TiCO^+$, both assumed to be linear. Only the valence electrons of Ti (ground state $4s^2\,3d^2$, ^3F with $4s\,3d^3$, ^5F 0.8 a.u. higher) and CO interact at all strongly. The ground state of TiCO is $\ldots 11\sigma^2\,4\pi\,1\delta$, $^2\Phi$ and that of $TiCO^+$ is $\ldots 11\sigma\,4\pi^2$, $^4\Sigma^-$. Fifteen excited states of TiCO are computed, mostly by SCF methods, and two of $TiCO^+$ by GVB. In the ground state, the main effects are: s–p hybridization of the Ti 4s, Ti–C binding by donation from CO 5σ, and some $3d\pi$ back donation from Ti into CO π^*. This indicates that the presence of the metal is just a minor perturbation on the bonding in CO. However, for the metal, $3d\sigma$ orbitals become involved in the bonding. The authors go on to generalize about carbonyls.

REFERENCES

1. R. S. Mulliken, *Rev. Mod. Phys.* **14,** 204 (1942); *Can. J. Chem.* **36,** 10 (1958); note also Refs. 3 and 5 in the latter article.
2. A. D. Walsh, *J. Chem. Soc.* p. 2260 (1953).
3. R. J. Buenker and S. D. Peyerimhoff, *Chem. Rev.* **74,** 127 (1974).
4. J. Kendrick and I. H. Hillier, *Mol. Phys.* **33,** 635 (1977); see also P. S. Bagus, G. del Conde, and D. W. Davies, *Faraday Discuss. Chem. Soc.* **62,** 321 (1977).
5. R. L. Martin and E. R. Davidson, *Mol. Phys.* **35,** 1713 (1978).
6. W. H. Gerbar and E. Schumacher, *J. Chem. Phys.* **69,** 1692 (1978).

6a. J. F. Harrison, R. C. Liedtke, and J. F. Liebman, *J. Am. Chem. Soc.* **101,** 7162 (1979).

7. E. L. Wagner, *Theor. Chim. Acta* **32,** 295 (1974).
8. A. Hinchliffe and J. C. Dobson, *Chem. Phys.* **8,** 166 (1975), on Li_2O; *Mol. Phys.* **28,** 543 (1974) and **29,** 1315 (1975), on Li_2S; *Chem. Phys.* **8,** 158 (1975), on Na_2S.
9. D. H. Liskow, C. F. Bender, and H. F. Schaefer, III, *J. Chem. Phys.* **56,** 5075 (1972).
10. J. Perić-Radić, J. Römelt, S. D. Peyerimhoff, and R. J. Buenker, *Chem. Phys. Lett.* **50,** 344 (1977).
11. J. Römelt, S. D. Peyerimhoff, and R. J. Buenker, *Chem. Phys. Lett.* **58,** 1 (1978).
12. J. Peslak, D. S. Klett, and C. W. David, *J. Am. Chem. Soc.* **93,** 5001 (1971).
13. J. Peslak, D. S. Klett, and C. W. David, *J. Chem. Phys.* **55,** 1993 (1971).
14. C. Thomson, *J. Chem. Phys.* **58,** 216 (1973).
15. B. Bak, E. Clementi, and R. N. Kortzeborn, *J. Chem. Phys.* **52,** 764 (1970).
16. E. Clementi, H. Kistenmacher, and H. Popkie, *J. Chem. Phys.* **58,** 2460 (1973).
17. V. Staemmler, *Chem. Phys.* **7,** 17 (1975).

17a. V. Staemmler, *Chem. Phys.* **17,** 187 (1976).

18. S. V. O'Neil, H. F. Schaefer, III, and C. F. Bender, *J. Chem. Phys.* **59,** 7 (1973).
19. C. Thomson, *J. Chem. Phys.* **58,** 841 (1973).
20. C. Thomson and B. J. Wishart, *Theor. Chim. Acta* **31,** 347 (1973).
21. A. A. Wu and C. Schlier, *Chem. Phys.* **28,** 73 (1978).
22. C. Thomson and B. J. Wishart, *Theor. Chim. Acta* **35,** 261 (1974).
23. J. A. Horsley and J. A. Hall, *Mol. Phys.* **25,** 483 (1973).
24. T. W. Archibald and J. R. Sabin, *J. Chem. Phys.* **55,** 1821 (1971).

24a. T. A. Claxton, R. E. Overill, and M. C. R. Symons, *Mol. Phys.* **26,** 75 (1973).

24b. T. J. Venanzi and J. M. Schulman, *Mol. Phys.* **30,** 281 (1975).
25. D. G. Hopper, *Chem. Phys. Lett.* **31,** 446 (1975).
26. A. D. McLean and M. Yoshimine, *IBM J. Res. Dev.* **2,** Suppl. (1967).
27. Å. Støgård, *Chem. Phys. Lett.* **40,** 429 (1976); see also Ref. 28.
28. E. Gilberg, W. Shätzl, and W. Schrenk, *Chem. Phys.* **13,** 115 (1976).
29. S. D. Peyerimhoff and R. J. Buenker, *J. Chem. Phys.* **49,** 2473 (1968).
30. N. W. Winter, *Chem. Phys. Lett.* **33,** 300 (1975).
31. W. Domcke, L. S. Cederbaum, J. Schirmer, W. von Niessen, C. E. Brion, and K. Tan, *Chem. Phys.* **40,** 171 (1979).
31a. T. Darko, I. H. Hillier, and J. Kendrik, *Mol. Phys.* **32,** 33 (1976).
32. M. P. S. Collins and B. J. Duke, *J. C. S. Dalton Trans.* p. 277 (1978).
33. G. V. Pfeiffer and L. C. Allen, *J. Chem. Phys.* **51,** 190 (1969).
34. P. Cremaschi and M. Simonetta, *Theor. Chim. Acta* **34,** 175 (1974).
35. J. Pacansky, U. Wahlgren, and P. S. Bagus, *Theor. Chim. Acta* **41,** 301 (1976).
36. M. A. Morrison and P. J. Hay, *J. Chem. Phys.* **70,** 4034 (1979).
37. C. Thomson and D. A. Brotchie, *Chem. Phys. Lett.* **16,** 573 (1972).
38. C. Thomson and D. A. Brotchie, *Theor. Chim. Acta* **32,** 101 (1973).
39. W. B. England, B. J. Rosenberg, P. J. Fortune, and A. C. Wahl, *J. Chem. Phys.* **65,** 684 (1976).
40. R. S. Mulliken, *Can. J. Chem.* **36,** 10 (1958), Table III.
41. W. B. England, D. Yeager, and A. C. Wahl, *J. Chem. Phys.* **66,** 2344 (1977).
42. W. B. England, W. C. Ermler, and A. C. Wahl, *J. Chem. Phys.* **66,** 2336 (1977).
43. W. von Niessen, G. H. F. Diercksen, and L. S. Cederbaum, *J. Chem. Phys.* **67,** 4124 (1977), Table IV.
44. L. S. Cederbaum, J. Schirmer, W. Domcke, and W. von Niessen, *J. Phys. B* **10,** L549 (1977).
45. W. B. England and W. C. Ermler, *J. Chem. Phys.* **70,** 1711 (1979).
46. R. N. Dixon, *Proc. R. Soc. London Ser. A* **275,** 431 (1966).
47. J. Schirmer, W. Domcke, L. S. Cederbaum, W. von Niessen, and L. Åsbrink, *Chem. Phys. Lett.* **61,** 30 (1979).
48. J. Pacansky and K. Hermann, *J. Chem. Phys.* **69,** 963 (1978).
49. S. Rothenberg and H. F. Schaefer, III, *J. Am. Chem. Soc.* **95,** 2095 (1973).
50. J. L. Gole, A. K. Q. Siu, and E. F. Hayes, *J. Chem. Phys.* **58,** 857 (1973).
51. J. L. Gole, *J. Chem. Phys.* **58,** 869 (1973).
51a. D. R. Yarkony, W. J. Hunt, and H. F. Schaefer, III, *Mol. Phys.* **26,** 941 (1973).
51b. P. Pulay, A. Ruoff, and W. Sawodny, *Mol. Phys.* **30,** 1123 (1975).
52. P. Pendergast and E. F. Hayes, *J. Chem. Phys.* **68,** 4022 (1978).
53. B. A. B. Seiders, W. L. Luken, and K. D. Jordan, *Int. J. Quantum Chem.* **14,** 741 (1978); B. A. B. Seiders, W. L. Luken, R. P. Blickensderfer, and K. D. Jordan, *Chem. Phys.* **39,** 285 (1979).
54. D. R. Yarkony and H. F. Schaefer, III, *Chem. Phys. Lett.* **15,** 514 (1972).
55. P. J. Hay, W. R. Wadt, L. R. Kahn, and F. W. Bobrowicz, *J. Chem. Phys.* **69,** 984 (1978).
56. H. Basch and S. Topiol, *J. Chem. Phys.* **71,** 802 (1979).
57. M. Boring, J. H. Wood, and J. W. Moskowitz, *J. Chem. Phys.* **63,** 638 (1975).
58. D. E. Ellis, A. Rosen, and P. F. Walch, *Int. J. Quantum Chem., Symp.* **9,** 351 (1975).
59. M. Krauss and D. Neumann, *Chem. Phys. Lett.* **14,** 26 (1972).
60. J. Almlöf, A. Lund, and K. Thomas, *Chem. Phys. Lett.* **28,** 179 (1974).
60a. C. Thompson and D. A. Brotchie, *Int. J. Quantum Chem., Symp.* **8,** 277 (1974).
60b. W. Bleicher and P. Botschwina, *Mol. Phys.* **30,** 1029 (1975).

61. S. Rothenberg and H. F. Schaefer, III, *Mol. Phys.* **21,** 317 (1971).
62. C. F. Jackels and E. R. Davidson, *J. Chem. Phys.* **63,** 4672 (1975).
63. P. J. Hay, *J. Chem. Phys.* **58,** 4706 (1973).
64. C. F. Jackels and E. R. Davidson, *J. Chem. Phys.* **64,** 2908 (1976).
65. C. F. Jackels and E. R. Davidson, *J. Chem. Phys.* **65,** 2941 (1976).
66. G. D. Gillispie, A. U. Khan, A. C. Wahl, R. P. Hosteny, and M. Krauss, *J. Chem. Phys.* **63,** 3425 (1975).
67. J. B. Coon, F. A. Cesarri, and F. P. Huberman, *J. Chem. Phys.* **52,** 1647 (1970).
68. P. A. Benioff, *J. Chem. Phys.* **68,** 3405 (1978).
69. P. J. Fortune, B. J. Rosenberg, W. B. England, and A. C. Wahl, *Theor. Chim. Acta* **46,** 183 (1977).
70. S.-K. Shih, S. D. Peyerimhoff, and R. J. Buenker, *Chem. Phys. Lett.* **46,** 201 (1977).
71. C. F. Jackels, *J. Chem. Phys.* **70,** 4660 (1979).
72. N. C. Handy, J. D. Goddard, and H. F. Schaefer, III, *J. Chem. Phys.* **71,** 426 (1979).
73. E. A. Andersen and J. Simons, *J. Chem. Phys.* **66,** 2427 (1977).
74. P. K. Pearson, H. F. Schaefer, III, J. H. Richardson, L. M. Stephenson, and J. I.
74a. L. A. Curtiss, S. R. Langhoff, and G. D. Carney, *J. Chem. Phys.* **71,** 5016 (1979).
75. K.-H. Thunemann, S. D. Peyerimhoff, and R. J. Buenker, *J. Mol. Spectrosc.* **70,** 432 (1978).
76. P. J. Hay and T. H. Dunning, Jr., *J. Chem. Phys.* **67,** 2290 (1977).
77. P. J. Hay, T. H. Dunning, Jr., and W. A. Goddard, III, *J. Chem. Phys.* **62,** 3912 (1975).
78. H. Basch, *J. Am. Chem. Soc.* **97,** 6047 (1975).
79. L. B. Harding and W. A. Goddard, III, *J. Chem. Phys.* **67,** 2377 (1977).
80. R. R. Lucchese and H. F. Schaefer, III, *J. Chem Phys.* **67,** 848 (1977).
81. S. Shih, R. J. Buenker, and S. D. Peyerimhoff, *Chem. Phys. Lett.* **28,** 463 (1974).
82. P. G. Burton, *J. Chem. Phys.* **71,** 961 (1979).
83. A. Devaquet and J. Ryan, *Chem. Phys. Lett.* **22,** 269 (1973).
84. S. Rothenberg and H. F. Schaefer, III, *J. Chem. Phys.* **53,** 304 (1970).
85. S. Huzinaga and M. Yoshimine, *J. Chem. Phys.* **68,** 4486 (1978).
86. B. Roos and P. Siegbahn, *Theor. Chim. Acta* **21,** 368 (1971).
87. U. Gelius, B. Roos, and P. Siegbahn, *Chem. Phys. Lett.* **4,** 471 (1970).
87a. L. S. Cederbaum, W. Domcke, and W. von Niessen, *Mol. Phys.* **33,** 1399 (1977).
88. C. W. Bauschlicher, Jr., H. F. Schaefer, III, and P. S. Bagus, *J. Am. Chem. Soc.* **99,** 7106 (1977).
89. V. Staemmler, *Theor. Chim. Acta* **35,** 309 (1974).
90. D. T. Clark, B. J. Cromarty, and A. Sgamellotti, *J. Electron Spectrosc.* **17,** 237 (1979).
91. T.-K. Ha, H.-U. Gremlich, and R. E. Bühler, *Chem. Phys. Lett.* **65,** 16 (1979).
92. S. P. So, *J.C.S. Faraday Trans. II* **75,** 820 (1979).
93. C. Thomson, *Theor. Chim. Acta* **32,** 93 (1973).
94. J. F. Harrison and C. W. Eakers, *J. Am. Chem. Soc.* **95,** 3467 (1973).
95. L. J. Lawlor, K. Vasudevan, and F. Grein, *J. Am. Chem. Soc.* **100,** 8062 (1978).
96. P. Rosmus, P. D. Dacre, B. Solouki, and H. Bock, *Theor. Chim. Acta* **35,** 129 (1974).
97. J. Kendrick, I. H. Hillier, and M. F. Guest, *Chem. Phys. Lett.* **33,** 173 (1975).
98. J. C. Cobb and A. Hinchliffe, *Chem. Phys. Lett.* **24,** 75 (1974); A. Hinchliffe and J. C. Cobb, *Chem. Phys.* **3,** 271 (1974).
99. P. J. Hay, *J. Am. Chem. Soc.* **99,** 1003 (1977).
100. B. Solouki, P. Rosmus, and H. Bock, *Chem. Phys. Lett.* **26,** 20 (1974).
101. E. F. Riedel and R. D. Willett, *Theor. Chim. Acta* **42,** 237 (1976).

102. S. R. Ungemach and H. F. Schaefer, III, *J. Am. Chem. Soc.* **98,** 1658 (1976).
103. M. F. Guest, M. B. Hall, and I. H. Hillier, *J.C.S. Faraday Trans. II* **69,** 1829 (1973).
104. B. D. Joshi and K. Morokuma, *J. Am. Chem. Soc.* **101,** 1714 (1979).
105. C. A. Coulson, *J. Chem. Phys.* **44,** 468 (1966).
106. P. S. Bagus, B. Liu, and H. F. Schaefer, III, *J. Am. Chem. Soc.* **94,** 6635 (1972).
107. P. S. Bagus, B. Liu, D. H. Liskow, and H. F. Schaefer, III, *J. Am. Chem. Soc.* **97,** 9216 (1975).
108. A. Rosén and D. E. Ellis, *Chem. Phys. Lett.* **27,** 595 (1974).
109. W. R. Wadt and P. J. Hay, *Appl. Phys. Lett.* **30,** 573 (1977); *J. Chem. Phys.* **68,** 3850 (1978); see also D. C. Lorentz, D. L. Huestis, M. V. McKusker, H. H. Nakamo, and R. M. Hill, *J. Chem. Phys.* **68,** 4657 (1978).
110. C. F. Bender and H. F. Schaefer, III, *Chem. Phys. Lett.* **53,** 27 (1978).
111. S. P. Walch and W. A. Goddard, III, *J. Am. Chem. Soc.* **98,** 7908 (1976).
112. P. S. Bagus and K. Hermann, *Solid State Commun.* **20,** 5 (1976).
113. D. T. Clark, B. J. Cromarty, and A. Sgamellotti, *Chem. Phys. Lett.* **55,** 482 (1978).
114. D. T. Clark, B. J. Cromarty, and A. Sgamellotti, *Chem. Phys. Lett.* **68,** 420 (1979).
115. A. P. Mortola and W. A. Goddard, III, *J. Am. Chem. Soc.* **96,** 1 (1974).

CHAPTER V

REACTION SURFACES

A. GENERAL CONSIDERATIONS

A large number of reaction surfaces involving many three-atom combinations have been computed using *ab initio* procedures at various levels of sophistication. Difficulties involved in the computation of such surfaces derive from the necessity of using a point-by-point procedure, i.e., a separate wave function calculation is required for each geometric configuration of the nuclei. In addition, the wave function must be suitable for describing the asymptotic regions of the surface such as the dissociation to an atom plus a diatomic molecule or to three atoms in appropriate electronic states. Consequently, for most reaction surfaces where bond cleavage or formation is involved, configuration mixing is imperative.

Whereas Chapters II–IV deal with *bona fide* triatomic molecules, many of the systems discussed here are metastable with respect to certain dissociation products. Properties of interest include the minimum-energy path for reaction, the location of barriers and saddle points on the surface and their heights and shapes (expressed in terms of force constants), and the exothermicity of reactions. The *ab initio* reaction surface often is subsequently employed, using methods of molecular dynamics, to yield reaction cross sections and other properties [1]. This subject is outside the scope of this book and will not be discussed. In most studies to date the *ab initio* reaction surfaces are published separately from the molecular dynamics studies, the latter utilizing, in general, these surfaces in parametrized forms.

The interaction of three hydrogen atoms and ions is discussed in the next section and the various hydrogen–helium interactions in the following section. The hydride systems AH_2, A_2H, and ABH are treated in Sections D–F and AB_2 and ABC surfaces in Section G. The last topic of this chapter deals with selected studies of "slices" of more complex reaction surfaces. There are a large number of studies of this nature, many of which are discussed in later chapters.

A review of *ab initio* potential energy surfaces by Bader and Gangi [2] contains an excellent analysis of procedures used for their calculations and analysis as well as a review of results to 1974. A somewhat more formal discussion with an application to the $O^+ + N_2$ reaction is given by Kaufman [3].

Peyerimhoff and Buenker [4] discuss the general problem of the calculation of potential energy surfaces using *ab initio* procedures. They make the following division of the general potential energy surface, which, although it is nonunique in that the parts may be overlapping, is useful for designing and analyzing computational approaches. The first region (a) corresponds to a small range of nuclear configurations in the vicinity of the equilibrium geometry of a stable molecular state. Such potential surfaces are useful for the prediction of spectroscopic constants, bond lengths, bond angles, and force constants. Calculations discussed in the earlier and following chapters of this book are mostly of this type in the sense that ground and excited state surfaces that provide the aforementioned constants as well as electronic transition energies are their primary objectives. The second region (b) may be identified by specifying that some particular path or slice of the total surface is singled out for study for the purpose of computing bonding energies, activation energies, and geometric properties of transition states. Calculations of these parts of potential surfaces are also discussed in the earlier and following chapters. The third region (c) is termed [4] the multidimensional surface. It is defined as the nearly complete locus of energy points corresponding to all possible nuclear configurations, from that of the system of noninteracting atoms infinitely separated to that of all nuclei forming a united atom. Although such a complete locus of points is unattainable, there are many computations recorded in the literature for three-atom systems that map out a large portion of the surface in all directions. The last region (d) is the most comprehensive and corresponds to several complete surfaces that are mutually interacting in many cases. Thus potential energy surfaces for several electronic states of a polyatomic system may be required for an accurate characterization of its chemical reactions. *Ab initio* studies of reaction surfaces of the categories (c) and (d) are reported for three-atom systems in this chapter. It should be noted that to date there are only a few systems for which surfaces of the type (d) have been studied. Devaquet *et al.* [5] give a discussion of the general problem of the interactions of electronic states on potential energy surfaces.

The results discussed in the remainder of this chapter as given in the following sections may be categorized as follows. Each of the next six sections deals with three-atom systems beginning with H_3 and proceeding to general ABC systems. Within each section the positive ionic, neutral, and negative ionic systems are discussed for the existing calculations. Attempts at the construction of potential energy surfaces appropriate for use in the applications of the methods of molecular dynamics [1] are more or less the basis for the studies reported in Sections B–G.

B. H_3

The potential energy surfaces computed for the various combinations of H atoms and ions are discussed here.

H_3^{2+} and H_3^+ both have stable equilateral triangular ground states (D_{3h}) with configurations 1a′ and $(1a')^2$, respectively. Potential energy surfaces for H_3^+ have been computed by several groups, the most accurate being by Carney and Porter [6], who characterized the surface primarily near equilibrium, and by Bauschlicher *et al.* [7].

Carney and Porter [6] carried out accurate CM calculations and determined the ground-state bond distance of H_3^+ to be $R = 0.914$ Å. They also computed vibrational energy levels and transition moments and gave charge density contour plots and dipole moment contours for geometries in the vicinity of equilibrium.

Bauschlicher *et al.* [7] treated the problem of the formation of H_3^+ along the two possible paths $H_2^+ + H$ and $H_2 + H^+$. In both cases the minimum energy path is for the atom or ion to approach the diatomic along a line perpendicular to that joining the nuclei in H_2 or H_2^+. The potential energy curves for the two processes undergo an avoided crossing, both being of $^1A_1$ symmetry. These authors examine the nature of the two surfaces using contour and perspective plots and compute the probability of surface hopping.

The reaction $H_2 + H \rightarrow H + H_2$ is of great interest to chemists since it

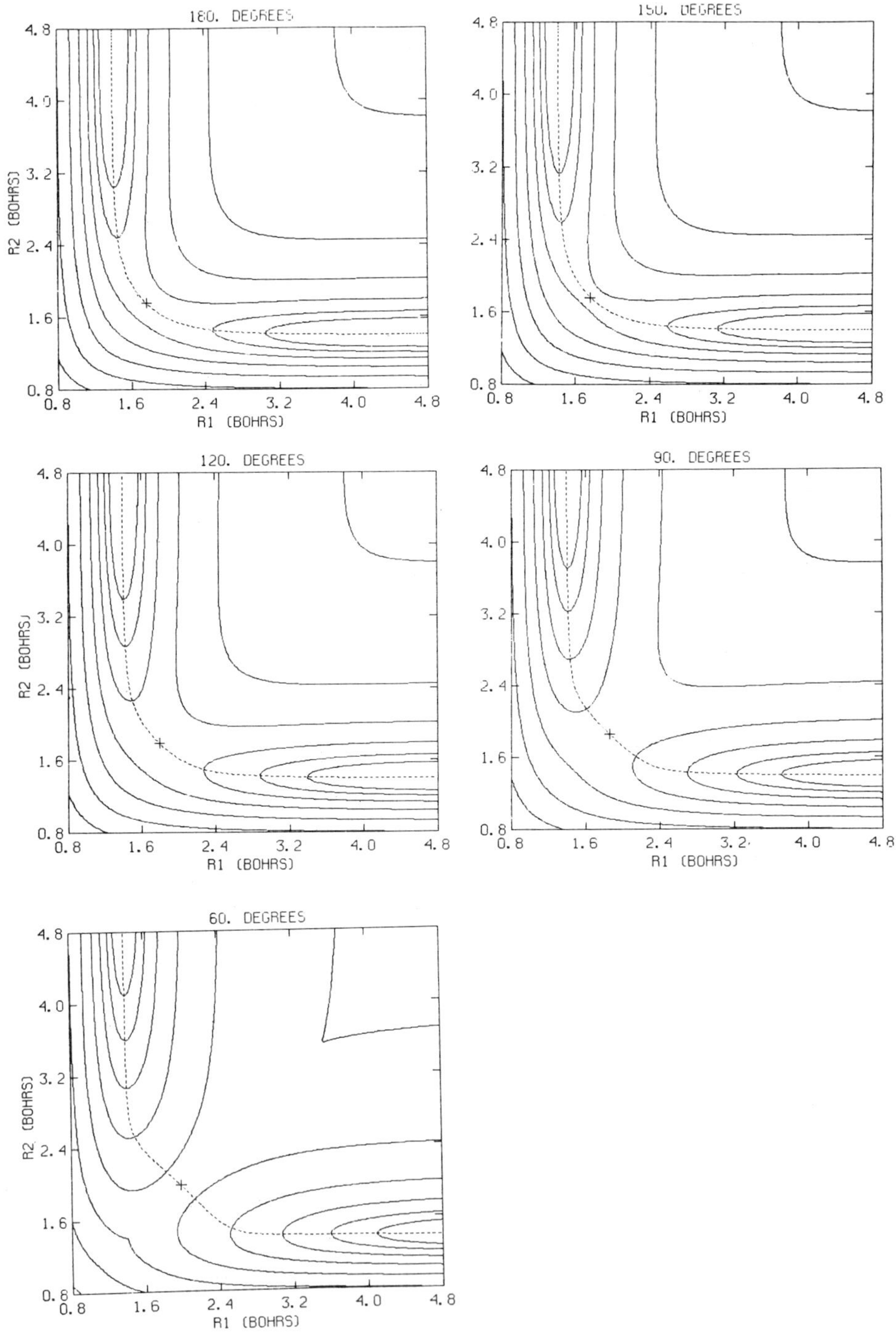

FIG. 1. Contours of the interaction potential for H + $H_2(R_1) \rightarrow H_2(R_2)$ + H. For each contour map, one internal bond angle is fixed. The fixed internal bond angle is indicated at the top of each figure, and the solid curves are contours of the interaction potential corresponding to 3, 6, 13, 24, 48, 96, 192, and 384 kcal/mol. The dashed curves are minimum-energy paths and the crosses are fixed-angle saddle points. From Truhlar and Horowitz [68].

is the simplest exchange reaction involving neutral species. The minimum-energy path was computed by Shavitt *et al.* in a landmark CM study [8], where it was confirmed to be a collinear one. Later calculations by Liu [9] using a large basis set of STFs and extensive CM for the collinear approach yielded a classical barrier height of 9.68 ± 0.12 kcal/mol at a symmetric saddle point geometry of 1.575 ± 0.001 a.u. Siegbahn and Liu [10], using an accurate basis of GTFs and complete CM using the direct CI method (see Section I.D), computed energies for 156 geometries on the three-dimensional potential surface. Their surface in the region $R_2 \leqslant R_1 \leqslant 3.5$ a.u., where R_1 and R_2 are the internuclear distances, is believed to be accurate to 0.1 kcal/mol. Truhlar and Horowitz [11] have fitted 267 points from the Liu [9] and Siegbahn and Liu [10] surfaces for H_3 to an analytical form, obtaining a root-mean-square deviation of 0.17 kcal/mol and a maximum deviation of 0.55 kcal/mol. The contours of the interaction potential for five fixed values of the internal angle are shown in Fig. 1. The H_3 surface calculated by Siegbahn and Liu provides the most accurate *ab initio* data for the complete characterization of any chemical reaction. Potential energy surfaces for H_3 computed prior to 1977 are evaluated in the context of their reliability for use in molecular dynamics applications by Truhlar and Wyatt [12].

Although there have not as yet been published accurate calculations of entire potential energy surfaces of excited states of H_3, two studies should be mentioned because of recent experimental interest [13]. Using a large basis set of GTFs in a UHF procedure, King and Morokuma [14] report energies, geometries, and transition moments for 15 Rydberg states of H_3. Carney [15] has calculated spectroscopic constants for these states.

The H_3^- system is found to be linear with a bond distance of 2.05 a.u. as determined from CM [16] and SCF calculations [17]. The energy barrier for the reaction of $H_2 + H^- \rightarrow H^- + H_2$ is found to be 11.55 kcal/mol using a CEPA procedure that corresponds to a lowering by 4.70 kcal/mol from the SCF value, an appreciable electron correlation effect [17].

C. HeH_2 AND He_2H

The HeH_2^{2+} system has been investigated by Zetik and Poshusta using a rather small basis set of GTFs with a 16-term CM procedure [18]. Contour plots are given for the two collinear reactions

$$H^+ + HeH^+ \rightarrow HHe^+ + H^+ \text{ and } He^+ + H_2 \rightarrow HeH^+ + H^+$$

They calculated activation energies of 6.5 and 4.1 eV for the first and second reactions, respectively.

An extensive open-shell SCF potential energy surface for the collinear HeH_2^+ system was first calculated by Brown and Hayes [19]. A minimum in the surface near $R_{HH} = R_{HeH} = 2.0$ a.u. of about 5 kcal/mol is calculated. They mapped a few points on the surface corresponding to angles smaller than 180° and observed a steady increase in energy, thus supporting the choice of a collinear reaction coordinate. More recently, Hopper has computed a potential energy surface for this system using CM [20]. The surface characteristics are quite similar to those of the earlier study with a minimum occurring at the same geometry and dissociation energies for the three asymptotes $He + H_2^+$, $H + HHe^+$, and $H + H^+ + He$ of 0.272, 1.028, and 3.032 eV, respectively.

Good SCF and CM calculations were reported for the HeH_2 potential surface as early as 1970 [21]. The SCF results are expected to be reasonable for certain reaction pathways in this system where closed-shell reactants and products are involved. Tsapline and Kutzelnigg [22] computed the collinear potential energy surface using the IEPA PNO method and a large basis set of GLFs. The inclusion of electron correlation results in a van der Waals minimum of about 20°K at a geometry of $R_{H_2} = 1.406$ a.u. and $R_{HHe} = 6.5$ a.u. A saddle point of 14°K having a force constant of 1.2×10^{-3} mdyn/Å is computed for the perpendicular approach of He at 6.29 a.u. to H_2 having $R = 1.406$ a.u.

Recently Romelt *et al.* [23] have reexamined the HeH_2 system using the MRD-CI method with an extended basis of GTFs. They examined collinear and orthogonal approaches of He in its ground ^1S state to H_2 in the $X\,^1\Sigma_g^+$, $b\,^3\Sigma_u^+$, $B\,^1\Sigma_u^+$, $a\,^3\Sigma_g^+$, $c\,^3\Pi_u$, and $C\,^1\Pi_u$ electronic states. All internuclear distances were varied for the surfaces for each of the states of H_2. Of particular interest is the discussion of the nature of the interactions of the excited state surfaces, especially the quenching of the B state fluorescence in HD by He atoms.

Poshusta and Siems reported *ab initio* valence bond calculations for the He_2H^+ system using a basis set of GLFs of approximately double-zeta plus polarization function quality [24]. Their wave functions included 30 valence bond configurations and predict a linear symmetric equilibrium geometry of 1.764 a.u. and a dissociation energy of 0.456 eV.

Milleur *et al.* computed potential energy surfaces for He_2H^+ and He_2H at the SCF level of approximation using an extended basis set of GTFs [25]. The He_2H^+ complex was found to have a stable collinear geometry in agreement with the earlier calculations [24].

D. AH_2

The first calculations in the LiH_2^+ potential energy surface were carried out by Lester using an SCF procedure [26]. Kutzelnigg *et al.* performed complementary SCF and IEPA-PNO calculations for the $Li^+ + H_2$ approach for both perpendicular geometries and certain nonperpendicular geometries [27]. Both studies used relatively large Gaussian basis sets including polarization functions. An absolute minimum of 0.22 eV relative to the separated atoms occurs for the perpendicular C_{2v} geometry for the H—H distance r = 1.4 a.u. and the Li to H_2 distance of R = 3.75 a.u. A saddle point was located at r = 1.4 a.u. and R = 4.7 a.u. in the collinear arrangement with a binding energy of 0.06 eV.

Mizutani *et al.* [28] report MCSCF plus CM calculations for the interactions of Li (^2S) and Li (^2P) with H_2. A primary motivation was the study of the phenomenon of the quenching of excited atomic states by diatomic molecules. They used a minimal (STO-3G) basis set and considered only the perpendicular approach of Li to the bond midpoint of H_2. The surface for the excited $^2B_2$ was found to possess two minima. They present a natural orbital analysis and tabulated the wave functions for selected nuclear configurations. Several interesting features were noted, suggesting further calculations using an extended basis set.

To explain features of molecular beam experiments involving the C^+ + H_2 ion–molecule reaction, Liskow *et al.* computed potential energy surfaces for six electronic states of this system [29]. They employed a double-zeta quality basis set of GTFs and included CM for each state. For the perpendicular approaches three reference configurations were used for the $^2A_1$ ground state, two for the $^2B_1$ state, and one for the $^2B_2$ state. For linear nonsymmetric geometries one reference configuration was used for the $^2\Sigma^+$ state and two for the $^2\Pi$ state. For the asymmetric geometries two reference configurations were used for the $^2A_1$ state. In each case the MOs for the reference configurations were obtained via an INO procedure. All single and double replacements with respect to the reference configurations were used in the CM wave functions for each symmetry. Saddle points for the perpendicular approaches were calculated to be 85.7 kcal/mol at r = 2.34 a.u. and R = 2.94 a.u. for the $^2A_1$ state and 62.8 kcal/mol at r = 2.33 a.u. and R = 3.00 a.u. for the $^2B_1$ state, where r is the H—H distance and R the C^+ to bond midpoint distance. The $^2B_2$ state was calculated to be repulsive at all geometries. The barriers given above proved to be in accord with the Woodward–Hoffmann rule for conservation of electron configuration. The linear CHH^+ calculations showed that for the $^2\Sigma^+$ surface there is a barrier of 28.4 kcal/mol for R_{CH} = 2.51 a.u. and R_{HH} = 2.11 a.u. No barrier was found for the $^2\Pi$ surface.

The calculations of Bender *et al.* [30] on the $N^+ + H_2$ system were discussed in Section II.E. A double-zeta plus polarization basis set of GTFs was used with extensive CM for the lowest $^3A_2$ and $^3B_1$ surfaces. The interaction of these two surfaces allows the reaction to proceed without having to surpass a barrier. The $^3A_2$ surface was found to have an absolute minimum of 60 kcal/mol at $R_{NA} = 1.26$ Å and an HNH angle of 43°. As in the earlier study of CH_2^+ [29], a comparison of the *ab initio* results with qualitative arguments based on Mulliken–Walsh correlation diagrams was generally favorable. Plots of contours of both surfaces showing the line of intersection are presented. An interesting feature of this reaction is that for nonsymmetrical approach (C_s symmetry) the states are both of $^3A''$ symmetry and the intersection becomes "avoided." The minimum pathway favors the C_{2v} approach on the $^3A_2$ surface.

Hirst has reported SCF calculations for the $^5\Sigma^-$, $^5A''$, and $^5A_2$ surfaces of NH_2^+ using a GTF basis set of double-zeta plus polarization function quality [31]. No barrier was observed for the collinear and near-collinear approaches to the formation of $NH^+(^4\Sigma^-) + H$, while the surface remained repulsive for C_{2v} approaches.

Bader and Gangi discussed the procedure for constructing potential surfaces for the interaction of oxygen atoms in their ^3P and ^1D states with ground state H_2 to form H_2O in the $^3B_1$ and $^1A_1$ states [2]. They attempted to generate reasonably accurate surfaces by augmenting the single determinant wave functions for the system with only those determinants that permit proper dissociation to the desired asymptotic limits. They caution against adding additional configurations indiscriminately. As we have discussed in Chapter I, the choice of CSFs to be used in the calculation of a molecular wave function must be "balanced" in the same sense that the atomic basis must be in order to avoid the spurious transfer of charge density onto the atom or atoms that are more accurately represented. Bader and Gangi also report UHF calculations for these surfaces.

Using basis sets of about double-zeta size (with and without polarization functions) for the symmetric insertion of $O(^1D)$ into H_2, proceeding to the linear HOH form they find a barrier of 33 kJ/mol above that of the equilibrium geometry. The C_{2v} approach was found to be that of lowest energy until the O atom reached about 3 a.u. from the H_2 bond midpoint. In the region from 2–3 a.u. a double mimimum character was observed with the minimum-energy approach being about 30° away from the C_{2v} axis of approach. For distances closer than 2 a.u. the C_{2v} approach is again favored. The insertion of $O(^3P)$ into H_2 along the C_{2v} axis was found to have a barrier of 80 kcal/mol using a similar procedure.

A comprehensive study of the $^1A'$, $^3A''$, $^1A''$, and B $^1A'$ surfaces of H_2O was carried out by Howard *et al.* using a basis set of GTFs of double-zeta plus polarization function quality and at the first-order CM INO level of approximation [32]. In addition, basis functions with small ζ values were included to represent the Rydberg character of the excited states. The results differed in many respects from those of Bader and Gangi. For the ground $^1A_1$ surface energies for 310 geometries were computed. No barrier to insertion of O(^{1}D) to ground-state H_2 was observed. The barrier for the continued insertion to form linear HOH was nearly the same as that of Bader and Gangi although the pathway decreased more steeply. However, the insertion of O(^{3}P) was found to proceed with a barrier of about one-half the magnitude of that computed by Bader and Gangi. A total of 307 geometries comprised this surface, which was found to be totally repulsive. The collinear approach of O to H_2 showed a saddle point at R_{OH} = 1.15 Å and R_{H_2} = 0.95 Å having a barrier of 14.0 kcal/mol, about one-half the value determined by Bader and Gangi. The shape of this saddle point is reflected in the computed frequencies of 569 and 787 cm^{-1} for the symmetric stretch and bend and the (imaginary) dissociative asymmetric stretch value of 6724 cm^{-1}.

Walch *et al.* report a complementary study of the O(^{3}P) + H_2 → OH + H surface using a somewhat larger basis set than that used by Howard *et al.* but including CM in nearly the same manner [33]. They compared the effects of choice of basis set, MOs, and reference configurations on the calculated barrier height and saddle point geometry for the collinear approach. Their most extensive calculations resulted in a barrier height of 12.5 kcal/mol at R_{OH} = 1.23 Å and R_{HH} = 0.92 Å.

Mahan *et al.* [34] have computed potential energy surfaces for F^+ + H_2 and F + H_2^+ using a double-zeta plus polarization basis set and CM with all single and double replacements with respect to one or two reference configurations. The computed correlation diagram for the reactants F(^{2}P) + H_2^+ and F^+(^{3}P) + H_2 for the C_{2v} approach to $D_{\infty h}$ HFH^+ indicates that all excited states are quite high (this is contrasted with the CH_2^+ and NH_2^+ cases) in relative energy, nearly 13 eV, near the equilibrium geometry of the ground state. They conclude that the important features of the surfaces occur considerably more toward the side of the reactants. The $^3A_2$ state for the approach of F^+ to H_2 showed only a small minimum of 0.8 kcal/mol at R_{HF} = 2.9 Å and an HFH angle of 15°. The $^3B_1$ and $^3B_2$ states for this approach were even less attractive, indicating the unlikelihood of it leading to reaction. For the F + H_2^+ approach the $^3B_1$ is predicted to be the lowest energy state with R_{FH} = 1.67 Å and an HFH angle of 31.6° and a dissociation energy of 17.5 kcal/mol. The $^3B_2$ was found to be strongly bound as well with bond distance, bond angle, and dissociation energy of 1.72 Å, 30.3°, and 14.5 kcal/mol, respectively.

Bender *et al.* carried out first-order CM calculations for 150 geometries on the collinear FH_2 system using a double-zeta plus polarization basis set of GTFs [35]. The large number of experimental studies on this system, in particular the F + H_2 reaction, make it an important case for testing the validity of *ab initio* potential surfaces. A comparison of results obtained at the SCF and correlated levels of approximation using the same basis set showed the importance of electron correlation for this surface. The SCF values for the barrier height, exothermicity, and saddle point geometry were 29.3 kcal/mol, 13.2 kcal/mol, and R_{FH} = 1.18 Å and R_{HH} = 0.84 Å, respectively. The corresponding values derived from the first-order CM surface were 1.66 kcal/mol, 34.4 kcal/mol, and R_{FH} = 1.54 Å and R_{HH} = 0.77 Å. Experimental values of 1.7 and 31.8 kcal/mol were reported for the barrier height and exothermicity. The failure of the SCF approximation is striking. In addition to the contour maps of the computed potential surface shown in Fig. 2, we reproduce the perspective plots of this surface in Fig. 3, in which the entrance (H_2 + F) and exit (H + HF) channels are clearly depicted [35]. Finally, to complete the detailed picture of this surface the energies and internuclear separations for the minimum-energy path are given in Table 1.

Subsequent studies on the FH_2 surface have been carried out. Jaffe *et al.* [36] examined the effects of spin–orbit coupling as well as the nature of the lowest excited state surfaces using SCF and CM wave functions. They investigated points on the surface corresponding to complex values of the nuclear coordinates. Rebentrost and Lester [37] calculated potential surfaces for the $1^2A'$, $2^2A'$, and $^2A''$ states of FH_2 using an extended basis set of GTFs, including polarization functions, at the SCF level of approximation. They included nonlinear geometries and, with knowledge of the poor quality of the SCF results in the region of close approaches of F to H_2 [35], concentrated on the longer-range interactions. Botschwinna and Meyer [38] carried out PNO-CEPA calculations using an extended basis including polarization functions.

Wadt and Winter [39] extended the calculations of Bender *et al.* [35] to nonlinear geometries. They used a basis set of double-zeta plus polariza-

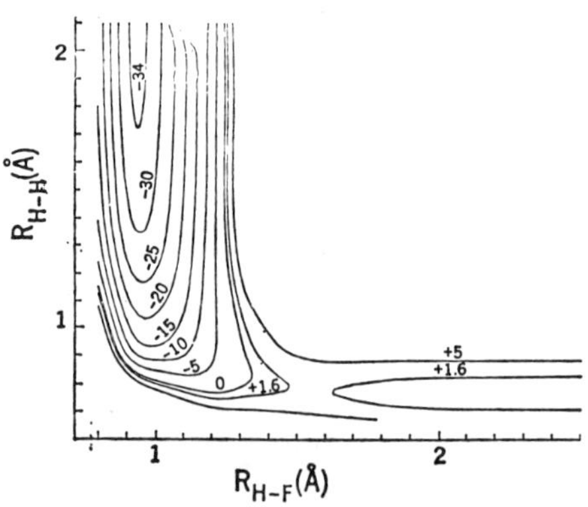

FIG. 2. Traditional contour map of the linear FH_2 potential energy surface. From Bender *et al.* [35].

TABLE 1
Minimum Energy (E) Path for $F + H_2 \rightarrow FH + H$

R_{HF}	R_{HH}	E (kcal/mol)	System
∞	1.42	0.00	Reactants
6.0	1.42	0.00	
4.0	1.42	0.25	
3.6	1.43	0.44	
3.4	1.43	0.75	
3.2	1.44	1.27	
3.0	1.44	1.61	
2.90	1.45	1.66	Saddle point
2.8	1.48	1.59	
2.63	1.5	1.18	
2.36	1.6	− 2.15	
2.03	1.8	−12.30	
1.85	2.0	−20.70	
1.80	2.2	−25.30	
1.78	2.5	−29.15	
1.77	3.0	−32.91	
1.76	∞	−34.44	Products

[a]Distances (R) are in Bohr radii; 338-configuration first-order wave functions were used.

tion quality and calculated wave functions using the POL-CI method in which MOs were obtained using the GVB procedure. Their results for collinear geometries were in close agreement with those of Bender *et al.* Their results indicate that the basis sets for the FH_2 should include functions with small ζ to describe the admixture of Rydberg [35] character to the wave function in the vicinity of the saddle point. The surface was found to be quite flat with respect to the noncollinear C_{2v} approaches of F to H_2 with a variation of only $\sim$2 kcal/mol for the energies of HFH at angles from 109° to 180°. The saddle point geometry was predicted to be R_{HF} = 1.08 Å and HFH angle of 106° with a barrier height of 44 kcal/mol. The near constancy of the saddle point for angles greater than 90° is illustrated in the perspective plot shown in Fig. 4 [39]. The solid line indicates the loci of the minima.

The H_2F^- system has been studied by Baybutt [40] in the SCF approximation using an extended basis set of GTFs including polarization functions. A stable collinear complex was determined with R_{HH} = 1.44 a.u. and R_{HF} = 3.66 a.u. The stabilization energies for the forward and reverse of the reaction $H^- + HF$ were calculated as 33.42 and 5.32 kcal/mol, respectively. The surface was extended to larger internuclear distances

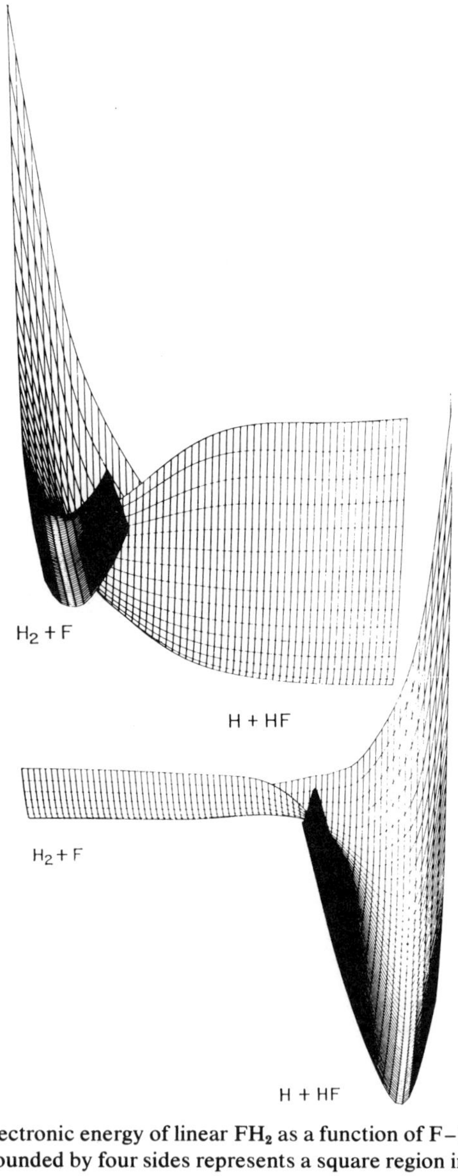

FIG. 3. (Top) Electronic energy of linear FH_2 as a function of F–H and H–H distances. Each small section bounded by four sides represents a square region in space, 0.05 bohr on a side. (Bottom) Same as (top) but viewed from the exit channel FH + H. From Bender *et al.* [35].

by Amos [41] using an MCSCF procedure with a basis set of GTFs of double-zeta plus polarization quality. Thirty-five geometries along the collinear $H^- + HF \rightarrow HF^- + H$ reaction path were computed. The equilibrium geometry of the complex was found to be $R_{HH} = 1.5$ a.u. and $R_{HF} = 3.5$ a.u. with a well depth of 7.5 kcal/mol with respect to the reverse reaction.

Two SCF studies of the NeH_2^+ surface have been published. Bolotin *et al.* [42], using a UHF procedure with minimal and double-zeta quality basis sets of STFs, predicted a collinear-bound NeH_2^+ complex. Energy surfaces for the linear symmetric, linear asymmetric, and triangular (C_{2v}) approaches are reported. Hayes *et al.* [43] reported open-shell SCF calculations for the NeH_2^+ system using a basis set of GTFs that was of slightly better than double-zeta plus polarization quality. A surface comprised of 47 collinear geometries was calculated and fitted using a diatomics-in-molecules procedure. They compute an endoergicity of 12 kcal/mol for the $Ne + H_2^+ \rightarrow NeH^+ + H$ reactions, which agress well with the experiment.

Birks *et al.* published a potential energy surface for NeH_2 for the colinear and perpendicular C_{2v} approaches [44]. A double-zeta quality basis set of GTFs was used in a two-configuration MCSCF-CI procedure. Of a total of 63 geometries considered, 42 were for the collinear NeH_2 system and 21 were for the C_{2v} approach of Ne to H_2. A fit of the surface to an analytical form was given. It was observed that for the C_{2v} approach Ne

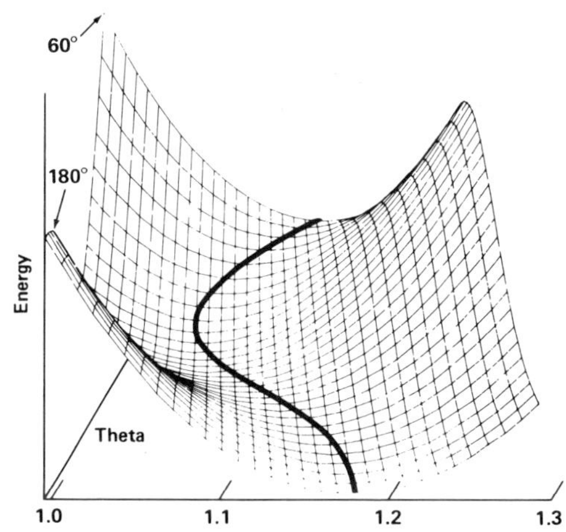

FIG. 4. POL (2) CI potential energy surface for the symmetric stretch and bending motions of the HFH transition state. From Walt and Winter [39].

exerts a contractive force on the H_2 bond for distance greater than about 3 a.u. Inside this distance a stretching force is exerted.

The Cl + $H_2 \rightarrow$ ClH + H reaction surface was computed by Rothenberg and Schaefer [45] using a GTF basis set of double-zeta plus polarization quality in the SCF approximation. From a total of 38 geometries it was determined that the minimum-energy path was collinear with an energy barrier of 26 kcal/mol at R_{ClH} and R_{HH} distances of 1.46 and 0.94 Å, respectively. Using a GTF basis of nearly the same quality, Dunning carried out calculations on the ClH_2 system using a GVB plus CM procedure that included single and double replacement configurations [46]. He obtained a barrier height of 8.1 kcal/mol for R_{ClH} and R_{HH} distances of 1.53 and 0.92 Å, respectively. This barrier height, considerably lower than that obtained using the SCF approximation, is expected to be more accurate. Also reported is the saddle point geometry (R_{HClH} = 1.52 Å) and barrier height (25.3 kcal/mol) for the collinear exchange reaction HCl + H $\rightarrow$ H + ClH. The calculated exothermicity for the abstraction reaction was 5.8 kcal/mol as compared with the experimental value of 3 kcal/mol.

E. HA_2

The simplest prototypical potential energy surface for A_2H systems for which chemical reactions come into play is that for H + $Li_2 \rightarrow$ LiH + Li A highly accurate study of this surface was carried out by Siegbahn and Schaefer [47]. A basis set comprised of three s- and one p-type contracted GTFs for the H and four s- and three p-type contracted GTFs for the Li was used. Both SCF and complete valence electron (frozen Li 1s shell) configuration mixing procedures were used for linear and nonlinear geometries. The absolute minimum of the potential surface was found to be for a C_{2v} arrangement with a bond angle of 95°, i.e., R_{LiLi} = 2.53 Å and a H to Li_2 bond midpoint distance of 1.16 Å. This geometry lies 22.4 kcal/mol below the LiH + Li asymptotic energies and is of $^2A_1$ symmetry. A small well of 4.2 kcal/mol was found for the collinear approach at R_{HLi} = 1.63 Å and R_{LiLi} = 3.32 Å. They estimate the position of the electron jump between the neutral and ion-pair surfaces (the latter lying below the former after this distance) at R_{HLi} = 5.5 a.u. and R_{LiLi} = 5.7 a.u. for the collinear approach. Contour maps and a natural orbital analysis of the correlated wave functions are reported.

England *et al.* [48] reported SCF and MCSCF plus CM results for the collinear Li + LiH and C_{2v} Li_2 + H surfaces. A basis set of nearly the same quality as that of Siegbahn and Schaefer [47] was used. SCF results were given for the 26 lowest lying states of the system, while MCSCF-CM results for the lowest states of $^2A_1$, $^2B_2$, and $^2\Sigma$ symmetries were reported. Electron jump geometries for six states are reported. All occur for C_{2v} approaches at angles between about 50° and 80° [48a]. SCF correlation diagrams are reported for low-lying linear and bent states [48b].

Open-shell SCF calculations have been published by Baird on the HN_2 surface using a 4-31G basis set [49]. An optimum geometry of R_{NN} = 1.18 Å, R_{HN} = 1.02 Å, and HNN angle of 118° was determined. The activation energy for the decomposition to H + N_2 was computed as 22.6 kcal/mol. The use of a small basis set as well as the restricted Hartree–Fock approximation reduce the credibility of this study to some degree.

Melius and Blint computed a potential surface for the nonlinear H + $O_2 \rightarrow HO_2^* \rightarrow$ OH + O reaction using an MCSCF-CM procedure [50]. A basis set of contracted GTFs of double-zeta plus polarization function quality was employed. The MCSCF procedure included the 20 configurations necessary to describe the orbitals of either the O_2, the OH, or the HO_2 molecule. Using a selection procedure similar to that of the MRD-CM procedure, a total of 7406 CSFs were used to define the wave functions. A total of 406 geometries were sampled and the resulting surface fitted to an analytical form. They find that the reaction is endothermic but that HO_2 is bound by 2.0 eV with respect to reactants and by 2.6 eV with respect to products. The equilibrium geometry of R_{OO} = 2.58 a.u., R_{OH} = 1.88, and HOO angle of 104.2° is in excellent agreement with experiment. A small barrier of less than 0.1 eV at 120° was computed for the incoming channel. Contour diagrams are given.

Bender, Schaefer, and co-workers have computed potential surfaces for the HF_2 system at several levels of approximation [51]. The careful SCF, INO, and first-order CM calculations, using a basis set of double-zeta plus polarization quality, indicated a collinear HFF minimum-energy pathway. A saddle point geometry of R_{HF} = 1.168 Å and R_{FF} = 1.50 Å was found. A barrier height of 4.1 kcal/mol and an exothermicity of 99.0 kcal/mol agree well with the corresponding experimental values of 2.4 ± 0.2 and 102.5 ± 2.8 kcal/mol, respectively. Calculations for the F + HF collinear surface yield a symmetrical saddle point geometry of 1.099 Å and a barrier height of 23.9 kcal/mol [51b].

Umeyama *et al.* reported SCF calculations for the collinear F^- + HF $\rightarrow$ FHF^- reaction using a 4-31G basis set [52]. The optimum bond distance in FHF^- was found to be 1.146 Å and no barrier from product to reactants was observed. An energy decomposition analysis was carried out which showed that the electrostatic component was predominant for three paths, including the one corresponding to the reaction coordinate. Keil and Ahlrichs [17] used the CEPA-PNO method to study the same reaction. A basis set of GLFs of considerably better than double-zeta plus polarization quality was used. The computed energy of formation was 44.32 kcal/mol.

Dedieu reports SCF calculations for H + Cl_2 using a basis set including polarization functions [53]. Baybutt *et al.* calculated surfaces for H + Br_2 in which GVB-CI methods were used with basis sets of contracted GTFs of double-zeta and double-zeta plus polarization function quality. In addition, *ab initio* effective core potentials were used to reduce the problem to one of 15 valence electrons. An interesting comparison with the earlier studies on H + F_2[51] and H + Cl_2 [53] is made. A classical barrier of 3 kcal/mol is computed for a linear geometry of R_{HBr} = 2.74 a.u. and R_{BrBr} = 4.51 a.u. Orbital contour plots are shown for several points along the collinear reaction path. Several noncollinear calculations were performed.

F. HAB

The interaction of He with LiH has been studied using three different basis sets of STFs and a diagrammatic Rayleigh–Schrödinger perturbation expansion procedure [55]. Comparisons with earlier SCF studies of this system are made. While the Li—H distance was kept fixed, the approach of He from several different angles was investigated and an analytical fit was presented. The object of the study was a collaborative theoretical and experimental effort to examine rotational energy transfer during inelastic collisions.

An earlier study of this problem was carried out by Gelb *et al.* on the Ne + H_2 and He + HF systems [56]. Basis sets of contracted GTFs of double-zeta quality were used in the SCF and CM (for selected geometries) approximations. Analytical representations of the surfaces were given.

Possibly the simplest HAB reaction that involves bond breaking is the Li + FH → LiF + H system. It was investigated as early as 1970 by Lester and Krauss in an SCF procedure [57]. They observed an attractive well attributed to the hydrogen bond. Balint-Kurti and Yardley used both semiempirical and *ab initio* procedures to study the linear and bent regions of the surface [58]. The most definitive study of this system is by Chen and Schaefer [59]. A basis set of contracted GTFs that included both small exponent p-type functions (for the description of F^-) and polarization functions was used in SCF and CM calculations. The latter included all single and double replacements with respect to the ground $^2A'$ configuration of LiHF. A total of 214 geometries for $C_{\infty v}$ and C_s approaches were chosen. Table 2 contains a summary of the computed re-

TABLE 2

Geometry, Energy, and Dipole Moment of LiFH at Stationary Points along the Minimum-Energy Pathway from Reactants to Products

System		ΔE (kcal/mol)	Dipole moment (D)
Reactants	Li + H—F		
	R_{HF} = 0.932 Å	0	1.9
Complex	LiFH		
	R_{FH} = 0.942 Å,		
	R_{LiF} = 1.947 Å	−4.6	3.4
	LiFH angle 114°		
Transition state	Li FH		
	R_{FH} = 1.291 Å		
	R_{LiF} = 1.699 Å	10.0	5.1
	LiFH angle 74°		
Products	LiF + H		
	R_{LiF} = 1.605 Å	2.9	6.7

[a] From Chen and Schaefer [59].

sults for points of interest along the reaction coordinate. A series of contour maps for several LiFH angles are presented as are contours for fixed Li–F distance and varying angle. Analysis of the surfaces, including zero-point vibrational corrections, resulted in a calculated threshold for reaction of 6.4 kcal/mol. The authors suggest that this value may be reduced to about 4 kcal/mol by additional electron correlation effects.

Schor *et al.* have published an extensive SCF and CM study of the collinear Be + FH → BeF + H surface using four different basis sets of STFs [60]. The best results derive from first-order CM calculations using an INO procedure with a medium-sized basis set to define the MOs. An exothermicity of 6 kcal/mol was computed for the overall reaction. The saddle point geometry corresponds to a point at which both internuclear separations are extended by 0.4 a.u. from their asymptotic equilibrium bond distances, and lies 28 kcal/mole above the reactants. The dynamics of the reaction were studied over a range of relative translational energies using a quasi-classical trajectory method.

Losonczy *et al.* have carried out SCF calculations for the interaction of Ne with HF [61]. A basis set of contracted GTFs of approximately double-zeta plus polarization quality was used for several distances and angles of the approaching Ne atom to a HF molecule with fixed bond length. A weak hydrogen bond of 0.23 kcal/mol along the collinear approach at a H—F distance of 6.375 a.u. was observed.

G. ABC

The F + $Li_2 \rightarrow$ FLi + Li reaction has been studied by Pearson *et al.* using a basis set of contracted GTFs of double-zeta plus polarization quality [62]. In addition, a small exponent p-type function was included to account for F^- character. SCF and two-configuration GVB calculations were carried out for collinear, symmetric (C_{2v}) and asymmetric (C_s) approaches of F to Li_2. Most of the reaction surfaces discussed to this point were "repulsive," i.e., a finite potential barrier to reaction exists and most of the exothermicity of the reaction is released as the products separate. The F + Li_2 reaction provides an example of an "attractive" surface, where the exothermicity is released as the reactants approach and there is no barrier to reaction. Thus, although the SCF approximation is not expected to be at all appropriate for bond breaking processes, Pearson *et al.* indicate that for attractive reaction surfaces it may lead to reasonable results. Although the minimum-reaction path was found to be for the collinear approach, it was observed that the most stable geometry for this approach was bound by 4 kcal/mol with respect to LiF + Li, whereas the C_{2v} Li_2F complex was bound by 34 kcal/mol for a bond distance of 1.79 Å and bond angle of 99°. Along the reaction coordinate (minimum-energy path) there appeared significant lengthening of the Li_2 bond until the F–Li distance reached about 5.8 a.u.; thereafter the Li_2 distance remained nearly constant over a considerable range. This was attributed to an electron "jump" from the covalent FLi_2 to the ionic $F^-Li_2^+$ surface. At a F–Li distance of about 3.2 a.u. the Li_2 distance begins to increase as the exit channel is mapped out. These authors point out the possible importance of interactions with excited electronic states of the FLi_2 system.

Schmidt *et al.* have reported "frozen orbital" SCF potential surfaces for the $NeHe_2$ system [63]. Closed-shell SCF calculations were performed for the neutral $NeHe_2$ system, and the ion surface was constructed through the use of Koopmans' theorem. Although this does not allow for SCF orbital relaxation in the ionic species, it was claimed tht the frozen orbital surfaces are nearly parallel to the open-shell SCF surfaces. A basis set of GTFs comprised of seven s-type and four p-type functions on Ne and four s-type functions on He was used. A total of 250 geometries corresponding to collinear and perpendicular (C_{2v}) approaches of Ne to He_2 were chosen. Five electronic states, defined in C_s symmetry, were investigated. Although this study is quite extensive from the standpoint of the large number of nuclear configurations chosen and the investigation of several electronic states, the use of such a small basis set combined with the frozen orbital approximation detracts from its ultimate reliability.

Potential energy surfaces for the Li + $F_2 \rightarrow$ LiF + F reaction have been published [64]. Both an *ab initio* multistructure valence bond and a so-called orthogonalized Moffit method were employed; the latter may be categorized as being semiempirical. A basis set of contracted GTFs of minimal quality was used together with two sets of composite valence bond functions made up of 123 and 204 functions. Calculations for the two linear cases LiFF and FLiF, the C_{2v} case of Li approaching F_2, and the case of Li approaching one F atom perpendicular to the F_2 bond are reported. This reaction is similar to the F + Li_2 reaction discussed above in that it is of the attractive type. While the accuracies of the potential surfaces are limited by the use of a minimal basis set, the extensive analysis given makes this study a worthwhile source for the investigation of more accurate procedures.

Thomas *et al.* have computed a CM potential energy surface for the Li^+ + CO reaction using a large basis set of contracted GTFs including polarization functions [65]. The CM calculations included all single and double substitutions with respect to an SCF-MO configuration for the three-atom system. In addition, the effects of quadruple substitutions were included using the Davidson–Langhoff procedure (Chapter I). A rigid rotor surface, i.e., fixed CO bond distance, was augmented with sets of geometries for two additional CO distances. Approaches of the Li^+ ion along five directions (0°, 45°, 90°, 135°, and 180°) with respect to the CO axis were investigated to yield a potential surface comprised of 83 geometries. This surface was then fitted to an analytical form and classical differential cross sections were computed. Calculations by Staemmler [66] on portions of the $LiCO^+$ as well as the $LiN_2{}^+$ surfaces should be noted (also see Chapter IV).

SCF calculations for the HeF_2 system have been published by Noble and Kortzeborn [67]. A basis set of contracted GTFs of slightly better than double-zeta quality was used. A few geometries were computed using a larger basis set including d-type functions on the F atoms. Energies for many nuclear configurations, mostly for the collinear approach of He, are tabulated. (SCF calculations of HF_2 and $HF_2{}^-$ are also reported in this study.)

Open-shell SCF calculations of states of the N_2O^+ system of ten (linear) symmetries have been carried out by Hopper [68]. A basis set of contracted GTFs of double-zeta quality was used. Bending curves for certain of the states are given. Particular emphasis is placed on the establishment of a mechanism for the reaction $O^+ + N_2 \rightarrow NO^+ + N$. A saddle point of about 0.1 eV was found for the geometry R_{NN} = 1.3 Å, R_{NO} = 1.3 Å, and NNO angle of 120° for the reaction proceeding on the $1\,{}^4A''$ surface. The lack of polarization functions in the basis set and the need for the inclusion of configuration mixing shed some doubt on the reliability of these calculations. Several aspects of these potential surfaces have also been discussed by Kaufman using configuration mixing with a minimal basis set of STFs [3].

The minimum-energy path for the reaction $N + O_2 \rightarrow NO + O$ has been investigated by Benioff *et al.* [69a] and Das and Benioff [69b]. They performed MCSCF-CM calculations using basis sets of contracted GTFs both with and without polarization functions. It was concluded [69a] that while the barrier height was not significantly altered by the inclusion in the basis set of polarization functions, there is the necessity to go beyond the valence CI approximation to at least a first-order level (see Chapter I). A barrier height of 21.7 kcal/mol for the surface of $^2A'$ symmetry was determined in the earlier study [69a]. The calculations were refined by a new method, termed the "configuration group approach," that permits the systematic selection of configurations to be included in the wave functions [69b]. Using a basis set augmented by a single d-type function on each center, the barrier height was computed as 13.8 kcal/mol at a geometry of $R_{NO} = 3.1$ a.u., $R_{OO} = 2.61$ a.u., and NOO angle of 109.8°. By a similar procedure a barrier of 18.1 kcal/mol was computed for the lowest $^4A'$ surface at a geometry of $R_{NO} = 3.3$ a.u., $R_{OO} = 2.64$ a.u., and NOO angle of 105.2°.

Kuntz *et al.* report first-order CM calculations on the $HeNe_2$ system [70]. A basis set of contracted GTFs of double-zeta quality including d-type polarization functions on Ne and p-type on He was employed. The particular reaction paths studied were $Ne_2^+ + He \rightarrow NeHe^+ + Ne$ and $Ne_2^+ + He \rightarrow Ne^+ + Ne + He$. A total of 91 collinear HeNeNe geometries comprised the final potential surface, which was fitted to an analytical form. Classical trajectory calculations were carried out.

Small basis sets of GTFs were used in SCF calculations by Novaro and Nieves to study the Ne_3 surface at short internuclear separations [71]. Collinear, isosceles, and asymmetrical nuclear configurations were chosen to generate a surface comprised of about 90 geometries. Comparisons are made with other studies of three- and four-atom rare gas systems.

Calculations of portions of the Ar_2F surface have been reported [72].

H. FOUR OR MORE ATOMS

Chemical reactions involving four or more atoms require potential energy hypersurfaces that need six or more internal coordinates to describe the relative positions of the nuclei. The expense of generating *ab initio* energy points for all possible nuclear positions on such surfaces, i.e., to qualify as being of types (c) or (d) as defined in Section A, is most often prohibitively large. It is, of course, possible in many cases to predict which degrees of freedom are relevant to a particular chemical reaction using knowledge gained from previous studies (in particular) and from "chemical intuition" (in general). This is discussed in some detail in Refs. 2–5. Peyerimhoff and Buenker set forth *ab initio* procedures for constructing surfaces involving free radicals [73a]. These include bimolecular interactions [73b] and electrocyclic reactions [73c]. They have applied SCF and CM methods to such systems as $C_2H_4 + NH_2$ [73b] and to the cyclobutene to butadiene isomerization and the cyclopropyl to allyl cation transformation [73c]. Potential curves for various slices along the reaction coordinate and the positions of extrema are particular properties discussed. The remainder of this chapter is devoted to brief descriptions of some high-quality studies of reaction surfaces involving more than three atoms.

The insertion of the methylidine radical in its lowest doublet and quartet states into molecular hydrogen to form $^1A_1$ and $^3B_1$ methyl radicals, respectively, has been studied by Brooks and Schaefer [74]. Basis sets of contracted GTFs of double-zeta and double-zeta plus polarization quality were used in a CM procedure that included all single and double replacements with respect to a two-configuration representation of the $CH + H_2$ surfaces. Approaches were constrained to conform to C_{2v} and C_s symmetries. A barrier for insertion on the doublet surface of 4 kcal/mol was found for a perpendicular HH midpoint to CH distance of 1.46 Å with $R_{HH} = 0.82$ Å. Abstraction on the quartet surface yielded a barrier of 13.8 kcal/mol. The HCH angle was constrained to be 134° in the latter case with distances $R_{HH} = 0.88$ Å and $R_{CH} = 1.40$ Å.

Nakatsuji *et al.* have proposed a procedure that makes use of the force density as derived based on the Hellman–Feynman theorem to compute the minimum-reaction path for the $NH_2 + H \rightarrow NH_3$ reaction [75]. Two-configuration MCSCF calculations employing STO-3G and STO-3G plus p-type polarization functions were carried out. The method shows promise and should be attempted with more adequate basis sets.

Dunning and co-workers have carried out POL-CM studies on the $O + H_2 \rightarrow H_2O$ [33], $OH + H_2 \rightarrow H_2O + H$ [76a], $O + CH_4 \rightarrow OH + CH_3$ [76b], and $H + CH_4 \rightarrow H_2 + CH_3$ [76b] reactions using basis sets of contracted GTFs of up to triple-zeta plus double-zeta polarization quality. In the second reaction a barrier of 6.2 kcal/mol was computed, which agrees well with the experimental value of 4–6 kcal/mol. The saddle point geometry of OH_3 was found to be coplanar. Barrier heights for the two methane reactions noted above were computed to be 14.4 and 15.5 kcal/mol, respectively. Calculations on the OH_3^+ surface have been published [77].

In a study of a variety of S_N2 reactions Keil and Ahlrichs report reaction energies and barrier heights for $ACH_3 + B^- \rightarrow A^- + H_3CB$ (A, B = H, F, Cl) and the $FSiH_3 + F^-$ exchange reaction using the CEPA-NO method [17]. The effect of electron correlation in energy barriers was found to be as much as 17 kcal/mol even for those reactions involving closed-shell species.

The insertion of singlet methylene into H_2 has been investigated by Kollmar and Staemmler [78] and by Jeziorek and Zurawski [79] using basis sets of GLFs, both including and excluding polarization functions in both the SCF and CEPA-PNO approximations. Both studies found the reaction path to decrease monotonically with no barrier to the formation of methane. This feature was not changed by the inclusion of electron correlation. Numerous cuts of the potential surface are displayed in contour maps in Ref. 79.

The photodissociation of formaldehyde $H_2CO \rightarrow H_2 + CO$ has been investigated by Jaffe and Morokuma using a 4-31G basis set [80]. Eighteen configurations defined in C_s and C_{2v} symmetries were used in an MCSCF procedure to define MOs for subsequent CM involving 52 configurations. A saddle point was found for the coplanar system at a height of 4.55 eV above the equilibrium formaldehyde molecule. A normal mode analysis for the system at the saddle point was carried out.

Garrison *et al.* computed potential surfaces for the interaction of He with formaldehyde as a rigid asymmetric top using two basis sets of contracted GTFs, one of which was of better than triple-zeta plus polarization quality [81]. Only a small minimum of less than 3°K was found for the coplanar approach at about 9 a.u. from the H_2CO molecule. The resulting surface was fitted to an expansion in spherical harmonics.

The dissociation of ketene into CH_2 and CO has been investigated by Prendergast and Fink using SCF and CI procedures with a small basis set of GLFs [82]. Truncation of the virtual space of MOs, the restriction to a class of "diagonal" double replacements, i.e., the excitation of two electrons from one occupied to one virtual MO, and the limited basis set detract from the reliability of this study. The ground-state and first-excited-state surfaces were examined.

Jonsson *et al.* computed a potential surface for the $H_2O + CO_2 \rightarrow H_2CO_3$ reaction using a double-zeta quality basis set of contracted GTFs in using SCF and CM procedures [83]. An energy barrier of 219 kJ/mol was computed using all single and double replacements from the ground MO configuration. This result is lower by 15 kJ/mol than that found in the SCF approximation.

The addition of methylene to ethylene to form cyclopropane has been studied in the SCF, CEPA, and PNO approximations by Zurawski and

Kutzelnigg using basis sets of GLFs of double-zeta and double-zeta plus d-type functions on the C's [84]. The energy was found to decrease monotonically along the reaction coordinate with no barrier. They found that the contribution to the computed reaction energy (~ 100 kcal/mol) due to the addition of polarization functions was about 10 kcal/mol and that due to electron correlation about 15 kcal/mol.

Kato and Morokuma used an energy gradient technique in UHF computations (4-31G) of the characteristics of the reaction surface for $H_2CCH_2F \rightarrow H_2C{=}CHF + H$ [85]. A transition state was located at about 40 kcal/mol above the reactant and about 6 kcal/mol above the products. The energy gradient method permits the tracing of the reaction path for this system of 15 degrees of freedom to a good approximation. A normal mode analysis is given, as is a calculation of the lifetime of the reactant complex in the Rice–Rampsberger–Kassel–Marcus theory.

The energy barrier for a proton jump from NH_4^+ to NH_3 has been computed by Delpuech *et al.* to be 2.5 kcal/mol in the SCF approximation using a large basis set of GTFs, including polarization functions on N and H [86]. The N—H—N path was constrained to be collinear and the NH_3 angles were varied in test calculations. Three distinct regions corresponding to association, proton transfer, and dissociation are observed.

REFERENCES

1. See, for example: W. H. Miller, ed., "Modern Theoretical Chemistry," Vols. 1 and 2. Plenum, New York, 1977; K. Lawley, ed., *Adv. Chem. Phys.* **42,** (1980).
2. R. F. W. Bader and R. A. Gangi, *in* "Theoretical Chemistry," Vol. 2. The Chemical Society, London, 1975.
3. J. J. Kaufman, *Adv. Chem. Phys.* **28,** 113 (1975).
4. S. D. Peyerimhoff and R. J. Buenker, "The New World of Quantum Chemistry," (B. Pullman and R. Parr eds.), Reidel, Dodrecht, 1976.
5. A. Devaquet, A. Seven, and B. Bigot, *J. Am. Chem. Soc.* **100,** 2009 (1978).
6. G. D. Carney and R. N. Porter, *J. Chem. Phys.* **65,** 3547 (1976).
7. C. W. Bauschlicher, Jr., S. V. O'Neil, R. K. Preston, H. F. Schaefer, III, and C. F. Bender, *J. Chem. Phys.* **59,** 1286 (1973).
8. I. Shavitt, R. M. Stevens, F. L. Minn, and M. Karplus, *J. Chem. Phys.* **48,** 2700 (1968).
9. B. Liu, *J. Chem. Phys.* **58,** 1925, (1973).
10. P. Siegbahn and B. Liu, *J. Chem. Phys.* **68,** 2457 (1978).
11. D. G. Truhlar and C. J. Horowitz, *J. Chem. Phys.* **68,** 2466 (1978); *ibid* **71,** 1514(E) (1979).
12. D. G. Truhlar and R. E. Wyatt, *Adv. Chem. Phys.* **36,** 141 (1977).
13. G. Herzberg, *J. Chem. Phys.* **70,** 4806 (1979).
14. H. F. King and K. Morokuma, *J. Chem. Phys.* **71,** 3213 (1979).
15. G. D. Carney, *J. Chem. Phys.* **72,** 4764 (1980).
16. A. Macias, *J. Chem. Phys.* **48,** 3464 (1968); **49,** 2198 (1969).

17. F. Kiel and R. Ahlrichs, *J. Am. Chem. Soc.* **98,** 4787 (1976).
18. D. F. Zetik and R. D. Poshusta, *J. Chem. Phys.* **52,** 4920 (1970).
19. P. J. Brown and E. F. Hayes, *J. Chem. Phys.* **55,** 922 (1971).
20. D. G. Hopper, Int. *J. Quantum. Chem.* **S12,** 305 (1978).
21. M. D. Gordon and D. Secrest, *J. Chem. Phys.* **52,** 120 (1970); M. D. Gordon, D. Secrest, and C. Llaguno, *ibid.* **55,** 1046 (1971).
22. B. Tsapline and W. Kutzelnigg, Chem. *Phys. Lett.* **23,** 173 (1973).
23. J. Römelt, S. D. Peyerimhoff, and R. J. Buenker, *Chem. Phys.* **34,** 403 (1978); **41,** 133 (1979).
24. R. D. Poshusta and W. F. Siems, *J. Chem. Phys.* **55,** 1995 (1971).
25. M. B. Milleur, R. L. Matcha, and E. F. Hayes, *J. Chem. Phys.* **60,** 674 (1974).
26. W. A. Lester, Jr., *J. Chem. Phys.* **53,** 1511 (1970); **54,** 3171 (1971).
27. W. Kutzelnigg, V. Staemmler, and C. Hoheisel, *Chem. Phys.* **1,** 27 (1973).
28. K. Mizutani, Y. Kuribara, H. Hayoshi, and S. Matsumoto, *Bull. Chem. Soc.* Japan **52,** 2184 (1979).
29. D. H. Liskow, C. F. Bender, and H. F. Schaefer III, *J. Chem. Phys.* **61,** 2507 (1974).
30. C. F. Bender, J. H. Meadows, and H. F. Schaefer III, *J. Chem. Soc. Faraday Trans.* **62,** 59 (1977).
31. D. M. Hirst, *Chem. Phys. Lett.* **53,** 125 (1978).
32. R. E. Howard, A. D. McLean, and W. A. Lester, Jr., *J. Chem. Phys.* **71,** 2412 (1979).
33. S. P. Walch, T. H. Dunning, Jr., R. C. Raffenetti, and F. W. Bobrowicz, *J. Chem. Phys.* **72,** 406 (1980).
34. B. H. Mahan, H. F. Schaefer III, and S. R. Ungemach, *J. Chem. Phys.* **68,** 781 (1978).
35. C. F. Bender, S. V. O'Neil, P. K. Pearson, and H. F. Schaefer III, *Science* **176,** 1412 (1972).
36. R. L. Jaffe, K. Morokuma, and T. F. George, *J. Chem. Phys.* **63,** 3417 (1975).
37. F. Rebentrost and W. A. Lester, Jr., *J. Chem. Phys.* **63,** 3737 (1975).
38. P. Botschwina and W. Meyer, *Chem. Phys.* **20,** 43 (1977).
39. W. R. Wadt and N. W. Winter, *J. Chem Phys.* **67,** 3068 (1977).
40. P. Baybutt, *Chem. Phys. Lett.* **30,** 293 (1975).
41. R. D. Amos, *Chem. Phys. Lett.* **61,** 597 (1979).
42. A. B. Bolotin, O. P. Bugaets, and D. A. Zhogolow, *Chem. Phys. Lett.* **37,** 9 (1976).
43. E. F. Hayes, A. K. Q. Siu, F. M. Chapman, Jr., and R. L. Matcha, *J. Chem. Phys.* **65,** 1901 (1976).
44. J. W. Birks, H. S. Johnston, and H. F. Schaefer III, *J. Chem. Phys.* **63,** 1741 (1975).
45. S. Rothenberg and H. F. Schaefer III, *Chem. Phys. Lett.* **10,** 565 (1971).
46. T. H. Dunning, Jr. *J Chem Phys.* **66,** 2752 (1977).
47. P. Siegbahn and H. F. Schaefer, III, *J. Chem. Phys.* **62,** 3288 (1975).
48a. W. B. England, N. H. Sabelli, and A. C. Wahl, *J. Chem. Phys.* **63,** 4596 (1975).
48b. W. B. England, N. H. Sabelli, A. C. Wahl, and A. Karo, *J. Phys. Chem.* **81,** 772 (1977).
49. N. C. Baird, *J. Chem Phys.* **62,** 300 (1975).
50. C. F. Melius and R. J. Blint, *Chem. Phys. Lett.* **64,** 183 (1979).
51. S. V. O'Neil, P. K. Pearson, H. F. Schaefer III, and C. F. Bender, *J. Chem. Phys.* **58,** 1126 (1973); S. V. O'Neil, H. F. Schaefer III, and C. F. Bender, *Proc. Natl. Acad. Sci. U.S.A.,* **71,** 104 (1974); C. F. Bender, C. W. Bauschlicher, Jr., and H. F. Schaefer III, *Jr. Chem. Phys.* **60,** 3707 (1974).
52. H. Umeyama, K. Kitaura, and R. Morokuma, *Chem. Phys. Lett.* **36,** 11 (1975); see also A. B. Isaacson and K. Morokuma, *J. Am. Chem. Soc.* **97,** 4453 (1975).
53. A. Dedieu, *J. Chim. Phys.* **71,** 615 (1974).
54. P. Baybutt, F. W. Bobrowicz, L. R. Kahn, and D. G. Truhlar, *J. Chem. Phys.* **68,** 4809 (1978); *erratum* **70,** 2050 (1979).

55. D. M. Silver, *J. Chem. Phys.* **72,** 6445 (1980).
56. A. G. Gelb, C. W. Wilson, R. Kaprol, and G. Burns, *J. Chem. Phys.* **7,** 3421 (1972).
57. W. A. Lester and M. Krauss, *J. Chem Phys.* **52,** 4775 (1970).
58. G. G. Balint-Kurti and R. N. Yardley, *Faraday Discussions Chem. Soc.* **62,** 77 (1977).
59. M. M. L. Chen and H. F. Schaefer III, *J. Chem Phys.* **72,** 4376 (1980).
60. H. Schor, S. Chapman, S. Green, and R. N. Zare, *J. Chem. Phys.* **69,** 3790 (1978).
61. M. Losonczy, J. W. Moskowitz, and F. H. Stillinger, *J. Chem. Phys.* **61,** 2438 (1974).
62. P. K. Pearson, W. J. Hunt, C. F. Bender, and H. F. Schaefer III, *J. Chem. Phys.* **58,** 5385 (1973).
63. H. M. Schmidt, H. Von Hirschhausen, and K. Helfrich, *Chem. Phys.* **29,** 219 (1978).
64. G. G. Balint-Kurti and M. Karplus, *Chem. Phys Lett.* **11,** 203 (1971); G. G. Balint-Kurti, *Molec. Phys.* **25,** 393 (1973).
65. L. D. Thomas, W. P. Kraemer, and G. H. F. Diercksen, *Chem. Phys.* **30,** 33 (1978).
66. V. Staemmler, *Chem. Phys.* **17,** 187 (1976); *ibid.* **7,** 17 (1975).
67. P. N. Noble and R. N. Kortzeborn, *J. Chem. Phys.* **52,** 5375 (1970).
68. D. G. Hopper, *J. Am. Chem. Soc.* **100,** 1019 (1978).
69a. P. A. Benioff, G. Das, and A. C. Wahl, *J. Chem. Phys.* **67,** 2449 (1977).
69b. G. Das and P. Benioff, *Chem. Phys. Lett.* **75,** 519 (1980).
70. P. J. Kuntz, M. Doran, and I. H. Hillier, *Chem. Phys.* **43,** 197 (1979).
71. O. Novaro and F. Nieves, *J. Chem. Phys.* **65,** 1109 (1976).
72. H. H. Michels, R. H. Hobbs, and L. A. Wright, *Chem. Phys. Lett.* **48,** 158 (1977).
73a. S. D. Peyerimhoff and R. J. Buenker, *Berichte der Bunsen-Gesellschaft für physikalische Chemie,* **78,** 119 (1974).
73b. R. J. Buenker, *in* "Chemical and Biochemical Reactivity, The Jerusalem Symposia on Quantum Chemistry and Biochemistry VI," The Israeli Academy of Sciences and Humanities, Jerusalem, 1974.
73c. S. D. Peyerimhoff, "*Ab Initio* Calculations for Electrocyclic Reactions," *ibid.;* P. Merlet, S. D. Peyerimhoff, R. J. Buenker, and S. Shih, *J. Am. Chem. Soc.* **96,** 959 (1974).
74. B. R. Brooks and H. F. Schaefer III, *J. Chem. Phys.* **67,** 5146 (1977).
75. H. Nakatsuji, T. Koga, K. Kondo, and T. Yonezawa, *J. Am. Chem. Soc.* **100,** 1029 (1978).
76a. S. P. Walch and T. H. Dunning, Jr., *J. Chem. Phys.* **72,** 1303 (1980).
76b. S. P. Walch and T. H. Dunning, Jr., *J. Chem. Phys.* (to be published) and Annual Report, Theoretical Chemistry Group, Argonne National Laboratory, Argonne, Illinois, 1978.
77. R. Daudel, C. Kozmitza, and E. Kapuy, *Chem. Phys. Lett.* **36,** 555 (1976).
78. H. Kollmar and V. Staemmler, *Theoret. Chim. Acta,* **51,** 207 (1979).
79. D. Jeziorek and B. Zurawski, *Int. J. Quantum Chem.* **16,** 277 (1979).
80. R. L. Jaffe and K. Morokuma, *J. Chem. Phys.* **64,** 4881 (1976).
81. B. J. Garrison, W. A. Lester, Jr., and H. F. Schaefer III, *J. Chem. Phys.* **63,** 1449 (1975).
82. P. Prendergast and W. H. Fink, *J. Am. Chem. Soc.* **98,** 648 (1976).
83. B. Jonsson, G. Karlstrom, H. Wennerstrom, S. Forsen, B. Roos, and J. Almlof, *J. Am. Chem. Soc.* **99,** 4628 (1977).
84. B. Zurawski and W. Kutzelnigg, *J. Am. Chem. Soc.* **100,** 2654 (1978).
85. S. Kato and K. Morokuma, *J. Chem. Phys.* **72,** 206 (1980).
86. J. J. Delpuech, G. Serratrice, A. Strich, and A. Veillard, *J. Chem. Soc. Chem. Commun.* p. 817 (1972).

CHAPTER VI

MOLECULES WITH TRIGONAL SYMMETRY

A. GENERAL CONSIDERATIONS

Many molecules have a threefold symmetry axis, belonging then to one of the symmetry groups C_{3v}, D_{3h}, or D_{3d}. The character tables for the several representations of these groups are as follows, with the coordinate z being taken along the trigonal axis.

C_{3v}	I	$2C_3(z)$	$3\sigma_v$	Coordinates
A_1	1	1	1	z
A_2	1	1	−1	
E	2	−1	0	x, y

D_{3h}	I	$2C_3(z)$	$3C_2$	σ_h	$2S_3$	$3\sigma_v$	Coordinates
A_1'	1	1	1	1	1	1	
A_1''	1	1	1	−1	−1	−1	
A_2'	1	1	−1	1	1	−1	
A_2''	1	1	−1	−1	−1	1	z
E'	2	−1	0	2	−1	0	x, y
E''	2	−1	0	−2	−1	0	

D_{3d}	I	$2S_6(z)$	$2S_6^2 \equiv 2C_3$	$S_6^3 \equiv S_2 \equiv i$	$3C_2$	$3\sigma_d$	Coordinates
A_{1g}	1	1	1	1	1	1	
A_{1u}	1	−1	1	−1	1	−1	
A_{2g}	1	1	1	1	−1	−1	
A_{2u}	1	−1	1	−1	−1	1	z
E_g	2	−1	−1	2	0	0	
E_u	2	1	−1	−2	0	0	x, y

The grouping of molecules by symmetry, as in this chapter, has as a consequence that molecules of a great variety of chemical types are included.

B. SURVEY

AH_3 molecules are planar (symmetry D_{3h}) or pyramidal (C_{3v}). For the first-row A atoms, the respective electron configurations are as follows:

$$\text{Planar:} \quad (1a_1')^2\,(2a_1')^2\,(1e')^m\,(1a_2'')^n$$

$$\text{Pyramidal:} \quad (1a_1)^2\,(2a_1)^2\,(1e)^m\,(3a_1)^n$$

The $1a_1'$ or $1a_1$ MO is nonbonding, essentially 1s of A, the $2a_1'$ or $2a_1$ and the 1e′ or 1e are bonding, and the $3a_1$ is a nearly nonbonding lone-pair MO which can act as an electron donor MO in coordination compounds.

In molecules containing $3a_1$ (if pyramidal) or $1a_2''$ MOs (if planar), the form of the MO is pure $(pz)_A$ or "$p\pi_A$" if the molecule is planar, but is a linear combination $a(pz)_A + b(s_A) - c(1s_H + 1s_H + 1s_H)$ if it is pyramidal. An enlightening discussion of the factors that decide the extent of pyramidalization has been given by Cherry and Epiotis [1]. They show that the relative energies of the D_{3h} and C_{3v} forms and the acuteness of the angle HAH depend on the closeness of approach of the energy of the LUMO (lowest unoccupied MO) to that of the HOMO (highest occupied MO, in this case $1a_2''$ or $3a_1$). As a result, by a small margin, BH_3, CH_3, and NH_3^+ are planar, but CH_3^-, NH_3, and OH_3^+ are pyramidal (respective apex angles 110° calculated, 107°, and 116°), while SiH_3, PH_3, SH_3^+, etc. are strongly pyramidal (e.g., apex angle 93° in PH_3).

There are very many references on AH_3 molecules, especially on NH_3, in the bibliography of Richards *et al.* [2]. Only a few of these, and more recent references, will be mentioned here.

C. BH_3, CH_3^+, BH_3^-, CH_3, NH_3^+, H_3O^+, NH_3, H_3O, $(CH_3)_3C$

For BH_3 there is little to be said. Everyone agrees that it is D_{3h} planar. Hall *et al.* [3], using a good basis set in an SCF calculation, report $E_h = -26.4014$ for BH_3. In an earlier, less accurate SCF calculation, Schwartz and Allen [4] give orbital energies and a population $B^{0.57}(H^{-0.19})_3$, also an overlap population of 0.794 *e*.

Millie and Berthier [5], using a good basis set with polarization functions on C and H, made SCF calculations on CH_3^+, CH_3, and CH_3^-. For CH_3^+ they find $E_h = -39.2459$; for CH_3, $E_h = -39.5715$; and for CH_3^-, $E_h = -39.5125$. They give orbital energies and atomic charges, corresponding to $C^{0.09}(H^{0.30})_3$ in CH_3^+, $C^{0.45}(H^{0.13})_3$ in CH_3, and $C^{1.09}(H^{0.03})_3$ in CH_3^-. An ionization potential (8.86 eV) for CH_3 is obtained. In slightly less accurate SCF calculations, Kari and Csizmadia [6] discuss CH_3^+ and CH_3^-, plot potential surfaces, and compute vibrational constants. They predict a C—H bond length of 1.08 Å. It is generally agreed the CH_3 is planar or at least very nearly so.

Kari and Csizmadia [7], using a basis set including polarization functions on both atoms, made SCF computations on planar CH_3 ($E_h = -39.5703$) and NH_3^+ ($E_h = -55.8945$). Orbital energies, multipole moments, field gradients, and potentials at the nuclei are given.

Ishikawa and Binning [8], using an unpolarized basis set in the generalized coupling operator method, made computations on three states of CH_3, the ... $(1e_1')^4\,(1a_2'')$, $^2A_2''$ ground state and the ... $(1e_1')^4\,(3a_1')$ Rydberg excited state, and the pyramidal $(1e)^3\,(3a_1)$, ^2E valence-excited state (not observed, transition forbidden), which should be the lowest excited state. For the latter they found $R_{CH} = 1.18$ Å, and the C—H bonds are inclined at 33° to a plane perpendicular to the symmetry axis (however, this neglects the Jahn–Teller effect). They give tables of atomic and overlap populations for the three states, and vertical transition energies.

Hinchcliffe [9] made unrestricted SCF calculations with wrong-spin annihilation on BH_3^-, CH_3, and NH_3^+, using basis sets including polarization functions. Varying the geometry, all are predicted to be planar with optimal bond lengths 1.22, 1.08, and 1.02 Å, respectively. Electron spin resonance hyperfine coupling constants were computed for A and H atoms for each radical in fairly good agreement with experiment. Computed E_h values were as follows: BH_3^-, -26.3614; CH_3, -39.5709; NH_3^+, -55.8953.

Perhaps the best wave functions for CH_3 and CH_3^- are from the SCF plus CM calculations of Marynick and Dixon [10]. A good basis set with polarization functions on C and H is used. The CM includes all single and double excitations from the valence shell except for five high-energy orbitals. For planar CH_3, 2026 determinants were included in the CM. Some computed one-electron operator quantities are listed. For CH_3 the computed E_h is -39.724; with estimated corrections for neglected items it comes to -39.808, as compared with the experimental -39.824. The optimized C—H bond length is 1.079 Å, in exact agreement with experiment. The out-of-plane bending frequency is computed as 612 cm^{-1} (experiment: 607 cm^{-1}).

For CH_3^- the computed bond length at the minimum is 1.100 Å, shortening to 1.081 in the planar form. The computed barrier to inversion is 1.95 kcal/mole with CM, 1.78 kcal/mole without. Marynick and Dixon conclude that CH_3^- is unstable with respect to $CH_3^- \rightarrow CH_3 + e$ by 2–8 kcal/mole. However, CH_3^- has now been found experimentally, with an electron affinity of 1.8 ± 0.7 kcal/mole [11].

Vying with Marynick and Dixon [10] in accuracy appears to be a paper by Dykstra *et al.* on CH_3^- [12]. A large and flexible basis set due to Duke [13] is used, together with ample CM, employing the theory of self-consistent electron pairs (SCEP). The computed barrier to inversion is 1.50 kcal/mole with CM, 1.72 kcal/mole without; here the effect of CM is the reverse of that reported by Marynick and Dixon. Dykstra *et al.* give a critical review of the best previous calculations on CH_3^-.

Jemmis *et al.* [14] have made approximate SCF calculations on the energy of CH_3^+, CH_3, and CH_3^- as a function of bond distances, one HCH angle, and the out-of-plane bend angle of the third CH. Walsh diagrams are given. The results were used to systematize data on large structural fragments having similar angular strains.

A GVB and POL-CI calculation on CH_3, CH_3^+, and CH_3^- by Surratt and Goddard [15], using a good polarized basis, explored energy changes on bonding, and gave $\nu_2 = 585$ cm^{-1} for CH_3 and 1570 cm^{-1} for CH_3^+. $E_h = 39.6347$ for CH_3 by POL–CI.

Chang *et al.* [16] and Ellinger *et al.* [17] have discussed proton and ^{13}C hyperfine splittings in CH_3 using SCF computations and perturbation theory. Almlöf *et al.* [18] have calculated these and also splittings in CH_3^+, using unrestricted SCF, with spin annihilation.

Of the many *ab initio* calculations on NH_3, there have been only a few good computations of SCF plus CM character. We mention first a non-variational calculation by Lee and Das [19], using a united-atom (i.e., single-center basis) many-body perturbation procedure. They calculated a correlation energy of 0.317 ± 0.01 a.u. This, added to their SCF energy of -56.151 ± 0.2 a.u., gives -56.468 ± 0.2 a.u. as compared with the experimental value -56.5601 a.u. However, if one combines the much better SCF energy -56.2154 a.u. of Ahlrichs *et al.* [22] with the Lee and

Das correlation energy, one arrives at a correlated energy of -56.532 ± 0.01 a.u.

The first SCF plus CM calculation of the conventional type was that of Kari and Csizmadia [20] who obtained $E_h = -56.3747$. The following SCF orbital energies (in a.u.) are reported: $1a_1$, -15.5294; $2a_1$, -1.1371; 1e, -0.6223; $3a_1$, -0.4240; further, out-of-plane angle 22.14°; R(NH) = 1.012 Å. For planar NH_3, $1a_1$, -15.4781; $2a_1$, -1.0990; 1e, -0.6254; $3a_1$, -0.5754; $R_{NH} = 0.996$ Å.

Stevens [21] in a study of the effect of CM on the calculated inversion barrier obtained $E_h = -56.4251$ for pyramidal NH_3, while Ahlrichs *et al.* [22], with the PNO-CI and CEPA-PNO methods, and using a large Gaussian basis set with one f and two d functions on N and two p functions on H obtained $E_h = -56.4428$ by CM and -56.4576 by CEPA, which, however, is not fully variational. Stevens [21] gives a table of computed one-electron operator expectation values. Ahlrichs *et al.* [22] give a critical review of calculations (including their own and that of Stevens) on the role of correlation energy in determining the inversion barrier in CH_3^-, NH_3, and OH_3^+. They conclude that correlation raises the barrier by about 0.4 kcal/mole in NH_3; the total barrier by the CEPA method is 5.6 kcal/mole, as compared with 5.8 experimental. For CH_3^- and H_3O^+ they find 0.9 and 1.5 kcal/mole, respectively, by the CEPA method.

For H_3O^+, Lischka and Dyczmons [23], with a large basis set, obtain $E_h = -76.5983$ by the IEPA method, which is not variational. Ahlrichs *et al.* obtain $E_h = -76.5445$ by CM, -76.5553 by CEPA, which again is not fully variational.

Runau *et al.* [24] made an SCF and CM study of the photodissociation of NH_3 using a good basis set. Their E_h by CM is -56.4241. Calculations were made of the vertical excitation energies to singlet states in which the 3a, MO is excited to the near-Rydberg MOs $4a_1$ (near 3s), 2e ($3p_{x,y}$), and $5a_1$ ($3p_z$); agreement with experiment (in which vertical excitation involves about six vibrational quanta in the upper, planar, state) is good. Direct calculations of the 0,0 vibrational transition from a planar ground state conformation are also made. The 1e → $4a_1$ transition is also computed; here, the excited state is pyramidal. Oscillator strengths for the transitions are also computed and agree well with experiment. SCF potential curves in C_{2v} symmetry are computed for the photodissociation process $NH_3^* \rightarrow NH_2 + H$ for the excited states $\tilde{A}\,{}^1A_2''$ ($1a_2'' \rightarrow 3s$), $\tilde{C}\,{}^1A_1'$ ($1a_2'' \rightarrow 3p_z$), and ${}^1E'$ ($1e' \rightarrow 3s$). In the $\tilde{A}$ state, there is a computed barrier of 0.57 eV for this process, assuming a 120° angle in NH_2; actually the angle is smaller and the barrier may well be lower. The 3s (or $4a_1$) MO of NH_3 is de-Rydbergized to become $1s_H$ as dissociation is approached. Runau *et al.* give further details in discussion of the photodissociation of the $\tilde{B}$ and $\tilde{C}$ states, and also give calculated SCF potential curves and discussion for the process $NH_3^* \rightarrow NH + H_2$.

Rianda *et al.* [25] also have calculated vertical excitation energies on the three lowest excited states of NH_3, using GVB plus CM methods. Their computations gave 6.14 and 6.37 eV for $3a_1 \rightarrow 3s$; 7.86 and 7.88 eV for $3a_1 \rightarrow 3p_{x,y}$; 7.87 and 8.15 eV for $3a_1 \rightarrow 3p_z$; the first number is for the triplet, the second for the singlet state. The following were observed by electron impact: 6.39 ($A\,^1A_2''$), 7.93 ($B\,^1E''$), and 8.26 eV ($C\,^1A_1'$). The smallness of the singlet–triplet splittings is in agreement with their Rydberg character.

Pulay and Meyer [26], using ground SCF wave functions, have computed force constants for NH_3 in rather good agreement with experiment, and also dipole moment derivatives. Signs for the latter, not known from experiment, are obtained, but only two (out of four) with certainty.

Using SCF wave functions, Laws *et al.* [27] computed diamagnetic shielding and susceptibility constants, dipole moment, and nuclear quadrupole coupling constants, and other one-electron properties of NH_3. They also examined vibrational averages of these properties.

Dejardin *et al.* [28] have used the NH_3 molecule as a test case in a study of MCSCF and CM calculations, with instructive discussion.

With good basis sets, Rupp and Ahlrichs [28a] made SCF and CEPA-PNO calculations on polymers of LiH, NaH, and LiF. For the trimers they found D_{3h} geometry. Energy data are given.

Wilhite and Whitten [29] have discussed SCF localized MOs for NH_3 and other molecules using essentially the method of Peters, and also the improvement of these by CM. Orbital energies for localized MOs are compared with the (very different) values for ordinary delocalized SCF MOs, with equal computed E_h. A modest polarized basis set is used. Percentages of delocalization for truncated MOs restricted to single atoms or bonds are computed (values are 1.2% for K shell, 2.5% for lone pair, 5.5% for NH bond MOs).

In a discussion of CM by pseudonatural orbitals, Meyer [30] has given some calculated results obtained by the PNO-CEPA method for the orbital energies, dipole moment, and isotropic polarizability of NH_3. Agreements with experiment are good.

Schwartz [31] made SCF computations on H_3O (not known experimentally) in the following way. He used a basis somewhat better than double zeta on the O atom plus diffuse functions but no polarization, and diffuse functions including p types on the H atoms. First he made SCF calculations on H_3O^+, whose virtual MOs should approximate the MOs of excited states of H_3O. He also made explicit SCF correlations on the lowest two excited states of H_3O; these turned out to be only slightly lower than the results from the H_3O^+ virtual orbitals. The resulting excitation energies were compared with experimental electron-impact data on excited states of K-shell–ionized and excited NH_3. As is reasonable, agreements are good. References are made to other related papers. Schwartz suggested that the observed absorption maximum at 1.73 eV for the hydrated electron may in part be due to the lowest-energy predicted absorption peak of H_3O (calculated: 1.82 eV).

Claxton *et al.* [31a] conclude from small-basis SCF calculations that the heavy-atom framework of the tert-butyl radical $(CH_3)_3C$ is planar. They calculate the isotropic hyperfine coupling to ^{13}C as +37G for the rigid molecule, or +69G on including the zero-point vibratory contribution (experiment: 46–51G).

D. AlH_3, SiH_3, PH_3^+, PH_3, OPH_3, SH_3^+

The amines A are well known for their formation of two-way charge-transfer oxides AO. PH_3 and its derivatives behave similarly. Demuynck and Veillard [32] have made SCF calculations on H_3PO. They give valence-electron gross and overlap populations showing $d\sigma$ bonding from P to O and $2p\pi$ (to $3d\pi_P$) back-bonding to be strong; d participation is much stronger than in $PH_3{\cdot}BH_3$ ($d\sigma_P$ and $d\pi_P$ populations 0.14 *e* and 0.22 *e*, respectively; overlap populations 0.44 *e* for $3p\pi_P - 2p\pi_O$, 0.19 *e* for $3d\pi_P - 2p\pi_O$). The σ charge transfer towards oxygen is about 1.2 *e* and the π back donation from oxygen is about 0.5 *e* (subject to the usual qualifications about population analysis).

Ahlrichs *et al.* [33] have made relatively accurate PNO-CI and CEPA-PNO calculations on AlH_3 (planar) and PH_3 (planar and pyramidal), obtaining $E_h = -243.7299$ for AlH_3, and -342.6403 for PH_3 (pyramidal) by CEPA. There are no experimental data on AlH_3. For R(AlH) they compute 1.59 Å. Kutzelnigg *et al.* [34] have made similar but slightly more accurate calculations on PH_3, including force constants and vibration frequencies for PH_3 and PD_3.

For PH_3 the computed barrier to inversion is 35.3 kcal/mole by CEPA [31]. In less accurate computations, Lehn and Munsch [35] in an SCF computation and Petke and Whitten [36] in a limited CM computation found similar values. Lehn and Munsch give detailed population analyses. Petke and Whitten compute dipole and quadrupole moments and field gradient at the P atom. Schmiedekamp *et al.* [37] also made an SCF computation in which they point out that the computed PH bond lengths decrease from 1.41 Å (experimental: 1.41) in actual pyramidal to 1.38 Å in calculated planar PH_3; corresponding behavior occurs in NH_3.

In a small-basis SCF calculation Aarons *et al.* [38] report an inversion barrier of 1.6 eV for PH_3 but 0.2 eV for PH_3^+, and angular inclinations of 35° for PH_3, 16° for PH_3^+, ionization potentials for PH_3 of 8.8 eV adiabatic and 9.8 eV vertical.

Domcke *et al.* [39] have used Green's function methods with a good polarized basis set to compute the valence-shell ionization spectrum of PH_3. Ionization out of the $5a_1$ MO gives one main line (calculated ionization potential 10.31 eV, experimental 10.58 eV) with a weak satellite; out of the 2e MO a single line is predicted (13.56 eV, observed 13.50 eV). Out of the $4a_1$ MO a group of ten lines of appreciable intensity is predicted, with 0.53 of the intensity in the first line at 19.4 eV and 0.13 of it in the second at 22.4 eV; all the rest are weak except 0.13 of the intensity in one at 28.0 eV; experiment shows two or three lines at the low-energy end.

Barone *et al.* [40] have made an SCF calculation on SiH_3 using a polarized double-zeta basis set (d on Si, p on H), supplemented by perturbation calculations to take account partially of CM, and have computed spin hyperfine structures. They find a C_{3v} structure (nearly tetrahedral) with $R(SiH) = 1.48$ Å and HSiH angle 111°. They have calculated bonding ($v = 0$–5) and stretching ($v = 0$–2) vibrational levels and corresponding hyperfine constants, and from these the temperature dependence of the latter; here there is good agreement with experiment (but only when a polarized basis is used).

Keil and Ahlrichs [41], using an extended basis set including d and f polarization, and using PNO CI and (preferably) CEPA methods for CM, have made computations on SiH_3^- and SiH_5^-. They sought first to illuminate the reaction $SiH_4 + H^- \rightarrow SiH_5^-$. The optimum geometry found is C_{3v} for SiH_3^- and D_{3h} for SiH_5^-. However, SiH_5^- is unstable with respect to $SiH_5^- \rightarrow SiH_3^- + H_2$, but SiH_5^- may represent a metastable minimum. For $R(SiH)$, 1.54 Å was found in SiH_5^- and 1.525 Å in SiH_3^-. For SiH_3^-, angle HSiH = 97.2° was found, along with an inversion barrier of 26 kcal/mole. For SiH_5^-, a barrier for pseudorotation (see under PH_5 in Section E) of 2.9 kcal/mole was found. E_h values and orbital energies are tabulated, for C_{3v} and D_{3h} in the case of SiH_3^-, and C_{4v} and D_{3h} in the case of SiH_5^-.

Dixon and Marynick [42] have made SCF plus CM calculations using a polarized double-zeta basis (d on S, p on H) on SH_3^+. They find $R(SH) =$ 1.35 Å and angle H—S—H = 94°, a barrier to inversion of 32 kcal/mole, and a proton affinity for H_2S of 196 kcal/mole. They also computed symmetric stretch ($\nu_1 = 2576$ cm^{-1}) and umbrella ($\nu_2 = 1099$ cm^{-1}) frequencies; the experimental values were 2520 and 1028 cm^{-1}.

E. BF_3, BCl_3, BF_3^-, $AlCl_3$, SO_3, NO_3^-, NF_3, NCl_3, PF_3, H_3NO, H_3CO^-, F_3NO, F_3CO^-, OPF_3, $CFCl_3$, SPF_3, PCl_3, $OPCl_3$, ScF_3, TiF_3, VF_3, CrF_3, MnF_3, FeF_3, CoF_3, NiF_3, PF_5, PH_5, VF_5, VF_5^-, $MnCl_5^{-2}$, $CuCl_5^{-3}$, UF_5, $LiBH_4$, $NaBH_4$, $LiAlH_4$, $NaAlH_4$, $B_3N_3H_6$, $B_3O_3H_3$

BF_3 and BCl_3 are experimentally planar. Several SCF calculations on BF_3 have been made, all with unpolarized basis sets (sp Gaussians only). The lowest reported energy, with no further details, was by Walker and Horsley [43]: $E_h = -323.178$. They also computed five states of BF_3^+; these gave BF_3 ionization potentials which do not agree well with the ESCA spectrum, doubtless partly because of the absence of CM. They also give computed energies for four states of NO_3. Goutier and Burnelle [44] compute $E_h = -322.733$ for BF_3. They also compute BCl_3 and give a detailed population analysis for both molecules. They list MO orbital energies for both. Schwartz and Allen [45] studied and compared BH_3, BH_2F, BHF_2, and BF_3 ($E_h = -322.902$ for BF_3).

Using an STO-3G (near minimal) plus a boron 3d basis, So [46] found BF_3^- to be pyramidal with $R(BF) = 1.40$ Å and angle FBF = 112°. The ^2E lowest excited state was found to be 4.0 eV above the $^2A_1$ ground state.

The assignment of peaks in the photoelectron spectrum of SO_3 has been controversial. von Niessen *et al.* [47] have resolved the controversy by a many-body Green's function calculation. Using a good polarized basis set (the polarization functions are unusually important), they give E_h (SCF) and a table of orbital energies and the last six vertical orbital ionization energies, as follows (experimental values in parentheses): $1a_2'$, 12.98 eV (12.82); $1e''(\pi)$, 14.23 (14.1); $5e'$, 15.10 (14.97); $4e'$, 18.37 (17.9); $2a_2''$ (π), 18.95 (18.3?); $5a_1'$, 21.14 (20.6). The vertical electron affinity of SO_3 was tentatively computed as 0.8 eV.

Connor *et al.* [47a], using a minimal basis set, obtain orbital energies for the valence-shell MOs of ClO_3^- and use them to interpret a photoelectron spectrum of $Li^+ClO_3^-$ solid.

The same authors [47a], using a double-zeta basis made a similar study of $CO_3^=$ (symmetry D_{3h}) with application to the Li_2CO_3 photoelectron spectrum.

NF_3 (pyramidal) apparently has been investigated only with a very limited unpolarized basis set [48]. Schmiedekamp *et al.* [37] investigated geometry changes during inversion of NH_3, NH_2F, NHF_2, NF_3, PH_3, PH_2F, PHF_2, and PF_3. In a small-basis SCF calculation [38] Aarons *et al.* reported an inversion barrier of 6.9 eV for PF_3, 3.0 eV for PF_3^+, angular inclinations of 28.5° for PF_3 and 20° for PF_3^+, ionization potentials of 10.6 eV adiabatic and 11.3 eV vertical for PF_3.

Serafini *et al.* [49] discuss bonding in PF_3 and F_3PO in an SCF study. They give valence-shell gross atomic and overlap populations, and discuss these in comparison with OPH_3.

Faegri and Kosmus [50], using a good basis set including d orbitals for both atoms, made SCF calculations on NCl_3. They found a charge distribution $N^{-0.19}(Cl^{0.06})_3$ and a dipole moment 0.73 D. An inversion barrier less than for NH_3 was found. Orbital energies are given for the valence MOs.

Hillier and Saunders [51] give a minimal-basis plus phosphorus 3d SCF calculation for PCl_3, also for $POCl_3$.

Serafini and LaBarre [52] in SCF calculations using a basis set of double-zeta quality, plus d functions on P and S, compared POF_3 and PSF_3, referring also to [49] as a more accurate calculation on POF_3. They give a complete tabulation of orbital energies for both molecules and a population analysis. The P atom has a charge + 1.50 in OPF_3 and + 1.37 in SPF_3. Computed electron transfers from PF_3 to O or S are, 1.02 e for σ and -0.52 e for π in POF_3 and 0.94 e for σ and -0.066 e for π in PSF_3.

Olsen and Howell [53] have made SCF computations with a 4-31G basis set, plus d orbitals, on H_3NO, H_3CO^-, F_3NO, and F_3CO^-. All are predicted to be of C_{3v} symmetry, with $R(YO)$ = 1.45 and $R(XY)$ = 1.00 Å, 1.36 and 1.12 Å (no d), 1.16 (experimental: 1.16) and 1.41 Å (experimental: 1.43 Å), 1.21 and 1.42 Å, angle OYX = 108°, 116°, 116° (experimental: 117°), and 117°, for H_3NO, H_3CO^-, F_3NO, and F_3CO^-, respectively. E_h values and population analyses are given.

Bonaccorsi *et al.* [54] have made SCF calculations on the geometry and charge distribution of $LiBH_4$, $NaBH_4$, $LiAlH_4$, and $NaAlH_4$, using a minimal and a double-zeta-quality basis set. For the first three they find the stablest to be a C_{3v} structure with the metallic atomic Y facing the *inside* of an XH_3 umbrella with the fourth H atom stretched out behind the peak of the umbrella. For $NaAlH_4$ a C_{2v} structure is found which may be formulated as $(YH_2)H_2$, the YH_2 forming a C_{2v} triangle, with the other two H in a perpendicular plane. Using population analysis, relations to the process $Y^+ + XH_4^- \rightarrow YXH_4$ are discussed.

On the other hand, Curtiss [54a], using a 6-31G basis set finds that the lowest-energy forms of $LiAlF_4$ have C_{2v} structure 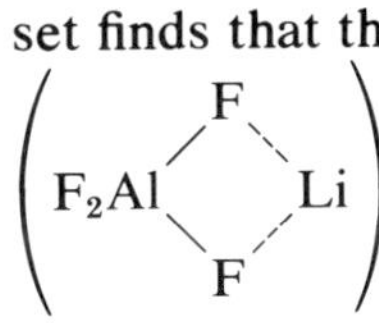, but with migration of the Li^+ in the complex relatively easy. The two possible C_{3v} structures seem to be definite minima in the energy surface.

Experimental evidence favors the C_{2v} structures, but is perhaps not conclusive. A table of orbital energies is given for the C_{2v} structure.

Hand *et al.* [55] made an essentially double-zeta SCF computation on FeF_3, and showed that it is planar or nearly so. The ground state is then $^6A_1'$. They also computed the excited low-spin state $^2A_1'$, finding it to be 7.66 eV higher in energy. They give MO orbital energies and detailed gross population analyses for both states, also multipole moments, potentials, fields, and field gradients at nuclei, and diamagnetic shielding.

Yates and Pitzer [56] have made SCF calculations on all the transition-metal trifluorides from ScF_3 to NiF_3, using for the metal atom a (14s 11p 5d) basis set contracted to (9s 7p 2d), and for the F atoms a (9s 5p) set contracted to (4s 3p). The results are summarized in Table 1, from their article. The computed dissociation energies are SCF and, of course, not very good, but the trends may be significant. As Table 1 shows, probably all the MF_3 have D_{3h} symmetry except CrF_3 (though MnF_3 could not be computed), and all are in high spin states. Computed excitation energies for two excited states are listed in each case. For two excited states of CoF_3, a Jahn–Teller interaction pattern is worked out and discussed. For the ground state of CrF_3, a dipole moment of 3.00 D is computed. Quadrupole moments, electric field gradients, out-of-plane vibrational frequencies, ionization potentials, and population analyses are given. Computed charges on the metal atoms vary from 1.72 to 1.90.

TABLE 1

Equilibrium Geometries, Ground States, and Binding Energies of the Transition Metal Trifluorides (MF_3)[a]

M	M—F (Å)	M—F (solid)[b]	Geometry	Ground state	Binding energy (eV)
Sc	1.88	*c*	D_{3h}	$^1A_1'$	10.7, (18.0 ± 0.9)[d]
Ti	1.83	1.97	D_{3h}	$(a_1')^1\ ^2A_1'$	9.5
V	1.81	1.95	D_{3h}	$(e'')^2\ ^3A_2'$	8.4
Cr	1.77	1.90	C_{3v}[e]	$e^2 a_1^1\ ^4A_2$	7.3 (15.4 ± 0.4)[f]
Mn	1.77[g]	1.90	—	$(e'')^2 (a_1')^1 (e')^1\ ^5E'$[g]	5.5[g] (12.7 ± 0.4)[f]
Fe	1.78	1.90	D_{3h}	$(e'')^2 (a_1')^1 (e')^2\ ^6A_1'$	7.9 (14.3 ± 0.7)[f]
Co	1.75	1.89	D_{3h}	$(e'')^2 (a_1')^2 (e')^2\ ^5A_1'$	6.7
Ni	1.73	—	D_{3h}	$(e'')^4 (a_1')^1 (e')^2\ ^4A_2'$	5.9

[a] From Yates and Pitzer [56].
[b] Experimental.
[c] Gaseous electron diffraction values are 1.91 ± 0.03 Å, 1.926 ± 0.005 Å.
[d] Mass spectrometric value.
[e] F—Cr—F angle of 117°.
[f] Mass spectrometric value.
[g] Assuming D_{3h} geometry.

Strich and Veillard [57] have made an SCF calculation on PF_5 with a medium basis (including d on P but not on F). Experimentally, PF_5 has symmetry D_{3h} (three equatorial and two axial fluorines) but the simplicity of the NMR spectrum suggests that the fluorines are all readily interchangeable. SV made calculations for D_{3h}, C_{2v}, C_{4v}, and C_s symmetries, for which E_h values -837.8366, -837.8323, -837.8289, and -837.8077 were found. From these values, the computed barrier to Berry's [58] pseudorotation (via C_{4v}) is 4.8 kcal/mole. Earlier work (with similar conclusions on PH_5) is reviewed. Orbital energies and a population analysis for the valence MOs for D_{3h} symmetry are given. The energies of "polytopal rearrangements" in phosphoranes in general are discussed. (For the "polytopal bond" in LiNC, see Section IV,C.)

With a large basis set except for the absence of polarization functions on the H atoms, Shih *et al.* [59] have made SCF plus MRD-CM calculations on the ground state of PH_5 in D_{3h}, C_{4v}, and C_s geometries. The computed barrier to Berry's pseudorotation via C_{4v} from the stable D_{3h} is only 1.0 kcal. For tilted turnstile rotation (via C_s structures) a barrier of at least 3 kcal is found. The extrapolated energy of the D_{3h} form for full CM is $E_h = -343.6643$, while corresponding optimal geometry is 1.49 Å for PH axial and 1.44 Å for PH equatorial; corresponding stretching force constants are given. The mechanisms for polytopal rearrangements in more

Kutzelnigg *et al.* [34] have made CEPA calculations on PH_5. They compute force constants, and find 1.455 and 1.404 Å for the axial and equatorial PH distances. The harmonic vibration frequencies of PH_5 and PD_5 are computed. For the barrier to Berry's pseudorotation they find 2 kcal/mole.

Wadt and Hay [60] have made SCF calculations, using relativistic core potentials for the inner electrons, for UF_5. A population analysis indicates a charge of +2.4 on the U, and a +3 configuration $6s^2\,6p^6\,5f^3$ is assumed for the outer electrons of the U. A double-zeta basis (4s 4p 3d 4f) contracted to (3s 3p 2d 2f) is used for these, and nonrelativistic (4s 5p) contracted to (2s 2p) for the fluorines. Using calculations which seem to assume only one 5f electron on the uranium, the authors after allowing for spin–orbit coupling arrive at a Γ_9 ground state of the double group (an unsplit doublet state) of D_{3h} symmetry, but with a C_{4v} state that is only 1 kcal/mole higher. The two are connected by a monotonic C_{2v} path, so that fluxional behavior (Berry pseudorotational) is expected. The authors conclude that this last conclusion is well established, in spite of the absence of CM, which of course could considerably affect all the energies. Oscillator strengths for transitions to a number of excited states are given. Contour plots of the various uranium f orbitals are shown.

Demuynck *et al.* [61] have made SCF calculations on VF_5, VF_5^-, $MnCl_5^{2-}$, and $CuCl_5^{3-}$ and report geometries, metal–ligand overlap populations, and σ and π charge transfers from metal to ligand. The symmetries as calculated are all D_{3h}; but the energy is only 2–5 kcal/mole higher for C_{4v}, and Berry pseudorotation is expected. Experimentally, $MnCl_5^{2-}$ is C_{2v} (distorted C_{4v}).

Peyerimhoff and Buenker [62] have made MRD-CI calculations on $CFCl_3$ and $CFCl_3^-$, also $CFCl_3^+$, with a double-zeta basis set. Potential curves are obtained for the dissociation of one Cl atom. The curve for $CFCl_3^-$ (which is more stable than $CFCl_3$) crosses the curve for $CFCl_3$ near its minimum, showing that dissociative Cl^- detachment (dissociative electron attachment) should occur with almost zero-energy electron impact. A table of orbital energies of $CFCl_3$ is given. The two lowest vertical ionization energies of $CFCl_3$, in good agreement with experiment, and vertical energies to the two lowest singlet and triplet excited states are computed and used to identify experimental observations.

Jemmis *et al.* [63] have made SCF calculations on the geometry and energy of $CH_3Li_2^+$, $CH_3(BeH)_2^+$, $CH_3Na_2^+$, and $CH_3(MgH)_2^+$, using STO-3G up to 6-31G* basis sets. For $CH_3Li_2^+$, the CH_3 lies unsymmetrically between the two Li (C_{3v} symmetry but near D_{3h}). In $CH_3(BeH)_2^+$, the symmetry is D_{3h}. In $CH_3Na_2^+$ and perhaps $CH_3(MgH)_2^+$, a C_s symmetry with both Na on MgH on one side of the CH_3 is preferred. Final conclusions may well await the use of CM. A discussion is given relative to reaction intermediates.

Serafini and LaBarre [64], using a basis of double-zeta quality, have made calculations on the semi-aromatic ring molecules borazine ($B_3N_3H_6$) and boroxine ($B_3O_3H_3$), of D_{3h} symmetry. The E_h values and orbital energies are listed, and a population analysis is given. The latter shows for borazine a σ charge transfer of 1.01 *e* from B to N, and (with transfers from the H atoms) a total σ charge on each N of -1.37 *e*; at the same time, a π charge transfer of 0.35 *e* from N to B is found. For boroxine a σ charge transfer from H and B to O of 1.14 *e* is found, and a reverse π charge transfer of 0.26 *e*.

F. ACH_3 [$LiCH_3$, H_3CBeF, H_3CMgF, NCH_3, OCH_3, OCH_3^-, CH_3X, $NiCH_3$, $Ni(CH_3)_2$], ACETYLENE DERIVATIVES; $CFCl_3$

Streitweiser *et al.* [65] have performed SCF calculations including polarization functions on methyllithium ($LiCH_3$), also on ethyl and vinyl lithium and on ethynyllithium (HC≡CLi). Geometries, population analysis, and a full discussion are given. It is concluded that the bonding is highly ionic. Geometry parameters for $LiCH_3$ are as follows: C—Li, 2.02 Å, C—H, 1.09 Å, HCH angle, 106°; for HC≡CLi, C—Li, 1.93 Å, C—C, 1.21 Å, C—H, 1.06 Å. E_h values are given (−47.0206 for $LiCH_3$). Dissociation energies and polymerization are discussed.

Hinchcliffe and Dobson [66], using a polarized basis set, made an SCF computation on $(LiNH_3)^+$, also on $LiNH_2$, LiNH, and Li_3N. In $LiNH_3^+$ they found $R(LiN) = 1.97$ Å in a nearly tetrahedral C_{3v} structure. With N as origin, they calculate a dipole moment of 3.42 D with N as origin, and give a population analysis $Li^{+0.94}N^{-0.35}(H^{+.14})_3$.

Yarkony and Schaefer [67], using an unpolarized basis set (9s 5p) contracted to (4s 2p) for C and O and (4s) contracted to (2s) for H, made SCF calculations on the ... $7a_1^2\,2e^4$, $^1A_1$ ground state and the $7a_1\,2e^4\,3e$, ^3E excited state of the acetyl cation CH_3CO^+. They also calculated or estimated the several states of the configuration ... $7a_1^2\,2e^3\,3e$. They found the ground state to be of C_{3v} symmetry, three other models of C_s symmetry having considerably higher energy. They list orbital energies for all these forms. From a population analysis, they report charge distribution $(CH_3)^{0.37} - C^{0.60} \equiv O^{0.02}$, changing on dissociation to CH_3^+ plus CO, during which the $7a_1$ MO of CH_3CO^+ becomes $5a_1$ of CO. They also discuss population analysis and dissociation for the C_s models, and they compare all four models with corresponding states of CO. In a subsequent paper, Yarkony *et al.* [68] give the results of accurate calculations of the energies of the several ... $7a_1^2\,2e^3\,3e$ states. The results differ little from the previous estimates.

Cremaschi and Simonetta [69] made 4-31G SCF calculations on the dication CH_3COH^{2+} (also on $HCOH^{2+}$) and on its dissociation path to CH_3COH^+ plus H^+.

Clark and Schleyer [70], using a 4-31G basis, made SCF calculations on several isomeric forms of CCl_3Li. The stablest form has symmetry C_{3v}, with Li, surprisingly, *under* the C_{3v} umbrella formed by CCl_3.

Baskin *et al.* [71] have made double-zeta quality SCF calculations but without polarization on CH_3BeF and on CH_3MgF. The latter molecules serve as a prototype for the well-known Grignard compounds RMgX. For CH_3BeF the predicted C—Be and Be—F distances are 1.70 and 1.40 Å, while the analogous distances for CH_3MgF are 2.08 and 1.78 Å. Dipole moments of 1.78 and 2.38 D are predicted for CH_3BeF and CH_3MgF, respectively. Orbital energies and atomic populations are tabulated, also quadrupole and second and third moments, field gradient tensor, and potential at each nucleus.

Yarkony *et al.* [72] have made double-zeta plus polarization function calculations on the three lowest-energy states of methyl nitrene (NCH_3).

For the $^3A_2$ ground state, they predict R(CH) = 1.09 Å, R(CN) = 1.47 Å, θ(NCH) = 110°, dipole moment, 2.11 D. The excited states have similar properties, with ^1E at 14,200 cm^{-1} up and $^2A_1$ at 27,700 cm^{-1} up. Analogies to NH are close. Detailed population analyses of the MOs are given. Orbital energies are given. Molecular properties are computed and tabulated *in extenso*.

Yarkony *et al.* [73] have made double-zeta quality calculations, without polarization, on the ground and first excited states of the radical OCH_3, and the ground state of OCH_3^-, the simplest derivatives of OH and OH^-. The following are the geometry predictions: OCH_3, ^2E ground R(CH) = 1.08 Å, R(CO) = 1.44 Å, θ(OCH), 109°; $^2A_1$, R(CH) = 1.08 Å, R(CO) = 1.65 Å, θ = 102°; OCH_3^-, $^1A_1$, R(CH) = 1.12 Å, R(CO) = 1.33 Å, θ = 114°. Population analyses are given and discussed. Of the MOs in the ground-state configuration ($1a_1^2\,2a_1^2\,3a_1^2\,4a_1^2\,1e^4\,5a_1^2\,2e$, ^2E) of OCH_3, $3a_1$, 1e, and $5a_1$ are bonding, $4a_1$ and 5e are antibonding. The ^2E state suffers a small Jahn–Teller distortion.

Rappé and Goddard [74] have made a GVB calculation on $NiCH_3$ (also on $NiCH_2$, see Section IX,H,4; see Section IV,I for a calculation on NiCO). As ground state they find a combination of Ni ... 4s $3d\sigma^2\,3d\pi^4\,3d\delta^3$ from ... 4s $3d^9$, ^3D with CH_3, involving a strong bond between 4s of Ni (with a little $3d\sigma$ mixed in) and a σ orbital of CH_3. Two excited states are also calculated, with Ni ... $3d\sigma^2\,3d\pi^3\,3d\delta^4$ and Ni ... $3d\sigma\,3d\pi^4\,3d\delta^4$, both up about 10 kcal/mole.

Åkermark *et al.* (74a) have made SCF plus valence-shell CM on several low-energy states of $Ni(CH_3)_2$, finding nearly always a linear arrangement of the C—Ni—C axis. They orient the CH_3 groups so as to produce D_{3h} symmetry, and find a $^3E''$ state lowest, followed closely by $^3E'$, $^3A_2'$, and $^3E''$ (the $^3E''$ and $^3E'$ states subject of course to Jahn–Teller splitting). Potential curves for the energies as functions of the C—Ni—C angle are shown. Some population analyses are given, and some density contours, and the chemistry of dialkylnickel compounds is discussed. A negative charge on the methyl groups is indicated. SCF calculations are also made on the complex $Ni(CH_3)_2(H_2O)_2$, when the C—Ni—C angle was taken to be 90°.

Brundle *et al.* [75], in an article primarily on the photoelectron spectra of the fluoromethanes, report Gaussian SCF calculations of double-zeta quality but without polarization functions. A table of computed ground-state properties is given (E_h, orbital energies, dipole, quadrupole, and second moments, diamagnetic shieldings). For CH_3F and CHF_3, the respective computed dipole moments are −2.597 and +2.245. Computed atomic charges for CH_3F are C −0.14; H 0.17, F −0.38; and for CHF_3 they are C 0.67, H 0.18, F, −0.28.

Schlegel *et al.* [76], using 4-31G basis SCF calculations, have computed force constants and vibrational frequencies for CH_3F and CH_3Cl. Agreements with experiment are moderately good. Skancke [77] has made double-zeta plus chlorine d SCF calculations on CH_3Cl, and gives a partial population analysis. She obtains a calculated C—Cl distance of 1.773 Å (experimental: 1.784) and a dipole moment of 1.93 D (experimental: 1.86).

Peyerimhoff and Buenker [77a], using MRD-CI calculations, obtained potential curves for $FCCl_3$ and its positive and negative ions in symmetry C_s for the separation of one of the Cl. The curve of lowest energy is one in which $CFCl_3^-$ dissociates to $CFCl_2$ plus Cl^-. As the distance from Cl^- to $CFCl_2$ is decreased, the $CFCl_3^-$ curve rises until it crosses the $CFCl_3$ curve (which is stable with respect to dissociation of a neutral Cl atom), close to the latter's minimum, thus showing that dissociative electron attachment can occur for $CFCl_3$ for electrons of nearly zero kinetic energy. A complete table of orbital energies for $CFCl_3$ is given. The Cl-dissociation curves for several states of valence-shell excitation and ionization (the first two ionized states vertically at 11.92 and 12.34 eV) are shown. All the results are in agreement with experimental evidence. The dissociative electron attachment results are relevant to the behavior of $CFCl_3$ and similar molecules in the upper atmosphere.

Goscinski *et al.* [78], using a rather small unpolarized basis set, computed potential surfaces for the $\nu_2(a_1)$ "umbrella" mode of distortion for the ground state and the F_{1s} hole state of CH_3F. It is concluded that the F at the first moment is unbound in the core-ionized state. The unusually broad observed width of the fluorine 1s ESCA line is explained. A similar explanation is given for the ESCA spectrum of CF_4 under the assumption (justified by calculations in other cases) that the 1s ionization is localized in one fluorine atom.

Roelandt *et al.* [78a], using an approximately double-zeta basis set, have made SCF calculations on SiH_3F (also SiH_4 and SiH_2F_2). They give tables of orbital energies and discuss the nature of the MOs, using a population analysis.

Semlow *et al.* [79] have made SCF and PNO-CEPA calculations on CH_3Cl and SiH_3Cl (also SiH_4, see Section VII,B,2). They used good basis sets (9s 5p 1d) for C, (11s 7p 1d) for Si, (5s 1p) for H, (?) for Cl. "All the basis sets are only slightly contracted in the core region." The five valence-electron vertical ionization energies were calculated in each of three ways: (a) by Koopmans' theorem using ground-state orbital energies; (b) by ΔSCF: the difference between SCF of ionic and neutral molecule state; (c) by ΔCEPA: the difference between ionic and neutral using CEPA method. Agreements with experiment for ΔCEPA were good. Core ionization energies for inner and outer cores of Cl and Si were computed by methods (a) and (b). (Perturbation theory calculations were also made.) Agreements with experiment for CH_3Cl core ionization potentials by ΔSCF were good.

del Conde *et al.* [80], using an essentially double-zeta basis set (with two 3d's for Br but no polarization for C or H), have made SCF calculations of vertical ionization potentials for most of the MOs of CH_3Br using methods (a) and (b) of the preceding paragraph. Three of the ΔSCF values for the valence electrons agree with experiment; a fourth value does not match a questionable experimental value.

Barthelot *et al.* [81], using a pseudopotential method, investigated various geometric models for C_2SiH_4 finding the ground state to be of structure $H_3Si{\equiv}CH$.

Kollman *et al.* [82] have made a (9s 5p/4s) basis SCF calculation on the rotational barrier for the molecule $H_3C{-}C{\equiv}C{-}CH_3$ and STO-3G calculations on those for $H_3Si{-}C{\equiv}C{-}CH_3$ and $F_3C{-}C{\equiv}C{-}CH_3$. The calculated barriers are 5.6, 3.3, and 3.8 kcal/mole, respectively, the eclipsed configuration being the more stable.

Von Niessen *et al.* have made Green's function calculations on ionization potentials of a number of CH_3 and CF_3 derivatives of acetylenic compounds; see under Bleri *et al.* in Section VIII,B,2.

G. ETHANE (C_2H_6) AND SILANE (Si_2H_6)

The ethane molecule at room temperature shows essentially free rotation between a more stable (D_{3d}) staggered and a 2.93 kcal/mole less stable (C_{3v}) eclipsed conformation. For the former, the electron configuration is ($1a_{1g}^2\ 1a_{2u}^2\ 2a_{1g}^2\ 2a_{2u}^2\ 1e_u^4\ 3a_{1g}^2\ 1e_g^4$, ^1A_g. Many SCF computations have been made on C_2H_6 [2]. Perhaps closest to the limiting accurate SCF is a computation using an extended basis and including d but not f basis orbitals, by Clementi and Popkie [83]. They give orbital energies and an atomic population analysis for three values of the R(CC) distance. They also make comparable calculations on C_2, C_2H_2, and C_2H_4, and discuss the relations of these molecules.

Much attention has been given to trying to understand in detail the nature and reason for the rotation energy barrier between D_{3d} and C_{3v} ethane. On the whole, this is perhaps rather useless: the main point is that several computations of the difference in energy agree well with experiment. However, the consensus is that exchange repulsion is the key factor; rather obviously this is greater for the eclipsed configuration. This is seen also in the computed equilibrium R(CC) values [83]: 1.551 Å for the staggered and 1.569 Å for the eclipsed form. For the HCH angle, the respective values are 107.0° and 107.3°.

A detailed analysis has been given by Morokuma and Umeyama [84] in terms of Morokuma's decomposition of the SCF interaction between two parts of a molecule into electrostatic, polarization, exchange, charge transfer, and mixed coupling components. The contributions of each of these to the staggered and eclipsed energies and to the barrier are evaluated, and by far the largest contribution to the barrier is found to result from exchange. In the same paper, the authors make similar analyses for the rotation barrier in CH_3NH_2 and CH_3OH.

Buenker and Peyerimhoff [85] have reported double-replacement CM calculations on the ground ($E_h = -79.18226$) and 15 approximately Rydberg-type excited states of ethane, and the lowest 4 states of its positive ion. The Gaussian basis set used is of essentially double-zeta quality without polarization functions, augmented by one-component s, p_x, p_y, and p_z functions, all with the same exponent to give approximately 3-quantum Rydberg character to the upper-state MOs of C_2H_6. These MOs are found to be not quite pure Rydberg, but to have some valence-shell character. Orbital energies for the ground state from SCF calculations are listed. Vertical transition energies to the excited states of C_2H_6 involving excitation from the $3a_{1g}$ and $1e_g$ occupied MOs to 3s, $3p\sigma$, and $3p\pi$ Rydberg MOs, and corresponding oscillator strengths, are reported.

A comparison with the observed absorption spectrum shows that the first strong region of absorption near 1400 Å contains two parallel type (to $^1A_{2u}$) and three perpendicular type (to ^1E_u) transitions ($3a_{1g} \rightarrow 3p$ and $1e_g \rightarrow 3p$ transitions). At longer wavelengths there is some much weaker absorption attributable to $3a_{1g} \rightarrow 3s$ and $1e_g \rightarrow 3s$ transitions.

Richartz *et al.* [86] have calculated potential surfaces for C_2H_6 and the three lowest-lying states of $C_2H_6^+$ using SCF and CM methods. For E_h of C_2H_6, -79.3933 was found for the largest secular equation solved; extrapolated to the limit, $E_h = -79.3976$. The authors use an essentially double-zeta basis augmented by polarization functions centered at the middle of the C—C bond. The authors discuss the photoelectron spectrum (11.5–12.6 eV) of ethane in detail, and conclude that the ^2B_g component of the Jahn–Teller distorted ^2E_g state of $C_2H_6^+$ from $1e_g$ ionization is responsible for the lowest-energy feature in the spectrum; at higher frequencies the $^2A_{1g}$ state from $3a_{1g}$ ionization mixed with the $^2A_{1g}$ component of the ^2E_g comes into the picture.

In another article using a similar basis set, Richartz *et al.* [87] make CM computations on the following: (a) vertical ionization energies for ionization out of the $1e_u$, $2a_{2u}$, and $2a_{1g}$ MOs of C_2H_6; (b) at R_e of $C_2H_6^+$ ($^2A_{1g}$), vertical excitation energies of several excited states including Rydberg states of the ion. With respect to (a), there is reasonable agreement with the photoelectron spectrum.

Clementi and von Niessen [88] have made a "bond energy analysis" of C_2H_6, decomposing the total energy into one-, two-, three-, and four-center contributions.

Hinchcliffe [89] has made SCF calculations on Si_2H_6 using a (12s 9p 2d) basis contracted to (7, 5, 2) for Si and (5, 1) contracted to (3, 1) for H. He computed a barrier height of 2.32 kJ/mole for the more stable staggered form to the less stable eclipsed form. He found for the Si—Si distance 2.36 Å for the staggered and 2.37 Å for the eclipsed forms, and used these and the corresponding HSiH angles (110.52° for the former, and 110.63° for the latter) in computing the barrier. The observed bond length is 2.33 Å. He lists orbital energies and E_h values for both staggered and eclipsed forms. He gives total charge $(Si^{0.1976})_2(H^{-0.0659})_6$ for staggered and $(Si^{0.1931})_2(H^{-0.0602})_6$ for eclipsed Si_2H_6.

Snyder and Wasserman [90], in a 4-31G SCF calculation, report $R(\text{SiSi}) = 2.35$ Å and $R(\text{SiH}) = 1.49$ Å computed. They list orbital energies, the highest four showing some agreement with photoelectron ionization potentials.

H. MOLECULAR COMPOUNDS

1. Survey

Electron donor molecules form rather stable *compounds*, and looser *complexes*,[1] with electron acceptor molecules [91]. Lone-pair donors form dative compounds with vacant-orbital acceptors; an example which is isoelectronic with ethane is borazane (H_3NBH_3). Looser complexes are formed when one or both partners are lacking in lone pairs or vacant valence-shell acceptor orbitals. All these compounds and complexes can be formulated in terms of a positive electron transfer from the donor to the acceptor molecule with a wave function which can be written in valence-bond theory as

$$\Psi = a\Psi(\text{D, A}) + b\Psi(\text{D}^+ - \text{A}^-) + \ldots$$

with $a > b$ (D = donor, A = acceptor). In MO theory, the description is rather different and more complicated, but a population analysis of the charge distribution shows partial electron transfer from D to A, though

[1] No sharp distinction between compounds and complexes can be made.

not necessarily from just one atom of D, e.g., N in NH_3 to one atom of A, e.g., B in BH_3.

A special characteristic of typical molecular compounds and complexes is that, because we start with closed-shell molecules, relatively little change is to be expected in the correlation energy when the molecule or complex is formed. In other words, the SCF MO wave function without CM should be a rather good approximation at all distances of approach of the two partners.

In molecular compounds and complexes, an instructive analysis of the interaction energy of a donor and an acceptor into several terms can be made by considering the energy of the complex computed under certain constraints. Earlier, perturbation theory was used to compute these terms. More recently, Morokuma and colleagues [92] have carried out the analysis using MO theory. One term is the classical Coulomb electrostatic energy (ES) obtained by subtraction of the computed SCF energy of the complex from that of the nonantisymmetrized product of the SCF wave functions of the two rigid partners. Next is the polarization energy (PL), the further energy obtained using the Hartree product of the two partner molecule wave functions with each individually optimized in the presence of the partner. ES is usually an attraction, but may be a repulsion; PL is always attractive. Next is the exchange interaction (EX), the further energy, always a repulsion, obtained if the product of the partner wave functions is now antisymmetrized. Then there is the additional stabilization energy, an attraction, obtained in the complete SCF wave function of the complex upon allowing charge transfer (CT) between the partners. Finally there are mixed terms (MIX).

A better procedure for obtaining the CT energy and separating it from the MIX term has been given by Kitaura and Morokuma [92]. In this, the forms of the MOs of the complex are compared with those of the donor and acceptor, and one finds changes in which there is partial transfer to previously unoccupied orbitals of each partner (PL), CT from occupied MOs of one partner to unoccupied MOs of the other, and other shifts due to EX.

In the above, changes in dispersion energy (which come in under CM) are not included, but in stable complexes this is not very important.

2. Molecular Compounds and Complexes of NH_3 (NH_3BH_3, NH_3BF_3, NH_3 + HCl, $LiNH_3$, $NaNH_3$, FNH_3, $NH_3 \cdot X_2$); also PH_3BH_3, $LiPH_3$, $CH_3 \cdot SiH_3$, and $N(CH_3)_3$ Complexes

With respect to rotation of BH_3 relative to NH_3 in borazane there are two configurations differing slightly in energy, like those in ethane, but the symmetry is C_{3v} for both. Moreau and Veillard [93] have made an SCF calculation on borazane including d functions for B and N but no p functions for H, assuming the angles tetrahedral. They found a population of polarity $(BH_3)^-(NH_3)^+$ due mainly to negative charges on the boron hydrogens and increased positive charges on the nitrogen hydrogens, without much transfer of charge between N and B (0.17 on B, −0.78 on N, but the absence of p polarization on the hydrogens might affect this). Orbital energies are listed. The B—N bond in borazane is weak (overlap population 0.13 *e*). In an improved computation including p polarization on the H atoms, and making allowance for the different equilibrium bond lengths for the staggered and eclipsed configurations, Veillard [94] reports 3.07 kcal/mole for the barrier to internal rotation.

Similar computations on methysilane [93, 94], another relative of ethane, gave a barrier of 1.44 kcal/mole, in good agreement with the experimental value of 1.71. Here no allowance was made for changed geometry between the two configurations.

Returning to borazane, there have been several other articles of interest, but these were mostly carried out with relatively poor basis sets. In an early article by Peyerimhoff and Buenker [95], a Gaussian-lobe basis set was used for B and C and a five-component expansion of $e^{-\zeta r}$ for the hydrogen s functions, in an SCF calculation. The BNH and NBH angles were both fixed at 110° for the computations. Because $H_3N \cdot BH_3$ is formed from two closed-shell molecules BH_3 and NH_3, little change in correlation energy is to be expected when borazane is formed. This expectation was confirmed by a fairly extensive CM computation, which showed only a 0.011 hartree lowering of the computed energy. The computed dipole moment also was little changed by CM (from 5.71 to 5.72 D; experimental: 4.9 D). Twelve excited states were computed in the CM treatment. The lowest-energy predicted singlet–singlet ($^1E \leftarrow {}^1A_1$) absorption is at 74,000 cm^{-1}, in the far UV like that for the isolectronic ethane. In the SCF calculations, comparisons were made with models of C_s and C_{2v} symmetry in connection with a possible resemblance to B_2H_6 (D_{2h} symmetry). However, the stable symmetry is C_{3v}. In the same paper, similar comparisons with B_2H_6 were made in extensive computations on $C_2H_6{}^{2+}$. Most of the MOs of borazane are found to be nearly unchanged NH_3 and BH_3 MOs. Only $4a_1$ and especially $5a_1$ (for which contour diagrams are shown) show weak B—N binding corresponding to partial charge transfer from N(2pσ) to B(2s).

Fujimoto *et al.* [96] have made an interesting analysis of the electronic structure of borazane by constructing its MOs as linear combinations of those of the two molecules NH_3 and BH_3. They assumed an HNH angle of 110°, an HBH angle of 105.4°. It is not clear how large a basis set they used. The major effect which they find is an extensive transfer from the $3a_1$ highest-occupied MO of NH_3 to the unoccupied $3a_1$ MO of BH_3. This result is in accord with the usual electron donor–acceptor theory. Changes in other MOs of NH_3 and BH_3 are relatively small.

In an SCF computation but using a minimal basis set, Dixon and Palke [97] have computed a number of molecular properties of borazane. In an earlier paper [98] Palke discusses the barrier to free rotation.

Umeyama and Morokuma [99] made SCF calculations on borazane and several other strong complexes using the split-valence 4-31G basis set. For borazane, computations were made both with planar BH_3 and with an angle HBZ of 106.3°, where Z is the C_{3v} axis. The new conclusions are as follows: (a) the binding of BH_3 to NH_3 is due more to the electrostatic forces than to charge transfer and polarization combined; exchange fights against all these; (b) there is a substantial contribution of the $H_3N \leftarrow BH_3$ back CT stabilization ($\cong$7 kcal/mole); (c) the barrier to internal rotation, as in C_2H_6, is due to the overlap repulsion exchange terms; (d) the stable C_{3v} structure is destabilized on axis tilting by a reduction in the electrostatic stabilization and an increase in exchange repulsion. Borazane is rather stable; the computed stabilization energy for pyramidal BH_3 is 44.7 kcal/mole, but 17.0 kcal/mole is needed to go from stable planar BH_3 to pyramidal BH_3, leaving a net stabilization energy of 27.7 kcal/mole. No experimental value is available.

The authors also discuss the complexes in which one to three methyl groups are substituted for H atoms in NH_3 in borazane.

They also discuss $H_3N \cdot BF_3$, where the role of the electrostatic forces is increased. Archibald *et al.* [100] have made double-zeta (no polarization) SCF calculations on the heats of formation of $H_3N \cdot BF_3$, also $Cl^- \cdot BF_3$ (stable), $Ar \cdot BF_3$, $Ne \cdot BF_3$, and $OHC \cdot BF_3$ (all unstable). Barber *et al.* [101] have also made double-zeta SCF calculations on $H_3N \cdot BF_3$, $CH_3CN \cdot BF_3$, and $C_2H_5NH_2 \cdot BF_3$, finding that the $2p\sigma_N$ population decreases, the $2p\pi_N$ population increases on complexing, with corresponding changes in BF_3.

Similar to borazane is phosphineborane (H_3PBH_3). Demuynck and Veillard [102] have made comparative SCF computations using an ample basis set on $H_3P \cdot BH_3$ and H_3PO, assuming tetrahedral angles in H_3P and BH_3. A population analysis including overlap shows P dσ populations of 0.08 e in H_3PBH_3 and 0.14 e in H_3PO, P dπ populations of 0.05 e in H_3PBH_3 and 0.22 e in H_3PO. Sabin [103] has reported an STO-3G calculation on phosphineborane, and as expected found the staggered configuration to be the more stable. He finds a charge transfer of 0.30 e for PH_3 to BH_3, at the expense of the phosphorus hydrogens; both B and P and the boron hydrogens are found to be slightly negative.

The most accurate computations on borazane and also on phosphineborine appear to be those of Ahlrichs and Koch [104], who with good polarized basis sets added electron correlation to SCF computations using the CEPA PNO method. They assumed staggered configurations. On optimizing the B—N and B—P distances, they found by CEPA $R(BN) = 1.65$ Å and $R(BP) = 1.95$ Å. Experimentally (microwave), $R(BP) = 1.937$ Å. They give SCF E_h values and CEPA correlation energies. For the dissociation energies they compute (CEPA) 27 kcal/mole for borazane, including about 4 kcal/mole from correlation, and 17 kcal/mole for phosphineborine, including about 6 kcal/mole from correlation.

Redmon *et al.* [104a] used many-body perturbation theory, which is size-consistent, to calculate the dissociation energy of borazane. They used a slightly poorer basis set than Ahlrichs and Koch [104]. They obtain 30 kcal/mole, of which 32% is from correlation.

Hillier *et al.* [105] have made a minimal basis (including 3d on P) SCF computation on $PF_3 \cdot BH_3$. They give a population survey, and several computed valence-shell orbital energies, which are compared with photoelectron ionization potentials.

Clementi [106] has made a comprehensive SCF computation of the system NH_3 + HCl all the way from separate molecules to the molecule NH_4Cl. This molecule has been shown to exist to a minor extent in the hot vapor, but on cooling immediately polymerizes to crystals of $NH_4^+Cl^-$ structure. In the isolated molecule, according to Clementi's calculations, the H of HCl has become bound to NH_3, and some charge transfer from NH_3 and H to Cl has occurred (already there was $H^{0.33}Cl^{-0.33}$ structure in HCl), but the structure of the NH_4Cl molecule at equilibrium is rather far from being $NH_4^+Cl^-$; it is approximately $NH_4^{+0.6}Cl^{-0.6}$. Clementi finds 0.84 eV (19.45 kcal/mole) for the energy of formation. In the process, the charge on the N atom remains nearly constant at $-1.1\ e$; the transfer of charge from the NH_3 to the Cl occurs at the expense of the H atoms of NH_3, whose charge goes from $+0.36\ e$ toward $+0.52\ e$ in NH_4^+. The H atom of HCl becomes more positive as it approaches the N. Clementi describes the changes in great detail (including the changes in the individual MOs) in his second paper.

His papers are of much interest not only for the later stages of the approach to form NH_4Cl, but also for the intermediate stages which correspond to normal hydrogen bonding. In NH_4Cl a stage far beyond ordinary H bonding is reached.

Using a larger basis set (12s 9p 1d) for Cl, (9s 5p 1d) for N, (5s 1p) for H, contracted, respectively, to (6s 4p 1d), (4s 2p 1d), and (2s 1p), and supplemented by full valence-shell CM, Raffenetti and Phillips [107] have greatly improved on Clementi's calculations. The authors compute vibration frequencies for the equilibrium NH_4Cl molecule, and compare with limited experimental evidence. They find a dissociation energy, after a zero-point energy correction of 3.1 kcal/mole, $D_0^0 = 5.9$ kcal/mole, much less than Clementi's value. A population analysis gives at equilibrium a charge transfer of 0.1 *e* from NH_3 to HCl, or 0.35 *e* from NH_4 to Cl. They show a good contour diagram of a potential surface for the molecule. E_h values are given.

Douglas *et al.* [108] have made an SCF calculation on $NH_3 \cdot SO_3$, using a 431-G basis set. They assumed an eclipsed C_{3v} configuration and an NH_3 structure as in the free molecule, and on optimizing with an STO-3G basis, they found an OSO angle of 117°. With R(NS) at 2.55 Å, they found a dissociation energy of 18 kcal/mole. They report also a Morokuma analysis of the charge distribution, finding more charge transfer than in the $NH_3 \cdot SO_2$ complex.

Lucchese and Schaefer [109] have made SCF calculations of double-zeta quality (with d polarization in the case of $NH_3 \cdot F_2$ and $NH_3 \cdot ClF$) on the H_3N complexes of F_2, Cl_2, and ClF, also on derivatives with $N(CH_3)_3$ instead of NH_3. These are of interest as prototypes for the well-known [91], much more stable I_2 complexes of these and similar molecules. C_{3v} symmetry is assumed for the $H_3N \cdot XY$ complexes, as seems reasonable. Predicted N—X distances are 3.08 Å in $H_3N \cdot F_2$, 2.93 Å in $H_3N \cdot Cl_2$, 2.65 Å in $NH_3 \cdot ClF$, with respective binding energies of 0.6, 2.4, and 7.7 kcal/mole. $NH_3 \cdot FCl$ is predicted to have a binding energy less than 0.1 kcal/mole. The predicted binding energies are larger for the NH_3 than for the corresponding $N(CH_3)_3$ complexes; this is explained by a greater negative charge on the N atom in the former case, as computed by population analysis.

In 1970, F. B. van Duijneveldt and R. S. Mulliken made much more accurate SCF calculations on the $H_3N \cdot Cl_2$ complex (including d functions on the Cl atoms), over a large range of distances. The individual MOs were analyzed, and we showed that charge transfer increases rapidly with closer approach of the two partners, becoming relatively more important than what we called induction. Unfortunately, the work is unpublished.

Umeyama *et al.* [110] have made 4-31G SCF calculations and from these, an energy decomposition analysis for the complexes $H_3N \cdot F_2$, $H_3N \cdot Cl_2$, $H_3N \cdot ClF$, and others. They give for $H_3N \cdot F_2$ and $H_3N \cdot ClF$ tables showing the relative importance of the several terms in the energy analysis at several distances of approach, also instructive figures showing the effects of the separate terms EX, PL, and CT (see Section H) on the charge density change from separate partners to complex; also they give the changes in the total distribution. They show that the most stable configuration is C_{3v}.

Trenary *et al.* [111] have reported computed interaction energies, geometries, and charge transfers in $LiNH_3$, $LiPH_3$, $NaNH_3$, $H_3N{\cdot}F$, and others. The largest stabilization energy found was for $LiNH_3$ (15 kcal/mole).

3. Borine Carbonyl ($OC{\cdot}BH_3$)

From the valence-bond point of view, borine carbonyl is an unusually stable two-way complex [91] or compound. Sigma charge is transferred from the 5σ donor MO of CO into the $3a_1$ vacant orbital of BH_3, while hyperconjugative π charge is transferred to a lesser extent from the quasi-π BH_3 MO into the previously vacant 2π MO of CO. Unlike BH_3CO, BF_3CO is unstable, both experimentally and computationally. This instability is explained in part by the unusual stability of planar BF_3 resulting from the migration of π charge from the fluorine orbitals into the boron vacant orbital.

Kato *et al.* [112] have constructed the MOs of $OC{\cdot}BH_3$ as linear combinations of those of CO and BH_3. They used a minimal basis set of STFs expanded in terms of Gaussians. The main effect which they find is a loss of 0.39 e from 5σ of CO and a gain of 0.48 e in the $3a_1$ MO of BH_3. Other intramolecular transfers (to previously unoccupied MOs), and intermolecular transfers, are much smaller. However, 2π of CO gains 0.13 e and 1 e of BH_3 loses 0.13 e.

Using a Gaussian basis of double-zeta quality in an SCF calculation, Ha [113] obtains 10.98 kcal/mole, and in a moderately inclusive CM treatment, 14.56 kcal/mole, as compared with an experimental value of 16.6 (corrected to 0°K from a value 18.8 kcal/mole at 273°K). The included conversion energy of BH_3 necessary from D_{3h} to C_{3v} is calculated as 15.97 kcal/mole. The calculated SCF energy was -139.04 a.u. at the calculated equilibrium geometry of $R(BC) = 1.58$ Å and HBH angle 113°. From a population analysis, complex formation causes a loss of σ charge from C of 0.23 e, from O of 0.06 e, and a gain by B of 0.25 e; a loss of charge from 3H of 0.12 e, of π charge from B of 0.01 e, and a gain of π charge by C of 0.20 e and a loss by O of 0.02 e. These figures, although not entirely comparable, are qualitatively but not quantitatively like those of Kato *et al.* Ha also lists computed electric field gradients for B in BH_3CO.

In an SCF calculation using a good basis set including polarization functions, Ermler *et al.* [114] obtain $E_h = -139.1723$ and estimate -139.210 ± 0.004 as the SCF limit. From this and an experimental value of -139.954 for BH_3CO they obtain a heat of formation of 25.1 kcal/mole; note that this differs considerably from the value 18.8 given by Ha; the difference may be due to uncertainty in the dissociation

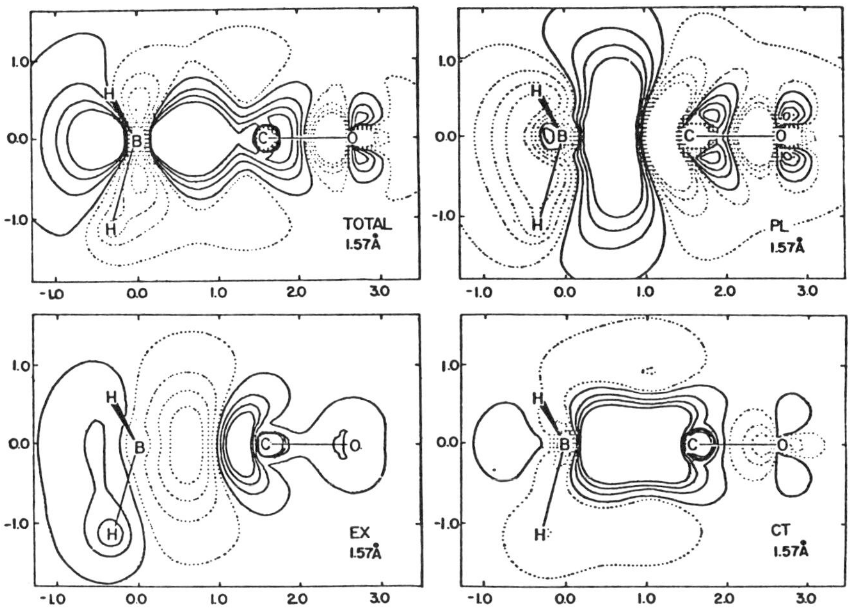

FIG. 1. Component electron density maps for the complex $OC \cdot BH_3$ at R(CB) = 1.57 Å in the C_{3v} approach. Contours indicated are, succesively, ±1, ±5, ±9, ±13 × 10^{-3} ($bohr^{-3}$). From Umeyama and Morokuma [99].

energy of B_2H_6 to $2BH_3$. Ermler *et al.* (see their Table III) find 6.4 kcal/mole as the SCF-estimated contribution to the heat of formation of BH_3CO from BH_3 and CO; subtracting from 25.1 leaves 18.7 kcal/mole as the contribution of correlation energy to the heat of formation. In view of the closed-shell character of both BH_3 and CO, this is unexpectedly large. This "molecular extra correlation energy" is much larger than the contribution of 3.58 kcal/mole found by Ha, though with a poorer basis set. In any event, it appears that the correlation energy contribution is larger here than in typical cases (cf., e.g., the discussion of borazane in Section I and of diborane in Section VIII,C,2). Ermler *et al.* give a table of orbital energies and a table of selected properties of the ground state of BH_3CO. In an extensive discussion, they obtain localized MOs and present contour diagrams for these. Population analyses are given.

Redmon *et al.* [104a] used many-body perturbation theory, which is size-consistent, to calculate the dissociation energy of borine carbonyl. They used a slightly poorer basis set than Ahlrichs and Koch [104]. They obtain 30 kcal/mole, of which 32% is from correlation.

Umeyama and Morokuma [99] give an energy and charge decomposition analysis (cf. Section H) for $OC \cdot BH_3$, using a split-valence 4-31G basis. Stability is found to be a maximum for C_{3v} symmetry. At the experimental B—C separation of 1.57 Å, the ES, PL, and CT contributions to the calculated formation energy of −27.9 kcal/mole for pyramidal BH_3 (−11.5 net for planar BH_3) are −70, −82, and −90 for ES, PL, and CT, respectively, as opposed to 119 for EX and 95 for MIX. An electron density map for the changes in charge distribution from BH_3 + CO to $OC{\cdot}BH_3$ is given, together with maps of the contributions of EX, PL, and CT to this total. These maps are reproduced here as Fig. 1.

I. TRIMETHYLENEMETHANE [$C(CH_2)_3$]

The diradical trimethylenemethane has been much discussed theoretically, but only in 1966 was it prepared experimentally by Dowd [115]. The ground state is a triplet, almost certainly planar, of symmetry D_{3h}. The MO electron configuration would then be

$$(1a_1')^2\,(1e')^2\,(2a_1')^2\,(3a_1')^2\,(2e')^4\,(4a_1')^2\,(3e_1)^4\,(1a_2')^2\,(4e')^4\,(1a_2'')^2\,(1e'')^2$$

The last four electrons are π electrons, of which $(1e'')^2$ is an open shell that should give rise to a $^3A_2'$ ground state and excited states $^1E'$ and $^1A_1'$. However, according to the Jahn–Taller theorem, the $^1E'$ must split into two nondegenerate states of lower symmetry, a $^1B_2$ and a $^1A_1$ of C_{2v} symmetry; in this symmetry, the ground state would become $^3B_2$. In valence bond theory, the ground state is a mixture of three equivalent resonance structures, each with a double bond from the central atom to one CH_2 group and an odd π electron on each of the other two CH_2 groups.

Yarkony and Schaefer [116] have carried out double-zeta level Gaussian SCF calculations assuming D_{3h} or C_{2v} symmetry. Several empirical calculations and other indications predict that in the $^1B_2$ component of $^1E'$ one CH_2 group is orthogonal to the other two, giving a $^1B_1$ C_{2v} state of somewhat lower energy. This is 20 kcal/mole above the planar $^3A_2'$ ground state according to GVB with CM calculations by Davis and Goddard [117]. They find the ($^1A_1$, $^1B_2$) planar C_{2v} component of $^1E'$ to be 26 kcal/mole above the ground state (see Fig. 2, taken from their paper). The $^1A_1'$ state of $(1e'')^2$ is much higher in energy (see Fig. 2). The nature of the two components of the $^1E'$ state has been very extensively discussed by Borden and Davidson [118], but the matter is too complicated for a brief review. Yarkony and Schaefer give tables of MO orbital energies for the D_{3h} and the orthogonal C_{2v} configurations. In another study [119] these authors report SCF ionization energies for the five states that arise upon removal of the 1e″ or 1a″ electron.

In connection with the interpretation of the photoelectron spectrum of

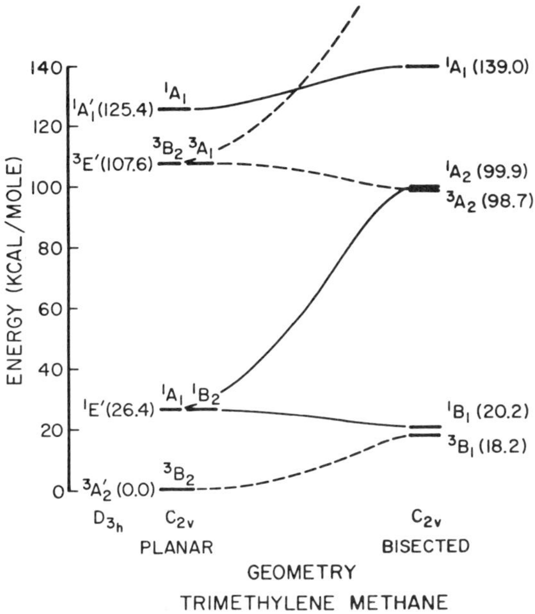

FIG. 2. From Davis and Goddard [117].

trimethylenemethane, Yarkony and Schaefer [120] have made SCF calculations on the ground state ... $1a_2''^2\ 1e''$, $^2E''$ and the four excited states $1a_2''\ 1e''^2$, $^4A_1''$, $^2A_1''$, $^2E''$, and $^2A_2''$ "of the ion." The lowest ionization potential is 8.7 eV by a frozen orbital approximation and 8.03 eV by a direct SCF calculation. The excited states show a wide energy spread from 10.15 to 18.24 eV by frozen orbital approximation, 8.99 to 16.62 eV by direct SCF calculation.

For an account of iron tricarbonyl trimethylenemethane, see Section XII,C,3.

J. CYCLOPROPANE [$(CH_2)_3$]

Another molecule of D_{3h} symmetry, a stable closed-shell hydrocarbon, is cyclopropane. The MO electron configuration is

$$(1a_1')^2\,(1e')^4\,(2a_1')^2\,(2e')^4\,(1a_2'')^2\,(3a_1')^2\,(1e'')^4\,(3e')^4,\ {}^1A_1'$$

Several SCF calculations have been published. The most accurate ($E_h = -117.05425$) appears to be one by Skancke [121], who used a (9, 3, 1) Gaussian basis including polarization (d) orbitals. She lists orbital energies and gives the following for atomic charges: -0.50 on the carbons,

+0.25 on the hydrogens. She has also done cyclopropene and cyclopropenone.

The other SCF calculations did not include polarization orbitals in the basis set. Basch *et al.* [122], obtaining $E_h = -117.0099$, computed second moments of electronic charge, quadrupole coupling constants, and diamagnetic shielding at the protons. They identify two strong absorption bands at 70,000 and 82,000 cm^{-1} as $^1E' \leftarrow {}^1A_1'$ valence-shell transitions $1a_2' \leftarrow 3e'$ and $4e' \leftarrow 3e'$, and other weaker bands as Rydberg transitions. They also discuss the photoelectron spectrum. In addition, they make calculations on and comparisons with ethylene oxide, ethylene imine, and diaziridine.

Kochanski and Lehn [123], with $E_h = -116.9918$, give a population analysis of MOs and bonds, along with contour diagrams of total density and of some MOs. They also make calculations on and comparisons with cyclopropene and diazirine.

Buenker and Peyerimhoff [124], with $E_h = -116.9164$, make calculations as a function of valence angles and display these by Walsh diagrams, and give contour diagrams for two MOs and the total density. They compare cyclopropane with the isoelectronic C_{2v} molecules O_3 and $HCOO^-$. They also discuss excited states and the electronic spectrum as well as reactivity and conjugative effects.

Stevens *et al.* [125], using a minimal SCF basis, found $E_h = -116.8322$. They give some difference density countour diagrams (comparisons with atoms).

Kao and Radom [126], using an STO-3G and 4-31G basis ($E_h = -116.882$), studied structural variations accompanying changes in the exocylic HCH angles.

For localized MOs see the article by Newton *et al.* [126a] (also for $CH_3—C{\equiv}CH$).

von Niessen *et al.* [127] with $E_h = -116.9835$, used a many-body Green's function approach for a good calculation, which agrees very well with experiment, of the five lowest vertical ionization potentials of cyclopropane.

For some molecules related to cyclopropane, see Section IX, C,1.

K. CYCLOPROPENYL CATION ($C_3H_3^+$)

Radom *et al.* [128] have made SCF calculations on the cyclopropenyl cation, the smallest aromatic molecule, with D_{3h} symmetry, using up to a 6-31G* basis set. (The latter contains d orbitals.) They report $E_h = -115.0037$. They calculated eight different isomeric structures, of which the lowest in energy is the cyclopropenyl cation. Next higher, by 34 kcal/mole, is the propargyl cation ($H—C{\equiv}C—C^+H_2 \leftrightarrow H—C^+{=}C{=}CH_2$).

Takuda and Ohno [129] also made calculations on the cyclopropenyl cation, including six excited singlet and six triplet states. They used a minimal basis set plus a d orbital on carbon (SCF $E_h = -114.7388$) and then supplemented this by valence-shell CM, obtaining $E_h = -114.9466$. They also obtained C—C distances (1.39 Å) and stretching force constants (7.92 mdyn/Å) in rather good agreement with experimental evidence.

REFERENCES

1. W. Cherry and N. Epiotis, *J. Am. Chem. Soc.* **98,** 1135 (1976); W. Cherry, N. Epiottis, and W. T. Borden, *Acc. Chem. Res.* **10,** 167 (1977).
2. See W. G. Richards, T. E. H. Walker, and R. K. Hinkley, "Bibliography of *ab initio* Molecular Wave Functions." Oxford Univ. Press (Clarendon), London and New York, 1971; W. G. Richards, T. E. H. Walker, L. Farnell, and P. R. Scott, "Supplement for 1970–1973," *ibid.,* 1974; W. G. Richards, P. R. Scott, E. A. Colburn, and A. F. Muchington, "Supplement for 1974–1977," ibid., 1978.
3. J. H. Hall, D. S. Marynick, and W. N. Lipscomb, *Inorg. Chem.* **11,** 3127 (1972).
4. M. E. Schwartz and L. C. Allen, *J. Am. Chem. Soc.* **92,** 1466 (1970).
5. P. Millie and G. Berthier, *Int. J. Quantum Chem. Symp.* **2,** 67 (1968).
6. R. E. Kari and I. G. Csizmadia, *J. Chem. Phys.* **50,** 1443 (1969).
7. R. E. Kari and I. G. Csizmadia, *Int. J. Quantum Chem.* **6,** 401 (1972).
8. Y. Ishikawa and R. C. Binning, Jr., *Chem. Phys. Lett.* **40,** 342 (1976).
9. A. Hinchcliffe, *J. Mol. Struct.* **27,** 329 (1975).
10. D. S. Marynick and D. A. Dixon, *Proc. Natl. Acad. Sci. U.S.A.* **74,** 410 (1977).
11. G. B. Ellison, P. C. Engelking, and W. C. Lineberger, *J. Am. Chem. Soc.* **100,** 2556 (1978).
12. C. E. Dykstra, M. Hereld, R. R. Lucchese, H. F. Schaefer, III, and W. Meyer, *J. Chem. Phys.* **67,** 4071 (1977).
13. A. J. Duke, *Chem. Phys. Lett.* **21,** 275 (1973).
14. E. M. Jemmis, V. Buss, P. von R. Schleyer, and L. C. Allen, *J. Am. Chem. Soc.* **98,** 6483 (1975).
15. G. T. Surratt and W. A. Goddard, III, *Chem. Phys.* **23,** 39 (1977).
16. S. Y. Chang, E. R. Davidson, and G. Vincow, *J. Chem. Phys.* **52,** 1740 (1970).
17. Y. Ellinger, A. Rasset, R. Suba, and G. Berthier, *J. Chem. Phys.* **62,** 1, 10 (1975).
18. Almlöf, A. Lund, and K.-A. Thnomas, *Chem. Phys.* **7,** 465 (1975).
19. T. Lee and T. P. Das, *Phys. Rev. A* **6,** 968 (1972).
20. R. E. Kari and I. G. Csizmadia, *J. Chem. Phys.* **56,** 4337 (1972).
21. R. M. Stevens, *J. Chem. Phys.* **61,** 2086 (1974).
22. R. Ahlrichs, F. Driessler, H. Lischka, V. Staemmler, and W. Kutzelnigg, *J. Chem. Phys.* **62,** 1235 (1975).
23. H. Lischka and V. Dyczmons, *Chem. Phys. Lett.* **23,** 167 (1973).
24. R. Runau, S. D. Peyerimhoff, and R. J. Buenker, *J. Mol. Spectrosc.* **68,** 253 (1977).
25. R. Rianda, R. P. Fruehalz, and W. A. Goddard, III, *Chem. Phys.* **19,** 131 (1977).
26. P. Pulay and W. Meyer, *J. Chem. Phys.* **57,** 3337 (1972).

27. E. A. Laws, R. M. Stevens, and W. N. Lipscomb, *J. Chem. Phys.* **56,** 2029 (1972).
28. P. Dejardin, E. Kochanski, A. Veillard, B. Roos, and P. Siegbahn, *J. Chem. Phys.* **59,** 5546 (1973).
28a. M. Rupp and R. Ahlrichs, *Theor. Chim. Acta* **46,** 117 (1977).
29. D. L. Wilhite and J. L. Whitten, *J. Chem. Phys.* **58,** 948 (1973).
30. W. Meyer, *in* "Modern Theoretical Chemistry, Vol. 3, Methods of Electronic Structure Theory" (H. F. Schaefer, III, ed.). Plenum, New York, 1977; see also, H. J. Werner and W. Meyer, *Mol. Phys.* **31,** 855 (1978).
31. M. E. Schwartz, *Chem. Phys. Lett.* **40,** 1 (1976); see also P. W. Deutsch and L. A. Curtiss, *Chem. Phys. Lett.* **39,** 588 (1976), for SCF computations on K-shell excited NH_3.
31a. T. A. Claxton, E. Platt, and M. C. R. Symons, *Mol. Phys.* **32,** 1321 (1976).
32. J. Demuynck and A. Veillard, *Chem. Commun.* p. 873 (1970).
33. R. Ahlrichs, F. Keil, H. Lischke, W. Kutzelnigg, and V. Staemmler, *J. Chem. Phys.* **63,** 455 (1975).
34. W. Kutzelnigg, H. Wollmeir, and J. Wasilewski, *Theor. Chim. Acta* **51,** 261 (1979).
35. J. M. Lehn and B. Munsch, *Mol. Phys.* **23,** 91 (1972).
36. J. D. Petke and J. L. Whitten, *J. Chem. Phys.* **59,** 4855 (1973).
37. A. Schmiedekamp, S. Skaerup, P. Pulay, and J. E. Boggs, *J. Chem. Phys.* **66,** 5769 (1977).
38. C. J. Aarons, M. F. Guest, M. B. Hall, and I. H. Hillier, *J.C.S. Faraday Trans. II* **69,** 643 (1973).
39. W. Domcke, L. S. Cederbaum, J. Schirmer, W. von Niessen, and J. P. Maier, *J. Electron Spectrosc.* **14,** 59 (1978).
40. V. Barone, J. Domady, Y. Ellinger, R. Subra, and F. Panzat, *Chem. Phys. Lett.* **65,** 542 (1979).
41. F. Keil and R. Ahlrichs, *Chem. Phys.* **8,** 384 (1975).
42. D. A. Dixon and D. S. Marynick, *J. Chem Phys.* **71,** 2860 (1979).
43. T. E. H. Walker and J. A. Horsley, *Mol. Phys.* **21,** 939 (1971).
44. D. Goutier and L. A. Burnelle, *Chem. Phys. Lett.* **18,** 460 (1973).
45. M. E. Schwartz and L. C. Allen, *J. Am. Chem. Soc.* **92,** 1466 (1970).
46. S. P. So, *J. Chem. Phys.* **67,** 2929 (1977).
47. W. von Niessen, L. S. Cederbaum, and W. P. Kraemer, *Mol. Phys.* **33,** 1415 (1977).
47a. J. A. Connor, I. H. Hillier, V. R. Saunders, and M. Barker, *Mol. Phys.* **23,** 81 (1972).
48. M. I. Umland, J. H. Letcher, and J. R. Van Wazer, *J. Chem. Phys.* **50,** 3214 (1969).
49. A. Serafini, J. F. Labarre, A. Veillard, and G. Vinot, *Chem. Commun.* p. 996 (1971).
50. K. Faegri, Jr. and W. Kosmus, *J.C.S. Faraday Trans. II* **73,** 1602 (1977).
51. I. H. Hillier and V. R. Saunders, *Chem. Commun.* p. 1510 (1970).
52. A. Serafini and J.-F. LaBarre, *Chem. Phys. Lett.* **27,** 430 (1974).
53. J. F. Olsen and J. M. Howell, *Theor. Chim. Acta* **47,** 39 (1978).
54. R. Bonaccorsi, E. Scrocco, and J. Tomasi, *Theor. Chim. Acta* **52,** 113 (1979).
54a. L. A. Curtiss, *Chem. Phys. Lett.* **68,** 225 (1979).
55. R. W. Hand, W. J. Hunt, and H. F. Schaefer, III, *J. Am. Chem. Soc.* **95,** 4517 (1973).
56. J. H. Yates and R. M. Pitzer, *J. Chem. Phys.* **70,** 4049 (1979).
57. A. Strich and A. Veillard, *J. Am. Chem. Soc.* **95,** 5574 (1973).
58. R. S. Berry, *J. Chem. Phys.* **32,** 933 (1960).
59. S.-K. Shih, S. D. Peyerimhoff, and R. J. Buenker, *J.C.S. Faraday Trans. II* **75,** 379 (1979).
60. W. R. Wadt and P. J. Hay, *J. Am. Chem. Soc.* **101,** 5198 (1979).

61. J. Demuynck, A. Strich, and A. Veillard, *Nouv. J. Chim.* **1,** 217 (1977).
62. S. D. Peyerimhoff and R. J. Buenker, *Chem. Phys. Lett.* **65,** 434 (1979).
63. E. D. Jemmis, J. Chandrasekkar, and P. von R. Schleyer, *J. Am. Chem. Soc.* **101,** 527 (1979).
64. A. Serafini and J.-F. LaBarre, *J. Mol. Struct.* **26,** 129 (1975).
65. A. Streitweiser, Jr., J. E. Williams, Jr., S. Alexandratos, and J. M. McElvey, *J. Chem. Soc.* **98,** 4778 (1976).
66. A. Hinchcliffe and J. C. Dobson, *Theor. Chim. Acta* **39,** 17 (1975).
67. D. R. Yarkony and H. F. Schaefer, III, *J. Chem. Phys.* **63,** 4317 (1975).
68. D. R. Yarkony, H. F. Schaefer, III, and C. F. Bender, *J. Chem. Phys.* **64,** 981 (1976).
69. P. Cremaschi and M. Simonetta, *Theor. Chim. Acta* **43,** 351 (1977).
70. T. Clark and P. von R. Schleyer, *J. Am. Chem. Soc.* **101,** 7747 (1979).
71. C. P. Baskin, C. F. Bender, R. R. Lucchese, C. W. Bauschlicher, Jr., and H. F. Schaefer, III, *J. Mol. Struct.* **32,** 125 (1976).
72. D. R. Yarkony, H. F. Schaefer, III, and S. Rothenberg, *J. Am. Chem. Soc.* **96,** 5974 (1974).
73. D. R. Yarkony, H. F. Schaefer, III, and S. Rothenberg, *J. Am. Chem. Soc.* **96,** 656 (1974).
74. A. K. Rappé and W. A. Goddard, III, *J. Am. Chem. Soc.* **99,** 3966 (1977).
74a. B. Åkermark, H. Johansen, B. Roos, and V. Wahlgren, *J. Am. Chem. Soc.* **101,** 5876 (1979).
75. C. R. Brundle, M. B. Robin, and H. Basch, *J. Chem. Phys.* **53,** 2196 (1970).
76. H. B. Schlegel, S. Wolfe, and F. Bernardi, *J. Chem. Phys.* **67,** 4194 (1977).
77. A. Skancke, *J. Mol. Struct.* **42,** 235 (1977).
77a. S. D. Peyerimhoff and R. J. Buenker, *Chem. Phys. Lett.* **65,** 434 (1979).
78. O. Goscinski, J. Muller, E. Poulain, and H. Siegbahn, *Chem. Phys. Lett.* **55,** 407 (1978).
78a. F. F. V. Roelandt, D. F. Van der Vondel, and G. P. Van der Kelen, *J. Mol. Struct.* **50,** 147 (1978).
79. A. M. Semlow, P. Rosmus, H. Bock, and P. Botschwina, *Chem. Phys.* **40,** 377 (1979).
80. G. del Conde, P. S. Bagus, and C. W. Bauschlicher, Jr., *Theor. Chim. Acta* **45,** 121 (1977).
81. J.-C. Barthelot, G. Trinquier, and G. Berthand, *J. Am. Chem. Soc.* **101,** 3785 (1979).
82. P. A. Kollman, C. F. Bender, and J. M. McKelvey, *Chem. Phys. Lett.* **28,** 407 (1974).
83. E. Clementi and H. Popkie, *J. Chem. Phys.* **57,** 4870 (1972).
84. K. Morokuma and H. Umeyama, *Chem. Phys. Lett.* **49,** 333 (1977).
85. R. J. Buenker and S. D. Peyerimhoff, *Chem. Phys.* **8,** 56 (1975).
86. A. Richartz, R. J. Buenker, P. J. Bruna, and S. D. Peyerimhoff, *Mol. Phys.* **33,** 1345 (1977).
87. A. Richartz, R. J. Buenker, and S. D. Peyerimhoff, *Chem. Phys.* **28,** 305 (1978).
88. E. Clementi and W. von Niessen, *J. Chem. Phys.* **54,** 521 (1971).
89. A. Hinchliffe, *J. Mol. Struct.* **48,** 279 (1978).
90. L. C. Snyder and Z. Wasserman, *Chem. Phys. Lett.* **51,** 349 (1977).
91. See R. S. Mulliken and W. B. Person, "Molecular Complexes." Wiley (Interscience), New York, 1969.
92. K. Morokuma, *J. Chem. Phys.* **55,** 1236 (1971); K. Kitaura and K. Morokuma, *Int. J. Quantum Chem.* **10,** 395 (1976); see also K. Morokuma, *Acc. Chem. Res.* **10,** 294 (1977); S. Nagase, T. Fueno, S. Yamabe, and K. Kitaura, *Theor. Chim. Acta* **49,** 309 (1978).
93. M. C. Moreau and A. Veillard, *Theor. Chim. Acta* **11,** 344 (1968).

94. A. Veillard, *Chem. Phys. Lett.* **3,** 128 (1969).
95. S. D. Peyerimhoff and R. J. Buenker, *J. Chem. Phys.* **49,** 312 (1968).
96. H. Fujimoto, S. Kato, S. Yamabe, and K. Fukui, *J. Chem. Phys.* **60,** 572 (1974).
97. M. Dixon and W. E. Palke, *J. Chem. Phys.* **61,** 2250 (1974).
98. W. E. Palke, *J. Chem. Phys.* **56,** 5308 (1972).
99. H. Umeyama and K. Morokuma, *J. Am. Chem. Soc.* **98,** 7208 (1976).
100. R. M. Archibald, D. R. Armstrong, and P. G. Perkins, *J.C.S. Faraday Trans. II* **69,** 1793 (1973).
101. M. Barber, J. A. Connor, M. F. Guest, I. Hillier, and M. Schwarz, *J.C.S. Faraday Trans. II* **69,** 551 (1973).
102. J. Demuynck and A. Veillard, *Chem. Commun.* p. 873 (1970).
103. J. R. Sabin, *Chem. Phys. Lett.* **20,** 212 (1973).
104. R. Ahlrichs and W. Koch, *Chem. Phys. Lett.* **53,** 341 (1978).
104a. L. T. Redmon, G. D. Purvis, III, and R. J. Bartlett, *J. Am. Chem. Soc.* **101,** 2856 (1979).
105. I. H. Hillier, J. C. Marriott, V. R. Saunders, M. J. Ware, D. R. Lloyd, and N. Lynaugh, *Chem. Commun.* p. 1586 (1970).
106. E. Clementi, *J. Chem. Phys.* **46,** 3842, 3851; **47,** 2323 (1967).
107. R. C. Raffenetti and D. H. Phillips, *J. Chem. Phys.* **71,** 4534 (1979).
108. J. E. Douglas, G. L. Kenyon, and P. A. Kollman, *Chem. Phys. Lett.* **57,** 553 (1978).
109. R. R. Lucchese and H. F. Schaefer, III, *J. Am. Chem. Soc.* **97,** 7205 (1975).
110. H. Umeyama, K. Morokuma, and S. Yamabe, *J. Am. Chem. Soc.* **99,** 330 (1977).
111. M. Trenary, H. F. Schaefer, III, and P. Kollman, *J. Am. Chem. Soc.* **99,** 3885 (1977).
112. S. Kato, H. Fujimoto, S. Yamabe, and K. Fukui, *J. Am. Chem. Soc.* **96,** 2024 (1974).
113. T.-K. Ha, *J. Mol. Struct.* **30,** 103 (1976).
114. W. C. Ermler, F. D. Glasser, and C. W. Kern, *J. Am. Chem. Soc.* **98,** 3799 (1976).
115. P. Dowd, *J. Am. Chem. Soc.* **88,** 2587 (1966); P. Dowd, *Acc. Chem. Res.* **5,** 242 (1972).
116. D. R. Yarkony and H. F. Schaefer, III, *J. Am. Chem. Soc.* **96,** 3754 (1974).
117. J. H. Davis and W. A. Goddard, III, *J. Am. Chem. Soc.* **98,** 303 (1976).
118. W. T. Borden, *J. Am. Chem. Soc.* **98,** 2695 (1976); E. R. Davidson and W. T. Borden, *J. Chem. Phys.* **64,** 663 (1976); *J. Am. Chem. Soc.* **99,** 2053 (1977).
119. D. R. Yarkony and H. F. Schaefer, III, *Chem. Phys. Lett.* **35,** 291 (1975).
120. D. R. Yarkony and H. F. Schaefer, III, *Chem. Phys. Lett.* **35,** 291 (1975).
121. A. Skancke, *J. Mol. Struct.* **30,** 95 (1976).
122. H. Basch, M. B. Robin, N. A. Kuebler, C. Bomker, and D. W. Turner, *J. Chem. Phys.* **51,** 52 (1969).
123. E. Kochanski and J. M. Lehn, *Theor. Chim. Acta* **14,** 281 (1969).
124. R. J. Buenker and S. D. Peyerimhoff, *J. Phys. Chem.* **73,** 1299 (1969).
125. R. M. Stevens, E. A. Laws, and W. N. Lipscomb, *J. Am. Chem. Soc.* **93,** 2603 (1971).
126. J. Kao and L. Radom, *J. Am. Chem. Soc.* **100,** 379 (1978).
126a. M. D. Newton, E. Switkes, and W. N. Lipscomb, *J. Chem. Phys.* **53,** 2645 (1970).
127. W. von Niessen, L. S. Cederbaum, and W. P. Kraemer, *Theor. Chim. Acta* **44,** 85 (1977).
128. L. Radom, P. C. Haribaran, J. A. Pople, and P. von R. Schleyer, *J. Am. Chem. Soc.* **98,** 10 (1976).
129. T. Takuda and K. Ohno, *Bull. Chem. Soc. Jpn.* **52,** 334 (1979).

CHAPTER VII

MOLECULES OF TETRAHEDRAL AND OCTAHEDRAL SYMMETRY

A. GENERAL CONSIDERATIONS

A number of important molecules have tetrahedral (T_d) or octahedral (O_h) symmetry. The group-theoretical character tables are as follows:

T_d	I	$8C_3$	$6\sigma_d$	$6S_4$	$3S_4^2 \equiv 3C_2$	Coordinates
A_1	1	1	1	1	1	
A_2	1	1	−1	−1	1	
E	2	−1	0	0	2	
T_1	3	0	−1	1	−1	
T_2	3	0	1	−1	−1	x, y, z

O_h	I	$8C_3$	$6C_2$	$6C_4$	$3C_4^2 \equiv 3C_2'$	$S_2 \equiv i$	$6S_4$	$8S_6$	$3\sigma_h$	$6\sigma_d$	Coordinates
A_{1g}	1	1	1	1	1	1	1	1	1	1	
A_{1u}	1	1	1	1	1	−1	−1	−1	−1	−1	
A_{2g}	1	1	−1	−1	1	1	−1	1	1	−1	
A_{2u}	1	1	−1	−1	1	−1	1	−1	−1	1	
E_g	2	−1	0	0	2	2	0	−1	2	0	
E_u	2	−1	0	0	2	−2	0	1	−2	0	
T_{1g}	3	0	−1	1	−1	3	1	0	−1	−1	
T_{1u}	3	0	−1	1	−1	−3	−1	0	1	1	x, y, z
T_{2g}	3	0	1	−1	−1	3	−1	0	−1	1	
T_{2u}	3	0	1	−1	−1	−3	1	0	1	−1	

B. METHANE (CH_4), SILANE (SiH_4), AND CF_4

The ground state of methane (CH_4) is $1a_1^2\,2a_1^2\,1t_2^6$, $^1A_1$, while that of CH_4^+ would be $1a_1^2\,2a_1^2\,1t_2^5$, $^2T_2$ but because of the Jahn–Teller effect it must split into nondegenerate components (six equivalent minima of C_{2v}, three of D_{2d}, and four of C_{3v} symmetry).

Many *ab initio* calculations (mostly only SCF) have been made on methane, but the most accurate including electron correlation seems to be that of Meyer [1], who has also treated CH_4^+. Meyer's best basis set includes carbon d and f orbitals and hydrogen p orbitals. He uses the PNO-CI (variational) and CEPA methods (not quite variational) for obtaining electron correlation (see Chapter I). He obtains $E_h = -40.4584$ (PNO-CI) and -40.4723 (CEPA), and corresponding binding energies of 0.6355 and 0.6451 a.u. (experiment: 0.671 a.u.). The PNO-CI method yields about 83% of the experimental correlation energy, whereas CEPA yields 89%. Correlation shifts the computed R_e(CH) from 1.08 to 1.09 Å, and the symmetrical vibration frequency ν_1 from 3149 to 3037 cm^{-1}. For the corresponding anharmonicity constant x_{11}, 13.6 cm^{-1} is obtained.

Using up to similar basis sets and PNO-CI and CEPA for correlation, Werner and Meyer [2] have computed static dipole polarizability for CH_4, in good agreement (for CEPA) with experiment.

Amos [3], using somewhat less accurate SCF plus CM wave functions, has made accurate calculations for CH_4 of octopole and hexadecapole moments, polarizability, hyperpolarizability, dipole–quadrupole polarizability, the diamagnetic contributions to the nuclear magnetic shielding constants and to the magnetizability, and the electric field gradients at the hydrogen nuclei. Excellent agreements with experiment are found when the latter is available.

Contrary to previous calculations, the lowest minimum of CH_4^+ is found to be of C_{2v} symmetry. The D_{2d} minimum is computed as 0.15 eV, and the C_{3v} minimum as 0.58 eV higher by CEPA. Detailed geometries for the three cases are given. Dissociation to CH_3^+ + H is computed to require 1.63 eV, giving 14.2 eV for the appearance potential of CH_3^+ on photoionization (experimental: 14.25 eV). For the adiabatic ionization potential of CH_4, a value of about 12.6 eV is obtained (experimental: 12.62 eV) for the beginning of a rather broad photoionization region. The computed vertical ionization energies (CEPA) for the respective MOs $1t_2$, $2a_1$, and $1a_1$ are 0.525, 0.859, and 10.683 a.u. (observed: 0.529, 0.840, and 10.68, respectively).

Similar calculations on SiH_4 (and other molecules) were made by

Ahlrichs *et al.* [4]. By PNO-CI, $E_h = -291.38144$; by CEPA, $E_h = -291.39323$. Orbital energies are tabulated. In a good SCF calculation, Rothenberg *et al.* [5] report computed values of octopole moment (the quadrupole moment is zero), diamagnetic susceptibility, potential at the Si nucleus, diamagnetic shieldings at nuclei, electric field gradients at H nuclei, charge densities at nuclei, and d orbital populations.

Returning to CH_4, in gaussian SCF computations Meyer and Pulay [6] obtain force constants F_{11}, F_{22}, F_{33}, F_{34}, and F_{44} that agree rather well with experiment; also dipole moment derivatives, but with large uncertainties. Pulay *et al.* [7] evaluated the same and also cubic (and two quartic) force constants in SCF and CEPA calculations. The effect of correlation is to diminish the force constants, but only moderately. Using the force constants to correct electron diffraction and spectroscopic values, they obtained an R_e of 1.0862 ± 0.0005 Å. This differs in an unexplained way from Meyer's [1] theoretical value of 1.090 Å. There is also a similar small but disturbing discrepancy in the case of the CH stretching force constants. Wiberg *et al.* [8] used up to a double-zeta level basis in SCF computations on the energy changes for four angle deformation modes, including large changes. They apply the results to the interpretation of the strain relations in various alkanes with unusual valence angles, for example cubane, propellane, and bicyclo (1, 1, 1) pentane.

Collins *et al.* [9], citing Shavitt, state that the best available *ab initio* calculations indicate that D_{4h} singlet planar methane should be about 150 kcal/mole less stable than the actual tetrahedral form. They also mention that computations on CH_4^{2+} show that if it is stable it would be planar. They go on, using up to 4-31G calculations, to show that donor substituents tend to produce planarity. Durmaz *et al.* [10] in a STF minimal basis SCF calculation report that the lowest energy state of planar methane is a triplet state about 7.8 eV above the ground state that is stable to all small distortions. This must be a Rydberg state.

Arrighini *et al.* [11] report proton magnetic shielding constants obtained by perturbed SCF calculations using an extended SCF basis.

Using a special procedure, Peters [12] obtains minimal-basis SCF MOs and transforms them to localized MOs (LMOs). The localized CH LMOs are not quite fully localized, but each contains a tail (σ hyperconjugation) consisting of AOs of the three H atoms other than the bonded one. Bond energies and bond dipole moments are discussed. In another paper [13], Peters discusses the transferability of the CH MOs of CH_4 to C_2H_6 (see also Newton *et al.* [13a]).

Clementi and Popkie [14], using a fairly extensive Gaussian basis for SCF computations, have analyzed the energy of formation of CH_4 starting from the nuclear field and adding one pair of electrons at a time, with variations of the CH distances, also by the removal of an electron from each MO. The energy and electron densities for CH_4^{8+}, CH_4^{6+}, CH_4^{4+}, CH_4^{2+}, CH_4^+, and CH_4 are discussed.

Bagus *et al.* [15] have made a fairly accurate SCF calculation (including polarization functions on the hydrogen) of the singlet transitions $2t_2 \leftarrow 1a_1$ and $3a_1 \leftarrow 1a_1$ (a vibronic transition mode allowed by interaction with $2t_2 \leftarrow 1a_1$ and $2t_2 \leftarrow 3a_1$). The latter is about 10% as strong as the former (estimated f about 6.7×10^{-4} for CH_4, 1.4×10^{-3} for CD_4). The former is identified with a strong observed transition at 288.4 eV computed, 288.3 eV observed, with f observed roughly 0.6×10^{-2}, computed roughly 1.9×10^{-2}. The latter is at 287.3 eV computed, 287.2 eV observed, with f computed about 10% as strong for CH_4 as the stronger transition, 20% as strong for CD_4. For the $1a_1^{-1}$, $^2A_1$ ion, the computed ionization energy is 290.8 eV, in agreement with Meyer [1] even though he included correlation.

Deutsch and Kunz [16] have made similar calculations on the excitation and ionization energies, including also the $2a_1$ and $1t_2$ ionizations. Agreement with experiment is good for the excitations and the $1a_1$ ionization, moderate for the others. In a previous paper [17] they made less accurate one-center calculations on ground and some excited states of CH_4 and SiH_4.

Ortenburger and Bagus [18] have used a large basis set, including polarization functions in an SCF computation, to obtain an interpretation of the KLL Auger spectra of CH_4. The Auger states (CH_4^{2+}) $1a_1^2\,2a_1\,2t_2^5$, $^1T_2$ and $^3T_2$, and $1a_1^2\,2a_1^2\,2t_2^4$, $^1A_1$, $^1T_2$, $^1E_1$, and $^3T_1$ are computed in good agreement with experiment (in the case of $1a_1^2\,2a_1^2\,2t_2^4$, the experimental observation is a blend of the computed states). This is the first theoretical analysis of the Auger spectra of a molecule.

Gordon [19] and Gordon and Caldwell [20], using a 4-31G basis set plus s and p Rydberg functions, and supplemented by single-substitution CM, have discussed the lower excited singlet states of CH_4 (mostly C_{2v} symmetry), and their dissociation products (mostly CH_2, $^1B_1$ plus H_2). Vertical energies were computed, and potential energy surfaces were evaluated out to dissociation.

Gordon [21], using up to a 4-31G silicon basis set plus diffuse s and p functions along with d functions, followed by single-substitution CM, has computed vertical excitation energies for a number of excited states of SiH_4, and compared them with those of CH_4. All the singlet states are primarily Rydberg states, but the triplet states have much more valence character.

Using 4-31G calculations, Gordon [22] has made SCF calculations on the species SiH_n ($n = 0$ to 4) and the corresponding positive ions, in particular, SiH_4^+. Small distortions from the T_d vertical ($^2T_2$) state lead to a C_{3v} ($^2A_1$) lowest state, with a C_{2v} state ($^2B_1$) and a D_{2d} state ($^2B_2$), respectively, 11 and 24 kcal/mole higher. The order of the three split states is the reverse of that in CH_4^+ and they are much farther apart. In the C_{3v} state of SiH_4^+, the lone Si—H bond is much lengthened, being almost dissociated to SiH_3^+ plus H. Consistent with experiment, SiH_4^+ is found to be much less stable to dissociation than CH_4^+.

Semkow *et al.* [23], with a basis set (12s, 2p, 1d) for Si, slightly contracted in the core region, and (5s, 1p) for H, made PNO-CEPA calculations on SiH_4. They obtained an optimized Si—C bond length of 1.486 Å. They also calculated symmetric harmonic stretching frequencies and anharmonicity constants. The same quantities, and E_h values, were also obtained for vertical ionization (T_d symmetry) to the $^2T_2(2t_2)$, $^2A_1(3a_1)$, $^2T_2(1t_2)$, $^2A_1(2a_1)$, and $^2A_1(1a_1)$ states of SiH_4^+, the first two with PNO-CEPA, the last three only with SCF. The computed first and second vertical ionization potentials are 12.68 ($^2T_2$) and 18.45 ($^2A_1$), as compared with about 12.8 and 18.2 observed. Vibration structures for photoelectron bands are computed for all the $^2A_1$ ionic states, and compared with experiment for the $3a_1^{-1}$ band.

Effects of fluorine 1s core ionization of CF_4 have been discussed in connection with those of CH_3F near the end of Section VI,B.

C. Be_4, Be_4^-, Mg_4, P_4, P_4^+, B_4H_4, B_4F_4, B_4Cl_4

Metal clusters are of considerable interest, especially in connection with adsorption and catalytic processes [24]. Brewington *et al.* [25] have made an SCF computation on Be_4 using basis sets including d functions. Theoretical predictions include $E_h = -58.3455$, bond distance 2.085 Å, dissociation energy 40.1 kcal/mole, assuming T_d symmetry. The electronic state is $1a_1^2\,1t_2^6\,2a_1^2\,2t_2^6$, $^1A_1$. Orbital energies and s, p, d populations (12.44, 3.44, 0.12) are given. In a later paper, Dykstra *et al.* [26] use extended CM in the SCEP/CEPA method, but with a smaller basis set (no d). As expected, the correlation energy is small, but the bond distance is increased by 0.02 Å by CM.

Jordan and Simons [27] have made SCF calculations using a moderate basis set on Be_4 and Be_4^- (also Be_2, Be_3, and their anions). An electron affinity of 1.0 eV is found for Be_4 (addition of a $3a_1$ MO). Be_2^- and Be_3^- but perhaps not Be_3 are stable. Charge density contour maps are given for the valence-shell MOs of Be_4.

When a Be_n cluster is formed, one can express the interaction energy as a sum of n-body interaction energies, e.g., in an equilateral Be_3, three 2-body interactions $E(2, 3) = 3E(2)$ plus a 3-body interaction $E(3, 3)$; and in tetrahedral Be_4, six 2-body interactions $E(2, 4) = 6E(2)$ plus four 3-body interactions $E(3, 4) = 4E(3, 3)$ plus a 4-body interaction $E(4, 4)$. Kolos *et al.* [28] discuss nonadditive effects in these interactions, using a (9s 2p) basis set in SCF calculations of total interaction energies $E(n)$ for Be_n. (Because of the closed-shell nature of the Be atom, $E(n)$ should be given rather well without CM.) They find that 2-body $E(2, n)$ and 4-body $E(4, n)$ terms, except for $E(2)$, give repulsive contributions to $E(n) = \Sigma_{k=2}^{n} E(k, n)$, but 3-body terms are attractive. However, the terms are such that Be_3 is unstable but Be_4 is stable if interatomic distances are favorable. Convergence of the $E(k, n)$ is slow, e.g., $E(3, 3)$ is -50% of $E(2, 3)$ at $R = 3.0$ a.u. in Be_3, and $E(4, 4)$ is 26% of $E(2,4)$ at $R = 3$ a.u. for Be_4. Be_2, Be_3, Be_4, and Be_5 are considered.

Daudey *et al.* [29] have made a similar study of Mg_n clusters, where the rate of convergence of $E(k, n)$ with k is found much faster than for Be_n.

Brundle *et al.* [30] have made SCF calculations on the tetrahedral molecule P_4 of white phosphorus and its ion, using a rather limited Gaussian basis (s, p only, single-ζ for the core, double-ζ for the valence-shell orbitals). They report $E_h = -1362.00$ and give a table of orbital energies, along with partial contour diagrams and overlap populations for the valence-shell MOs. The Koopmans' theorem ionization energies when reduced by 8% agree well with the observed photoelectron spectrum. Successive ionization potentials correspond to ionization of 2e giving a ^2E state with Jahn–Teller splitting of 0.35 eV, of $6t_2$ to give $^2T_2$ with a small Jahn–Teller splitting, $5a_1$ to give $^2A_1$, and $5t_2$ to give $^2T_2$ with a strong (about 2 eV) Jahn–Teller splitting. The 2e, $6t_2$, and $5a_1$ MOs are composed largely of 3p AOs, but the $5t_2$ of 3s.

Supplementing an earlier paper [31] through improved computational accuracy and some CM, the ultraviolet absorption spectrum of P_4 vapor is discussed. The only dipole-allowed transitions are of the $^1T_2 \leftarrow {}^1A_1$ type. Strong CM was found among four such transitions, and it was concluded that the observed transitions at 5.6 eV ($f = 0.13$), 6.3 eV ($f = 0.13$), 7.1 eV ($f = 1.30$), and 8.4 eV ($f = 0.40$) correspond to such mixtures (observed and calculated energies and oscillator strengths correspond rather well), with the possibility of participation also by $3d(^1T_2) \leftarrow 2e$ transitions. The 6.3 eV band could possibly be a $^1T_1 \leftarrow {}^1A_1$ transition mode allowed vibronically.

Venanzi and Schulman [31a] in a double-zeta SCF calculation showed that N_4 in tetrahedral symmetry has a $^1A_1$ local minimum with energy 275 kcal/mole greater than the energy of $2N_2$. Force constants were computed; R(NN) = 1.47 Å.

Hall *et al.* [31b] on B_4H_4 and Hall and Lipscomb [31c] on BF_4 and BCl_4 have made SCF calculations with minimal basis sets of STFs, using Slater AOs and assuming T_d symmetry. Only B_4Cl_4 is known experimentally. They give orbital energy tables. They calculate σ and π populations in valence-shell e and e* MOs of B_4F_4 and B_4Cl_4 (σ and π defined from a localized viewpoint) and, including other evidence, conclude that there is more π back charge transfer in B_4F_4. They give other details concerning charge density, and the forms of the e and e* MOs.

D. ClO_4^-, SO_4^{2-}, AND PO_4^{3-}

The isoelectronic ions ClO_4^- and SO_4^{2-} are well known in their salt compounds. The ground-state electron configuration is $1a_1^2\, 2a_1^2\, 1t_2^6\, 3a_1^2\, 2t_2^6$ (core) $4a_1^2\, 3t_2^2\, 5a_1^2\, 4t_2^2\, 1e^4\, 5t_2^6\, 1t_1^6$, $^1A_1$. A number of semiempirical and *ab initio* computations have been made on these ions.

Johansen [32] has made an SCF calculation on the ClO_4^- ion using a moderately large Gaussian basis including d orbitals on the Cl but not the O. She gives a table of orbital energies and a population analysis of the MOs; she obtains $E_h = -757.9364$ and a charge distribution $Cl^{1.539}(O^{-0.635})_4$, with s, p, and d populations 5.26, 9.32, and 0.88 on the Cl and s and p populations 3.94 and 4.70 on the O; also a Cl—O total overlap population 0.62 e. The valence-shell orbital energies agree moderately well with corresponding photoelectron spectroscopy peaks (on solid salts, of course).

A similar SCF calculation on the SO_4^{2-} ion has been made by Gelius *et al.* [33]. Total energy ($E_h = -696.142$) and orbital energies and a population analysis are given. The S—O overlap population is 0.86 e. The charge distribution found is $S^{1.34}(O^{-0.84})_4$. A contour diagram of total electron density in a plane through the S and two O atoms is shown. Several molecular properties are computed (diamagnetic susceptibility, diamagnetic shielding, potential at the S nucleus, fields and field gradients at the O nuclei, quadrupole coupling constant for ^{17}O, etc.).

Connor *et al.* [33a], in a minimal-basis SCF calculation supplemented by d orbitals on the heavy atom, investigated ClO_4^-, SO_4^{2-}, and PO_4^{3-}. They give orbital energies and atomic percentages from population analyses for the valence MOs, and use their results to interpret the photoelectron spectra of the lithium salts.

E. MnO_4^-, MnO_4, MnO_4^{2-}, MnO_4^{3-}, CrO_4^{2-}, VO_4^{3-}, $MnCl_4^{2-}$, $CoCl_4^{2-}$, $CuCl_4^{2-}$, $NiCl_4^{2-}$, MH_4 (M=Ti ... Ni)

Many calculations, mostly semiempirical, have been made on MnO_4^- and its optical spectrum. Johansen [34] has made an SCF calculation on the ground state of the ion, using an extensive Gaussian basis set not including d functions on O. She gives a table of orbital energies of core MOs, valence MOs, and four virtual MOs. The computed E_h is -1448.7571. Computed charges correspond to $Mn^{0.99}(O^{-0.50})_4$, with respective s, p, and d populations 6.3, 12.7, and 4.9 on Mn and s and p populations 3.9 and 4.6 on O, and an Mn—O overlap population 0.65 *e* (total, not for each bond).

Hsu *et al.* [35] made a similar calculation with nearly the same results, and also calculated the energies of many ionized (MnO_4) states, with $E_h = -1448.510355$ for the ground state (electron affinity 6.71 eV), along with the energies of many triplet and singlet excited states of MnO_4^- and two states of MnO_4^{2-} ($E_h = -1448.738595$, up 0.50 eV from MnO_4^- for the ^2E ground state). Observed excitations in MnO_4^- correspond well to the computed values but are low by 0.4–1.2 eV, which is understandable because of the lack of CM. Previous identifications of the first two transitions at 2.1 and 2.3 eV as $2e \leftarrow 1t_1$, $^1T_1$ (forbidden) and $^1T_2$ are accepted. Similarly the fourth band at 4.0 eV is identified as $2e \leftarrow 6t_2$, $^1T_2$. Bands at 5.5 and 6.6 eV are identified as $2e \leftarrow 5t_2$, $^1T_2$ or $^1T_1$, and $7t_2 \leftarrow 6a_1$, $^1T_2$. The computed values are averages over several states of a configuration. Calculations on magnetic circular dichroism A/D ratios for the $2e \leftarrow 1t_1$, $2e \leftarrow 6t_2$, and $7t_2 \leftarrow 1t_1$ transitions are reported; the first of these agrees with experiment.

Another SCF calculation by Mortola *et al.* [36], but using a somewhat smaller (more contracted) basis set, deserves mention ($E_h = -1447.1931$). Magnetic circular dichroism A/D ratios are computed; the first and third agree with experiment. The major atomic compositions of the MOs are listed and discussed.

Jasinski and Holt [37] have made SCF Xα scattered wave computations on MnO_4^{2-} and MnO_4^{3-}, but one cannot expect these to be as satisfactory as full-basis Gaussian calculations.

Connor *et al.* [37a], using a near double-zeta basis, made SCF calculations on MnO_4^-, CrO_4^{2-}, and VO_4^{3-}, and give orbital energies, along with population analyses. They use their data to interpret the X-ray emission spectra of the ions.

Demuynck *et al.* [38] have made SCF calculations on $CuCl_4^{2-}$ in D_{4h}, D_{2d}, and T_d symmetries, using moderately large Gaussian basis sets. The T_d structure ($E_h = -3470.6057$) is found to be more stable than D_{4h} by 18 kcal/mole, but a flattened D_{2d} structure, with a value of 120° for the Cl—Cu—Cl angle, which is close to the experimental 124° in Cs_2CuCl_4, has a slightly lower computed energy. It is believed that $CuCl_4^{2-}$ in solution is nearly tetrahedral. Orbital energies for the valence MOs in D_{4h} and T_d symmetry are given, as well as population analyses. In the latter, the computed charge on the Cu is found to vary strongly with the size of the basis set. Ionization energies as well as excited states and excitation energies are calculated.

In SCF calculations using ample polarized basis sets for both atoms, Smit *et al.* [39] have examined charge distribution and ligand field splitting in $CuCl_4^{2-}$ and $NiCl_4^{2-}$. Ground-state E_h values are given. For ground (t_2^4 e^4, ^1E) and excited (t_2^6 e^2, $^3A_2$; t_2^5 e^3, $^3T_2$; and t_2^4 e^4, $^3T_1$) states of $NiCl_4^{2-}$, E_h values are given, and Racah parameters B and C are obtained. Population analyses and comparisons with energies of several states of Cu^{2+} and Ni^{2+} are given.

Barber *et al.* [40] have made Xα and unrestricted SCF calculations with moderate basis sets on $MnCl_4^{2-}$, $CoCl_4^{2-}$, and $CuCl_4^{2-}$ assuming T_d symmetry (very probable for $MnCl_4^{2-}$ and $CoCl_4^{2-}$). Orbital energies are tabulated, with different values for α and β spin in the cases of $CoCl_4^{2-}$ and $CuCl_4^{2-}$. Comparisons are made with ESCA and with X-ray emission spectra.

Hood *et al.* [41] have made SCF calculations for the ground and up to four excited states of MH_4, *assumed* to be tetrahedral, where M is Ti, V, Cr, Mn, Fe, Co, and Ni, using a (14s 11p 6d) basis contracted to (10s 8p 3d) for M and (5s 1p) contracted to (3s 1p) for H. Of the MH_4 molecules, only TiH_4 is known experimentally, but even so, it seems rather unlikely that TiH_4 is more stable than Ti + $2H_2$. The other molecules are still less stable, but it is suggested that VH_4 and CrH_4 may be capable of existence. The electron configurations considered are $1a_1^2\,2a_1^2\,1t_2^6\,3a_1^2\,2t_2^6\,4a_1^2\,3t_2^6\,1e^m\,4t_2^n$. Predicted SCF dissociation energies (relative to M + 4H) and M—H bond distances are as follows: TiH_4, 132 kcal/mole and 1.70 Å; VH_4, 86 and 1.64; CrH_4, 65 and 1.59; MnH_4, −36 and 1.58; FeH_4, 0 and 1.58; CoH_4, 27 and 1.61; NiH_4, 18 and 1.75. Of course, CM would increase the dissociation energies, by a roughly estimated 92 kcal/mole. Ground states are TiH_4, $^1A_1$ ($m = n = 0$); VH_4, ^2E ($m = 1$, $n = 0$); CrH_4 (e^2, $^3A_2$); MnH_4 (t_2e^2, $^4T_1$); FeH_4 (t_2^2 e^2, $^5T_2$); CoH_4 (t_2^3 e^2, $^6A_1$); NiH_4 (t_2^4 e^2, $^5T_2$). Excited states with equal $m + n$ values, in each case, are predicted to be from 19 to 122 kcal/mole higher. E_h values and M—H distances are listed for all states. Orbital energies and some population analyses are given for the ground states. The $4t_2$ MO is antibonding and its occupancy increases the M—H distances.

F. SF_6, SF_6^+, XeF_6, UF_6, $Co(CN)_6^-$, Cubane

The hexafluorides of S and Xe have octahedral symmetry. For SF_6, with a moderate basis set, Hay [42] has made SCF calculations, and GVB calculations followed by CM including single and double excitations, with the core orbitals held doubly occupied. Orbital energies from the SCF calculations are given, along with charge distributions (0.945, 1.918, and −0.952 for s, p, and d on S; 0.044 and −0.382 for s and p on each F; total charges $S^{1.91}(F^{-0.32})_6$. $E_h = -994.0287$ is found from SCF and −994.1907 from GVB-CM. Calculations are made also on SF_2 and SF_4. Analysis of the GVB orbitals indicates that the stability of SF_4 and SF_6 is largely due to charge transfer configurations, with sulfur d functions being less important.

Excited states and positive ions of SF_6 are calculated by Hay [43] using the same methods as in [42] for the ground state of SF_6, with the addition of diffuse s, p, and d functions to deal with Rydberg orbitals. The SCF calculations on the states of the ion (neglecting Jahn–Teller splitting) give results about 0.5 eV lower than the Koopmans' theorem values (i.e., the orbital energies). Results are given for $1t_{1g}$, $5t_{1u}$, $1t_{2u}$, $3e_g$, $1t_{2g}$, and $4t_{1u}$ ionization (increasing in that order). Results by GVB-CM are about 1.5 eV lower, and experimental results from photoelectron spectra about 1 eV lower still.

Excitation energies are given for singlet states from six occupied to seven unoccupied MOs, and for triplet states from the highest two occupied to the seven unoccupied MOs. Allowed transitions are indicated. The excitation energies are calculated by taking differences between orbital energies of occupied and virtual MOs, corrected by the difference between calculated and experimental ionization potentials. Dipole transition matrix elements are computed. The lowest excited states are the $^3T_{1g}$ (9.6 eV) and $^1T_{1g}$ (9.9 eV) from $6a_{1g}$ $(\sigma) \leftarrow 1t_{1g}$. Next are the $^3T_{1u}$ (10.8 eV) and $^1T_{1u}$ (11.8 eV) from $6a_{1g} \leftarrow 5t_{1u}$. The dipole-allowed transition to the $^1T_{1u}$ state is identified with the first major UV experimental peak at 11.73 eV and the 11.54 eV peak in the electron-impact energy loss spectrum. Chutjian's electron loss peaks at large scattering angles, at 9.8 and 12.9 eV, are tentatively identified with the symmetry-forbidden calculated

$^1T_{1g} \leftarrow {}^1A_1$ at 9.9 and 13.1 eV. A table gives a number of further identifications between calculated UV and electron-impact spectra. The lowest excited state MOs are in increasing order $6a_{1g}$ (σ^*, an antibonding valence orbital), $7a_{1g}$ (4s), $6t_{1u}$ (4p), $2t_{2g}$ (3d), $4e_g$ (3d), $8a_{1g}$ (5s), and $7t_{1u}$ (5p); all but σ^* are Rydberg orbitals. Finally, orbital energies of occupied ($1s_S$, $1s_F$, $2s_S$, and $2p_S$) and virtual orbitals for core excitations in SF_6 are listed.

Von Niessen *et al.* [44], using Green's function methods with a fairly good basis set, have determined vertical ionization potentials of SF_6 for comparison with ESCA data. In another paper, von Niessen *et al.* [45], with an improved basis set, (12s 9p 2d 1f) contracted to (6s 4p 2d 1f) for S and (9s 5p 1d) contracted to (4s 2p 1d) on F, made an SCF calculation and corrected the Koopmans' theorem results (orbital energies) to correspond to the CM involved in the previous paper, but with somewhat improved corrections here. They thereby obtained improved ionization energies as follows (experimental values in parentheses): $1t_{1g}$, 16.22 eV (15.7); $5t_{1u}$, 17.43 (17.0); $1t_{2u}$, 17.52 (17.0); $3e_g$, 18.44 (18.6); $1t_{2g}$, 20.32 (19.8); $4t_{1u}$, 23.02 (22.9); $5a_{1g}$, 27.41 (27.0). Their SCF E_h (−993.8546, previously −993.7867) was only moderately improved because of an inadvertent change in the contraction of the s-type functions in the basis, but this should not seriously affect the valence-shell ionization potentials. The orbital energies are of course listed.

Basch *et al.* [46] have made an SCF calculation on XeF_6 (also XeF_4 and XeF_2), with a single-zeta basis set for the core and a double-zeta basis for the valence orbitals. Orbital energies are tabulated for the core, valence, and first unoccupied MOs; $E_h = -7814.18$. Approximate populations of valence orbitals are tabulated. The chemical shifts (from Xe) are discussed. Overlap populations for one Xe—F bond are given. Comparisons with the photoelectron spectrum are discussed. Calculated energies for two excited states are given. Some one-electron properties (diamagnetic shielding, $\langle r^2 \rangle$, diamagnetic shielding at the nuclei, field gradients at the nuclei) are computed.

In UF_6, outside a core comprised of 78 U electrons $\cdots 5s^2\,5p^6\,4f^{14}\,5d^{10}$ plus 12 K-shell F electrons, there are 56 valence-shell electrons corresponding to $14(6s^2\,6p^6\,5f^3\,6d\,7s^2)$ of U and 42 $(6x\,2s^2\,2p^5)$ of F. In calculations on UF_6, UF_6^+, and UF_6^-, Hay *et al.* [47] first built up a relativistic core potential (ECP) for the five s, p, d, f, and g components corresponding to a 78-electron core suitable for a U^{+3} ion. (The ECP were found to be only slightly different for a U^0 or a U^{+3} core, but since a population analysis of their final results gave a charge distribution $U^{2.4}(F^{-0.4})_6$, the U^{+3} derived core was used.) The resulting U atom ECP was expanded in Gaussian functions. The nonrelativistic ECP for the F atom was similarly expanded. SCF calculations for the 56 valence-shell electrons in the field of the total core were then carried out, using a (4s 4p 3d 4f) basis contracted to (3s 3p 2d 2f) for U and (4s 5p) contracted to (2s 2p) for F.

The ground state of UF_6 may be described as

$$1a_{1g}^2\ 1t_{1u}^6\ 1e_g^4\ 2a_{1g}^2\ 2t_{1u}^6\ 2e_g^4\ 1t_{2g}^6\ 3t_{1u}^6\ 3a_{1g}^2\ 1t_{2u}^6\ 1t_{1g}^6\ 4t_{1u}^6\ {}^1A_{1g}$$

outside the core electrons. Corresponding orbital energies and a population analysis are given. In UF_6^+, ionization occurs most readily from $1t_{1g}$ or $4t_{1u}$ with respective orbital energies -0.6819 and -0.6772 a.u. Six excited states of the ion are considered, and the effects of a moderate amount of CM added; SCF calculations are also made on the ground and two excited states. But now spin–orbit coupling must also be considered; here O_h' double-group representations are needed. Spin–orbit splitting is small for $1t_{1g}$, which, like most of the valence-shell MOs, consists mainly of fluorine 2p. However, $4t_{1u}$ contains 7% of U 2p (and 5% of U 4f) and the spin–orbit splitting amounts to about 1.3 eV, giving two components, one on each side of the slightly split $1t_{1g}$. Comparison with the experimental photoelectron spectrum shows satisfactory agreement with theory for the last seven MOs after allowing for the wide splitting in the $4t_{1u}$ (splittings are small for the other six). For R_{UF} in UF_6, 1.96 Å is computed; the experimental value is 1.999 Å.

In the ${}^2A_{2u}$ ground state of UF_6^- in O_h symmetry a $1a_{2u}$ electron is added, where $1a_{2u}$ is primarily one of the 5f U atom orbitals. Two excited states, ... t_{2u}, ${}^2T_{2u}$ and ... t_{1u}, ${}^2T_{1u}$ are obtained by using other components of the 5f U atom orbital. When spin–orbit coupling is included (O_h' symmetry), the sequence of last-electron MSOs is γ_{7u} (ground state), γ_{8u}, γ_{7u}, γ_{8u}, γ_{6u}. An electron affinity of about 7 eV is found versus an experimental value of about 5 eV.

Although we consider the MS Xα method to be semiempirical and do not include its results, it seems desirable to mention a comprehensive relativistic calculation of this kind on UF_6 by Boring and Wood [48].

In a preliminary report on an SCF computation on $Co(CN)_6^{-3}$, assumed octahedral, Sano *et al.* [49] give orbital energies for the valence shell MOs and population analyses. They used two basis sets, a double-zeta set of (12s 8p 4d) contracted to (8s 6p 2d) for Co and (8s, 5p) sets contracted to (4s 2p) for C and N, and the same set with the most diffuse 4s and 4p orbitals omitted. The population analyses differed surprisingly much in the two cases, even though they are well known to be somewhat lacking in meaning.

Using a minimal basis set (also semiempirical methods) Schulman *et al.* [50] made an SCF calculation on cubane (symmetry O_h). They list valence-shell orbital energies. They also discuss how the MOs could be attained by the interaction of eight localized CH MOs.

REFERENCES

1. W. Meyer, *J. Chem. Phys.* **58,** 1017 (1973).
2. H.-J. Werner and W. Meyer, *Mol. Phys.* **31,** 855 (1976).
3. R. D. Amos, *Mol. Phys.* **38,** 33 (1979).
4. R. Ahlrichs, R. Keil, H. Lischka, W. Kutzelnigg, and V. Staemmler, *J. Chem. Phys.* **63,** 455 (1975).
5. S. Rothenberg, R. H. Young, and H. F. Schaefer, III, *J. Am. Chem. Soc.* **92,** 3234 (1970).
6. W. Meyer and P. Pulay, *J. Chem. Phys.* **56,** 2109 (1972).
7. P. Pulay, W. Meyer, and J. E. Boggs, *J. Chem. Phys.* **68,** 5077 (1978).
8. K. B. Wiberg, G. B. Ellison, and J. J. Wendoloski, *J. Am. Chem. Soc.* **98,** 1212 (1976).
9. J. B. Collins, J. D. Dill, E. D. Jemmis, Y. Apelorg, P. von R. Schleyer, R. Seeter, and J. A. Pople, *J. Am. Chem. Soc.* **98,** 5419 (1976).
10. S. Durmaz, J. N. Murrell, and J. B. Pedley, *Chem. Commun.* p. 933 (1972).
11. G. P. Arrighini, M. Maestro, and R. Moccia, *J. Chem. Phys.* **52,** 6411 (1970).
12. D. Peters, *J. Am. Chem. Soc.* **51,** 1566 (1969).
13. D. Peters, *J. Am. Chem. Soc.* **94,** 707 (1972).

13a. M. D. Newton, E. Switkes, and W. N. Lipscomb, *J. Chem. Phys.* **53,** 2645 (1970).

14. E. Clementi and H. Popkie, *J. Am. Chem. Soc.* **94,** 4057 (1972).
15. P. S. Bagus, M. Krauss, and R. E. LaVilla, *Chem. Phys. Lett.* **23,** 13 (1973).
16. P. W. Deutsch and A. B. Kunz, *J. Chem. Phys.* **62,** 4069 (1975).
17. P. W. Deutsch and A. B. Kunz, *J. Chem. Phys.* **59,** 1155 (1973).
18. I. B. Ortenburger and P. S. Bagus, *Phys. Rev. A* **11,** 1501 (1975).
19. M. S. Gordon, *Chem. Phys. Lett.* **52,** 161 (1977).
20. M. S. Gordon and J. W. Caldwell, *J. Chem. Phys.* **70,** 5503 (1979).
21. M. S. Gordon, *J. Chem. Phys.* **69,** 4955 (1978).
22. M. S. Gordon, *Chem. Phys. Lett.* **59,** 410 (1978).
23. A. M. Semkow, P. Rosmus, H. Bock, and P. Botschwina, *Chem. Phys.* **40,** 377 (1979).
24. C. W. Bauschlier, Jr., D. H. Liskow, C. F. Bender, and H. F. Schaefer, III, *J. Chem. Phys.* **62,** 4815 (1975); C. W. Bauschlier, Jr., C. F. Bender, H. F. Schaefer, III, and P. S. Bagus, *Chem. Phys.* **15,** 227 (1976); J. F. Hamilton and R. C. Baetzold, *Science* **205,** 1213 (1979).
25. R. B. Brewington, C. F. Bender, and H. F. Schaefer, III, *J. Chem. Phys.* **64,** 905 (1976).
26. C. E. Dykstra, H. F. Schaefer, III, and W. Meyer, *J. Chem. Phys.* **65,** 514 (1976).
27. K. D. Jordan and J. Simons, *J. Chem. Phys.* **67,** 4027 (1977).
28. W. Kolos, F. Nieves, and O. Novaro, *Chem. Phys. Lett.* **41,** 431 (1976); O. Novaro and W. Kolos, *J. Chem. Phys.* **67,** 5066 (1979).
29. J. P. Daudey, O. Novaro, W. Kolos, and M. Berrondo, *J. Chem. Phys.* **71,** 4297 (1979).
30. C. R. Brundle, N. A. Kuebler, M. B. Robin, and H. Basch, *Inorg. Chem.* **11,** 20 (1972).
31. R. R. Hart, M. B. Robin, and N. A. Kuebler, *J. Chem. Phys.* **42,** 3631 (1965).

31a. T. J. Venanzi and J. M. Schulman, *Mol. Phys.* **30,** 281 (1975).

31b. J. K. Hall, Jr., I. R. Epstein, and W. N. Lipscomb, *Inorg. Chem.* **12,** 915 (1973).

31c. J. K. Hall, Jr. and W. N. Lipscomb, *Inorg. Chem.* **13,** 710 (1974).

32. H. Johansen, *Chem. Phys. Lett.* **11,** 466 (1971).
33. U. Gelius, B. Roos, and P. Siegbahn, *Theor. Chim. Acta* **23,** 59 (1971).

33a. J. A. Connor, I. H. Hillier, V. R. Saunders, and M. Barber, *Mol. Phys.* **23,** 81 (1972).

34. H. Johansen, *Chem. Phys. Lett.* **17,** 569 (1972).
35. H.-L. Hsu, C. Peterson, and R. M. Pitzer, *J. Chem. Phys.* **64,** 791 (1976).
36. A. P. Mortola, H. Basch, and J. W. Moskowitz, *Int. J. Quantum Chem.* **7,** 725 (1973).
37. J. P. Jasinski and S. L. Holt, *J. C. S. Faraday Trans. II* 1304 (1976).
37a. J. A. Connor, I. H. Hillier, V. R. Saunders, M. H. Wood, and M. Barber, *Mol. Phys.* **24,** 497 (1972).
38. J. Demuynck, A. Veillard, and U. Wahlgren, *J. Am. Chem. Soc.* **95,** 5563 (1973).
39. J. J. M. Smit, C. Haes, and W. C. Nieuwpoort, *Theor. Chim. Acta* **43,** 277 (1977).
40. M. Barber, J. D. Clark, A. Hinchcliffe, and D. S. Urch, *J. C. S. Faraday Trans. II* 681 (1978).
41. D. M. Hood, R. M. Pitzer, and H. F. Schaefer, III, *J. Chem. Phys.* **71,** 705 (1979).
42. P. J. Hay, *J. Am. Chem. Soc.* **99,** 1003 (1977).
43. P. J. Hay, *J. Am. Chem. Soc.* **99,** 1013 (1977).
44. W. von Niessen, L. S. Cederbaum, G. H. F. Diercksen, and G. Hohlneicher, *Chem. Phys.* **11,** 399 (1975).
45. W. von Niessen, W. P. Kraemer, and G. H. F. Diercksen, *Chem. Phys. Lett.* **63,** 65 (1979).
46. H. Basch, J. W. Moscowitz, C. Hollister, and D. Hankin, *J. Chem. Phys.* **55,** 1922 (1971).
47. P. J. Hay, W. R. Wadt, L. R. Kahn, R. C. Raffenneti, and D. R. Phillips, *J. Chem. Phys.* **71,** 1767 (1979).
48. M. Boring and J. H. Wood, *J. Chem. Phys.* **71,** 32 (1979); see also *J. Chem. Phys.* **71,** 392 (1979), on UF_6 and UO_2^{++}.
49. M. Sano, H. Yamatera, and Y. Hotano, *Chem. Phys. Lett.* **60,** 257 (1979).
50. J. M. Schulman, C. R. Fischer, P. Solomon, and T. J. Venanzi, J. Am. Chem. Soc. **100,** 2949 (1978).

CHAPTER VIII

ALIPHATIC AND INORGANIC MOLECULES WITH FOUR OR MORE ATOMS: LINEAR, D_{2h}, AND D_{2d} MOLECULES

A. GENERAL CONSIDERATIONS

In this chapter first a few linear molecules will be discussed, then numerous molecules with a dominant plane of symmetry and finally molecules of D_{2d} symmetry. Here as always, we are classifying molecules according to their ground-state symmetries, but note that many excited and ionized states assume different symmetries. Linear molecules, like diatomic molecules [1], belong to symmetry $D_{\infty h}$ or $C_{\infty v}$ and have the same species symbols. Group theoretical tables for other symmetries encountered in this chapter are given in Table 1.

Among additional linear and planar molecules are the triatomics treated in Chapters II–IV. Several planar molecules of D_{3h} symmetry and with more than three atoms were treated in Chapter VI.

As noted in Chapter II,A (and see [1] there), the choice of B_1 or B_2 labels for C_{2v} depends on how the axes x, y, z are located. We shall always place the x axis perpendicular to the dominant symmetry plane. We make the same choice for C_{2h} and D_{2h}. For D_{2h}, where possible, we use the long axis of the molecule as z axis, in order to fix the identification of the B_i labels.

TABLE 1
Character Tables of Representations for Various Symmetries

C_{2v}	E	$C_2(z)$	σ_{xz}	σ_{yz}	Coordinates
A_1	1	1	1	1	z
A_2	1	1	−1	−1	
B_1	1	−1	1	−1	x
B_2	1	−1	−1	1	y

C_2	E	$C_2(z)$	Coordinates
A	1	1	z
B	1	−1	x, y

C_s	E	σ_{yz}	Coordinates
A′	1	1	y, z
A″	1	−1	x

C_{2h}	E	$C_2(x)$	$\sigma_h(y, z)$	i	Coordinates
A_g	1	1	1	1	
A_u	1	1	−1	−1	x
B_g	1	−1	−1	1	
B_u	1	−1	1	−1	y, z

D_{2h}	E	$\sigma(xy)$	$\sigma(xz)$	$\sigma(yz)$	i	$C_2(z)$	$C_2(y)$	$C_2(x)$	Coordinates
A_g	1	1	1	1	1	1	1	1	
A_u	1	−1	−1	−1	−1	1	1	1	
B_{1g}	1	1	−1	−1	1	1	−1	−1	
B_{1u}	1	−1	1	1	−1	1	−1	−1	z
B_{2g}	1	−1	1	−1	1	−1	1	−1	
B_{2u}	1	1	−1	1	−1	−1	1	−1	y
B_{3g}	1	−1	−1	1	1	−1	−1	1	
B_{3u}	1	1	1	−1	−1	−1	−1	1	x

D_{2d}	E	$2S_4(z)$	$S_4^2 \equiv C_2''$	$2C_2$	$2\delta_d$	Coordinates
A_1	1	1	1	1	1	
A_2	1	1	1	−1	−1	
B_1	1	−1	1	1	−1	
B_2	1	−1	1	−1	1	z
E	2	0	−2	0	0	x, y

B. LINEAR MOLECULES

1. Acetylene (H—C≡C—H)

The ground state of acetylene has the structure

$$1\sigma_g^2\, 1\sigma_u^2\, 2\sigma_g^2\, 2\sigma_u^2\, 3\sigma_g^2\, 1\pi_u^4\, 1\pi_g^0\, 3\sigma_u^0,\ {}^1\Sigma_g^+$$

The most accurate SCF computation for the ground state is that of McLean and Yoshimine [2], who used 46 STFs including d and f orbitals. Their $E_h = -76.8540$ is estimated to be within 0.001 a.u. of the SCF limit. They give detailed forms of the MOs, along with the following orbital energies (in eV): 305.958, 305.857, 28.016, 20.948, 18.579, and 11.166 for $1\sigma_g$, $1\sigma_u$, $2\sigma_g$, $2\sigma_u$, $3\sigma_g$, and $1\pi_u$, respectively. For the last four, the photoelectron values are 23.4, 18.6, 16.7, and 11.4. With a slightly less accurate basis set, giving $E_h = -76.8482$, Fischer-Hjalmars and Siegbahn [3] also give orbital energies and population analyses, along with some good contour diagrams for the valence-shell MOs, which they compare with corresponding MOs of C_2H_4 (see Fig. 1, from their article).

Hay *et al.* [4] reported generalized valence bond (GVB) SCF wave functions for acetylene, using a double-zeta basis set with d polarization. They show contour maps of GVB orbitals.

Using a basis set up to 6-31G*, Pople *et al.* [5] have tested the coupled-cluster theory with double substitutions on 1.2 hydrogen shifts for C_2H_2 (to HC(H)C and H_2CC).

Botschwina *et al.* [6], using a moderate polarized basis set, have calculated harmonic and anharmonic force constants for C_2H_2. The diagonal force constants are in rather good agreement with experiment.

Clark and Colling [7], using an STO-31G basis, have compared calculated bond lengths and stretching force constants for C_2H_2 with those of (1s) core-ionized $C_2H_2^+$ and computed overall band profiles for the core-ionized levels.

Using the MRD-CI procedure with a basis set of 72 contracted Gaussians, Barth *et al.* [8] have studied core-ionized and various core-excited and shake-up states of C_2H_2. Some oscillator strengths were computed of which $1s \rightarrow \pi^*$ ($f = 0.17$) was by far the greatest; next is $1s \rightarrow 3p\pi$ with $f = 0.008$; for $1s \rightarrow 3s$ and $1s \rightarrow 3p\sigma$, $f = 0.001$; identifications with observed optical and electron-impact spectra are discussed. Four shake-up states ($1s \rightarrow \infty$ with $\pi \rightarrow \pi^*$, $\pi \rightarrow 3p\pi$, $\sigma \rightarrow \sigma^*$, and $2\sigma_u \rightarrow 3s$) are computed and identified in the photoelectron spectrum.

Nagase and Kerr [9], using a 4-31G basis set, have calculated the paths for the addition reaction $C_2H_2 + H \rightarrow CH_2{=}\dot{C}H$ and the subtraction

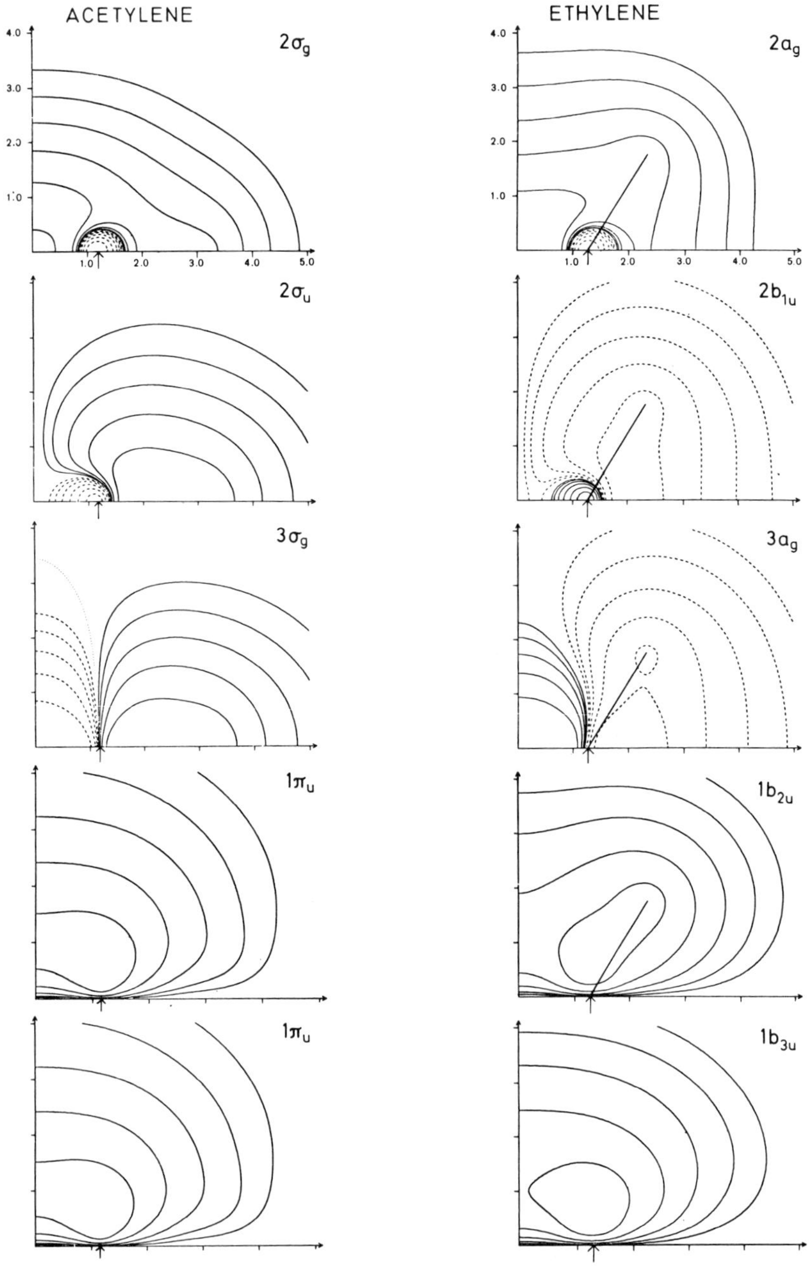

FIG. 1. Contour maps in the molecular plane of pairs of corresponding orbitals. The arrows point towards the position of the carbon nucleus on the z axis. The solid line in the ethylene maps is drawn along the CH bond. From Fischer-Hjalmars and Siegbahn [3].

reaction $C_2H_2 + H \rightarrow CH{\equiv}\dot{C} + H_2$. For the addition reaction, an approach is found in C_s symmetry with a 6 kcal/mole barrier to the transition state and a net reaction energy of 34 kcal/mole. The added H approaches trans to the orientation of the odd electron in the radical which is formed, as this orientation would be if the electron were forming a bond with a further added H atom as in C_2H_4. For the subtraction reaction, only a straight line approach along the C_2H_2 symmetry axis was investigated. For this approach a barrier of 11 kcal/mole was encountered, followed by a descent of 32 kcal/mole to the final products. Details of the two processes are discussed in terms of Morokuma energy component analysis and population analysis.

Clementi and Popkie [10], using a moderately accurate SCF computation, have analyzed each orbital energy into a sum of terms associated with each atom-pair or atom-triple, and also have discussed atomic populations at three CC distances corresponding to equilibrium in C_2H_2, C_2H_4, and C_2H_6. These three molecules are discussed and compared. For localized MOs, see the first article of Newton *et al.* [38] and Section VIII, C, 1 on ethylene for details of methods.

Much attention, experimentally and theoretically, has been given to the excited states of acetylene. For the lower valence-shell excited states, both triplet and singlet, the molecule is predicted to become planar, either cis (C_{2v}) or trans (C_{2h}), and for the higher ones, to stabilize in a gauche (C_2) conformation. For a linear molecule, these valence-shell states of $\pi_u^3\,\pi_g$ configuration would be $^3\Sigma_u^+$, $^3\Delta_u$, $^3\Sigma_u^-$ and $^1\Sigma_u^-$, $^1\Delta_u$ and (higher) $^1\Sigma_u^+$. In the nonlinear molecule, $^3\Delta_u$ and $^1\Delta_u$ are each split. The degenerate MOs $1\pi_u$ and $1\pi_g$ are split as follows:

$$\textit{cis:}\quad 1a_1^2\, 1b_2^2\, 2a_1^2\, 2b_2^2\, 3a_1^2\, 4a_1^m\, 1b_1^n\, 3b_2^p\, 1a_2^q\, 4b_2^r$$

$$\textit{trans:}\quad 1a_g^2\, 1b_u^2\, 2a_g^2\, 2b_u^2\, 3a_g^2\, 3b_u^m\, 1a_u^n\, 4a_g^p\, 1b_g^q\, 4b_u^r$$

$$\textit{gauche:}\quad 1a^2\, 1b^2\, 2a^2\, 2b^2\, 3a^2\, 4a^m\, 3b^n\, 4b^p\, 5a^q\, 5b^r$$

where $m + n = 3$, p, q, or $r = 1$.

The energies of the MOs as a function of the CCH angle can be represented by a Walsh diagram [11]; for C_{2v} and C_{2h} see Fig. 2, taken from an article by Wetmore and Schaefer [12] with group theory symbols added. In Fig. 2, cis and trans energies are not distinguished, but their relatively small differences must be determined by computation.

Wetmore and Schaefer have made large-scale computations on the four lowest triplet states using a double-zeta plus polarization (d for C, p for H) basis with extensive (about 6000) single- and double-excitation CM, at various CCH angles θ. To obtain better estimates of the excitation energies, they have added a Davidson [14] correction for higher than double CM excitations. For the lowest triplet, in agreement with an earlier article

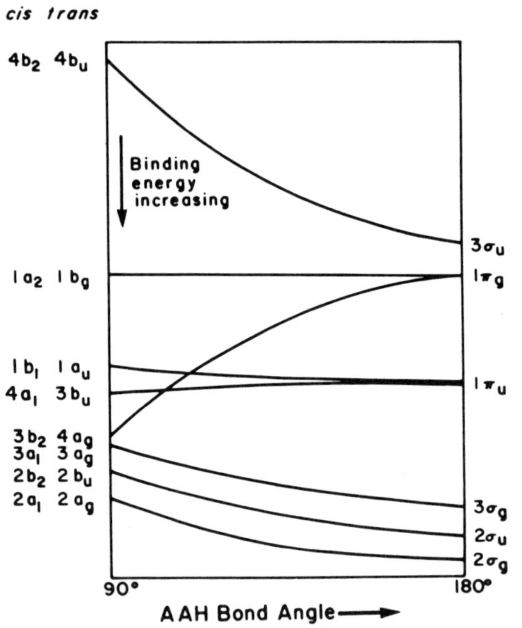

FIG. 2. Walsh diagram for HAAH molecules, with group theory symbols added. From Wetmore and Schaefer [12].

of Kammer [13], they find a cis $^3B_2$ state ($m = 1, n = 2, p = 1$) with excitation energy 3.49 eV, R(CC) = 1.343 Å, R(CH) = 1.090 Å, θ = 127.8°. Next is the trans ^3B_u state ($m = 1, n = 2, p = 1$) at 3.84 eV with R(CC) = 1.346 Å, R(CH) = 1.083 Å, θ = 130.7°. Then comes a trans ^3A_u state ($m = 2, n = 1, p = 1$) at 4.18 eV, with R(CC) = 1.390 Å, R(CH) = 1.094 Å, and θ = 120.6°, and finally a cis $^3A_2$ state ($m = 2, n = 1, p = 1$) at 4.56 eV, R(CC) = 1.363 Å, R(CH) = 1.094 Å, θ = 130.8°. (The Davidson corrections have been included here for the excitation energies.) For the $^3B_2$ and $^3A_2$ states, Wetmore and Schaefer predict dipole moments of 1.32 and 2.36 D, respectively. They also give natural orbital populations, with $1a_2$ being the largest CM population. They also give s, p, and d populations (about 0.06 for d) for carbon, and s and p populations (about 0.02) for hydrogen.

Similar results on energies and orientations have been obtained by Demoulin [15] but using a Gaussian-lobe basis without polarization functions and restricted CM. Demoulin also computed the corresponding singlet states (each above its triplet) as well as higher states with $q = 1$, and also Rydberg states (see below). The results of Wetmore and Schaefer and of

Demoulin are in general agreement with existing experimental evidence, in particular, spectroscopic ^1A_u and probably ^1B_u trans states at 5.23 and 6.71 eV, respectively, while electron impact evidence indicates triplet states at 5.3, 5.9, and 8.05 eV, and recently an electronic absorption spectrum of the excited $^3B_2$ states has been observed [16]. The results of Wetmore and Schaefer on the low triplet states are in contradiction to the Walsh diagram in that $1b_1$ comes below $4a_1$ (or $1a_u$ below $3b_u$), but for the corresponding singlet states, Demoulin finds Walsh's order confirmed.

With CM for the ground state, Wetmore and Schaefer obtain $E_h = -77.0824$, or $E_h = -77.1061$ with Davidson's correction included. $R(CC) = 1.213$ Å, $R(CH) = 1.066$ Å are found, agreeing rather well with the experimental values 1.203 and 1.060 Å.

Demoulin and Jungen [15] studied vertical electron transitions. They used a frozen core approximation (core is the $C_2H_2^+$ ground-state ion), and numerous diffuse functions for the excited states. For C_2H_2, their SCF E_h was -76.8118, and for $C_2H_2^+$, -76.4454. Demoulin, in extending the calculations to a variety of nuclear configurations, used a somewhat less extended basis set, giving an SCF E_h of -76.7113 for C_2H_2. The $^2\Pi_u$ ion is found to be linear as expected. In calculating excitation energies, "the difference in total energy between an excited state at the given geometry and the molecular ion at the molecular ground-state geometry is subtracted from 11.40 eV, the experimental first ionization potential of acetylene. This amounts to reporting all energies to that of the ion." This procedure is justified for Rydberg states, but probably involves some error for valence states.

For the valence excited states with $q = 1$ ($m, n = 2$, 1 or 1, 2), Demoulin found HCH angles of about 170° if the molecule was held planar (cis or trans), but found that the energy decreased upon introducing gauche symmetry, to a minimum for HCCH angle (not θ) = 90°; θ was not accurately determined, but might even approach 180°. Potential curves and a Walsh diagram for HCCH twisting (C_2 symmetry) from cis (0°) to trans (180°) at $\theta = 135°$, $R_{CC} = 2.521$ a.u. are given for the valence-shell states.

One such state, the $1\pi_u^3\, 1\pi_g$, $^1\Sigma_u^+$ (the V_π state) is found to be linear and of Rydberg dimensions. Calculations were also made in planar C_s geometry, but they gave no energy minima.

Three Rydberg series are known experimentally for acetylene. There ought to be four allowed Rydberg MOs, starting with 3s, 3dσ, 3dπ, and 3dδ and giving, respectively, $^1\Pi_u$, $^1\Pi_u$, $^1\Sigma_u^+$, and $^1\Pi_u$ states. Apparently either the ndσ or the ndπ series is missing, although their computed oscillator strengths are about equal. Calculations were made on all four series. All Rydberg states were found to be linear ($R_{CC} = 1.26$ Å for these and the ion; experimental: 1.255 Å for the ion, 1.265 and 1.257 Å for two of the Rydberg states).

Computed stretching frequencies agree well with experiment (ground state: 2166 cm^{-1}, experimental: 1974; for the $^2\Pi_u$ ion, 1890 cm^{-1}, experimental: 1830, with similar results for Rydberg states). But the agreement is much poorer for the valence states.

Approximate CM wave functions for selected excited states are given. Further details are contained in the articles just discussed, including orbital sizes and oscillator strengths for optical transitions, and discussion of orbital forms as they change with geometry.

In earlier work by Kammer [13] using a minimal Gaussian lobe basis plus extensive valence-MO CI, covering many excited (mostly Rydberg) states, some conclusions similar to those above were reached. In a later article with improved basis [17] he computed a large number of vertical transition energies.

Cederbaum *et al.* [18] discuss details of the photoelectron spectra of acetylene (and several other hydrocarbons) using computations by a many-electron Green's function method, using a basis set including polarization (SCF $E_h = -76.8406$). The computed photoelectron spectrum shows the $1\pi_u$ ionization as essentially a single line, the $3\sigma_g$ with one main line and weak satellites at about twice the energy, the $2\sigma_u$ as a strong main line with four very weak satellites at higher frequencies. The $2\sigma_g$ ionization is split into two main components, one at 23.9 eV (60% of the intensity) with CM of $2\sigma_g^{-1}$ (dominant) and $(2\sigma_u)^{-1}(1\pi_u)^{-1}(2\pi_g)$, and the second at 28.9 eV (25% relative intensity), whose main CM components are $(2\sigma_u)^{-1}(1\pi_u)^{-1}(2\pi_g)$, $(2\sigma_g)^{-1}$, and $(1\pi_u)^{-2}(4\sigma_g)$; there is a very weak third component at about 35 eV. The spectrum is shown in Fig. 3, taken from [18]. Experiment shows structures, partially unresolved, corresponding to the more intense of the predictions.

Barth *et al.* [8] have used the large-scale MRD-CI method to calculate vertical transition energies for 1s ionization and 1s excitation to π^* and Rydberg MOs (3s, 3p, 3d, and some $n = 4$) in C_2H_2 and C_2H_4. Localized and delocalized CM treatments are compared. (Localization here means use of unsymmetrical 1s ionization; in SCF, this gives best results; use of sufficient CM in the delocalized treatment brings agreement.) Shake-up states are also calculated; the lowest ($1s^{-1}$, $\pi^* \leftarrow \pi$) are 6–7 eV above the K edge. Agreements with experiment, including shake-up data, are good for C_2H_2. Term values are given and oscillator strengths for excitations computed.

Coplan *et al.* [19], making use of the wave functions of [2], computed momentum distributions for the MOs of acetylene. Their results are in

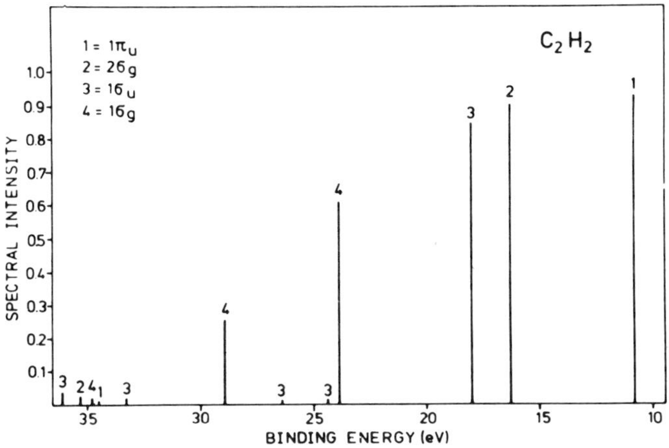

FIG. 3. Spectral intensity as a function of binding energy for acetylene. From Cederbaum *et al.* [18].

very good agreement with experiments in a separation energy spectrum obtained by them.

In SCF calculations using up to a large basis set, Chu *et al.* [20] have explored the geometry changes in going to $C_2H_2^+$ and $C_2H_2^-$. They find $E_h = -76.7927$ for C_2H_2, with $R_{CC} = 1.203$ Å and $R_{CH} = 1.059$ Å, $E_h = -76.4299$ for $C_2H_2^+$, with $R_{CC} = 1.250$ Å and $R_{CH} = 1.072$ Å, and $E_h = -76.6957$ for trans $C_2H_2^-$ with $p = 1$, -76.6859 for cis $C_2H_2^-$ with $p = 1$ ($R_{CC} = 1.282$ Å and $R_{CH} = 1.059$ Å for both); for $C_2H_2^-$ with $q = 1$, the ion should be linear and 1.36 eV higher in energy than for trans (1.09 eV higher than for cis). These figures lead to a trans electron affinity of -2.64 eV, or if an estimate of 1.4 eV is made for extra correlation energy of the ion, an estimate of -1.2 eV, indicating that $C_2H_2^-$ is unstable. However, the authors in the abstract of a second article [21] state that "the anion is predicted to possess stable cis and trans isomers." This seems unjustified. The text is confusing in that a negative electron affinity ("vertical electron affinity" of 4.2 eV) is there called positive.

2. Cyanogen (N≡C—C≡N), Carbon Suboxide (O=C=C=C=O), Ethylene Dione (O=C=C=O), Imidoborane, Haloacetylenes, and Cyanoacetylenes

A very good SCF wave function for cyanogen has been obtained by McLean and Yoshimine [2]; $E_h = -184.6568$. They give orbital energies and forms of MOs. They also give the results of similar calculations for

HC≡C—CN, LiC≡CH, FC≡CH, and ClC≡CH. From the wave functions of McLean and Yoshimine, Bonaccorsi *et al.* [22] compute the electric field gradient at the N nucleus in C_2N_2 and HCCCN. They also derive and give density contour maps for the "exclusive" localized MOs of HC≡C—CN, and calculate nuclear quadrupole coupling constants for this and C_2N_2.

Cederbaum *et al.* [23], using a Green's function method with a double-zeta plus polarization basis, compute the four lowest ionization potentials (IPs) $1\pi_u$, $4\sigma_u$, $5\sigma_g$, and $1\pi_g$ of C_2N_2 as 15.56, 14.80, 14.40, and 13.20 eV, respectively, in good agreement with the experimental values (15.6, 14.86, 14.49, and 13.36 eV). Their article also reports, for the first time for a polyatomic molecule, an *ab initio* calculation of corresponding vibrational structures, for the most part in very good agreement with the experiment; thus their predictions give evidence on the bonding properties of the MOs. Their IPs differ considerably from those predicted at the SCF level. In addition to those observed, they made calculations on the $4\sigma_g$, $3\sigma_u$, and $3\sigma_g$ IPs. For the $4\sigma_g$, they show a figure giving in great detail the predicted structure of the $4\sigma_g$ band.

Botschwina *et al.* [6], using moderate polarized basis sets, have computed force constants for FCCH and ClCCH. Off-diagonal constants are all very small.

In experimental work on photoelectron spectra and Green's function (in part Tamm–Dancoff) theoretical calculations, Bieri *et al.* [24] have studied the following twelve acetylenic derivations: CH_3—C≡C—X, CF_3—C≡C—X, N≡C—C≡C—X with X = Cl, Br, I and N≡C—C≡C—X with X = F, CH_3, CF_3. In most calculations a (9s 5p) basis set contracted to (4s 2p) was used for C, N, and F, and (4s) contracted to (2s) for H. Tables of orbital energies are given, but the ordering of the ionization potentials often differs from the predictions of Koopmans' theorem, and in any event the computed Green's function ionization potentials are much better and agree remarkably well with experiment.

Haddon *et al.* [25] have made an SCF calculation on ethylene dione (O═C═C═O) using a split-valence (4-31G) basis. The molecule is unstable relative to two CO. Its lowest state is probably $^3\Sigma_g^-$.

In an SCF calculation without polarization functions Bardo and Ruedenberg [26] conclude that carbon suboxide is probably stably linear, and give a table of orbital energies. In an ordinary SCF calculation Jonkman [27] finds 12.51 eV for the $^3\Pi_u$, 12.86 eV for the $^1\Pi_u$ $n\pi^*$ ($5\sigma_u \to 2\pi_g$) and 12.75 eV for the $^3\Pi_g$, 13.15 eV for the $^1\Pi_g$ $n\pi^*$ ($6\sigma_g \to 2\pi_g$) excited states of C_3O_2, but on relaxing the inversion symmetry constraints the electron hole becomes localized and the excitation energies of the $^1\Pi$ and $^3\Pi$ states drop to 10.21 and 9.89 eV, respectively. The effect is similar to one for O_2 reviewed in Mulliken and Ermler [1, pp. 116–118]. On introducing CM between pairs of unsymmetrical functions with localized MOs, the symmetry can, of course, be restored, but with only small energy effects.

Botschwina [28] has used PNO-CI and CEPA methods on the linear molecule imidoborane (HNBH), which is isoelectronic with C_2H_2. Very short and strong BH and NH bonds are predicted, with the calculated NH equilibrium bond length of 0.993 Å being probably the shortest known for a ground-state molecule. Force constants and vibration frequencies are obtained, in good agreement with experimental evidence; $E_h = -80.597$.

Summers and Tyrell [29] have given interatomic distances, E_h and valence-electron orbital energies, and total electronic populations, using SCF computed with a 4-31G basis set for the linear 14-electron molecules $HBCH^-$, HBNH, $HCOH^{2+}$, $HNNH^{2+}$, $HCNH^+$, $HBOH^+$, HCCH, and $HBFH^{2+}$. In another paper [30], using the same basis sets, they have computed minimum energy reaction paths for the reactions $HCO^+ + H^+ \rightarrow HCOH^{2+}$, $H^+ + COH^+ \rightarrow HCOH^{2+}$, $H^+ + HNN^+ \rightarrow HNNH^{2+}$, together with energy barriers for these reactions. The minimal-energy approach paths are all linear.

C. MOLECULES WITH D_{2h} SYMMETRY

1. Ethylene ($H_2C{=}CH_2$) and Si_2H_4

Ethylene in its planar ground state may be described as follows (see Section A for the symmetries of the MOs and the choice of coordinate axes):

$$1a_g^2\ 1b_{1u}^2\ 2a_g^2\ 2b_{1u}^2\ 1b_{2u}^2\ 3a_g^2\ 1b_{3g}^2\ 1b_{3u}^2\ 1b_{2g}^0,\ {}^1A_g$$

The MOs $1b_{3u}$ and $1b_{2g}$ are the well-known π and π^* MOs.

Figure 4 (taken from a review by Merer and Mulliken [31a]) shows the energies (from photoelectron spectra) and roughly the forms of the MOs, along with the approximate characteristics of the MOs when (except for the K-shell orbitals) the nuclei are all united (semi-united atom). Transitions to the V state and the first Rydberg singlet state are also indicated. Many higher Rydberg states are also known from spectroscopy.

Figure 5 (also from [31a]) shows an early version of how the energies of the most important electronic states of C_2H_4 behave upon twisting the molecule. The curves are based on good experimental data for the N, R, and I ($C_2H_4^+$) states near their minima, supplemented by approximate

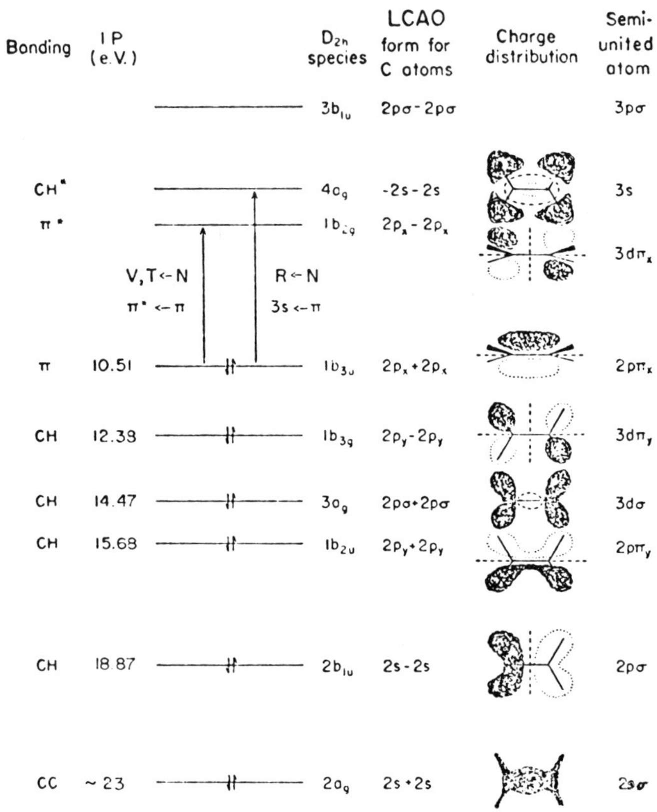

FIG. 4. Molecular orbitals of the ethylene molecule. From Merer and Mulliken [31a].

data on the T state [31b], but need modification on the basis of theoretical calculations (see below). Note that the symmetry is reduced to D_2 on twisting, permitting g and u states to mix, but at 90° the symmetry becomes D_{2d}, leading to orbital degeneracies and new species labels. With regard to the ^{1}E states formed at 90° by R and R′ and the ^{2}E states of $C_2H_4^+$ formed by I and I′, note that each of these is subject to a Jahn–Teller splitting into two states of reduced symmetry. The resulting surfaces would, of course, have equal energies at 0° and 180°.

At 90° the wave function of the ground state becomes a 50:50 mixture of what at 0° are the two ^1A_g states N and Z (see Fig. 6). Helpful for understanding are the correlations between 90° twisted ethylene and O_2 (see Fig. 6). In Fig. 5, theoretical calculations now indicate (see Table 2 below) that at 90° the T($^3A_2$) state is slightly *above* the N($^1B_1$) state. They also

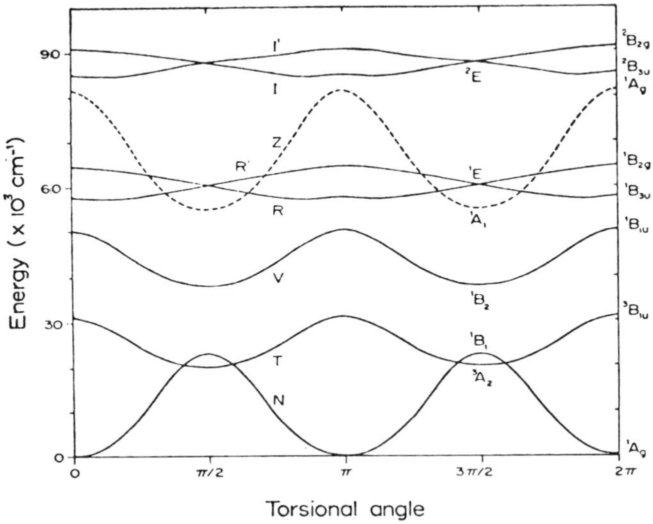

FIG. 5. Potential energies of some important electronic states of ethylene as functions of the torsional angle. From Merer and Mulliken [31a].

show that at 90° the Z state is slightly *below* the V state, instead of being high above it as in Figs. 5 and 6.

Very many *ab initio* computations have been made on the ground (N) and excited states, especially the $\pi\pi^*$ ($b_{3u} \rightarrow b_{2g}$) $^1B_{1u}$ or V state. The most accurate SCF computation for the ground state appears to be one by Fischer-Hjalmars and Siegbahn [3], using a double-zeta plus polarization (one d) Gaussian basis set on the carbons and s plus one p function on the hydrogens. They obtain $E_h = -78.0623$ and give orbital energies (in eV) as follows: -305.63, -305.58, -28.07, -21.78, -17.59, -16.01, -14.01, and -10.13 for $1a_g$, $1b_{1u}$, $2a_g$, $2b_{1u}$, $1b_{2u}$, $3a_g$, $1b_{3g}$, and $1b_{3u}$, respectively. These are somewhat smaller than the corresponding photoelectron ionization energies (cf. Fig. 3). They have obtained a better set of theoretical values for the latter by making SCF computations on various

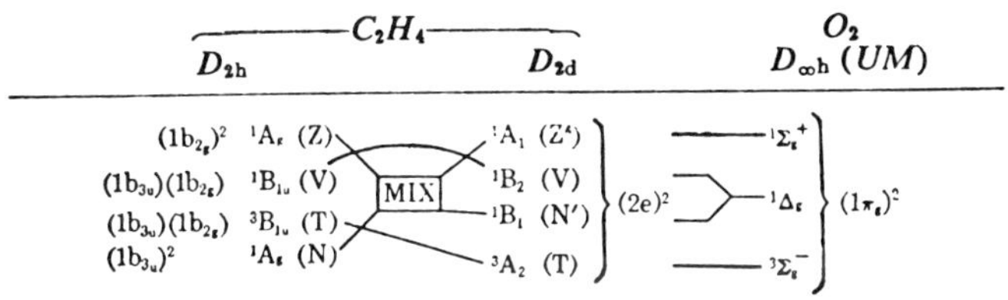

FIG. 6. Correlation of the states of planar and perpendicular ethylene with those of the united molecule, O_2. From Merer and Mulliken [31a].

TABLE 2
Computed Energies for 90° Twisted Ethylene

State (D_{2d})	R(CC) (Å) BP[a]	R(CC) (Å) BS[b]	0,0 excitation energy (eV) BP[a]	0,0 excitation energy (eV) BS[b]
N $^1B_1$	1.48	1.49	2.71	2.59
T $^3A_2$	1.48	1.49	2.79	2.67
Z $^1A_1$	1.41	1.40	6.14	5.76
V $^1B_2$	1.43	1.40	6.22	5.92

[a] From Buenker and Peyerimhoff [36].
[b] From Brooks and Schaefer [37].

states of the ion, and subtracting the ground-state SCF energy; they obtain 298.59 eV for a 1s hole, and 20.99, 16.76, 14.41, 13.31, and 8.9 for $2b_{1u}$, $1b_{2u}$, $3a_g$, $1b_{3g}$, and $1b_{3u}$, respectively. Remaining discrepancies with experiment are explainable by CM. The same authors give contour diagrams of the valence-shell MOs, and compare them with those for corresponding MOs of C_2H_2. These contour diagrams are reproduced in Fig. 1. They also give a comparison of overlap populations.

Using up to 4-31G plus in part bonding function basis sets (for out-of-plane bending) in SCF computations, Wiberg and Wendoloski [32] have computed dipole moment derivatives relevant to vibrational intensities, along with force fields and charge shifts in the atoms on vibrational excitation. Agreement of the force fields with experiment is fairly good. Agreement with earlier calculations of Pulay and Meyer [33], using larger but unpolarized basis sets, is good (but not much dependent on basis set size) except for out-of-plane bending, where Wiberg and Wendoloski find that polarization functions are needed.

Various CM calculations on the ground state have been reported in connection with calculations (see below) on the V state. The lowest directly computed E_h value, at a threshold T of 10^{-5} a.u. (10 microhartrees), is $E_h = -78.3365$, obtained by Buenker *et al.* [34]. Extrapolation to $T = 0$ yields $E_h = -78.3530$. Allowance for the imperfections of the MO basis set leads to an estimate $E_h = -78.373$. An interesting GVB calculation by Hay *et al.* [4], equivalent to SCF plus partial CM, gave $E_h = -78.1332$. Another GVB calculation has been published by Levin *et al.* [35]. Contour maps of the GVB orbitals are shown.

Incidental to a CM discussion of excited states (see below), Buenker and Peyerimhoff [36] give for the planar (0°) ground state the chief CM components (coefficients 0.95, −0.16, and −0.07 for the $\pi^2 \rightarrow \pi^{*2}$ and

$\sigma\pi \rightarrow \pi^*\sigma^*$ excitation). They also give what is probably the best computed value of the rotation barrier (90° excitation energy) as 2.71 eV or 62 kcal/mole, as compared with the experimental value of 65 kcal/mole. Recent calculations by Brooks and Schaefer yielded similar results [37].

Localized MOs (LMOs) have been discussed both by the procedure of Edmiston and Ruedenberg minimizing the exchange component of the total energy, and by the Peters SCF procedure starting with a set of qualitatively localized orbitals. Using the Edmiston–Ruedenberg procedure with approximate SCF wave functions, Newton *et al.* [38] obtain LMOs for C_2H_4, and discuss bond moments, hybridization, and the effect of truncation to remove nonlocal portions of the otherwise somewhat extended (σ hyperconjugation, see Section VII,B on CH_4) but still orthogonal MOs. They find a C—H local bond moment of 1.92 D, increased to 2.02 by truncation. They find an increase in energy of 0.046 a.u. by truncation. They also examine the degree of preference for the bent bonds which replace the usual σ–π separation in the LMOs of a double bond, and find 0.064 a.u. less exchange energy using LMOs; they find a similar effect in 1,3-butadiene but much less in benzene.

Wilhite and Whitten [39] have used the Peters method, including comparison with conventional SCF (perfect agreement) using each of two rather moderate Gaussian bases "frozen" and larger ("split"). The percent of delocalization in truncated orbitals, defined as $d = 100(1 - \langle\Phi'|\Phi''\rangle)^{1/2}$, is computed, where Φ' and Φ'' are untruncated and truncated LMOs. They do not delocalize the double-bond MOs to mix σ and π. With the "split" basis, d is about 1% for the 1s LMOs, 7.05% and 7.61% for the CC and the CH LMOs. A table of orbital energies is given for the LMOs (ε values, in a.u., are -0.7954 for CC σ and -0.7012 for CH LMOs, as compared with values ranging from -1.0327 to -0.5073 for the corresponding conventional MOs). An increase of energy of 0.029 a.u. by truncation is found for the "split" basis. This differs somewhat from the value given by Newton *et al.*, perhaps because the latter have not delocalized the double bond into bent bonds. Besides their SCF studies, they made fairly extensive CM calculations, using localized virtual orbitals. An energy lowering of 0.125 a.u. is obtained by CM; of this, intraorbital contributions give 77%, interorbital 23%.

Kollman [40], using up to a 431G SCF basis, obtained LMOs for the valence electrons of ethylene and 1,1-difluorethylene, using the Edmiston–Ruedenberg method. He shows that the anomalously small XCX angle in the latter is not due to lone-pair attractions but is an electronegativity effect. Comparisons are made with bond angles in molecules like Li_2O.

In a study of optical activity, Rauk *et al.* [41] have calculated the rotational strength of 10° twisted ethylene (and 1,3-butatriene). They used a minimal STF basis in STO-3G form.

Liskow and Segal [42], using a (9s 5p) basis contracted to (4s 2p) for carbon, plus s and p diffuse functions, report a series of CM computations (up to 23,322 configurations) on various excited states of ethylene twisted by 10°, the computations being directed toward an elucidation of the circular dichroism spectra of *trans*-cyclooctene. The energies, rotatory strengths, and oscillator strengths, along with $\langle x^2 \rangle$ values, of the excited states (including $\pi^* \leftarrow \pi$ and $\pi^* \leftarrow \sigma_{CH}$ and several Rydberg states) were calculated for ethylene in the planar, 10° twisted (D_2), and in the *trans*-cyclooctene (C_2) double-bond geometry. The results are used to account for the observed CD spectrum of cyclooctatetrene.

Using it as a model for torsionally strained olefins, Volland *et al.* [43] have calculated the effect of carbon atom pyramidalization on the bonding in ethylene. They used a (9s 5p 1d) basis set contracted to (4s 2p 1d) for the carbons, and (5s) contracted to (3s) for the hydrogens. They calculated the energy changes on bending both pairs of hydrogens equally out of plane either in the same direction (*syn*, C_{2v} symmetry) or in opposite directions (*anti*, C_{2h} symmetry). They found *anti* somewhat easier than *syn* bending. They discuss the changes in the MOs and give overlap integrals and bond orders.

They quote Snyder [44] as having found that Si_2H_4 prefers a nonplanar equilibrium geometry, with *anti* pyramidalization. On the other hand, Daudel *et al.* [45] report SCF calculations on planar π^2 Si_2H_4 and 90° twisted (D_{2d}) $\pi\pi^*$ triplet Si_2H_4. They used a (15s 10p) basis set contracted to (7s 5p) for Si, (10s 5p) contracted to (5s 3p) for carbon, and (5s) contracted to (3s) for H. At the end of the paper in a preliminary study, they flapped the hydrogens 2° out of plane in a trans manner and found a slight *increase* in total energy. In view of Snyder's report, however, it appears doubtful whether Si_2H_4 is stably planar. As a measure of the strength of the π bond, they took the difference between the energies of the 90° triplet state and the planar ground state, for which they found 7 kcal/mole, although correlation should considerably increase the difference. Optimization of the Si—Si distances gave 2.16 Å for the planar singlet and 2.33 Å for the twisted triplet. In addition to the calculations on Si_2H_4, they made similar calculations on silaethylene $H_2Si{=}CH_2$. Here they found the planar ground state to be 8 kcal/mole more stable than the twisted triplet.

Cheung *et al.* [46] have used an eight-configuration MCSCF calculation to follow the dissociation of ethylene in its equilibrium state ($\sigma^2\pi^2$, ^1A_g) into two ($\sigma\pi$, $^3B_1$) ground-state methylene (CH_2) radicals. They use an unpolarized (9s 6p) carbon basis set contracted to (3s 2p), with (4s) contracted to (2s) for the H atoms. For the MOs they use an orthonormal set of symmetrized LCAO forms of the simplest kind, and build their eight configurations from these localized MOs. During dissociation, similarly localized MO configurations of CH_2 result. Following the least-motion reaction path, with the HCH bond angle optimized all the way, dissociation proceeds without a barrier. The π bond breaks before the σ bond. The calculated dissociation energy agrees with experiment within 6 kcal/mole. A similar calculation and analysis is made for the dissociation of the $\sigma^2\,\pi^{*2}$ lowest ${}^1A_g$ excited state into two σ^2, ${}^1A_1$ excited methylenes.

The T (triplet) $\pi\pi^*$ state of ethylene has been the subject of numerous *ab initio* computations. Perhaps the most accurate for vertical excitation is that of Brooks and Schaefer [47]. They use a standard double-zeta basis plus one d function set ($\alpha = 0.75$) on the carbons and p functions ($\alpha = 1.0$) on the hydrogens. They then do valence-shell CM (all singles and doubles) and finally apply Davidson's correction for higher substitutions. For the excitation energy they obtain 4.33 eV, or 4.60 eV with the correction, or slightly different values when diffuse orbitals are included. These results agree well with experimental values of about 4.3 eV [31b] or about 4.5 eV [48]. For 90° twisted ethylene, Buenker and Peyerimhoff [36] in an extended CM calculation on excited states find the T state 0.08 eV (2 kcal/mole) *above* the N state. In Fig. 5, the T state at 90° is slightly *below* the N state, a relation for which there is some experimental evidence [31c]; but the correct result is uncertain. All calculations on the size of the MO in the T state show it to be much smaller (valence-like) than in the corresponding singlet state, the V state.

Walsh [11] suggested that in the T state, the molecule may be distorted by wagging or flapping of the hydrogens. Baird and Swenson [49] made minimal basis and then double-zeta SCF calculations with flapped methylene groups, both C_{2v} (cis) and C_{2h} (trans), as well as C_2 (twisted 90° and then flapped). With the minimal basis, a stabilization of 4–5 kcal/mole below the planar triplet is found, but with the double-zeta basis, the calculated stabilization is only about 1 kcal/mole. A better basis and CM would be needed to give a final answer. However, on consideration of NH_3, the authors estimate that inclusion of polarization functions in the basis would give (by flapping) a net stabilization of 1–5 kcal/mole for the planar conformation, and 0–1 kcal/mole for the 90° twisted form. For the C—C distance in the planar form, 1.55 Å is found, and 1.48 Å for 90° twist.

The $1b_{2g}$ (π^*) MO in the T state is relatively small: the contrast is shown in Fig. 7, taken from an article by Buenker *et al.* [50]. This figure shows, for planar geometry, contour maps of the π^* MO in the ${}^3B_{1u}$ and the corresponding ${}^1B_u$ V state. The ${}^3B_{1u}$ π^* is valence-like, while on ${}^1B_{1u}$ it is partly Rydberg-like. These maps are for SCF MOs. However, for the V state especially, one ought to use NOs (natural orbitals). To do so would make

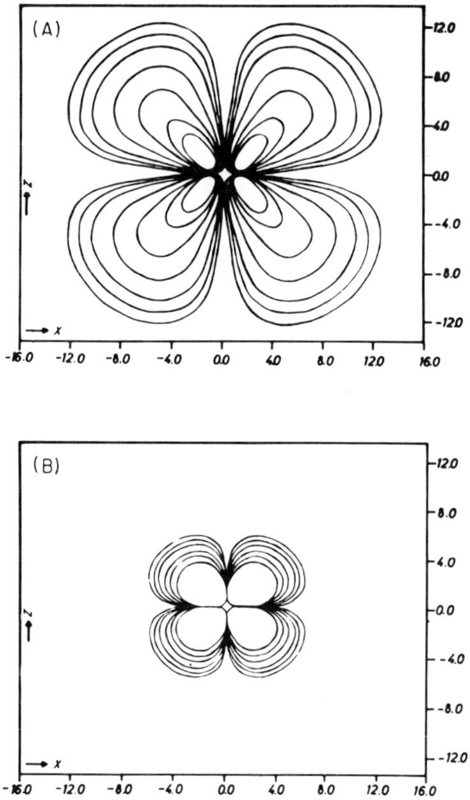

FIG. 7. Charge density contours for the $1b_{2g}$ MO (united-atom $3d\pi$) from the SCF calculation for the lowest $1b_{2g} \leftarrow 1b_{1u}$ ($\pi^* \leftarrow \pi$): (A) excited singlet, (B) excited triplet. From Buenker *et al.* [50].

the $1b_{2g}$ MO for the V state much smaller but still large. For other MOs, the difference would not be so great.

Relevant to Fig. 5, Table 2, taken from [36] and [37], shows dimensions and excitation energies (relative to the planar ground state) of states for 90° twisted ethylene. It now appears that the T $^3A_2$ state unexpectedly is probably *above* the N $^1B_1$ state.

Characterization of the V state has proved to be a difficult problem for theoreticians for a number of years. Experimentally also, the V ← N transition has been difficult to interpret [31], but its behavior in the condensed state indicates valence-like character for the V state. However, other experimental evidence [51] indicates that it also has significant Rydberg character. The best theoretical calculations in planar geometry give for a vertical VN transition a higher energy than for the intensity maximum of the observed transition at 7.65 eV.

Buenker *et al.* [52] have suggested that the observed transition is a nonvertical one and, further, have calculated that for a vertical transition, a $^1B_{1g}$ Rydberg state ($3p_y \leftarrow \pi$) lies somewhat below the ^1B_u V state. Thence upon twisting, both become $^1B_1$ so that they can mix. Later more accurate calculations by Buenker and Peyerimhoff [36] placed the vertical $^1B_{1g}$ state at about 7.6 eV. McMurchie and Davidson [53] agree that the vertical $^1B_{1g}$ is probably below the vertical $^1B_{1u}$. Experimental evidence probably also favors this conclusion (see discussion below).

The foregoing and other theoretical calculations are in agreement with the theoretical prediction that (aside from CM) the π^* MO for vertical excitation is partially Rydbergized [54] toward its united-atom form of a $3d\pi$ Rydberg orbital, while at larger C—C distances it largely retains its $2p\pi-2p\pi$ (but with reduced exponents) valence-type LCAO form. Nonverticality in the experimental VN transition favors the LCAO form, and experiment confirms its predominance for the observed transition.

The V states of N_2 (a $^1\Sigma_u^+$ $\pi\pi^*$ state) and O_2 (a $^1\Sigma_u^-$ $\pi\pi^*$ state) have been shown to contain a considerable amount of $\sigma\sigma^*$, where σ and σ^* are roughly $2p\sigma + 2p\sigma$ and $2p\sigma - 2p\sigma$ in LCAO form, where the excitations $\pi\pi^*$ and $\sigma\sigma^*$ are relative to the ground state. Mulliken has suggested [55] that $\sigma\sigma^*$ is relatively important also in C_2H_4. For the analogy to N_2 and O_2, $\sigma\sigma^*$ would mean $3a_g \rightarrow 3b_{1u}$; other non-π MOs commonly referred to as σ and σ^* should not be counted. In a 1977 article, McMurchie and Davidson [53] in an extended CM study find $3a_g \rightarrow 3b_{1u}$, using NOs, to be the most important CM contribution in the V state (coefficient -0.11 as compared with 0.97 for the main NO). In qualitative agreement, Buenker *et al.* in a 1979 article [34] give 0.006 as the square of this coefficient; the next largest squared coefficient, 0.004, corresponds to simultaneous $\pi\pi^*$ and $\sigma\sigma^*$ excitation. Buenker *et al.* [34, 36] seem to argue that $\sigma\sigma^*$ excitation is unimportant; perhaps they mean relative to the main $\pi\pi^*$ configurations, but in any event the importance of the coefficients just cited can speak for itself.

The size of the π^* MO in the vertical V state, related to the degree of Rydbergization, has been the subject of considerable discussion. McMurchie and Davidson give for $\langle\pi^*|x^2|\pi^*\rangle$ the relatively small value of 8.8 a.u. for the first π^* natural orbital, while Buenker *et al.* provide strong evidence for a considerably larger value. A critical discussion of three articles [34,47,53] with respect to the size of the π^* MO and the VN transition oscillator strengths is given by Buenker *et al.* [34] at the end of an extended article on the V state, using their MRD-CI method (see also

Buenker *et al.* [52]). More recently Buenker *et al.* [56] have made a very thorough study in which they find $\langle\pi^*|x^2|\pi^*\rangle = 10.36$ a.u.

The computed vertical excitation energy of the V state seems to have settled down: Buenker *et al.* [34], 7.95 eV; McMurchie and Davidson, 7.96 eV. It is still higher than the VN peak of 7.65 eV, a fact explainable by nonverticality in the latter, as mentioned above. Without the inclusion of Rydberg-like 3dπ in the basis set, the excitation energy would be 0.6 eV higher [34]. For the oscillator strength of the vertical VN transition, Buenker *et al.* [34] compute $f = 0.34$, which may be compared with experiment (0.34, accidentally in agreement), with a pure LCAO valence-type π^* (0.43), and with a pure Rydberg-type (3dπ) π^* (about 0.01); evidently the valence-type component has a predominant influence. Brooks and Schaefer [47] found $f = 0.36$.

The most recent and accurate articles on the V state ([47,50], and perhaps especially [34] supplemented by [52] and [56]) all used good polarized Gaussian basis sets with d functions on the carbons and p functions on the hydrogens, and with extensive CM, but differ in some ways. All included in the basis set "diffuse," i.e., large Rydberg-scale, d functions in addition to the smaller polarization d functions. In [50] polarization d's with orbital exponent $\alpha = 0.45$ were used, [47] used $\alpha = 0.75$, while [34] used two polarization d's with $\alpha = 0.35$ and 1.20. For the diffuse d's, [50] used two sets with $\alpha = 0.013$ and 0.026 centered at the inversion center, [47] used diffuse 2p functions (2pπ_x–2pπ_x) with $\alpha = 0.034$ and 0.012, and [36] tried four sets of pairs of d functions varying from $\alpha =$ 0.015, 0.008 to $\alpha = 0.040$, 0.030, of which 0.030, 0.022 proved to be slightly the best. In [50] the need for CM describing all correlation effects involving the π or π^* electrons was emphasized. Single and double excitations from the V as well as the ground-state configurations into the virtual space were included. For the V state, [47] obtains $E_h = -78.0273$, [34] obtains -78.0227 at $T = 10\ \mu$h, or -78.0640 upon extrapolation to $T = 0$, or -78.086 as an estimate of the complete CM energy.

Calculations on the TN and VN transitions by the equations-of-motion method are in fairly good agreement with the results in the articles just discussed. In this method, quantities of interest are computed directly without obtaining wave functions for the two states involved. Rose *et al.* [57] find 4.1 eV for the vertical TN transition, and for VN 7.9 eV with $f =$ 0.40. For the T state, they find $\langle\pi^*|x^2|\pi^*\rangle = 2.7$ a.u. and for the V state 9.0 a.u. They used a somewhat restricted basis set (no polarization) but included three diffuse pπ functions with exponents 0.0365, 0.0116, and 0.0037.

In a later article with a somewhat poorer basis set, Coughran *et al.* [58] computed the T and V excitation energies and the VN oscillator strengths for a range of R(CC) values from 1.24 to 1.80 Å. They give potential curves for the N, T, and V states showing them dissociating to two $^1A_1$ CH_2 radicals (actual dissociation should involve $^3B_1$ CH_2).

Using one version (RPA) of the equations-of-motion method and an STO-nG (n = 4,5) basis, Bouman and Hansen [59] have made computations of ordinary and rotatory intensities of low-lying electronic transitions of twisted ethylene and twisted *trans*-butene. The computations confirm that a localized ethylenic chromophore can account for the low-lying CD spectrum in monoolefins. Further, the computed rotatory strengths agree in sign and magnitude with experimental results on *trans*-cyclooctene.

In a careful study of the flapping (pyramidalization) of one CH_2 group in 90° twisted ethylene (symmetry D_{2d}), Brooks and Schaefer [37], using a double-zeta basis without polarization but followed by single and double valence-shell excitation CM, compute that as flapping proceeds, the V and Z states rapidly develop a dipole moment corresponding to a partially $CH_2{}^+CH_2{}^-$ wave function. (Note that the V and Z states in 90° ethylene are 50:50 mixtures of $CH_2{}^+CH_2{}^-$ and $CH_2{}^-CH_2{}^+$ wave functions.) This is the "sudden polarization"—a better term would seem to be "rapid polarization"—that has been evidenced in experimental work on certain substituted ethylenes [60]. Thus for a flapping angle of 5°, the Z and V states develop dipole moments of + 1.25 and − 1.18 D respectively, where a positive sign indicates that the flapped CH_2 carries the negative charge, and vice versa. For a C—C separation of 1.416 Å, this corresponds to the two CH_2 groups having opposite charges of 0.18 e. They further conclude that, while the higher, V, state (cf. Table 2) is stablest in the unflapped D_{2d} conformation, the lower, Z, state in equilibrium is flapped by 30.7° with 2.8 kcal/mole stabilization energy as compared with D_{2d}. In their calculations, they note that it is necessary to proceed with extreme caution to avoid spurious results sometimes obtained by previous authors.

In a later article [61], Bonačić-Koutecký *et al.*, using an MRD-CI procedure down to very low thresholds, have further examined the sudden polarization effect. They tested three AO basis sets, two double-zeta and one with d polarization on the carbons, and also examined the effect of varying the MO basis set in CM, including the use of SCF MOs or NOs. If they used NOs, their results agreed well with those of Brooks and Schaefer, but using MOs gave polarization about half as large for 5–10° flapping although they approached equality at 20°. Bonačić-Koutecký *et al.* differ from Brooks and Schaefer in not absolutely requiring zero dipole moments at 0°, but hopefully this is not serious. However, their work reinforces the need of extreme caution emphasized by Schaefer.

Numerous Rydberg states of ethylene are well known [31]. The lowest singlet Rydberg state is well known spectroscopically (the RN transition in Fig. 5). The bands of the RN transition are superposed on the VN transition. From their structure it has been deduced that the R state has minima at ±25° (±1°) of twist, but with an energy only 0.01 eV below that for a vertical transition [62]. The photoelectron spectrum shows that the ground state of $C_2H_4^+$ has a very similar twist [31]. The same must then be true of the Rydberg states in general. These results are in accord with early semiempirical calculations. The $4a_g$ Rydberg MO of the R state is essentially of 3s type with a quantum defect [53] of 1.00 [31]. Two contour maps of it, as obtained by McMurchie and Davidson, are shown as Figs. 8 and 9. One map shows the MO as obtained from an SCF computation [50] and the other the corresponding natural orbital (NO) as obtained in a later article [63].

Although a number of good papers on the Rydberg states of ethylene have been published, it may suffice to refer to an article by McMurchie and Davidson [63]. This article covers singlet states involving excitation of the π ($1b_{3u}$) MO to all $n = 3$ Rydberg MOs, including the R state just mentioned. Their paper also contains a critical review of earlier work. However, some of the articles [36,50,64] of Buenker, Peyerimhoff, *et al.*, who have investigated all low-lying states that can be obtained from a single excitation out of $3a_g$, $1b_{3g}$, and $1b_{3u}$ orbitals, should be consulted. Also, Fischer-Hjalmars and Kowalewski [65] present useful calculations

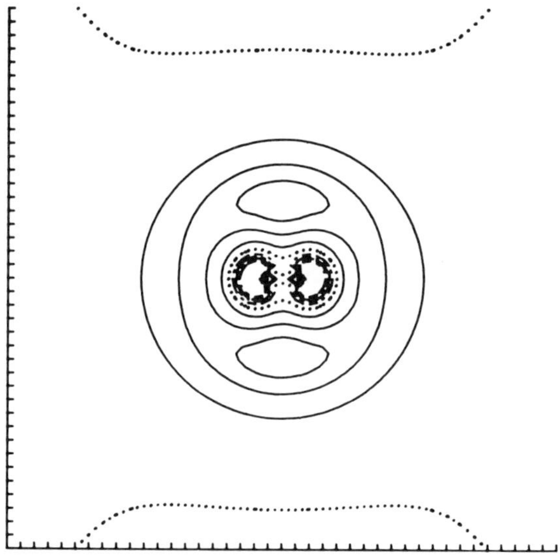

FIG. 8. $4a_g$ SCF MO for $^2B_{3u}$ in the molecular plane. From McMurchie and Davidson [53].

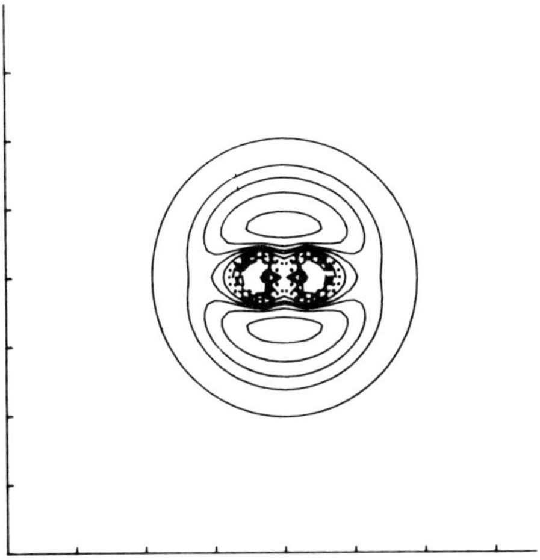

FIG. 9. 3s CM NO from $^1(\pi, 4a_g)$ in the molecular plane. From McMurchie and Davidson [63].

for excitations from π ($1b_{3u}$) to ns, np, and nd Rydberg MOs for several values of n, and for both singlet and triplet Rydberg states; they also compute oscillator strengths for absorption to these states. As expected, singlet–triplet separations decrease with increasing n.

McMurchie and Davidson used a fairly good basis set not including polarization functions, plus diffuse functions to represent the Rydberg MOs (two s, four d, and four p), with CM including all double excitations of the correct symmetry including the Rydberg MO and one valence MO (except for $1a_g$ and $1b_{1u}$), along with all single excitations from Rydberg and valence orbitals of the parent configuration of the excited state. Next, the resulting energies were subtracted from a singles-only calculation on the positive ion to obtain an MO term value for each state. Finally, to obtain realistic excitation energies, these computed term values were subtracted from the experimental ionization energy of 10.51 eV. All calculations were done at the experimental equilibrium configuration of the ground state. Use of the experimental ionization energy takes care of the fact that the Rydberg states like the ion would in equilibrium be twisted by about 25° but the correction to verticality would be only about 0.01 eV (see above).

Table 3 gives descriptions of excited states and probable assignments of the upper levels of experimentally observed features to theoretical states as given by Mulliken [31b, 31c]. These include modifications of several previously proposed assignments by various people. Spectroscopic transitions are observed to states V and 2R and to all the listed states begin-

TABLE 3

Excited States of Ethylene

State	Excited MO (0°)	n^* [j]	UAO [e]	State (0°)	Excitation energy: Experiment	Excitation energy: Theory [k]	Polarization [f] (dipole)
V	$1b_{2g}$ ($2p\pi - 2p\pi$, $3d\pi$)	2.18 (peak)	$3d\pi_x$	$^1B_{1u}$	5.5 (90°) [g]	6.2 [h]	
					7.65 (peak)	7.96	z
BLA [a]	$3b_{1u}$ ($3p\sigma$)	1.96	$4p\sigma$	$^1B_{2g}$	7.2	8.01	Forbidden
2R [b]	$4a_g$ (3s)	2.00	$3d\sigma$	$^1B_{3u}$	7.10	7.26	x
VV [c]	$2b_{2u}$ ($3p\pi_y$)	2.25	$3p\pi_y$	$^1B_{1g}$	7.83	7.93	Forbidden
3R′ [d]	$2b_{3u}$ ($3p\pi_x$)	2.46	$3p\pi_x$	^1A_g	8.26	8.36	Forbidden
3R″ [d]	$5a_g$ ($4d\sigma$)	2.68	4s	$^1B_{3u}$	8.62	8.80	x
3R [d]	{ $1b_{1g}$ ($3d\delta_{sin}$)	2.91	$3d\delta_{sin}$	$^1B_{2u}$	8.90	9.04	y }
	$6a_g$ ($3d\delta_{cos}$)		$3d\delta_{cos}$	$^1B_{3u}$		8.99	x }
	$2b_{2g}$ ($4d\pi_y$) }		$4d\pi_y$	^1A_u		8.99	}
4R‴ [d]	$7a_g$ (4s)	3.06	$4d\sigma$	$^1B_{3u}$	9.04	9.12	x
4R′ [d]	$2b_{2g}$ ($4d\pi_x$)	3.45	$4d\pi_x$	$^1B_{1u}$	9.36	9.31	z
4R″ [d]	$8a_g$ ($5d\sigma$)	3.70	5s	$^1B_{3u}$	9.51		x
4R [d]	{ $2b_{2g}$ ($4d\delta_{sin}$)	3.92	$4d\delta_{sin}$	$^1B_{2u}$	9.62		y }
	$9a_g$ ($4d\delta_{cos}$) }		$4d\delta_{cos}$	$^1B_{3u}$			x }

[a] Magnetic circular dichroism: from Brith-Lindner and Allen [65a].

[b] From Merer and Schoonfeld [62].

[c] Threshold electron-impact spectrum: from van Veen [65b]. This assumes a vertical transition. Note, however, that, on twisting, this state becomes $^1B_1$ and probably mixes with the V state (see below Table 2).

[d] Optical spectra: from Wilkinson [65c] and McDiarmid [31c].

[e] United-atom orbital.

[f] For optical transition from ground state.

[g] From Foo and Innes [65d].

[h] See Table 2.

[j] Effective principal quantum number of MO [31b].

[k] From McMurchie and Davidson [63].

ning with 3R′. The transition to 3R′ is agreed to be an electric-dipole forbidden one. Mulliken's assignment for 3R′ at first differed from that of McMurchie and Davidson, but now we accept the latters' assignment. For all the higher levels for allowed transitions, McMurchie and Davidson concurred with Mulliken [31b]. Two additional states to which electric-dipole transitions are forbidden are included in Table 3 [31c]. It should be mentioned that others, notably Buenker and Peyerimhoff (personal communication), disagree with the assignment of the BLA state, on the ground that its observed energy differs by an unacceptable amount from the results of their theoretical calculations. However, there seems to be no reasonable doubt as to its reality, and it seems very difficult to find a plausible alternative assignment.

McMurchie and Davidson give a table of computed values of $\langle x^2 \rangle$, $\langle y^2 \rangle$, and $\langle z^2 \rangle$ for the Rydberg MOs. They show that there is considerable mixing between s and d Rydberg MOs. This can be seen from the $\langle x^2 \rangle$ etc. values. Table 3 also shows that s and dσ change places as the united atom is approached (see UAO column); this effect results from avoided crossings.

von Niessen *et al.* [66], using a many-body Green's function approach, have computed vertical ionization potentials for ethylene. They tried five basis sets, four of them polarized, and got fairly good agreements with experiment with all. However, for the lowest (π, $1b_{3u}$) ionization potential the definitely best agreement was obtained when an abnormally large ("diffuse") d AO was used in the basis set for polarization. Thunemann *et al.* [67] have sought to explain this peculiarity by making CM calculations of the usual kind on the ground states of the molecule and its positive ion to obtain the π ionization potential for two of the basis sets of von Niessen *et al.*, one with a normal (exponent 0.8) and one with a diffuse (exponent 0.3) d AO. The results agree with those from the Green's function method. They conclude that the anomaly should be attributed to a poorer correlation energy in the ion than in the molecule, so that the use of an abnormal d AO accidentally improves the agreement with experiment; and so that much larger basis sets would be needed for more reliable calculated ionization potentials by either Green's function or conventional CM methods.

Cederbaum *et al.* [18] made calculations to predict details of the satellite structure of the photoelectron spectrum of ethylene. The method and results are similar to those for acetylene, discussed in Section B,1. As mentioned there, the satellite peaks are unusually strong in the inner valence shell region where the MOs are built up largely of carbon 2s AOs. The $2b_{1u}$ (2s − 2s) ionization is predicted to have a main 19.2 eV peak that may be identified with the experimental peak at 19.2 eV, followed by one moderately strong satellite near 25.7 eV, then one weaker and three very weak. The $2a_g$ (2s + 2s) is predicted to be split into two strong components close together at about 23.1 and 23.9 eV (experimentally, a strong peak at 24.6 eV) with five satellites, of which the strongest are two at about 27.2 and 28.2 eV. The experimental peak at 27.4 eV may be identified with a composite of these and the predicted $^2B_{1u}$ peak near 25.7 eV.

In an earlier article, Martin and Davidson [68] have discussed the nature of satellite peaks as corresponding to correlation states of the ion. They make a limited CM calculation for the $2a_g$ ionization using $(2a_g)^{-1}$, $(2b_{1u})^{-1}\,{}^3(\pi,\pi^*)$ and $(2b_u)^{-1}\,{}^1(\pi,\pi^*)$ to obtain three ^2A_g peaks with differing CM composition depending on the couplings assumed. For the $2b_{1u}$ ionization they similarly obtain three $^2B_{1u}$ peaks that are mixtures of $(2b_{1u})^{-1}$, $(2a_g)^{-1}\,{}^3(\pi\pi^*)$ and $(2a_g)^{-1}\,{}^1(\pi\pi^*)$. Additional satellites are obtainable by more complicated CM.

Barth *et al.* [8] have made core-ionized MRD-CI calculations on C_2H_2 and C_2H_4, most of which are discussed in Section B under C_2H_2 (see above). For C_2H_4, potential curves for C—C stretching, CH_2 torsion, and CH stretching and associated Franck–Condon factors for the transition $\pi^* \leftarrow 1s$ show that the observed fine structure is a progression in the CH stretching vibration.

By double charge transfer spectroscopy, Benoit and Horsley [69] measured double ionization potentials of C_2H_4 corresponding to the seven lowest states of $C_2H_4^{2+}$, and also made SCF and CM calculations, using SCF MOs of the ground state of the ion for all the states. They used a moderate Gaussian lobe basis set without polarization functions; for CM they considered all single and double excitations for each parent state and by a perturbation theory criterion selected the most important up to about 2000 configurations for each state. The states computed were $(1b_{3u})^{-2}$, ^1A_g (the ground state); $(1b_{3g})^{-1}(1b_{3u})^{-1}$, ^1A_u; $(3a_g)^{-1}(1b_{3u})^{-1}$, $^1B_{3u}$; $(1b_{3g})^{-2}$, ^1A_g; $(1b_{2u})^{-1}(1b_{3u})^{-1}$, $^1B_{1g}$; $(3a_g)^{-1}(1b_{3g})^{-1}$, $^1B_{3g}$; and $(3a_g)^{-2}$, ^1A_g. They obtained excellent agreement between measured and computed ionization energies, which lay between 29.4 ± 0.5 eV (ground state) and 40 ± 1 eV experimental versus 29.76 to 39.12 eV computed.

In an early SCF computation on C_2H_4, $C_2H_4^+$, and $C_2H_4^-$, Kaldor and Shavitt [70] obtained −6.69 eV for the electron affinity of ethylene.

The rearrangement of ethylidene (CH_3—CH) into ethylene has received considerable theoretical attention. In computations using a sesquizeta basis, Altmann *et al.* [71] conclude that the transformation takes place in the ground state via a cyclic transition state.

For results on ethylene derivatives, see Section IX,H,1.

2. Diborane (B_2H_6), Al_2F_6, and Al_2Cl_6

In diborane, two BH_2 groups lie in a plane like the CH_2 groups in C_2H_4, with the other two bridging H atoms directly above and below the midpoint. Although the two molecules are isoelectronic and both of D_{2h} symmetry, the $1b_{3u}$ MO is much more firmly bound in diborane than in ethylene, obviously because it has become strongly bonded to the bridge hydrogens.

$$1a_g^2\ 1b_{1u}^2\ 2a_g^2\ 2b_{1u}^2\ 1b_{3u}^2\ 1b_{2u}^2\ 3a_g^2\ 1b_{3g}^2\ 1b_{2g}^0,\ {}^1A_g$$

The numerical subscripts of the b MOs depend on how the axes x, y, z are located in the molecule (see Section A). Here the x axis is taken perpendicular to the plane determined by the two BH_2 groups, hence through the bridging hydrogens. Some but not all authors of articles on B_2H_6 have used the same choice. In other cases, there are permutations of the numerical subscripts as compared with the axis choice made here. Here the b MO labels agree with those for ethylene.

Elbert *et al.* [72] have made MRD-CI computations on the ground and a number of singlet and triplet excited states of diborane, along with six states of $B_2H_6^+$, using a large double-zeta basis plus one s and two pπ bond functions for polarization, and diffuse functions for Rydberg states. In the CM, all single and double excitations for the valence-shell occupied and virtual MOs are considered, except for the K-shell MOs and their virtual complements. For SCF they obtain $E_h = -52.8014$ for the equilibrium geometry, for CM they obtain -52.9569 at threshold $T = 20$ (microhartrees), with an extrapolated value -52.9604 for $T = 0$. Adding the Davidson estimated correction to include quadruple excitations gives $E_h = -52.972$.

Elbert *et al.* compute vertical excitation energies for numerous states (Table 9 in [72]) including the 3-quantum Rydberg states, along with orbital expectation values $\langle x^2 \rangle$, $\langle y^2 \rangle$, and $\langle z^2 \rangle$ for several of these, and oscillator strengths for transitions to eight of the singlet states. They also compute the energies of the six lowest states of $B_2H_6^+$ (Table 10 in [72]). In Tables 9 and 10 in [72] Elbert *et al.* say the zero of energy is -56.00 hartrees. To make sense, this figure apparently needs to be amended to -52.00 hartrees. They identify the lowest-frequency, weak, peaks in the absorption spectrum of diborane at 7.3 and 8.4 eV, with the forbidden transitions ${}^1A_g$ to $1b_{3g} \rightarrow 1b_{2g}$, ${}^1B_{1g}$ and ($3a_g \rightarrow 1b_{2g}$), ${}^1B_{2g}$, while transitions at higher frequencies are in part valence-allowed and in part Rydberg.

Regrettably in their calculations they have interchanged the x and y axis labels as compared with the present choice, but *have not so stated.* (One can deduce this interchange of axes, which leads to a replacement of 2 by 3 and vice versa in the numerical subscripts of the b MOs, by comparison with SCF orbital energy tables of other authors who used the present axis choices.)

In their Table 3 listing orbital energies, Elbert *et al.* write $1b_{2u}(\pi y)$, $1b_{3u}(\pi x)$, $1b_{2g}(n)$, $1b_{3g}(\pi y^*)$, where the symbols in parentheses indicate the form of the MOs, except that πx^* is replaced by n. The descriptive symbols πy, πx, πx^*, and πy^* are correct in terms of their choice of axes. With the present choice of axes, the subscripts 2 and 3 must be exchanged, and simultaneously the subscripts x and y. If these exchanges are made, $1b_{2u}(\pi y)$ becomes $1b_{3u}(\pi x)$, $1b_{3u}(\pi x)$ becomes $1b_{2u}(\pi y)$, $1b_{2g}(n)$ becomes $1b_{3g}(\pi y^*)$, and $1b_{3g}(\pi y^*)$ becomes $1b_{2g}(\pi x^*)$, with the orbital energies in the same order as in several earlier articles by authors who used the present axis labels (e.g., an early article by Buenker *et al.* [50]). A designation such as π_x or π_{x*} means $(2p\pi_x)_A \pm (2p\pi_x)_B$ for the participation of the two boron atoms A and B.

In an early SCF article using the present choice of axes, Buenker *et al.* [73] explored the varying energy of B_2H_6, also of C_2H_6, as the conformation is varied between D_{2h} and D_{3d} symmetry. They depict the energy variation of the $1b_{2g}$ MO of B_2H_6, "which is occupied in C_2H_6 (after becoming $3a_{1g}$) but not in B_2H_6" as a function of conformation in the transformation from D_{2h} to D_{3d} (see Fig. 2 in [73]). They give numerous maps of MO charge densities.

In a Green's function treatment, Cederbaum [74] reports calculated values of the six lowest excitation energies of B_2H_6. He predicts the occurrence of satellite lines of about 10% the intensity of the main lines. He uses an axis choice different both from the present and from Elbert *et al.* (x, z, y for our z, y, x). He gives computed ionization energies to reach the six lowest states of $B_2H_6^+$. All are within 0.2 eV of the corresponding observed photoelectron energies, with the exception of the highest IP ($2a_g$), which is computed to be about 1 eV too large.

With an extended STO basis set including polarization orbitals, Laws *et al.* [75] in an SCF computation find $E_h = -52.8331$, appreciably lower than Elbert *et al.*'s SCF results. They use the same choice of axes as Elbert *et al.* They give contour diagrams and lists of computed one-electron properties.

Blom and Mueller [76], using a 4-31G SCF basis, have determined the complete GVFF of B_2H_6, and computed vibrational frequencies after a scale factor adjustment. They obtained agreements with an average deviation of about 10 cm^{-1} or 0.7% with experimental wave numbers. They also computed a number of Coriolis coupling constants in good agreement with experiment, along with centrifugal distortion constants. In a comparison of calculated force constants with corresponding ones of C_2H_4, they find a close parallelism.

Ahlrichs [77] has discussed the dimerization energy of BH_3 to form B_2H_6, using a large polarized basis set. He computes the SCF contribution as 20.7 kcal/mole. Using the coupled electron pair (CEPA) approximation (which is not quite variational), he computes the correlation energy contribution as 15.9 kcal/mole. The total is 36.6 kcal/mole for the dimerization energy. Redmon *et al.* [78] have used many-body perturbation theory, which is size consistent, for a calculation of the dimerization energy of BH_3, using an ample polarized basis set. They obtain 35.0 kcal/mole. After correcting for zero-point vibration and temperature effects, there is agreement within 5% with what is probably the best experimental value of $\Delta H(300°C) = -34$ kcal/mole. It is seen that the change in correlation energy from $2BH_3$ to B_2H_6 is unusually large for a reaction involving only closed-shell molecules.

Minimal-basis calculations by Switkes *et al.* [79] and by Dixon *et al.* [80] using localized MOs should perhaps be mentioned. Their axis labeling differs from that used here.

Minimal-basis calculations by Curtiss [81] show that Al_2F_6 and Al_2Cl_6 have bridged D_{2h} structures like B_2H_6.

3. Dinitrogen Tetroxide (N_2O_4) and B_2F_4

Experimentally, both molecules are planar with a very weak bond between the N or B atoms. Two *ab initio* calculations of double-zeta quality that appeared in 1974 are especially relevant. One, by Howell and Van Wazer [82], is an SCF calculation using seven s-type and three sets of p-type Gaussians for each atom, and covering both N_2O_4 and B_2F_4 (with incidental discussions of B_2Cl_4). The other, by Ahlrichs and Keil [83], is restricted to N_2O_4. It uses Gaussian-lobe functions in a contracted 8,4 (for s,p) basis, supplemented by one Gaussian-lobe d function on the nitrogens only, but also supplemented by a pseudonatural orbital CM treatment of the principal ($6a_g$) σ N—N bonding electron pair. This article gives a table of computed orbital energies for all the MOs, and also compares them with a similar table for the hypothetical molecule C_2O_4.

The two articles are in essential agreement about N_2O_4, including the conclusion that the N—N bond is of σ character, and an explanation of the stability of the planar form as a result of a balance of the stabilizing effects of the σ bond (see below) and exchange repulsions resulting from the interactions of the closed-shell MOs of the two NO_2 groups when N_2O_4 is formed.

Howell and Van Wazer compute a rotation barrier to the 90° twisted D_{2d} conformation of 11.6 kcal/mole, and Ahlrichs and Keil with an improved wave function find 6.0 kcal/mole; the experimental barrier is 2.9. For B_2F_4, Howell and Van Wazer compute a preference of 0.3 kcal/mole for the twisted form, but experiment favors the planar form; in any event, the difference is evidently small. Both pairs of authors give charge-density diagrams (profile maps or contour diagrams) and discuss the forms of the valence-shell MOs, Howell and Van Wazer using population analyses.

Ahlrichs and Keil replace the structure of the weak σ bond in N_2O_4 by an expansion in terms of PNOs (pair natural orbitals). This bonding pair is found to be of the form $0.989(6a_g)^2 - 0.142(2b_{1u})^2 + \cdots$, giving an occupation number 0.0202 for the $2b_{1u}$ MO. They point out that this bond tends to stabilize the planar D_{2h} structure because the binding is not only N—N but partly also between opposite O atoms of the two NO_2 parts. They find the SCF binding energy -2 kcal/mole, which becomes $+5.0$ when the σ bond pair correlation is included; an experimental value is 12.7. The net effect of the N—N bonding and antibonding MOs which result from the interaction of doubly occupied MOs of the two NO_2 in forming N_2O_4 favors the 90° twisted D_{2d} structure, but not enough to cancel the effect of the bonding σ pair in stabilizing the planar structure.

Both sets of authors find 1.67 Å for the equilibrium N—N distance, as compared with an electron diffraction value of 1.78 Å. For the N—N force constant Ahlrichs and Keil calculate 1.32 mdyn/Å, and an NO_2—NO_2 frequency $\nu_3(a_g)$ of 307 cm^{-1}, as compared with an experimental value 283 cm^{-1} for the solid. Obviously all calculated results might be considerably altered if a complete CM were carried out.

von Niessen *et al.* [84], using a Green's function method, have computed the vertical ionization energies for N_2O_4. They find 11.23, 12.57, 12.80, 13.09, and 13.90 eV for the first five ionization potentials $6a_g$, $4b_{2g}$, $1a_u$, $1b_g$, and $4b_{3u}$, as compared with experimental values of 11.4, 12.35, 12.35 (?), 13.0, and 13.5 eV. Above 15 eV they conclude that the usual simple MO picture is no longer at all valid, because of correlation effects, and above 21 eV the intensity due to each MO is distributed over numerous lines, leading to a dense line structure from about 21 to about 44 eV in the photoelectron spectrum.

4. Cyclobutadiene and 1,3-Dimethylenecyclobutadiene

Several *ab initio* calculations on the interesting molecule cyclobutadiene

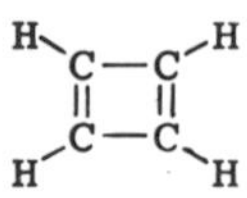

have been made, resulting in a consensus that the ground state is a rectangular singlet (^1A_g of D_{2h} symmetry), with a square (D_{4h}) triplet lying somewhat higher. The absence of an EPR signal for the matrix-isolated molecule gives experimental support to the singlet character of the ground state. With our choice of axes (x axis perpendicular to the molecular plane, z axis parallel to the longer side of the rectangle), the ground state is

$$1a_g^2\ 1b_{1u}^2\ 1b_{2u}^2\ 1b_{3g}^2\ 2a_g^2\ 2b_{1u}^2\ 2b_{2u}^2\ 3a_g^2\ 2b_{3g}^2\ 3b_{1u}^2\ 4a_g^2\ 1b_{3u}^2\ 3b_{2u}^2\ (c_1b_{1g}^2 + c_2b_{2g}^2),\ ^1A_g$$

Various authors have used various labels for the three axes, making the interpretation of their MO symbols difficult.

Early articles by Buenker and Peyerimhoff [85] use an axis choice that differs from the present one only in the interchange of x and z axes, which for D_{2h} leads only to an exchange of B_1 to B_3 and B_3 to B_1 labels. Using a minimal basis set of s and p Gaussian-lobe functions, they give energies and contour maps of all valence-shell SCF MOs and of total charge densities. They follow the SCF treatment with a π-electron CM calculation that gives a rectangular ^1A_g ground state and a higher energy almost square (D_{4h}) triplet state, in violation of Hund's rule. However, by SCF, constraint to the square geometry brings the triplet state ($^3A_{2u}$) below the singlet state ($^1B_{1g}$), in agreement with Hund's rule. By CM, however, the triplet is always above the singlet. Relations to $2C_2H_2$ and to other compounds are discussed. The rectangular singlet is found to be 11 kcal/mole stabler than the square singlet.

Later articles reach similar conclusions. Kollmar and Staemmler [86], using a double-zeta basis set plus d orbitals on the carbon atoms, followed by π-electron CM using IEPA calculations, estimate that the rectangular singlet lies 14 kcal/mole below the square singlet. The reason for the extra stabilization of the singlet versus the triplet they call "dynamic spin polarization." They compute long single C—C bonds (1.57 Å) and normal length double bonds. In a later article [87], the same authors calculate the frequencies and intensities of the IR bands to be expected for rectangular cyclobutadiene. They make plausible a reinterpretation of the fragmentary observed spectrum, previously interpreted to favor a square geometry, but now consistent with the rectangular geometry.

Borden *et al.* [88], using an STO-3G and a double-zeta quality basis plus full π-space CM, discuss the potential surfaces of the lowest singlet and triplet states. They find the rectangular singlet below the square singlet by 4.2. kcal/mole. On adding partial σ CM, this difference is increased to 8.3 kcal/mole. A square geometry is found to represent a minimum for the triplet and a transition state between two rectangular geometries for the singlet. They describe the last two π electrons in the manner indicated in the electron configuration above, with $c_1 > c_2$, corresponding to emphasis on the longer π bonds (b_{1g} MOs). (The other two π electrons are in the $3b_{2u}{}^2$ closed shell.)

Jafri and Newton [89] examined the four low-lying singlet and triplet states for square, rectangular, rhomboid, or trapezoidal geometry using a flexible 6-31G* basis (which includes d orbitals) with 2-configuration SCF followed by CM calculations including all single and double excitations from important reference configurations (both σ and π excitations: "limited $\pi + \sigma$"). Regarding the separation of the square singlet and triplet, the authors arrive at 8.2 kcal/mole using the 6-31G* basis (7.3 kcal/mole found in [87]), but this is increased to 11 kcal/mole when σ CM is included. For the energy difference between the square and the rectangular singlet they find 6 kcal/mole at the π-electron CM level, but come to an estimate of 12 kcal/mole when σ correlation is included, not far from the estimate of Kollman and Staemmler. With reference to the low-lying singlet manifold of the square this paper yields a much smaller energy spread than before. Rhomboidal geometries were found to give higher energy than square geometries.

Haddon and Williams [90] compute the excitation energies and oscillator strengths for 14 excited states of cyclobutadiene using a 4-31G basis by the SCF and by the equations of motion method.

Krogh-Jespersen *et al.* [91] in 6-31G* calculations on the cyclobutadiene dication, which is predicted by Hückel theory to be aromatic and planar, is computed to be strongly puckered, with D_{2d} symmetry (C_{2v} on methyl substitution).

Davidson *et al.* [92] discuss the structure of the diradical 1,3-dimethylenecyclobutadiene (triplet ground state, two low singlets) and the potential surface, using an STO-3G basis and complete π-space CM. They make similar computations on tetramethylenemethane.

5. Butatriene

Brogli *et al.* [93] present a photoelectron spectrum of butatriene ($H_2C{=}C{=}C{=}CH_2$) and STO-G and 4-31G SCF computations in explanation thereof. They also report on tetramethyl- and tetra-*t*-butyl butatriene. With axis choices as in ethylene (Brogli *et al.* exchange the subscripts 2 and 3, because they exchange x and y axes), the electronic structure of butatriene is

$$(1a_g)^2\,(1b_{1u})^2\,(2a_g)^2\,(2b_{1u})^2\,(3a_g)^2\,(3b_{1u})^2\,(4a_g)^2\,(4b_{1u})^2\,(1b_{2u})^2$$
$$(5a_g)^2\,(1b_{3g})^2\,(1b_{3u})^2\,(2b_{2u})^2\,(1b_{2g})^2,\ {}^1A_g,$$

where $1b_{2u}$ is π and $1b_{2g}$ is π^*. The orbital energies of the last seven orbital energies can be identified with corresponding major photoelectron peaks; however, between the two lowest-energy peaks is a "band" of several vibrational peaks that presents a problem.

In a later article, von Niessen *et al.* [66], using a many-body Green's function method, compute the valence-shell vertical ionization potentials of butatriene. They also give a table of SCF orbital energies for various basis sets, the best being an ample polarized set with exponent 0.8 for the carbon d AO. (As discussed above in Section C,2, von Niessen *et al.* [66] for ethylene found that an anomalously low exponent 0.3 gave much better agreement with experiment, but this was explained by Thunemann *et al.* [67].) They use the same axis choices and MO symbols (except for a shift in numbering) as Brogli *et al.* They suggest that the extra "band" found by Brogli *et al.* is a vibronic phenomenon. Calling it the "mystery band," Cederbaum *et al.* [94] then prove that the peaks of the mystery band can be explained by a complex interaction of the lowest two ($^2B_{2g}$ and $^2B_{2u}$) electronic states of the ions with the twisting (A_u) vibration and one of the symmetrical vibrations.

Tantardini and Simonetta [95], using in part a 6-31G basis set, discuss the ground and several vertical valence ionized states of butatriene using a valence-bond method.

6. Dimers of Salts (LiH, LiF, etc.) and SN Molecules

Kollman *et al.* [96] in an SCF calculation on $(LiH)_2$ with contracted (9s 5p) Gaussians on Li and (5s 2p) on H found a D_{2h} symmetry with H—H distance 2.74 Å and Li—Li distance 2.36 Å. With z axis along H—H and x axis perpendicular to the molecule plane, the ground state is $(1a_g)^2\,(1b_{2u})^2\,(2a_g)^2\,(1b_{1u})^2$, ^1A_g. The orbital energies are listed. A complete CM calculation including all triple and quadruple excitations is carried out. The vibrational frequencies, quadrupole moment, and thermodynamic functions were computed, and the dimer found to be bound by 44 kcal/mole at 1000°K.

Baskin *et al.* [97] in an SCF calculation on $(LiF)_2$ and $(NaH)_2$ found D_{2h} geometry at minimum energy. They used basis sets like Kollman *et al.* for Li and H, a contracted (9s 5p 2d) set for F, and a contracted (11s 7p) for Na. The computed dimerization energy of LiF at a minimum energy geometry of 2.25 Å for Li—Li and 2.65 Å for F—F was 63.5 kcal/mole as compared with an experimental value of 62.8 (both corrected for vibration energy).

Ahlrichs [77], using a large polarized basis set, has computed the dimerization energy of LiH, first by SCF (47.3 kcal/mole) and then with electron correlation by CEPA (48.3 kcal/mole). Thus, in contrast to B_2H_6 (see Section C,2) the correlation contribution is minor. The computations on Li_2H_2 agree well with those of Kollman *et al.*

Rupp and Ahlrichs [98] studied the structure and stability of Li_nH_n, Na_nH_n, and Li_nF_n with $n = 1 \ldots 4$, also NaF and Na_2F_2 using good polarized basis sets for Li_nH_n and unpolarized basis sets for F but with the addition of low-exponent p AO's (because F in MF is largely F^-). They used SCF and CEPA-PNO computations. Geometrical and energy data are listed for each case. They assumed D_{2h} structures for the dimers. For the trimers, cyclic D_{3h} geometry was found. For the tetramers, T_d geometry was found when a considerable correlation energy contribution was included; for D_{4h} symmetry, correlation energy is small.

Potts and Lee [99] measured the photoelectron spectra of LiX dimers (X = Cl, Br, I) and made minimal-basis calculations to interpret them, with consideration of spin–orbit splitting. Theory and experiment give higher ionization potentials to the dimers than to the monomers.

Jordan [100], using a fairly good basis set, has made calculations on Li_2H_2 and its negative ion. The latter is found to be stable in a linear configuration $(LiH—LiH)^-$ in which the negative charge is largely spread out to the left of the first Li, and which possesses a dipole moment of 15 D. The electron affinity of Li_2H_2 is estimated as about 0.2 eV.

Jafri *et al.* [101] have made SCF and CM calculations on S_2N_2 using up to double-zeta plus midbond and d polarization functions. They find a D_{2h} almost square singlet ground state and compute a number of excited singlet and triplet states, the lowest a $^3B_{3u}$ state at 3.43 eV according to their choice of axes, which we can deduce to be as follows: x along the S—S axis, y along the N—N axis, z perpendicular to the plane. With x perpendicular to the plane as we prefer, $^3B_{3u}$ becomes $^3B_{1u}$. For the lowest allowed transitions and their oscillator strengths f, they find, in our B notation, $^1B_{3u} \leftarrow {}^1A_g$ (5.38 eV, $f = 0.004$), $^1B_{2u} \leftarrow {}^1A_g$ (6.50 eV, $f = 0.34$), and $^1B_{1u} \leftarrow {}^1A_g$ (6.69 eV, $f = 0.15$). Observed is a transition at a 5.0 eV peak with an intermediate f value. The SCF orbital energies are listed ($b_1 \leftrightarrow b_3$ with our choice of axes). $E_h = -903.7901$ for the best basis set in the SCF calculation, -903.9139 with CM.

Karpfen *et al.* [102], with a large Gaussian basis set followed by CEPA-PNO CM, made calculations on the ground state of S_2N_2, obtaining $E_h = -903.6240$ (SCF?) for the best basis set. They then computed the complete six-dimensional harmonic force field at the SCF level. They found all four SN distances to be equal (D_{2h} symmetry), and computed the force constants and normal frequencies of vibration. Relations to experimental frequencies are discussed. For the NSN angle, 90° is found (experimental: 89.6°).

7. TTF and TCNQ and Their Complex

The molecules TTF (tetrathiofulvalene) and TCNQ (tetracyanoquinodimethane) both may be assumed to probably have D_{2h} symmetry. Cavallone and Clementi [103] have made minimum-basis calculations on these and on the 1:1 complex formed between them. They show orbital energies for the most easily ionized σ and π MOs and give values for the lowest peaks of the π electronic spectrum. For TCNQ these range from 7.8 to 12.6 eV and beyond; they also state that the σ electronic spectrum should start at about 19 eV. For TTF the corresponding values are 11.3 to 14.4 and beyond for π, and about 17 eV for σ. Using Koopmans' theorem, ionization potentials of 8.05 eV for TTF and 11.24 eV for TCNQ are found. The first virtual MO of TCNQ has a negative value, suggesting a considerable electron affinity. For the complex, three models are considered, each at three separation distances. One corresponding to a sandwich structure with a negative charge on TCNQ (the discussion is confusing) is found to be the most stable. Stacking of the sandwiches in the crystal is discussed.

More accurate calculations on TCNQ, also $TCNQ^+$, $TCNQ^-$, and $TCNQ^{2-}$, have been made by Johansen [103a]. He uses a (9s 5p) basis contracted to (4s 2p) for C and N, (4s 1p) contracted to (2s 1p) for H. A complete table of orbital energies is given for TCNQ. The lowest excited states are calculated as $^3B_{3u}$ at 1.05 eV up and $^1B_{3u}$ at 4.23 eV up (experimental: 3.1 eV). Several excited states of $TCNQ^-$ and $TCNQ^{2-}$ are calculated, $TCNQ^-$ is very stable; the electron affinity of TCNQ, from the calculated SCF energies of TCNQ and $TCNQ^-$, is 2.63 eV. Contour maps of the highest occupied π MOs $1a_u$, $2b_{2g}$, $2b_{3g}$, $3b_{1u}$ of TCNQ are shown. The three lowest levels of $TCNQ^+$ result from ionization of the last three of these MOs, and from their energies the corresponding vertical ionization energies are obtained as 8.90, 10.81, and 12.04 eV. Seven higher (σ and π) ionization energies are similarly obtained.

D. MOLECULES WITH D_{2d} SYMMETRY

1. Allene ($H_2C{=}C{=}CH_2$) and 1,3-Dimethylallene

The ground state of allene is nonplanar, of D_{2d} symmetry (see Table 1), with structure

$$(1a_1)^2\,(1b_2)^2\,(2a_1)^2\,(3a_1)^2\,(2b_2)^2\,(4a_1)^2\,(3b_2)^2\,(1e)^4\,(2e)^4,\ {}^1A_1$$

Several *ab initio* calculations have been published, largely on the magnitude of the barrier to rotation in the ground state. The lowest reported ground-state E_h appears to be from an article by Seeger *et al.* [104], with $E_h = -115.9948$ from 4-31G SCF followed by single- and double-substitution CM. It has often been assumed that the rotation barrier corresponds to rotation to planar D_{2h} symmetry (state ^1A_g). However, the calculations show that a planar C_{2v} structure, with the CCC axis bent, gives a lower energy, for a ... $(1b_1)^2 (1a_2) (6a_1)$, $^3A_2$ state. This is calculated to be 48 kcal/mole above the ground state. The corresponding planar (D_{2h}) energy is at 54 kcal/mole up for ^3A_u. Corresponding singlet transition states are estimated to be at 50 and 56 kcal/mole up. References to other articles are given by Seeger *et al.*

von Niessen *et al.* [66], using a many-body Green's function method, compute the five lowest valence-shell vertical ionization potentials, which compare well with experiment. Experimentally, the first band (to ^2E) shows a 0.6-eV Jahn–Teller splitting. They give a list of SCF orbital energies.

Tantardini and Simonetta [95], using in part a 6-31G basis set, use a valence-bond method to obtain several vertical valence ionized states.

Rauk *et al.* [104a] make double-zeta-basis calculations on excited states and optical activity of allenes: allene, 1,3-dimethylallene, and 1,2-cyclononadiene.

2. Spirononatetraene and Peroxochromate Ion

A minimal-basis SCF plus fairly extensive CM calculation has been made by Craig *et al.* [105] on the ground and low excited states of spirononatetraene $C(CH{=}CH{-}CH{=}CH)_2$. With an electronic structure ... $(1a_2)^2 (1b_1)^2 (7e)^4$, $^1A_1$ in the D_{2d} ground state, the lowest singlet excited state is calculated as a nondegenerate $^1B_1$, transition to which is forbidden. Next would come a ^1E state, but, with a twisting to symmetry D_2, this is split into two states $^1B_2$ and $^1B_3$ by the Jahn–Teller effect. An f value of 0.055 is computed for the transitions to these. In these, the structure is found to be twisted by about 2.5° from D_{2d} and stabilized by about 110 cm^{-1}. There follow two D_{2d} states of species $^1A_1$ and $^1B_2$ (f values 0 and 0.035), then another ^1E state, which becomes twisted and split into $^1B_2$ and $^1B_3$ (f value 0.156). The states of D_2 symmetry are chiral and should give rise to circular dichroism spectra. The rotatory strength of the lowest-frequency optically active transition is found to be negative for the R (or M) enantiomer. Further details are given.

Dacre and Elder [106] have made a minimal-basis SCF calculation on the peroxochromate amion $[CrO_4(O_2)_2]^{-3}$, experimentally of D_{2d} symmetry. They give a complete set of orbital energies for the $^2B_1$ ground state, a population analysis, and data on two excited states. They discuss the MO structure with the help of contour diagrams of the MOs and a total density contour diagram.

3. Cyclobutane

In 6-31G* calculations on cyclobutane, Cremer [107] finds the puckered D_{2d} form to be 0.9 kcal/mole more stable than the planar D_{4h} form. (Experimental values of the barrier height range from 1.1 to 1.5 kcal.) The puckering amounts to 0.23 Å; detailed geometry is given.

4. Beryllium borohydride

Marynick and Lipscomb [108] made minimal-basis STF followed by CM calculations on beryllium borohydride in 11 different configurations. They conclude that a D_{2d} configuration is the most stable.

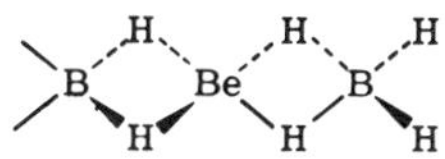

REFERENCES

1. R. S. Mulliken and W. C. Ermler, "Diatomic Molecules: Results of *ab initio* Computations." Academic Press, New York, 1977.
2. A. D. McLean and M. Yoshimine, *IBM J. Res. Dev.* **2,** 190 (1967).
3. I. Fischer-Hjalmars and P. Siegbahn, *Theor. Chim. Acta* **31,** 1 (1973).
4. P. J. Hay, W. J. Hunt, and W. A. Goddard, III, *J. Am. Chem. Soc.* **94,** 8293 (1972).
5. J. A. Pople, R. Krishnan, H. B. Schlegel, and J. S. Binkley, *Int. J. Quantum Chem.* **14,** 545 (1978).
6. P. Botschwina, K. Srinavasan, and W. Meyer, *Mol. Phys.* **35,** 1177 (1978).
7. D. T. Clark and L. Colling, *J. Electron Spectrosc.* **13,** 317 (1978).
8. A. Barth, R. J. Buenker, S. D. Peyerimhoff, and W. Butscher, *Chem. Phys.*, **46,** 149 (1980).
9. S. Nagase and C. W. Kerr, *J. Am. Chem. Soc.* **101,** 2544 (1979).
10. E. Clementi and H. Popkie, *J. Chem. Phys.* **57,** 4870 (1972).
11. A. D. Walsh, *J. Chem. Soc.* p. 2288 (1953).
12. R. W. Wetmore and H. F. Schaefer, III, *J. Chem. Phys.* **69,** 1648 (1978).
13. W. E. Kammer, *Chem. Phys. Lett.* **6,** 529 (1970).
14. S. R. Langhoff and E. R. Davidson, *Int. J. Quantum Chem.* **8,** 61 (1974); E. R. Davidson and D. W. Silver, *Chem. Phys. Lett.* **52,** 403 (1978).
15. D. Demoulin and M. Jungen, *Theor. Chim. Acta* **34,** 1 (1974); D. Demoulin, *Chem. Phys.* **11,** 329 (1975).
16. H. R. Wendt, H. Hippler, and H. E. Hunziger, *J. Chem. Phys.* **70,** 4044 (1979).
17. W. E. Kammer, *Chem. Phys.* **5,** 408 (1974).
18. L. S. Cederbaum, W. Domcke, J. Schirmer, W. von Niessen, G. H. F. Diercksen, and W. P. Kraemer, *J. Chem. Phys.* **69,** 1591 (1978).

19. M. A. Coplan, J. H. Moore, and J. A. Tossell, *J. Chem. Phys.* **68,** 329 (1978).
20. S. Y. Chu, I. Ozkan, and L. Goodman, *J. Chem. Phys.* **60,** 1268 (1974).
21. S. Y. Chu and L. Goodman, *J. Am. Chem. Soc.* **97,** 7 (1975).
22. R. Bonaccorsi, E. Scrocco, and J. Tomasi, *J. Chem. Phys.* **50,** 2940 (1969).
23. L. S. Cederbaum, W. Domcke, and W. von Niessen, *Chem. Phys.* **10,** 459 (1975).
24. G. Bieri, E. Heilbronner, V. Hornung, E. Kloster-Jensen, J. P. Maier, F. Thommen, and W. von Niessen, *Chem. Phys.* **36,** 1 (1979).
25. R. C. Haddon, D. Poppinger, and L. Radom, *J. Am. Chem. Soc.* **97,** 1645 (1975).
26. R. D. Bardo and K. Ruedenberg, *J. Chem. Phys.* **60,** 932 (1974).
27. H. T. Jonkman, Ph.D. Thesis, Univ. Groningen, Groningen, 1975.
28. P. Botschwina, *Chem. Phys.* **28,** 231 (1978).
29. N. L. Summers and J. Tyrell, *J. Am. Chem. Soc.* **99,** 3960 (1977).
30. N. L. Summers and J. Tyrell, *Theor. Chim. Acta* **47,** 223 (1978).
31a. A. J. Merer and R. S. Mulliken, *Chem. Rev.* **69,** 639 (1969).
31b. R. S. Mulliken, *J. Chem. Phys.* **66,** 2448 (1977).
31c. R. McDiarmid, *J. Chem. Phys.* **71,** 556 (1979); see also T. Cvitaš, H. Güsten, and L. Klasinc, *J. Chem. Phys.* **70,** 57 (1979), for improved photoelectron spectra.
32. K. B. Wiberg and J. J. Wendoloski, *J. Am. Chem. Soc.* **98,** 5465 (1976); see also A. Komornicki and J. W. McIver, Jr., *J. Chem. Phys.* **70,** 2014 (1979).
33. P. Pulay and W. Meyer, *J. Mol. Spectrosc.* **40,** 59 (1971); G. Jalsonsky and P. Pulay, *J. Mol. Struct.* **26,** 277 (1975).
34. R. J. Buenker, S.-K. Shih, and S. D. Peyerimhoff, *Chem. Phys.* **36,** 97 (1979).
35. G. Levin, W. A. Goddard, III, and D. L. Huestis, *Chem. Phys.* **4,** 409 (1974).
36. R. J. Buenker and S. D. Peyerimhoff, *Chem. Phys.* **9,** 75 (1975).
37. B. R. Brooks and H. F. Schaefer, III, *J. Am. Chem. Soc.* **101,** 307 (1979).
38. M. D. Newton, E. Switkes, and W. N. Lipscomb, *J. Chem. Phys.* **53,** 2645 (1970); M. D. Newton and E. Switkes, *J. Chem. Phys.* **54,** 3179 (1971).
39. D. L. Wilhite and J. L. Whitten, *J. Chem. Phys.* **58,** 948 (1973).
40. P. Kollman, *J. Am. Chem. Soc.* **96,** 4363 (1974).
41. A. Rauk, J. O. Jarvie, H. Ichimura, and J. M. Barriel, *J. Am. Chem. Soc.* **97,** 20 (1975).
42. D. H. Liskow and G. A. Segal, *J. Am. Chem. Soc.* **100,** 2945 (1978).
43. W. V. Volland, E. R. Davidson, and W. T. Borden, *J. Am. Chem. Soc.* **101,** 533 (1979).
44. L. Snyder, personal communication.
45. R. Daudel, R. E. Kari, R. A. Poirier, J. D. Goddard, and I. G. Czismadia, *J. Mol. Struct.* **50,** 115 (1978).
46. L. M. Cheung, K. R. Sundberg, and K. Ruedenberg, *Int. J. Quantum Chem.* **16,** 1103 (1979).
47. B. R. Brooks and H.F. Schaefer, III, *J. Chem. Phys.* **68,** 4839 (1978).
48. S. Trajmar, J. R. Rice, and A. Kupperman, *Adv. Chem. Phys.* **18,** 15 (1970).
49. N. C. Baird and J. R. Swenson, *Chem. Phys. Lett.* **22,** 183 (1973).
50. R. J. Buenker, S. D. Peyerimhoff, and W. E. Kammer, *J. Chem. Phys.* **55,** 814 (1971).
51. M. Krauss and S. R. Mielczarek, *J. Chem. Phys.* **51,** 5241 (1969).
52. R. J. Buenker, S. D. Peyerimhoff, and H. L. Hsu, *Chem. Phys. Lett.* **11,** 65 (1971); S. D. Peyerimhoff and R. J. Buenker, *Theor. Chim. Acta* **27,** 243 (1972).
53. L. E. McMurchie and E. R. Davidson, *J. Chem. Phys.* **66,** 2959 (1977).
54. R. S. Mulliken, *Acc. Chem. Res.* **9,** 7 (1976).
55. R. S. Mulliken, *Chem. Phys. Lett.* **25,** 305 (1974); see also S. Iwata and K. F. Freed, *J. Chem. Phys.* **61,** 1500 (1974).

56. R. J. Buenker, S. D. Peyerimhoff, and S.-H. Shih, to be published. The value for $\langle\pi^*|x^2|\pi^*\rangle$ is from a personal communication. The authors report the $\langle\Psi\Sigma_i x_i^2|\Psi\rangle$ value as in the 19 – 20 a.u. range, nearly double the value for $^3B_{1u}$.
57. J. Rose, T.-I. Shibuya, and V. McKoy, *J. Chem. Phys.* **58,** 74 (1973).
58. W. Coughran, J. Rose, T. Shibuya, and V. McKoy, *J. Chem. Phys.* **58,** 2699 (1973).
59. T. D. Bouman and A. E. Hansen, *J. Chem. Phys.* **66,** 3460 (1977).
60. For review, see L. Salem, *Acc. Chem. Res.* **12,** 87 (1979).
61. V. Bonačić-Koutecký, R. J. Buenker, and S. D. Peyerimhoff, *J. Am. Chem. Soc.* **101,** 5917 (1979).
62. A. J. Merer and L. Schoonfeld, *Can. J. Phys.* **47,** 1731 (1969).
63. L. McMurchie and E. R. Davidson, *J. Chem. Phys.* **67,** 5613 (1977).
64. U. Fischbach, R. J. Buenker, and S. D. Peyerimhoff, *Chem. Phys.* **5,** 265 (1974).
65. I. Fischer-Hjalmars and J. Kowalewski, *Theor. Chim. Acta* **27,** 197 (1972); **29,** 345 (1973).
65a. M. Brith-Lindner and S. D. Allen, *Chem. Phys. Lett.* **47,** 32 (1977).
65b. E. H. van Veen, *Chem. Phys. Lett.* **41,** 540 (1976).
65c. P. G. Wilkinson, *Can. J. Phys.* **34,** 643 (1956).
65d. P. D. Foo and K. K. Innes, *J. Chem. Phys.* **60,** 4582 (1974).
66. W. von Niessen, G. H. F. Diercksen, L. S. Cederbaum, and W. Domcke, *Chem. Phys.* **18,** 469 (1976).
67. K.-H. Thunemann, R. J. Buenker, S. D. Peyerimhoff, and S.-K. Shih, *Chem. Phys.* **35,** 35 (1978).
68. R. L. Martin and E. R. Davidson, *Chem. Phys. Lett.* **51,** 237 (1977).
69. C. Benoit and J. A. Horsley, *Mol. Phys.* **30,** 557 (1975).
70. U. Kaldor and I. Shavitt, *J. Chem. Phys.* **48,** 191 (1968).
71. S. L. Altmann, *Proc. R. Soc. London, Ser. A* **210,** 343 (1951).
72. S. T. Elbert, S. D. Peyerimhoff, and R. J. Buenker, *Chem. Phys.* **11,** 25 (1975).
73. R. J. Buenker, S. D. Peyerimhoff, L. C. Allen, and J. L. Whitten, *J. Chem. Phys.* **45,** 2835 (1966).
74. L. S. Cederbaum, *J. Phys. B* **8,** 290 (1975).
75. E. A. Laws, R. M. Stevens, and W. N. Lipscomb, *J. Am. Chem. Soc.* **94,** 4461 (1972).
76. C. E. Blom and A. Mueller, *J. Chem. Phys.* **69,** 3397 (1978).
77. R. Ahlrichs, *Theor. Chim. Acta* **35,** 59 (1974).
78. L. T. Redmon, G. D. Purvis, III, and R. J. Bartlett, *J. Am. Chem. Soc.* **101,** 2856 (1979).
79. E. Switkes, R. M. Stevens, W. N. Lipscomb, and M. D. Newton, *J. Chem. Phys.* **51,** 2085 (1969).
80. D. A. Dixon, I. M. Pepperberg, and W. N. Lipscomb, *J. Am. Chem. Soc.* **96,** 1325 (1974).
81. L. A. Curtiss, *Int. J. Quantum Chem.* **14,** 709 (1978).
82. J. M. Howell and J. R. Van Wazer, *J. Am. Chem. Soc.* **96,** 7902 (1974).
83. R. Ahlrichs and F. Keil, *J. Am. Chem. Soc.* **96,** 7615 (1974).
84. W. von Niessen, W. Domcke, L. S. Cederbaum, and J. Schirmer, *J. C. S. Faraday Trans. II* **8,** 1550 (1978).
85. R. J. Buenker and S. D. Peyerimhoff, *J. Chem. Phys.* **48,** 354 (1968); *J. Am. Chem. Soc.* **91,** 4342 (1969).
86. H. Kollmar and V. Staemmler, *J. Am. Chem. Soc.* **99,** 3583 (1977).
87. H. Kollmar and V. Staemmler, *J. Am. Chem. Soc.* **100,** 4304 (1978); see also L. J. Schaad, B. A. Hess, Jr., and C. S. Ewig, *J. Am. Chem. Soc.* **101,** 2281 (1979).
88. W. T. Borden, E. R. Davidson, and P. Hart, *J. Am. Chem. Soc.* **100,** 388 (1978).

89. J. A. Jafri and M. D. Newton, *J. Am. Chem. Soc.* **100,** 5012 (1978).
90. R. C. Haddon and G. R. J. Williams, *J. Am. Chem. Soc.* **97,** 6582 (1975).
91. K. Krogh-Jespersen, P. von R. Schleyer, J. A. Pople, and D. Cremer, *J. Am. Chem. Soc.* **100,** 131 (1978).
92. W. T. Borden and E. R. Davidson, *J. Am. Chem. Soc.* **99,** 4587 (1977); E. R. Davidson, W. T. Borden, and J. Smith, *J. Am. Chem. Soc.* **100,** 3299 (1978).
93. F. Brogli, E. Heilbronner, E. Kloster-Jensen, A. Schmelzer, A. S. Manocha, J. A. Pople, and L. Radom, *Chem. Phys.* **4,** 107 (1974).
94. L. S. Cederbaum, W. Domcke, H. Koppel, and W. von Niessen, *Chem. Phys.* **26,** 169 (1977).
95. G. F. Tantardini and M. Simonetta, *Int. J. Quantum Chem.* **15,** 655 (1979).
96. P. Kollman, C. F. Bender, and S. Rothenberg, *J. Am. Chem. Soc.* **94,** 8016 (1972).
97. C. P. Baskin, C. F. Bender, and P. A. Kollman, *J. Am. Chem. Soc.* **95,** 5868 (1973).
98. N. Rupp and R. Ahlrichs, *Theor. Chim. Acta* **46,** 117 (1977).
99. A. W. Potts and E. P. F. Lee, *J. C. S. Faraday Trans. II* p. 941 (1979).
100. K. D. Jordan, *Chem. Phys. Lett.* **40,** 441 (1976); **42,** 610 (1976).
101. J. A. Jafri, M. D. Newton, T. A. Pakkanen, and J. L. Whitten, *J. Chem. Phys.* **66,** 5167 (1977).
102. A. Karpfen, P. Schuster, J. Petkov, and H. Lischka, *J. Chem. Phys.* **68,** 3884 (1978).
103. F. Cavallone and E. Clementi, *J. Chem. Phys.* **63,** 4304 (1975).
103a. H. Johansen, *Int. J. Quantum Chem.* **9,** 459 (1975).
104. R. Seeger, R. Krishnan, J. A. Pople, and P. von R. Schleyer, *J. Am. Chem. Soc.* **99,** 7103 (1977).
104a. A. Rauk, A. F. Drake, and S. F. Mason, *J. Am. Chem. Soc.* **101,** 2284 (1979).
105. D. P. Craig, P. J. Stiles, P. Palmieri, and C. Zauli, *J. C. S. Faraday Trans. II* p. 97 (1979).
106. P. D. Dacre and M. Elder, *J. C. S. Dalton Trans.* p. 1426 (1972).
107. D. Cremer, *J. Am. Chem. Soc.* **99,** 1307 (1977).
108. D. S. Marynick and W. N. Lipscomb, *J. Am. Chem. Soc.* **95,** 7244 (1973).

CHAPTER IX

ALIPHATIC AND INORGANIC MOLECULES WITH FOUR OR MORE ATOMS: MOLECULES WITH ONE OR TWO SYMMETRY PLANES

A. GENERAL CONSIDERATIONS

Ab initio computations are available on numerous molecules with one (C_s or C_{2h}) or two (C_{2v}) symmetry planes. These will be discussed here. The relevant group theory tables are given at the beginning of Chapter VIII. Note especially the necessity of specifying one's choice of axis labels in assigning species labels to b MOs and B states of C_{2v} symmetry.

Although not all examples belong in Chapter IX, it is convenient here to refer to a valuable article by Schaefer [1], where examples of the 1,2-hydrogen shift and its *ab initio* calculation are reviewed. The examples relevant to Chapter IX are vinylidene → acetylene (Section B,1), hydroxycarbene → formaldehyde (Section H,1), the AlC_2H_2 isomers (Section H,3), and the isomers of H_2CN^+, of which the linear form H—C ≡ N$^+$—H is the most stable.

B. VINYL AND ALLYL RADICALS AND IONS; VINYLIDENE

1. Vinyl Ions and Radicals and Vinyllic Compounds

For the vinyl cation (protonated acetylene), two major structures have been discussed: classical (1) and bridged (2), both of C_{2v} symmetry.

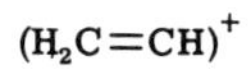

1

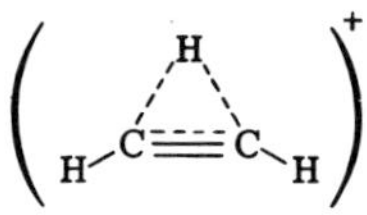

2

Weber *et al.* [2] made calculations using a double-zeta plus polarization basis followed by CM consisting of single and double excitation from the ground-state SCF function. They conclude that the two structures have almost the same energy ($E_h = -77.3449$), with the path between them planar with a very low barrier (less than 1–3 kcal/mole). Charge distributions from a population analysis for the two forms are given. Two singlet–singlet excitation energies are computed for each form; these differ greatly for the two forms. With more extended CM (PNO-CI and CEPA-PNO), Lischka and Köhler [3] find the bridged cation to be more stable than the classical one by 4.0 kcal/mole, with an almost negligible energy barrier between them.

For the vinyl radical, Millie and Bertheir [4] have made an SCF computation using a Gaussian basis including polarization functions on C and H. With planarity and C=C and C—H bond lengths (1.34 and 1.00 Å) and HCH angle of 120° assumed, a trans structure (3) with C—C—H angle of

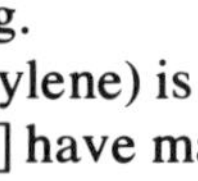

3

137° is found, in a $^2A'$ state (C_s symmetry). Orbital energies and total energy ($E_h = -77.3121$) are given. Using the same model, Ellinger *et al.* [5] calculated the hyperfine splitting.

Vinyl anion (deprotonated ethylene) is similar in geometry to vinyl radical. Williams and Streitwieser [6] have made up to double-zeta SCF calculations ($E_h = -77.3345$) and given ionization potential, electron affinities (−1.2 eV for the radical), proton affinities, and electron density diagrams. Dykstra *et al.* [7] have made SCF ($E_h = -77.3372$) and SCEP ($E_h = -77.5273$) calculations, and have listed orbital energies. They also compute excited transition structures (linear and C_{2v}) in connection with a discussion of the reaction path for the reaction $C_2H_2 + H^- \rightarrow C_2H_3^-$. For this they find a formation energy of 26 kcal and an activation energy of 16 kcal.

Davis *et al.* [8] have made GVB and CM calculations on the $^1A_1$ ground state and low excited states of the vinylidene radical ($H_2C = C$:) using a modest Gaussian basis supplemented by d polarization orbitals on each carbon. They discuss the $1a_1^2\,1b_2^2\,2a_1^2\,2b_2^2\,3a_1^2\,1b_1^2\,4a_1^2$, $^1A_1$ ground state (C_{2v}), the ... $4a_1\,3b_2$, $^1B_2$ and $^3B_2$ and the ... $1b_1\,4a_1{}^2\,3b_2$, $^3A_2$ and $^1A_2$ excited states (z and y axes in the molecular plane).

Dykstra and Schaefer [8a], using a polarized double-zeta basis in SCF followed by use of the SCEP method for correlation, have computed the energies of the 1,2-hydrogen shift in the transformation from the ground state of vinylidene to that of acetylene (H—C = C—H). They find for the transition state a planar (C_s) structure with the migrating H atom roughly on a perpendicular to the CC line, at a barrier height of 8.6 kcal/mole. They also make calculations on the interaction of Li^+ with C_2H_2.

In another article, Conrad and Schaefer [8b] make similar calculations on the transformation from the $^3B_2$ lowest excited state of vinylidene to the ^3B_u trans or the $^3B_2$ cis lowest excited states of acetylene. They find the transition state to be nonplanar with the migrating H atom in a plane (with the two C atoms) that is nearly perpendicular to the original vinylidene plane. For this transition state they find an energy barrier of about 55 kcal/mole, indicating a long natural lifetime for the $^3B_2$ state of vinylidene. This suggests likewise a long lifetime for the analogous 1,2-hydrogen shifts of methylcarbene and nitrene, which have triplet ground states.

Csizmadia *et al.* [9] have made SCF calculations of unpolarized double-zeta quality on $C_2H_2F^+$. They found the open structure (fluorovinyl cation) to be 31 kcal/mole more stable than the bridged structure (fluorenium ion), which in fact was found to represent an energy maximum. At the electron-deficient center, the H atom is found to be slightly (4.5°) bent away from the fluorine.

Csizmadia *et al.* [10] have made SCF calculations of up to polarized double-zeta quality (77 GTF) on $C_2H_2SH^+$. They found the bridged (thiirenium) structure to be more stable by 15 kcal/mole than the open (β-thiovinyl) structure with a considerable barrier between them. (With an unpolarized 53 GTF set, they found a difference of only 4 kcal/mole; as usual, polarization is important.) For pyramidal inversion at the sulfur atom, they calculated 73 kcal/mole.

Lister and Palmieri [11] have made an SCF study of vinyl mercaptan (C_2H_3SH) using a 4-31G basis. The potential energy is calculated as a function of torsion around the S—H bond. The syn and gauche rotamers are found to be more stable than the anti, with the syn rotamer more stable by 2 kJ/mole than the *gauche*, with a considerable barrier between them. Vibrational levels are computed. Obviously polarization and CM could change the picture. Kao [11a] has reached similar conclusions, and has also made calculations on methylvinyl sulfide and methylallenyl sulfide.

2. Allyl Ions and Radicals

Molecular orbital theory and calculations indicate that the allyl cation, radical, and anion have a C_{2v} geometry

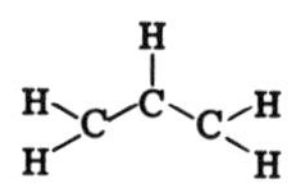

with a C—C—C angle of about 120° in the cation and somewhat greater values in the radical and anion (but see below).

Peyerimhoff and Buenker [12] studied all three using SCF and CM computations with an extensive Gaussian-lobe basis. They also make comparisons between the allyl cation and the higher energy isoelectronic molecules cyclopropyl cation and cyclopropene, with respectively 6,4,1,0; 6,3,1,1; and 6,3,2,0 a_1, b_2, b_1, a_2 MOs. They compute SCF energies of the cation for a variety of conformations (see Fig. 2 in [12], and discuss the structures in detail, also giving computed vertical excitation energies for a number of excited states. In computations on the radical, they also discuss excited states and spectra. Orbital energies are listed for the cation; $E_h = -166.1300$ is obtained. Computed CCC angles are 120°, 127°, and 134° for the cation, radical, and anion, respectively.

In a later article, Radom *et al.* [13] give computations on and discuss the allyl cation and the cyclopropyl cation and other isoelectronic forms, using up to an 6-31G* basis. Köhler and Lischka [14] have used very extensive CM in a CEPA calculation (almost variational) of the allyl cation, obtaining $E_h = -116.4617$. Like Radom *et al.*, they also compute the energies of seven higher energy isomeric forms of $C_3H_5^+$. In an unrestricted SCF calculation with a good basis set, Hinchcliffe [15] obtains energy and hyperfine coupling constants for the allyl radical.

Levin *et al.* [16] made GVB and partial CM calculations on the allyl cation and two triplet and two singlet excited states. Their excitation energies are considerably lower than those of [8]. Levin and Goddard [17] made GVB and CM calculations on the allyl radical and on its π-electron excited states. They found that the ground and first excited states are well described as the resonant and antiresonant states

with a resonance energy of 11.4 or (later) 14 kcal, in good agreement with thermochemical estimates (11.6 ± 2 kcal). The higher states involve excitation of Rydberg orbitals.

Several writers have called attention to the fact that in an SCF computation an energy minimum can be found that is only a local minimum in the sense that a lower minimum can be found if the symmetry is reduced. Following the earlier work of others, Paldus and Veillard [18], in a careful treatment using up to polarized double-zeta basis sets, find this to be true of the allyl radical if the symmetry is reduced from C_{2v} to C_s by making the two C—C bond lengths unequal (one more C═C, the other less so), giving a double-minimum potential with two equal minima. If the average bond length could be reduced somewhat, a single-minimum potential of C_{2v} symmetry would be restored. Although this has not been discussed, it seems likely that the same effect occurs for the allyl cation and anion (and in many other cases). However, it is conceivable that CM could change matters here.

C. SOME THREE-MEMBERED RING COMPOUNDS

1. Molecules Related to Cyclopropane

Cyclopropane has been discussed in Section VI,J. SCF calculations on it and the isoelectronic ethylene oxide or oxirane (4), ethyleneimine (5), and diaziridene (6), have been made by Basch *et al.* [19] using a Gaussian

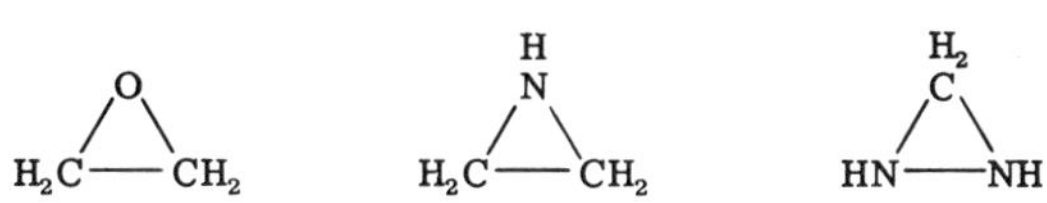

4 5 6

basis set (details not given). Orbital energies and total energies for the four molecules are given; the respective E_h values are −117.0099, −152.8012, −132.9726, −148.8430. The respective symmetries should be D_{3h}, C_{2v}, C_s, and C_2. Figure 1 in [19] is a diagram correlating the orbital energies of the four molecules. The authors compute the first three ionization potentials of each, assuming Koopmans' theorem, along with dipole moments, quadrupole coupling constants, average diamagnetic shielding at the protons, and second moments $\langle x^2 \rangle$, $\langle y^2 \rangle$, $\langle z^2 \rangle$ of the electric charge. Computations on the electronic spectrum and the orbital energies are compared with experiment.

Hopkinson *et al.* [24], using a (9s 5p 4s) basis set contracted to (4s 2p 2s), have made SCF calculations on the oxiranium ion (7) and the

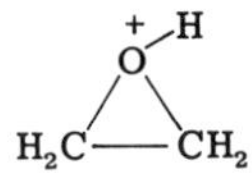

7

2-hydroxyethyl cation $H_2C^+—CH_2OH$. The geometries (C_s symmetry) were first optimized using a 4-31G basis set. Orbital and total SCF energies are listed, and atomic charges are given. The barrier for rotation of the CH_2^+ group in the 2-hydroxyethyl cation is computed as 15 kcal/mole and its profile is pictured. In the oxiranium ion, the OH proton was computed to be 48° out of the COC plane, with an inversion barrier of 5 kcal/mole.

von Niessen *et al.* [20], using a polarized double-zeta basis set in a many-body Green's function method, have made a good computation of the eight lowest vertical ionization potentials of ethylene oxide and of ethyleneimine that agrees very well with experiment.

Using an STO-3G minimal basis followed by limited CM, Bigot *et al.* [21] have computed potential energy curves for the transformation of each of the (about seven) lowest states of ethylene oxide into one or the other of the diradical forms

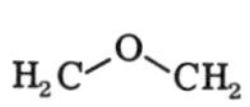 or 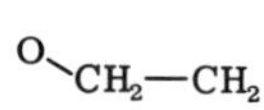

and discuss how the MOs change in each process. In a second article [22] they make calculations on various modes of scission (to $C_2H_4 + O$, $H_2CO + CH_2$, and indirectly to CH_3CHO). The calculation was made in order to illuminate the photochemical behavior of oxiranes, of which ethylene oxide is the simplest.

In relatively accurate double-zeta quality SCF calculations without polarization ($E_h = -152.8111$), Mezey *et al.* [23] considered angular distortion of ethylene oxide in its ground and six low excited states. For rotations (presumably of the H_2 groups) around the C—C bond, they found for the ground and lowest triplet states minimum energy for zero rotation, maximum for 90°. For distortions of the C—C—O angle

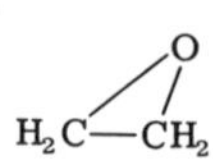

they found for minimum energy no distortion for the ground state and about 40° for nearly all the excited states. For all the excited states, they found a strong maximum of energy for the undistorted C_{2v} conformation.

Oxaziridene (8) is not known experimentally, but derivatives are famil-

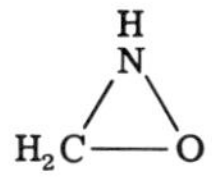

8

iar. With the H atom of the NH out of plane, the molecule should have no symmetry, and so belongs in Chapter X, but will be discussed here. With a basis set of double-zeta quality plus diffuse functions, Oliveros *et al.* [25] make SCF plus CM (CIPSI) calculations on the ground and low excited (including partially Rydberg) states. In the first singlet excited states, there should be photochemical transformation to formamide. The authors discuss a two-step transformation in which the N—O bond is first broken and then one H is transferred from C to N. They conclude that this should occur readily via an avoided crossing after the molecule has been activated to its first excited singlet state, whereas in the ground state a barrier must be overcome.

Bigot *et al.* [26], using a minimal basis set, have made SCF plus CM calculations on the transformation of oxaziridene by various paths to nitrone (9) or to formamide.

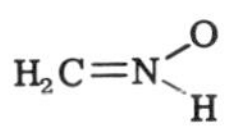

9

Cimiraglia *et al.* [27] make *ab initio* calculations on "alternative paths in the ring opening of oxadiaziridene: the diimide *N*-oxide versus the oxodiimide reearrangement." (The original reference was not available.)

Skancke and Boggs have made SCF calculations, using a basis set of double-zeta quality, on several cyclopropane derivations of mostly C_{2v} or C_s symmetry. They have determined geometry and computed dipole moments in each case: cyclopropyl chloride and 1,1-dichlorocyclopropane (Skancke [28]); methylcyclopropane, cyclopropylamine, and cyclopropyllithium (Skancke and Boggs [29]); vinylcyclopropane and cyclopropyl cyanide (Skancke and Boggs [30]); *cis*- and *trans*-1,2-difluorocyclopropane (Skancke and Boggs [31]). In the last case, the cis isomer is the more stable.

2. Cyclopropene, Diazirine, and Related Molecules

The most accurate SCF calculations on cyclopropane, cyclopropene (10) and cyclopropenone (11) are apparently calculations by Skancke [32]

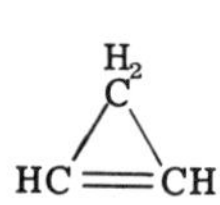

10

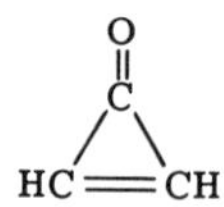

11

using a (9s 5p 1d) basis set, giving $E_h = -117.0543$, -115.8178, and -189.5240 for the three molecules. Orbital energies are tabulated for all three. Dipole moments are computed. The lowest three ionization potentials (Koopmans' theorem values) are compared with experiment.

In a previous article [33], using a (9s 5p) basis set, Skancke reported similar results on methylenecyclopropene (12), cyclopropenimine (13), and cyclopropenone. A population analysis is given, but its validity (quite properly) is questionable.

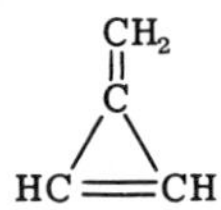

12

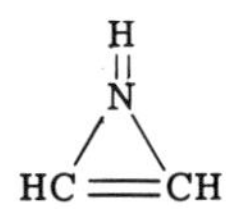

13

Hehre and Pople [34], in up to 6-31G* calculations on C_4 hydrocarbons, find that methylenecyclopropene is the most stable ring system within the C_4H_4 manifold, lower in energy by perhaps 10 kcal/mole than 1,3-cyclobutadiene (about which see Section VIII,C,4). Other noncyclic isomers (butatriene, Section VIII,C,5, and lowest but-1-yn-3-ene) are lower still in energy.

An early SCF and CM study by Peyerimhoff and Buenker [35] of cyclopropene and its C_4H_4 isomers, using a Gaussian-lobe expansion of STFs, should be mentioned.

Using a double-zeta Gaussian basis plus small-exponent s and p Gaussians to represent 3s and 3p Rydberg MOs, Basch [36] makes SCF calculations on the low excited states of cyclopropene. He identifies the observed level at about 7.24 eV as a V state like that of ethylene ($\pi^* \leftarrow \pi$, $1a_2 \leftarrow 2b_1$ with our choice of axes, s perpendicular to the molecule plane). A weaker transition to a level at about 6.52 eV is dubiously identified as $\pi^* \leftarrow \sigma$, $1a_2 \leftarrow 3b_2$ or perhaps in part $3s \leftarrow \pi$, $7a_1 \leftarrow 2b_1$. Ionization (Koopmans' theorem) to $^2B_1$ at 8.84 eV and $^2B_2$ at 10.32 eV corresponds to photoelectron peaks at 9.86 and 10.89 eV.

An early article by Kochanski and Lehn [37] gives an SCF treatment of cyclopropane, cyclopropene ($E_h = -115.7572$), and diazirine (14) with

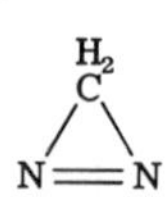

14

$E_h = -147.6980$, using unpolarized (9s 5p/4s) basis sets. Instructive electron density contour diagrams are given.

Using a basis set of double-zeta quality for an SCF calculation followed by extensive CM at several thresholds T, Vasudevan and Kammer [38] made calculations on the ground and nine excited states (some Rydberg or partially so) for diazirine and seven excited states for difluorodiazirine. They give tables describing the prinicpal CM configurations and their coefficients. For diazirine, they obtain $E_h = -147.7268$ (SCF) and -147.9445 (CM extrapolated to $T = 0$); for the difluor compound, the values were -345.3989 (SCF) and -345.5279 (extrapolated CM). Comparisons with the earlier SCF work of Robin *et al.* are given.

Bigot *et al.* [39] have made minimal basis STO-3G SCF and limited CM calculations on the potential surfaces of the ground and low-lying excited states of the diazirine–diazomethane ($H_2C{=}N{-}N{:}$) system. Experimentally, excited diazirine can isomerize into diazomethane or break to give $N_2 + CH_2$. Calculations on various paths for these processes are made. In another article [40] similar calculations on the photochemical behavior of 1-azirine (15) are made.

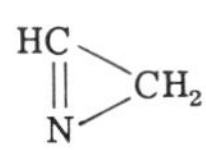

15

In an early SCF and limited CM computation on cyclopropene, 3,3-dimethylcyclopropene, and difluorodiazirine, Robin *et al.* [41] calculate a dipole moment of 0.364 D (observed: 0.455 D) for cyclopropene, with the calculated dipole negative at the methylene end, contrary to chemical intuition. They also compute frequencies and f values for transitions of cyclopropane, diazirine, and difluorodiazirine, to low excited states, with agreements with observed spectra.

In a 6-31G calculation, Radom *et al.* [42] report that the calculated dipole moment of cyclopropene is 0.54 D with the dipole moment again negative at the CH_2 end. Hehre and Pople [43] discuss the matter and, in an article on C_4 hydrocarbons, calculate a moment of 0.90 D, in agreement with experiment, for 1-methylcyclopropene. Since the CH_3 group tends to be positive, their result confirms that for cyclopropene. In further confirmation, Wiberg *et al.* [44], using minimal STO-G and 4-31G SCF bases, have computed the dipole moments of cyclopropene (0.513 D), 1,1-dimethylcyclopropene (methyls at the double bond, 0.864 D), and 3,3-dimethycyclopropene (0.387 D, experimental: 0.287 D). Wiberg *et al.* also compute the quadrupole moment of cyclopropene.

Kao and Radom [45] used STO-3G and 4-31G SCF calculations on the energetic and dipole moment effects of deformation of HCH angles in cyclopropane, cyclopropene, cyclobutene, and cyclopentadiene as models to study angular deformations for higher hydrocarbons. They found Walsh–Mulliken diagrams of orbital energy as a function of the orbital angle to be useful.

Davis *et al.* [46], using a d-polarized double-zeta Gaussian basis in GVB computations, have investigated the role of vinylmethylene as an intermediate in the ring opening of cyclopropene. They find the ground state of vinylmethylene to be a $^3A''$ state (16) either syn or anti. At higher levels they find singlet states $^1A'$ (17) and $^1A''$ (18). The $^3A''$ syn and anti states

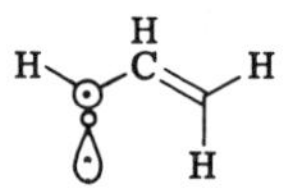

16

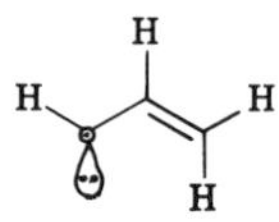

17

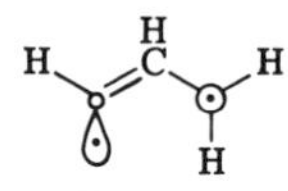

18

are identified with results of ESR observations.

Hase *et al.* [47], using a double-zeta basis for C, O, and H, and near double-zeta plus d for S, have made SCF calculations on cyclopropane, cyclopropenone, cyclopropenethione (19), thiirene 1-oxide (20), and

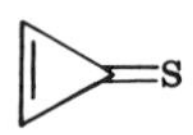

19

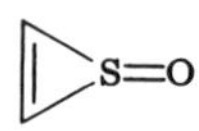

20

thiirene 1,1-dioxide. They give diagrams of valence-shell orbital energies, along with dipole moments, atomic charges, π charge transfers, and ionization energies, and discuss the MO structures.

The carbene cyclopropenylidene (21) is isoelectronic with the cyclo-

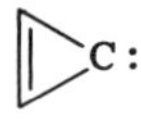

21

propyl cation, and like the latter, its $^1A_1$ ground state satisfies the $4n + 2$ rule for aromaticity. Shepard *et al.* [47a] have made a double-zeta-basis MCSCF calculation on the $^1A_1$ (six configurations) and the excited $^3B_1$ (four configurations) states of this molecule. The $^3B_1$ goes over a barrier on its way to dissociation to C_2H_2 plus $C(^3P)$, while $^1A_1$ dissociates without a barrier to C_2H_2 plus $C(^1D)$.

Levi *et al.* (47b), using a 4-31G basis set, have investigated the potential surface of $C_4H_7^+$ in the neighborhood of the bisected cyclopropylcarbinyl cation structure. For understanding, reference should be made to the original article.

3. Unusual Geometries in Compounds Containing BBC Rings

Van't Hoff predicted geometries in cumulenes concerning the relative orientations of the two CH_2 (or CR_2) groups; experiment has generally verified these. However, Krogh-Jesperson *et al.* [48] in *ab initio* computations have found that in several instances the stable form is rotated 90°

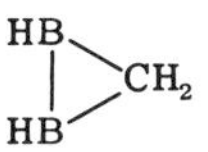

22a

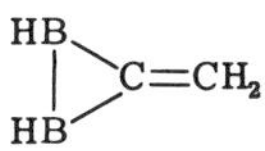

22b

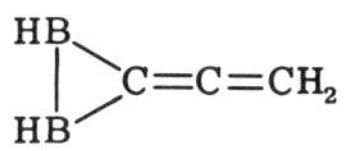

22c

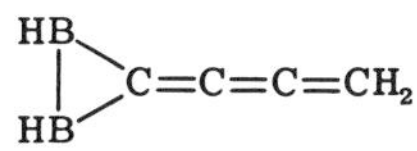

22d

from the van der Waals prediction. For example, 22a is found to be planar. Calculations on both the van der Waals predicted forms and stable forms are compared. The authors used 4-31G or 6-31G* basis sets supplemented by Møller-Plesset second-order perturbation theory estimates of correlation effects. Bond lengths, net charges and overlap populations are reported. Diagrams of MOs are shown; rotation barriers and other interesting details are discussed.

D. UREA, THIOUREA, AND FORMAMIDE

1. Urea [$O{=}C(NH_2)_2$] and Thiourea

Urea is well-known to be planar (symmetry C_{2v}). Using unpolarized Gaussian-lobe functions in an SCF and an IC ("internally consistent") SCF calculation [49], followed by extensive CM, Elbert and Davidson [50] give orbital energies and instructive contour maps of valence orbitals and some natural orbitals for urea. For the ground state, $E_h = -224.3254$ when double excitations leading to 2586 configurations were included. The configurations chosen were based on a cutoff at computed perturbation energies of 10^{-4} a.u. A table of the 30 most important excitations to natural orbitals is given. CM energies are also computed for the 8 lowest triplet and singlet excited states of each symmetry.

The lowest excited state is predicted to be the $\pi^* \leftarrow n$, $3b_1 \leftarrow 5b_2$, $^3A_2$, followed by $\pi^* \leftarrow \pi$, $3b_1 \leftarrow 2b_1$, $^3A_1$, then $3b_1 \leftarrow 5b_2$, $^1A_2$; $3b_1 \leftarrow 1a_2$, $^3B_2$; $\pi^* \leftarrow \pi$, $3b_1 \leftarrow 2b_1$, $^1A_1$; $3b_1 \leftarrow 1a_2$, $^1B_2$; $3b_1 \leftarrow 8a_1$, $^3B_1$; $3b_1 \leftarrow 8a_1$, $^1B_1$. The lowest singlet calculated excited state is the $\pi^* \leftarrow n$, $^1A_2$ at 7.2 eV (presumably vertical), in agreement with the frequency of the one spectroscopically observed (although dipole-forbidden) transition. The $\pi^* \leftarrow \pi$, $^1A_1$ is about 2 eV higher. Earlier interpretations are thereby revised, although indications from other molecules (formaldehyde and formamide) strongly favor the present conclusions. However, since the basis set used did not include Rydberg orbitals, allowance for this fact indicates that the calculated valence states above $^1A_2$ should be mixed with Rydberg states.

To evaluate the ionization potential, SCF calculations were made on the $(2b_1)^{-1}$, $^2B_1$ and $(5b_2)^{-1}$, $^2B_2$ states of the ion. The inclusion of extrapolated correlation energies gives an ionization energy of 0.33 a.u. for the $5b_2$ and 0.34 a.u. for the $2b_1$ electron, showing that Koopmans' theorem (0.41 a.u. for the $5b_2$, 0.34 a.u. for the $2b_1$) does not work well here.

Tables of numerous computed properties, including the dipole moment, are given for both singlet and triplet states. For the ground state, the computed dipole moment is -4.68 D. In this paper, the B notation agrees with ours, but seems to be inconsistent with their stated choice of axes.

Kapur *et al.* [51], using a minimal basis STO-3G, have made calculations on the potential surfaces of the ground and first excited triplet states of thiourea. The ground state is found, unlike urea, not to have C_{2v} symmetry, but the NH_2 hydrogens are somewhat bent out of plane, symmetrically by about 12° in the lowest energy (cissoid) form, or somewhat more and in opposite directions in the 3 kcal/mole higher-energy (transoid) form. The triplet state lacks any symmetry.

2. Formamide (O═CHNH$_2$): Symmetry C_s; Formamido Radical; (OCHNH); *N*-Methylformamide, Acetamide, and *N*-Methylacetamide

Whether formamide is planar or slightly nonplanar has long been uncertain. A recent polarized double-zeta SCF calculation by Carlsen *et al.* [52] indicated that there is only a few hundreths kcal/mole energy difference between the planar form and a nonplanar form with the NH_2 H atoms bent about 10° out of plane. The experimental evidence appears to be not quite conclusive, so that one can only say that the moleule is at least very nearly planar.

The formamide absorption spectrum is of much interest from the fact that it represents the basic chromophore in polypeptides. The observed optical transitions [53] include the weak W at 5.7 eV, medium strong R_1 ($f = 0.06$) at 6.7 eV, the strong V_1 ($f = 0.37$) at 7.3 eV, medium strong R_2 at 7.9, and medium strong Q at 9.2 eV. Some of these have been interpreted as composites [53]. Transitions to triplet excited states have been found at vertical energies of 5.30 and 6.60 eV by the use of trapped electron spectra [54]. Assuming planarity, the ground state of formamide is $(1a')^2 \ldots (9a')^2 (1a'')^2 (10a')^2 (2a'')^2$, $^1A'$, where $10a'$ is the oxygen atom lone-pair MO (n) and $2a''$ is the second π MO. Excited MOs needed are

11a′, 12a′, 13a′, and 14a′, which are Rydberg 3s, $3p_y$, $3p_z$, and 3d MOs (z axis along C═O); 3a″ and 4a″ are π^* and $3p\pi$.

In a GVB plus CM computation for vertical transitions Harding and Goddard [55] compute $\pi^* \leftarrow n$, $^3A''$ at 5.39 eV, in agreement with the observed triplet transition at 5.30 eV, $\pi^* \leftarrow n$, $^1A''$ at 5.65 eV in agreement with W, and 6.19 eV for the transition to $\pi^* \leftarrow \pi$, $^3A'$. They compute dipole moments of each state. The result for the ground state (3.79 D) agrees well with experiment (3.71D), and also agrees in direction. They also compute barriers to rotation about the C—N bond for all four states.

Stenkamp and Davidson [56], using a polarized Gaussian basis, for each state separately except for $\pi\pi^*$, $^1A_1$, but not including Rydberg orbitals, computed the ground and four excited states. They used the SCF and ICSCF methods plus extensive CM. They found vertical excitation energies of 5.34 eV for $^3n\pi^*$, 6.12 eV for $^3\pi\pi^*$, 5.63 eV for $^1n\pi^*$, and 8.45 eV for $^1\pi\pi^*$. Their results and those of Harding and Goddard agree well. Stenkamp and Davidson give extensive tables of computed properties for the several states, including, of course, dipole moment components (μ_y = 0.46 D and μ_z = 1.55 D for the ground state). They also tabulate orbital energies and dominant CM configurations for the ground state, and give contour maps of the valence and some excited MOs.

They were dissatisfied with their calculations on the $^1\pi\pi^*$ state, and Nitzsche and Davidson [57] published another article using perturbation methods to study this state, which they concluded to be mixed with the $\pi 3s$, $\pi 3p_x$, $\pi 3p_y$, and $\pi 3p_z$ Rydberg states. They find computed vertical energies in the range 6.96–8.21 eV, identifiable with transitions to the R_1, V_1, and R_2 upper states. They emphasize that (as for other T and V states) the π^* MO is very different in triplet and singlet states.

Oliveros *et al.* [58], using a double-zeta plus diffuse AO basis and the partially variational CM method CIPSI, calculated four vertical ionization potentials of formamide as follows (experimental values in parentheses): η_0 9.85 eV (10.32); π_2, 9.93 (10.52); π_1, 13.82 (14.18); σ_1, 14.09 (14.75).

Baird and Kathpal [59] made STO-3G and 4-31G calculations on the *N*-formamido radical OCH · NH. They find that the two lowest states $^2A'$ and $^2A''$ are almost degenerate, both having at least two potential minima in their planar conformations, corresponding to localization of the odd electron at the oxygen or the nitrogen atom.

For further enlightenment on the chromophore in polypeptides, Nitzsche and Davidson [60] made SCF plus CM computations on ground and excited states of a planar (C_s) conformer of *N*-methylacetamide, using a double-zeta basis plus diffuse orbitals to represent low-lying Rydberg states. The computed vertical excitation energies found were 5.61 eV for

$^3n\pi^*$, 5.85 eV for $^1n\pi^*$, and 6.11 eV for $^3\pi\pi^*$ after somewhat limited CM supplemented by a perturbation theory extrapolation for all double substitutions. These results are similar to those for formamide. A similar procedure gave 9.66 eV for the $^2A''$ and 10.11 eV for the $^2A'$ ionic states (experimental: 9.68 and 9.85 eV, respectively). Computations on $\pi\pi^*$, $^1A'$ and ten $^1A'$ Rydberg states are reported, together with oscillator strengths. The $^1\pi\pi^*$, $^1\pi 3p\pi$, and ^1n3p configurations are predicted to be strongly mixed, giving three states of large oscillator strength in the region of the broad V band of the absorption spectrum.

Fogarasi *et al.* [61], using a new gradient program, but with only a near-double-zeta basis, have obtained fully optimized geometries of formamide, *N*-methylformamide, acetamide, and *N*-methylacetamide.

E. OTHER NITROGEN-CONTAINING COMPOUNDS

1. Nitrous Acid (HNO_2) Isomers

Using a (7, 3) Gaussian basis, Skaarup and Boggs [62] have made an SCF computation on the experimentally rather well-known cis and trans isomers of HNO_2. Their computed geometries, and the barrier between

cis *trans*

the two isomers, agree fairly well with experiment. The same is true of harmonic force fields and vibration frequencies, vibrational isotope shifts (D, N^{15}), and centrifugal distortion constants. The isomers are almost equally stable, in agreement with experiment. Graphs of the energy as a function of rotation angle from cis to trans are given.

2. The Isomers of CHNO

Using up to a 6-31G* basis, Poppinger *et al.* [63] in SCF calculations have made a detailed explanation of the singlet potential surface of HCNO. Their results indicate that the most stable isomer is a cyanic acid (23a). They find cyanic acid (23b) 21 kcal/mole higher, formonitrile oxide (23c) 80 kcal/mole up, its tautomer carboxime (23d) slightly lower, next a $^1A'$ excited state of formylnitrene (23e) 88 kcal/mole up and a number of others higher.

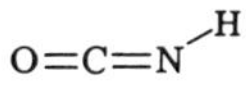
23a

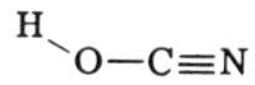
23b

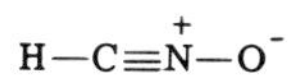
23c

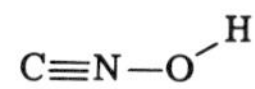
23d

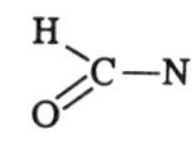
23e

Previously, Harrison and Shalhoub [64] had used a 9s, 5p/4s) basis in an SCF plus limited CM calculation on formylnitrene, finding a $^3A''$ ground state.

Later, Poppinger and Radom [65] made a detailed study of RCNO molecules using a 4-31G basis, with a great variety of substituents R, including Li, F, and Cl.

3. Cyanocarbenes

Lucchese and Schaefer [66] have made a polarized double-zeta Gaussian SCF calculation on the ground and lowest excited states of dicyanocarbene (compare Section II,E for carbene, CH_2). They conclude that the ground state is $^3B_1$ with a 133° bent C_{2v} structure, $R(CC) = 1.41$ Å, $R(CN)$ 1.15 Å, and the excited $^1A_1$ state has an angle of 115°, with $R(CC) = 1.42$ Å, $R(CN) = 1.16$ Å. The C—C bonds are notably long for single bonds. With an unpolarized (9s, 5p) Gaussian basis set but followed by modest CM, Harrison *et al.* [67] conclude that the ground state of HCCN when CM is included is neither a carbene (HC—C≡N) nor a nitrene (HC═C—N:) but an allene-like triplet H—Ċ═C═Ṅ. They suggest that a CM calculation on dicyanocarbene might similarly favor an allene-like structure Ṅ═C═C═C═Ṅ.

However, Zandler *et al.* [68], in a more recent calculation on HCCN using a double-zeta plus polarization basis and extensive CM, conclude that the bent carbene structure is lower in energy by about 4.4 kcal/mole than the allene-like structures. Details concerning the geometries of potential minima for alternative structures, also computed force constants, are given.

Baird and Tayler [69] have made STO-3G minimal-basis calculations on the lowest triplet and singlet states of cyanocarbene, dicyanocarbene, and also numerous other substituted carbenes; they give geometries.

Mezey *et al.* [70] discuss "the instability of the planar structure of the carbanion $(CH_2CN)^-$" (article not available).

4. Dinitrogen Dioxide (N_2O_2)

Skaarup *et al.* [71] studied the geometry of $(NO)_2$ using up to a (10, 6, 1) basis set of Gaussian-lobe functions. They made calculations on several energy surfaces but found only two with minima, the planar cis and trans forms. Although the cis form is experimentally the more stable, in agree-

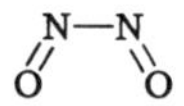

cis

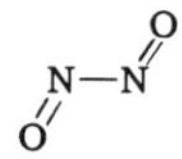

trans

ment with the theoretical calculation for a (7, 3) basis set, the (10, 6, 1) basis set makes the trans form more stable. Apparently CM is needed. Force constants and vibrational frequencies for both isomers are computed.

5. Diimide (N_2H_2) and N_2F_2

Diimide (diazene, diimine), isoelectronic with C_2H_4, OCH_2, and O_2, is known experimentally to have a C_{2h} trans structure. The ground state is

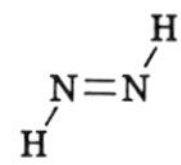

$(1a_g)^2 (1b_u)^2 (2a_g)^2 (2b_u)^2 (3a_g)^2 (3b_u)^2 (1a_u)^2 (4a_g)^2$, ^1A_g, where $2b_u$, $1a_u$, and $3a_g$ may be called n_+, π, and n_-. A higher-energy isomer of considerable interest is 1,1-diazine or aminonitrene (H_2NN, essentially $H_2N^+{=}N^-$). We select here from some of the more recent of the published literature, to display what are probably the most reliable results.

Vasudevan *et al.* [72] made SCF calculations using a double-zeta quality Gaussian-lobe basis augmented by Rydberg-type orbitals in order to study the ground and low-lying valence and Rydberg states ($E_h = -109.9345$ for the ground state). Thereafter, excluding only K-shell correlation, they made extensive CM calculations at several cutoff thresholds and extrapolated to $T = 0$ ($E_h = -110.1555$ for the ground state). They show computed SCF potential curves, as a function of the NNH angle going from cis (111°) to linear H—N=N—H (180°) to trans (111°), for the 12 lowest states including 5 Rydberg states. One of these favors a linear conformation, all the rest favor either trans or cis. The excited states correspond to excitations from $4a_g$, $1a_u$, and $3b_u$ to $1b_g$, $5a_g$, $4b_u$, and $2a_u$. They also show an SCF angular orbital energy correlation for MOs of the ground state, from cis to trans. They show contour diagrams of the $3b_u$ and $4a_g$ MOs and their cis counterparts, from SCF calculations. They estimate the barrier between trans and cis via the linear form as roughly 100 kcal, but via a singly bent form it is much lower (see below).

They show Koopmans' theorem potential curves as a function of angle for $N_2H_2^+$ and $N_2H_2^{2+}$, the latter isoelectronic with C_2H_2 and having a linear conformation. The CM calculations were carried out at the geometry of the ground state giving vertical excitation energies. Vasudevan *et al.* give energies and dominant configurations for the excited states, and discuss them in detail. They show a figure comparing the excited states with those of H_2CO and C_2H_4.

They identify the lowest-energy computed singlet transition $\pi^* \leftarrow n_-$ ($1b_g \leftarrow 4a_g$), ^1B_g-^1A_g with the weak observed ultraviolet transition, but note that their computed excitation is about 0.6 eV below the maximum of the observed transition. This they attribute to vibronic coupling in the otherwise forbidden transition.

In a later article in which vibrational frequencies and transition energies are computed for the ^1A_g and ^1B_g states of N_2H_2 and N_2D_2, Peric *et al.* [73] give a detailed explanation in which excitation of one quantum of the ν_5 vibration occurs in the ^1B_g state, making it a ^1B_u vibronic state. This article led the authors of an experimental work on N_2H_2 and N_2D_2 [73a] to revise their analysis. In a subsequent paper [74] Peric *et al.* improved their calculations on the vibrational frequencies in the ^1A_g and ^1B_g states by taking anharmonicity into account.

Using the calculated data of [72] and [73], Craig and Levin calculate vibration frequencies and make revised vibrational assignments for *trans*-diazene [75].

Pople *et al.* [76] discuss and apply Cizek's coupled-cluster theory with double substitutions (CCD) for extensive electron correlation. They obtain a (nonvariational) value $E_h = -110.2882$ for the ground state; see Section F,1 for some details of the method. They give preliminary results on the potential surfaces for 1,2 hydrogen shifts in C_2H_2, HCN, CH_2O, and N_2H_2.

Ahlrichs and Staemmler [77] in a polarized double-zeta calculation (d on N, p on H) obtained SCF data on the geometry of the trans and cis ground states and the transition state barrier, also on the ground and first excited states of H_2NN. The details are listed (SCF $E_h = -109.9531$ for *trans*-N_2H_2). With these geometries, they made rather extensive valence-shell CM calculations using PNO CM and CEPA methods, recovering about 80% of the valence-shell correlation energy. They found *cis*-N_2H_2 to be about 6 kcal/mole higher and $H_2N^+{=}N^-$ about 26 kcal/mole higher in energy than *trans*-N_2H_2. For the activation energy of the transition state from trans toward cis they find 55 kcal/mole. For the reaction N_2H_2 (*trans*) $\rightarrow$ H + N_2 + H they find 59 kcal/mole to be required. This result is relevant to the photochemically induced expulsion of N_2 from R—N=N—R.

The lowest energy states of H_2NN are a planar singlet (C_{2v}) and a triplet (C_s), for which they find H_2 bent 42° out of plane but with a small inversion barrier (about 3 kcal/mole). They find the triplet state about 3 kcal/mole below the singlet, with a rather long N—N bond (1.34 Å compared with an N=N double bond of 1.22 Å length in the singlet). Comparisons with CH_2 are instructive.

On the other hand, in a very thorough GVB plus CM study of aminonitrene Davis and Goddard [78] find the singlet state (largely $H_2N^+{=}N^-$, with R_{NN} = 1.25 Å and a dipole moment of 4.04 D) to be the ground state, with the $\pi\pi^*$ triplet state 15 kcal/mole higher, with a dipole moment of 2.35 D and a practically zero inversion barrier and a 21° nonplanarity angle. Five higher excited states are also discussed, along with heats of formation and five ionization potentials.

von Niessen *et al.* [79], using a Green's function method with a good polarized basis set (SCF E_h = −110.0313 for *trans*-N_2H_2, −110.0208 for *cis*-,N_2H_2 and −110.0021 for H_2NN) have computed ionization potentials for N_2H_2 and H_2NN. They list SCF orbital energies and ionization energies for all three isomers, those for H_2NN being for the singlet state. For the *trans*-diazene ionization potentials they find 10.26, 14.24, 15.87, 17.38, and 23.59 eV for $4a_g$ (n_+), $1a_u$ (π), $3b_u$ (n_-), $3a_g$, and $2b_u$, respectively. They compute and display the vibrational structure of the photoelectron spectrum of *trans*-HNNH and *trans*-DNND, in good agreement with experiment.

Parsons and Dykstra [80] in 1979 reported calculations by the SCEP CM method with up to 176,000 configurations, using a polarized triple-zeta basis set, on aminonitrene, *trans*-diimide, and *cis*-diamide, and the barriers between them. They considered only the lowest singlet state of H_2NN. They found the following energies (in kcal/mole) relative to *trans*-diimide: 7.2 for *cis*-diimide, 55.1 for the trans to cis barrier, 24.5 for H_2NN, and 82.6 for the *trans*-diimide to H_2NN barrier. The first three of these results agree almost exactly with those given by Ahlrichs and Staemmler [77], who do not give the fourth. Optimized geometries for the four cases are given. Contour plots are given showing the forms of the highest occupied MO in all the ground and barrier states. They also give the H_2NN dipole moment (3.64 D) and N and H populations (7.2 on amino N, 7.3 on nitrene N, 0.77 on each H). They give a concise survey of existing literature on the N_2H_2 isomers.

Pasto [81] has made minimal-basis computations on minimal-energy pathways and transition states for numerous disproportion reactions of one or a pair of diimide molecules (cis or trans) to form $N_2 + H_2$ or N_2H_4.

Using a basis set of double-zeta quality for an SCF calculation followed by extensive CM at several thresholds, Vasudevan and Kammer [38] made calculations on the ground and eight excited states (some Rydberg or partially so) of *trans*-N_2F_2. They give a table describing the principal CM configurations and their coefficients. For the ground state... $6b_u^2\,2a_u^2\,7a_g^2$, ^1A_g, the most important CM term involves

$2a_u^2(\pi) \rightarrow 2b_g^2\ (\pi^*)$. The MOs $6b_u$ and $7a_g$ are combinations n_- and n_+, i.e., $n_A \pm n_B$, of nonbonding N atom orbitals ($E_h = -307.5073$).

6. Diazomethane (CH_2NN)

In a GVB plus CM study Walsh and Goddard [82] conclude that the structure of diazomethane (symmetry C_{2v}) is basically that of a singlet diradical (24) rather than the usually accepted resonance structure (25).

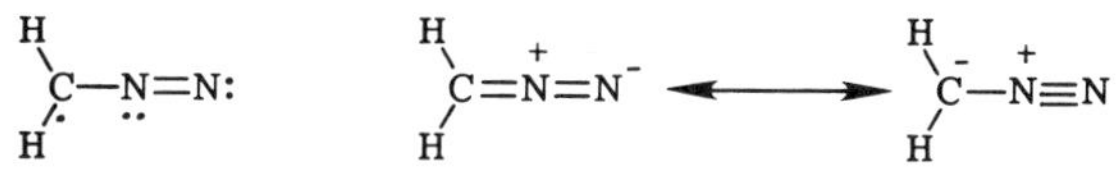

24 25

They use a double-zeta basis plus diffuse orbitals to deal with Rydberg states, and make calculations on the $^1A_1$ ground state, the $^3A_2$, $^1A_2$, $^3A_1$, and second $^1A_1$ valence-shell states, and several Rydberg states. They compute dipole moments for the ground and $^1A_2$ states in terms of bond and MO contributions and population analysis. They show contour maps of the valence-shell SCF MOs and of the GVB MOs. The lower excited states $^3A_2$, $^1A_2$, $^3A_1$, and 2^1A_1 ($\pi\pi^*$) are computed to lie at energies 2.65, 2.93, 3.66, and 5.90 eV, respectively. The oscillator strength for absorption to 2^1A_1 (which shows partial Rydberg character) is computed as 0.376, and the absorption is identified with the observed strong absorption peak at 5.70 eV. The dipole forbidden transition to $^1A_2$ is identified with a diffuse weak absorption with maximum at 3.14 eV. Contour diagrams of the calculated Rydberg MOs (3s, 3p, 3d, 4p) are shown, and corresponding transitions are related to observed spectra.

Lievin and Verhaegen [83], using a double-zeta basis plus polarization, studied the association paths for the reactions $CH_2\ (^3B_1) + N_2\ (^1\Sigma_g^+) \rightarrow CH_2N_2\ (^3B_1)$ and $CH_2\ (^1A_1) + N_2\ (^1\Sigma_g^+) \rightarrow CH_2N_2\ (^1A_1)$. For the latter, they concluded that the lowest path is one without potential barrier in which the N_2 approaches almost perpendicular (93°) to the CH_2 plane with the C and N atoms almost colinear (N—N—C angle 203°). They also obtained vertical excitation energies for the $^3A_2$ and $^1A_2$ states of diazomethane (67 and 73 kcal/mole, the latter in good agreement with the observed transition at 72 kcal/mole).

7. Nitrosomethane

Ha and Wild [84], using an unpolarized basis set of double-zeta quality followed by limited CM, have made computations on the ground and first three singlet and triplet excited states of the eclipsed (C_s symmetry) conformer (26) of nitrosomethane (CH_3NO), and have computed dipole mo-

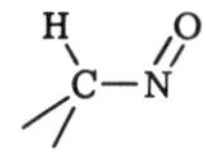

26

ments, along with oscillator strengths for the allowed spectroscopic transitions. They give a list of molecular orbital energies (though not the usual ones). The ground state is... $(8a')^2 (1a'')^2 (9a')^2 (2a'')^2 (10a')^2$, $^1A'$, where 9a', 2a'', and 10a' are designated n_+, π, and n_-; the unoccupied 3a'' is π^*. In their CM, they keep the first 6 (out of 12) MOs doubly occupied, and use the lowest 6 (out of 18) virtual MOs after a transformation toward natural MOs.

The first allowed transition, to $(\pi^* \leftarrow n_-)$, $^1A''$ at 2.17 eV computed, with $f = 1.2 \times 10^{-3}$, corresponds to an observed spectrum. Next comes one to $(\pi^* \leftarrow n_+)$, $^1A''$ computed at 7.14 eV with $f = 4.8 \times 10^{-3}$ (not yet observed), and then one to $(\pi^* \leftarrow \pi)$, $^1A'$ with $f = 0.63$ (not yet observed). The $\pi\pi^*$ transition is found to involve strong charge transfer from the CH_3 to the NO (dipole moment 6.55 D, almost twice that for the ground state). Major CM contributions for each state are listed. The LCAO coefficients are listed for the n_- MO 10a', the n_+ MO 9a', and the π^* MO 3a''. The MOs n_- and n_+ are not really nonbonding as the symbol n implies.

In a similar but apparently more accurate SCF plus CM computation, Ernsting *et al.* [85] have found that the lowest $^1A''$ state differs from the ground state by having a staggered conformation, lower by 325 cm^{-1} than for the eclipsed form. (In the ground state, the barrier, experimentally, is 400 cm^{-1}.) They reanalyze the observed spectrum accordingly.

8. Nitramine ($H_2N \cdot NO_2$) and Dimethylnitramine

Duke [86] in minimal-basis SCF calculations finds nitramine to be nonplanar with an out-of-plane angle of 52° for the NH_2 hydrogens, but for dimethylnitramine he finds a planar C_2NO skeleton. These results agree with experiment.

9. Glycine ($H_3N^+CH_2CO_2^-$) and Glycylglycine ($H_3N^+CH_2CONHCH_2CO_2^-$)

Ryan and Whitten [87] used an SCF procedure somewhat better than a minimal Slater treatment plus limited CM to study the zwitterion forms of the molecules glycine and glycylglycine. They give tables of orbital energies and discuss the charge distributions. The molecular skeletons are assumed to be planar.

10. Hydrazoic Acid (HN_3) and Isomers of HNCO

Lievin *et al.* [88] have made 6-31G plus single- and double-substitution CM (up to about 600 and 1300 configurations) on HN_3. They find the terminal N of NNN bent up by 10° and the H bent down by 69°, in a planar (C_s) configuration. Previously NNN was usually presumed to be linear. They also computed force constants and vibrational frequencies in fair agreement with experiment. They investigated the dissociation path to NH plus N_2, and found a relatively small dissociation energy but a barrier to dissociation of about 30 kcal/mole. They compared this with the dissociation of the isoelectronic azomethane. An instructive partial population analysis is given.

McLean *et al.* [89] have made *ab initio* computations on the four isomeric molecules HNCO (stablest), HOCN 22 kcal/mole higher using a DZP (polarized double-zeta) basis, HCNO 83 kcal/mole up by a DZP basis, and HONC 73 kcal/mole up by a DZ basis (expect somewhat higher by DZP). The HNCO and HCNO molecules are known spectroscopically, and HNCO has been detected in interstellar clouds. Particular attention has been devoted to the optimization of HOCN in order to predict its rotational constants with the idea that it may be found in interstellar clouds. To this end, extensive CM has been added for this isomer. The HCNO molecule is very nearly linear. In the other isomers, the lineup of three heavy atoms is bent by a few degrees, trans to the much more extensive bending of the H atom. A table of gross atomic populations for the four isomers is shown, thus giving an indication of charge distributions. Force constants and dipole moments are computed and compared with experiment when available.

11. Miscellaneous Calculations (Mostly SCF Only)

Kitaura *et al.* [90]. Ground and ($\pi^* \leftarrow n$) state of methylenimine (CH_2NH). STO-6G basis plus CM.

Osamura *et al.* [91]. Lowest triplet state ($\pi^* \leftarrow \pi$) of methylenimine.

Hopkinson and Csizmadia [92]. XNH_2 with X = CH_3, OH, F, CN, CHO, NO_2. Contracted 8s 3p basis for X.

Yamaguchi *et al.* [93]. Nitrene peroxide (HNOO). 4-31G basis.

Almlöf and Ischenko [94]. $LiNO_3$. Smallish unpolarized basis but MCSCF.

Ozias and Julg [95]. Thiocarbonates and thionitrates. STO-6G basis.

Ha [96]. Nitroethylene. Planar and perpendicular NO_2. Double-zeta basis.

Howell *et al.* [97]. NH_2CN, NF_2CN, PF_2CN. Moderately large s, p basis, plus d for P. RCN angles bent slightly.

Hinchcliffe and Dobson [98]. $LiNH_3^+$, $LiNH_2$, Li_2NH, Li_3N. Unpolarized basis.

Olsen [99]. Four planar conformers of 2-propene-1-imine ($H_2C{=}CH{-}C{=}NH$). 4-31G basis.

Barber *et al.* [100]. Core binding energies of complexes of several nitrogen bases with BF_3. Minimal basis for 1s and 2s, double-zeta for valence orbitals.

Gropen and Seip [101]. Aminoborane (H_2BNH_2). Stably planar; rotation barrier 33 kcal/mole with pyramidal N at barrier top. Moderately good basis.

So [101a]. Minimal-basis geometry and double-zeta valence orbital energies of NH_2Cl and geometry of NH_2Cl^+ (planar ground state).

Hinchcliffe [101b]. Energies and orbital energies of CH_2CN and CH_2NC with good polarized basis set.

Kollman *et al.* [101c]. Double-zeta calculations on the methyleneimmonium ion ($H_2C{\cdots}NH_2^+$), with orbital energies and population analysis; same on methyleneimine, CH_2O, CH_2OH^+, CH_3OH and CH_3NH_2, CH_3F and CH_2F^+ for comparison.

Merényi *et al.* [101d]. Isomerization of azomethine ($H_2C{=}NH$), $H_2C{=}NCH_3$, $NH{=}NH$, $NH{=}NCH_3$ and their protonated derivatives. Inversion barriers for all these are computed. Double-zeta basis.

Olsen [101e]. STO-3G calculations on the four planar conformers of 2-propane-1-imine.

Fogarasi *et al.* [101f]. Force constants and geometry of HNF_2 and H_2NF, using moderate basis with one s bond function.

Ozias and Julg [101g]. Effect of single and multiple substitution of oxygen by sulfur in carbonate and nitrate ions. Unpolarized STO-3G basis.

Bouma and Radom [101h] have made SCF calculations on nitrosoacetaldehyde ($O{=}N{-}CH_2CH{=}O$, 1) and its enol ($O{=}HCH{=}CHOH$, 2) and oxime ($HON{=}CH{\cdot}CH{=}O$, 3) isomers. Stablest is 3, followed by 2, then 1. A low-energy transition path for the 1,5-sigmatropic hydrogen shift from 2 to 3 has been found.

F. ALDEHYDES, KETONES, AND CARBOXY MOLECULES

1. Formaldehyde, Hydroxycarbene, and Protonated Formaldehyde

Formaldehyde (27) is isoelectronic with C_2H_4, N_2H_2, B_2H_6, and O_2. Its

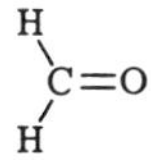

27

ground state is $1a_1^2\,2a_1^2\,3a_1^2\,4a_1^2\,1b_2^2\,5a_1^2\,1b_1^2\,2b_2^2$, $^1A_1$, where $1b_1$ is a C=O π and $2b_2$ a roughly nonbonding O atom n orbital, but somewhat antibonding between O and CH_2. Many *ab initio* calculations have been made on this molecule and its lower excited states, in particular, the spectroscopically well-known [102]... $2b_2\,2b_1$ ($\pi^* \leftarrow$ n), $^1A_2$ and the corresponding $^3A_2$. In these two states, the spectrum shows that the molecule deviates somewhat from planarity, assuming C_s symmetry, so that these states are $^1A''$ and $^3A''$. Numerous Rydberg states are known.

The best SCF calculations to date [103] on the ground and the lowest excited states $^3A''$ and $^1A''$ appear to be those of Garrison *et al.* (E_h = −113.9149, −113.8347, and −113.8248), who used a Gaussian (9s 5p 1d) on C and O, (5s 1p) on H as their best basis set, contracted in the case of the excited states. Their estimate of the SCF limit is $E_h = -113.925$. They give orbital energies for all three states, along with dipole, quadrupole, and octupole moments, and electric field gradients at the nuclei, and report having computed a number of other properties. The calculations were made using experimental equilibrium geometries for each state. The dipole moments are all C^+O^- oriented. For the ground state, $\mu_{calc} = 2.86$ D, 0.5 D higher than the experimental value. For $^1A''$, the computed μ_z is 1.470 D, $\mu_x = 0.513$ D, making $\mu = 1.557$ D; μ_z is in good agreement with experiment (1.56 ± 0.07 D). For $^3A''$, the computed μ_z and μ_x are 1.321 and 0.604 D, the experimental μ_z is 1.29 ± 0.03 D. Earlier articles are cited.

In an early article using a less accurate basis set, Dunning and Winter [104] give contour and profile diagrams of charge density, both total and relative to the atoms and individually for five of the valence MOs. In another early article, Buenker and Peyerimhoff [105] give contours for the $1b_1$ and $2b_2$ MOs themselves (see Figs. 1 and 2, reproduced from their article). Calculations are made for varying geometries and for several excited states, and effects of very limited CM are studied.

For the ground state, Meyer and Pulay [106], using an SCF calculation with a rather good basis set ($E_h = -113.9030$), compute harmonic force constants and vibrational frequencies (also for D_2CO and HDCO) in good agreement with experiment after some corrections for anharmonicity in the case of the frequencies. They also give the dipole moment and results from which its derivatives can be obtained.

In an early article using a minimal basis set Newton *et al.* [107] have obtained localized MOs and bond moments for the CH and CO bonds and

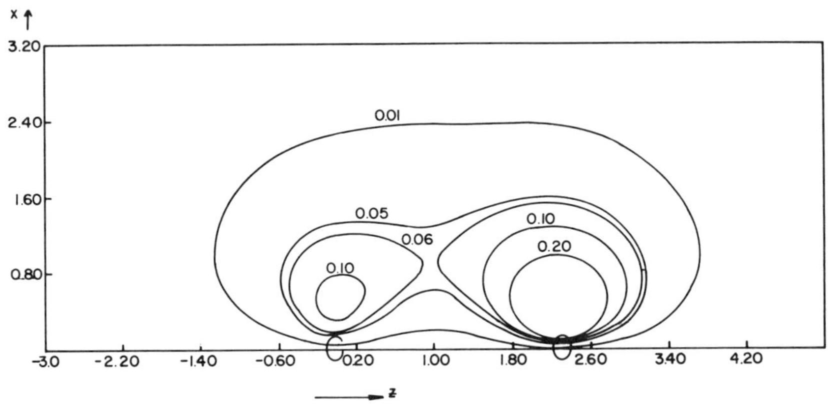

FIG. 1. Charge-density SCF contours for the $1b_1$ MO in the ground state. From Buenker and Peyerimhoff, [105].

the O atom lone pair using the Edmiston–Ruedenberg procedure (cf. Section VIII,C,1 for details). The resulting orthogonal sets of localized MOs have delocalization "tails" varying from 1 to 12%. Contour diagrams of electron density are shown for the CO bond MO and the O atom lone-pair Mo. The CH, CO, and O lone-pair bond moments of 2.06, 1.54, and 2.20 D are reported. The CH moment is much larger than what is generally inferred (about 0.4 D) from intensities of infrared absorption.

Wilhite and Whitten [108] obtain localized MOs by the Peters method

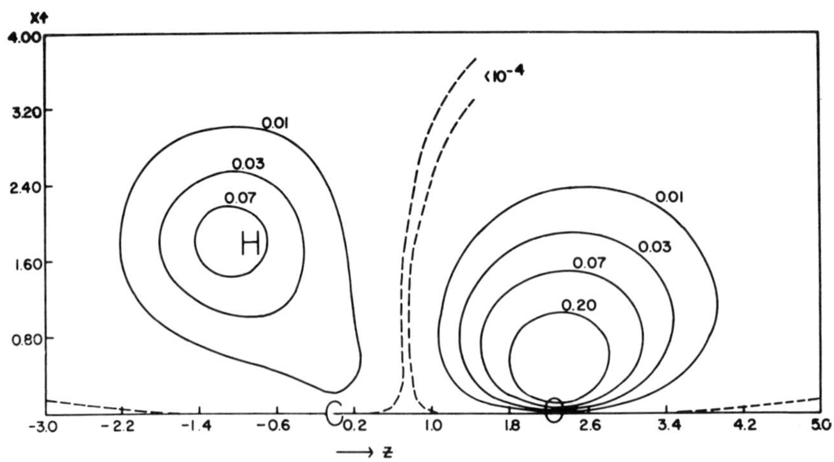

(a)

FIG. 2. Charge-density SCF contours for the $2b_2$ MO in the ground state. From Buenker and Peyerimhoff, [105].

(see Section VIII,C,1 for details). They list orbital energies for the various localized and unlocalized MOs, and percentage delocalizations for the localized MOs. They make a limited CM calculation using localized MOs.

There seem to be no ground-state calculations by conventional methods which approach complete CM.

An article by Langhoff *et al.* [109] reports the attainment of $E_h = -114.1366$, corresponding to about 50% of the correlation energy. A double-zeta basis of contracted Gaussian lobes was used, followed by CM. They compute 2.22 D (C^+O^-) for the dipole moment (experiment: 2.33 ± 0.02 D). Their computation of the dipole moment is probably the best to date. They also report computed values of several other properties, and also for the $n\pi^*$, $^2A_2$ and $\pi\pi^*$, $^3A_1$ excited states.

A lower E_h (-114.1795 with frozen core) has been obtained, using an 6-31G* basis, by Pople *et al.* [76] with methods that are not variational, namely the Moeller–Plesset perturbation method to most of fourth order and the nearly equivalent CCD method (coupled-cluster method) with double substitutions used by Cizek and others. In their article, Pople *et al.* apply the method to the potential surfaces for 1,2 hydrogen shifts.

Bell [110], using a polarized double-zeta basis followed by CM (4214 configurations), obtained $E_h = -114.1809$ at the equilibrium configuration, the lowest computed value yet reported.

In an article by Tawil and Langhoff [111], with 40% of the correlation energy ($E_h = -114.0708$), Compton profiles and the effect (a broadening) of correlation on these are calculated. A double-zeta plus polarization basis is used.

The most extensive discussion of magnetic susceptibilities and ^{13}C and ^{17}O magnetic shielding tensors is one by Tokuhiro *et al.* [112], using a double-zeta basis including both diamagnetic and paramagnetic contributions, using perturbation theory for the latter. Comparisons with experiment (not always accurately available) are made, and also with the earlier calculations of Neumann and Moscowitz [113], who used a better (polarized double-zeta) basis set, in the case of the diamagnetic contributions.

Yeager and McCoy [114] have used the equations-of-motion method, with a rather limited unpolarized basis plus diffuse functions, to compute excitation energies and oscillator strengths for transitions to twelve excited states, including seven Rydberg transitions, out of the $2b_2$ MO.

Harding and Goddard [115] have made GVB calculations using a polarized double-zeta basis, followed by a little CM, on the ground and low excited states. With GVB, $E_h = -113.9162$, then with CM, $E_h = -113.9409$, for the ground state. For the $^3A''$ ($n \rightarrow \pi^*$) and $^1A''$ ($n \rightarrow \pi^*$) they compute adiabatic excitation energies 3.03 and 3.42 eV, in excellent agreement with experimental values of 3.12 and 3.50 eV. They also compute vertical excitation energies of 3.62 and 4.09 eV for these states, in agreement with electron impact values of 3.5 and 4.1 eV, and 3.75 eV for $^3A''$ from trapped-electron experiments. They compute 5.95 eV for the vertical excitation energy of the $2b_1 \leftarrow 1b_1$ ($\pi\pi^*$), $^3A_1$ states, in agreement with the value (5.9 eV) from electron impact experiments and that (6.20 eV) from trapped-electron experiments [54]. For $\pi\pi^*$, $^1A_1$ they calculate an excitation energy of 10.77 eV with a diffuse π^* MO. Dipole moments are computed for each state. For the ground state, the GVB-CM result of 2.59 D is much closer to experiment (2.32 D) than the SCF value (3.10 D, but see above for the value 2.86 D obtained by Garrison *et al.*). For the $^3A''$ and $^1A''$ states, the differences between SCF and GVB-CM are much smaller, for reasons which are explained. For the $^3A_1$ state, GVB gives 1.33 D.

Bell [110] has used a polarized double-zeta basis followed by rather extensive CM in calculations on the $^3A''$ and $^1A''$ states, obtaining low E_h values, but only in planar geometry. However, adiabatic excitation potentials are also predicted: 2.66 eV for $^3A''$ and 2.95 eV for $^1A''$, as compared with the experimental values 3.12 and 3.50 eV.

Özkan *et al.* [116], in SCF calculations of moderate accuracy on the $^1A_1$ ground state ($E_h = -113.8847$), the $^1A''$ and $^3A''$ excited states, the lowest two ($^1B_1$ and $^1B_2$) cation states, and the $^2B_1$ anion state, study "potential surface distortion and orbital reorganization upon change of electronic state." They discuss changes in energy, in the forms of MOs, and in potential surfaces. Only the $^3A''$, $^1A''$, and the anion state are predicted to be nonplanar. Many interesting details are given, including first and second moments of several MOs.

Computations on the spin–spin component of the zero-field splitting in the $^3A''$ and the $\pi\pi^*$ $^3A_1$ states are presented by Langhoff *et al.* [109] (see above). The values $D = 0.539$ cm^{-1} and $E = 0.031$ cm^{-1} were obtained for the $^3A''$ (*treated as planar*), $D = -0.588$ cm^{-1} and $E = 0.058$ cm^{-1} for $\pi\pi^*$, $^3A_1$.

Further computations, on the spin–orbit component, were carried out for the $^3A''$ state (*assumed planar*) by Langhoff and Davidson [117], the resulting contribution to D and E being $\times 0.224$ cm^{-1} and 0.009 cm^{-1}, giving total values of $D = 0.314$ cm^{-1} and $E = 0.04$ cm^{-1}; experimental values are $D = 0.141$ cm^{-1}, $E = 0.02$ cm^{-1}. Allowance for nonplanarity might improve the agreement. The calculations involved a second-order perturbation treatment of an infinite sum of terms involving excited states. Incidental to this, vertical transition energies to 13 excited states were computed using essentially SCF plus CM for each state (see below for discussion). The authors also make extensive calculations on the phosphorescence lifetime of the (planar) $^3A''$ state, but there are difficulties.

van Dijk *et al.* [118] have used an unpolarized double-zeta basis in calculating the strengths of the observed vibronic transition of the otherwise forbidden absorption to the assumed planar $^1A_2$ (really $^1A''$) state. Unfortunatley they did not seem to realize that the $^1A''$ state is (moderately) nonplanar (inversion barrier only about 650 cm^{-1}) with an out-of-plane frequency ν_4 of about 700 cm^{-1}, but assumed that the observed inversion splitting [82] of 124 cm^{-1} for the lowest level of $^1A''$ is ν_4. This appears to deprive their computed oscillator strengths of significance, although they seem to run parallel to experimental values. Strangely, they try the use of the acceleration operator as an alternative to the dipole operator in their calculations of absorption strengths, whereas it is generally known that the velocity operator is preferable. In a companion article [119] the same authors make computations on the radiationless transition from the $^1A_2$ to the ground state for H_2CO and D_2CO, with the same assumptions. In a later article [120] Kemper *et al.* use the material from the first paper to calculate fluorescence emission spectra and absolute radiation lifetimes of single vibronic levels; agreement with experiment is fairly good.

In a communication, Lucchese and Schaefer [121] make SCF plus fairly extensive CM calculations ($E_h = -114.1338$ for the ground state of formaldehyde) on the lowest excited states of the formaldehyde isomer hydroxycarbene (28), beleived to be an intermediate in the decay of the $^1A''$

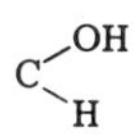

28

excited state of formaldehyde. Relative to the ground state of formaldehyde, they compute the following excitation energies: $^3A''$ of formaldehyde, 68.9 kcal/mole (experiment: 72.0); hydroxycarbene, $^1A'$, trans, 51.7, cis, 57.5 kcal/mole; ^3A (gauche) of hydroxycarbene, 73.0 kcal/mole. Thus it appears that hydroxycarbene is a possible intermediate for the decay of $^1A''$ but not of $^3A''$ of formaldehyde.

This time in contrast to [118] and [119] (see above) apparently recognizing the nonplanar character of the $^1A''$ (S_1) state, Kemper *et al.* [122], using an unpolarized double-zeta basis set in an SCF computation plus some CM, report computations on a reaction path via hydroxycarbene for the decay of the $^1A''$ state. They report a barrier of about 100 kcal/mole for the unimolecular transformation of $^1A''$ via hydroxycarbene, lowered only to about 95 kcal/mole in a bimolecular transformation (they make computations for the double molecule).

Photochemical dissociation of H_2CO to CO plus H_2 begins at the $1A''$ level (3.124 eV) and to H + CHO at 3.73 ± 0.04 eV. Goddard and Schaefer [123] have made very extensive SCF + CM calculations using up to polarized double-zeta basis sets and correcting the single- and double-substitution CM calculations for higher substitutions, on the two dissociation processes for the ground state. They beleive that the observed photochemical dissociation goes largely via this state. They have calculated the barriers to dissociation, including geometry optimization, with force constant evaluation and correction for zero-point vibration energy. For the molecular dissociation they find a barrier of 4.06 eV, which after corrections for zero-point vibration gives an approximate activation energy of 3.76 ± 0.2 eV. (The total reaction energy is 0.02 eV exothermic.) For the radical dissociation energy, they calculate 3.48 eV, as compared with experimental values from 3.7 to 3.8 eV.

They also make improved calculations on hydroxycarbene, which perhaps may be an important intermediate, and arrive at energies above formaldehyde of 52.6 kcal/mole (2.28 eV) for trans and 57.6 kcal/mole (2.50 eV) for cis, with a barrier to rearrangement (though a nonplanar structure) from trans to cis of about 29 ± 3 kcal/mole. For the barrier to rearrangement (in a plane throughout) from formaldehyde to *trans*-hydroxycarbene, they find 3.86 ± 0.22 eV (89 ± 5 kcal/mole). For the reverse reaction, the barrier is 1.56 eV (36 kcal/mole). Harmonic force fields for both isomers were calculated.

Several authors have made calculations on excited states. Langhoff and Davidson [117] list results on 13 excited states, including comparisons with results of other authors. Some earlier, rather good calculations [105, 114, 124–126] are cited. All agree that the observed absorption near 8 eV is a Rydberg transition to ($3pb_2 \leftarrow 2b_2$), $^1A_1$ and not, as first thought, the $\pi\pi^*$ transition to ($2b_1 \leftarrow 1b_1$), $^1A_1$. The observed absorption at 7.1 eV is identified as a Rydberg transition to ($3sa_1 \leftarrow 2b_2$), $^1B_2$, and that at 9.0 eV as a $\sigma\pi^*$ transition to ($2b_1 \leftarrow 5a_1$), $^1B_1$.

Several authors have discussed the whereabouts of the $^1(\pi\pi^*)$ transition, concluding that it is probably mixed with Rydberg transitions. Perhaps most recently, Langhoff *et al.* [127], after reviewing previous work, predict in an SCF plus limited CM calculation that the $\pi\pi^*$, $^1A_1$ state would be at 11.2 eV, slightly above the minimum ionization potential, so that it should be autoionized. A similar conclusion has been reached by Mentall *et al.* [128] in SCF calculations on Rydberg and mixed Rydberg-valence states. In a special calculation for this state, Peyerimhoff and Buenker [129] place it at 10.96 eV.

Gouyet and Prat [130] have made calculations on the ground and eight excited states by the "SPIN-CIPSI" method using STO-3G without d polarization but with very extended CM. The results are presented graphically. For the ground state, E_h is about -114.1. The results seem to be in general agreement with other comparable calculations.

Harding and Goddard [131], using GVB plus CM methods, have calculated the excitation energies and oscillator strengths of spectroscopic transitions from the $1b_1$ and $2b_2$ MOs to the various 3s-, 3p-, and 3d-like Rydberg MOs. There is good agreement with experiment on frequencies and fairly good agreement on oscillator strengths. Two states ($\pi\pi^*$, $^1A_1$ and $3s \leftarrow \pi$, $^1B_1$) are calculated to be between 10.7 and 10.8 eV, corresponding to a broad unassigned peak (10.5–11.0 eV) in the electron-impact spectrum. Other peaks in that spectrum are assigned to $3p \leftarrow \pi$ and $3d \leftarrow \pi$ transitions.

Cederbaum *et al.* [132] have used their Green's function approach, with a polarized double-zeta basis set, to compute the four lowest vertical ionization energies of formaldehyde. The results are 10.84 eV for the removal of a $2b_2$ MO, 14.29 eV for $1b_1$, 16.36 eV for $5a_1$, and 17.13 eV for $1b_2$. Excitation probabilities for normal modes ν_1, ν_2, and ν_3 are also reported for H_2CO and D_2CO. Detailed structures for the $5a_1$ and $1b_2$ ionizations are portrayed; they differ strongly for H_2CO and D_2CO. Domcke and Cederbaum [133] give a somewhat improved and more detailed discussion of the vibrational structures, and add a fifth computed ionization potential, 21.57 eV for $4a_1$, and for both isotopes show observed and computed structures for the first four ionizations.

Langhoff *et al.* [134], using a polarized double-zeta basis set plus diffuse functions, first make SCF plus modest CM computations, using ground-state and $^3A_2$ MOs for CM, on discrete vertical excitation energies and oscillator strengths for the neutral, and for the lowest of each species except $^2A_2$ for the ionized molecule. Then to deal with the ionization continuum they use the moment-theory methods of Stieltjes and Tchebychoff to compute vertical photoionization profiles for each of the allowed types of transitions ($^1A_1$, $^1B_2$, and $^1A_1$). They obtain graphs of photoexcitation and photoionization cross sections, and make comparisons with experiment.

The first excited state $^2B_1$ of H_2CO^+ at about 14 eV above the ground-state H_2CO is observed to dissociate both to $H_2 + CO^+$ (requiring 13.98 eV) and $HCO^+ + H$ (requiring 11.92 eV); the $^2B_2$ ground state of H_2CO^+ is at 10.87 eV [135]. Pires *et al.* [136], using SCF calculations with a good polarized basis set near equilibrium geometries, and minimal basis plus moderate CM for the reaction paths, have investigated the paths for the two modes of dissociation. They conclude that for the production of CO^+ ions, a conical intersection with the next higher, $^2A_1$ state of H_2CO^+ is involved: the intersection which occurs for C_{2v} symmetry is avoided by flapping of the H_2, $^2B_1$ and $^2A_1$ both become $^2A'$ of C_s symmetry and can mix, so that dissociation to $CO^+(^2\Sigma^+)$ plus $H_2(^1\Sigma_g{}^+)$ can proceed adiabatically on the lower of the two resulting surfaces. Rather similarly, HCO^+ may be produced by an unsymmetrical stretching of one HCO bond to remove all symmetry, so that the molecule can proceed adiabatically to dissociation via the ground state of H_2CO^+. Or without further work one cannot exclude an alternative route by predissociation on the way toward dissociation, from the $^2B_1$ (now $^2A''$ under C_s symmetry) into the dissociative ground state. In the adiabatic processes, the molecule has to go through various distortions.

Using up to a polarized near double-zeta basis, Saebø [137] has calculated the potential surface for the reaction $H_2COH^+ \rightarrow H_2CO^+ + H_2$.

Strausz *et al.* [138] have discussed vertical proton affinities of H_2CO and H_2CO^+H in their ground singlet, excited triplet, and ionized doublet states (article not available).

Cremaschi and Simonetta [139], using a 4-31G basis, have optimized the geometries of $HCOH^{2+}$ (protoformyl dication) and plotted the dissociation path, over a barrier, to HCO^+ or COH^+ plus H^+; a linear arrangement for the ions and the path is found. Similar calculations are given for the dissociation of the CH_3COH^{2+} (protoacetylium) dication.

According to calculations using a polarized basis set and some CM, Johansen and Nilssen [140] find H_3CO^+ to be 30 kcal/mole less stable than $(H_2COH)^+$. Dill *et al.* [141] also make calculations on these and $H_2\cdots HCO^+$.

Del Bene and Radovich [142], using up to a 4-31G basis, have made SCF calculations to find proton affinities for R_2CO molecules to form R_2COH^+, where R = CH_3, NH_2, OH, F. Proton affinities are found to increase in the order $NH_2 > CH_3 > OH > H > F$, the same order as for RCHO. Del Bene and Vaccaro [143] have studied the geometries and energies of these compounds.

As an example of a generalization of the direct CM method, Siegbahn [144] makes calculations on dioxymethane (CH_2O_2), finding a $^1A_1$ ground state with a $^3B_2$ state 2 kcal/mole higher.

2. Thioformaldehyde ($H_2C{=}S$) and Some Derivatives

Bruna *et al.* [145] have made SCF plus limited CM calculations on the ground and 16 excited states of $H_2C{=}S$, along with the ground state of the ion, using a double-zeta basis of Gaussian-lobe s and p functions plus CS-bond-centered s and $p\pi$ functions for polarization and diffuse functions to provide for Rydberg MOs. They report orbital energies ($1a_1$ to $7a_1$, $1b_1$, $2b_1$, $1b_2$, $2b_2$ occupied). They make comparisons with H_2CO and find that the CH_2 in the $n\pi^*$ ($3b_1 \rightarrow 3b_2$), $^3A''$ and $^1A''$ are only slightly flapped (about 16° versus 32° in H_2CO); they also find that the $\pi\pi^*$ ($3b_1 \leftarrow 2b_1$), $^3A'$ and $^1A'$ are probably very slightly flapped. They show calculated total and orbital energies as a function of the flapping angle, and the energy barrier for inversion in the $n\pi^*$ states (only about 0.01 eV versus 0.07–0.09 eV in H_2CO). They list excitation energies, and make identifications with experiment similar to those in H_2CO. They find that the computed $^1(\pi\pi^*)$, $^1A_1$ at 8 eV is well below the first ionization energy at 8.5 eV and thus is not autoionized as in H_2CO. The first spectroscopically allowed transition to (Ry s $\leftarrow 2b_2$), $^1B_2$ at 7.7 eV is analogous to one in H_2CO. Dipole moments are computed for the ground and $n\pi^*$ states. Charge density contours are depicted for the $2b_1$ (π), $3b_2$ (n), and $3b_1$ (π^*) MOs. Some of the predicted excitation energies have later been found to agree well with experiment (74a): $^1(n, \pi^*)$, 1.84 eV (1.80); $^3(n, \pi^*)$, 2.17 eV (2.03); (4s $\leftarrow$ n), 5.83 eV (5.83); (4p $\leftarrow$ n), 6.62 eV (6.59); experimental values in parentheses.

Kapur *et al.* [146] made STO-3G minimal-basis calculations to obtain potential surfaces for the ground and first excited triplet states of several derivatives of H_2CS. Their main result is that the flapping angle and the inversion barrier increase steadily from 24° and 108 cm^{-1} in H_2CS through Cl_2CS, HFCS, ClFCS, to 42° and 2707 cm^{-1} in F_2CS. They also compute out-of-plane bending frequencies for Cl_2CS and F_2CS; there is good agreement with the experimental frequencies.

Rosmus and Bock [147] have made a PNO-CEPA study of H_2CS^- using a polarized double-zeta basis and found a $^2B_1$ ground state. The last electron is in a $3b_1$ (π) antibonding MO, which causes a lengthening of the C—S bond. The anion is found to be stable by 0.3 eV. Correlation energy is small.

Yamabe *et al.* [148] have made 4-31G SCF calculations which indicate that protonation of $(CH_3)_2S$ very likely leads to a carbon-protonated derivative $(CH_3)_3S^+$ of C_{3v} symmetry, in contrast to the protonation of $(CH_3)_2O$, which gives $(CH_3)_2OH^+$ (see end of Section F,1). Possible reasons are explored.

3. Acetone, Thioacetone, Acetaldehyde, Vinyl Alcohol, and Glycolaldehyde

For Acetone ($(CH_3)_2CO$) Hess *et al.* [149] make an SCF plus CM calculation on the ground and 32 excited states and the first 2 ionic states, using a double-zeta unpolarized basis, plus diffuse functions for Rydberg states. They place the molecule so that it has C_{2v} symmetry (one H of each CH_3 in the principal plane determined by the heavy-atom skeleton). The CM was carried out according to the usual Buenker and Peyerimhoff MRD-CI method, but with a core of seven MOs always left doubly occupied. Excited configurations for CM were obtained using all single and double excitations from 2–4 main configurations. Five of the six highest occupied MOs correspond more or less closely to the five highest occupied MOs of H_2CO, as shown in a figure (no list of orbital energies is given). The sixth ($8a_1$) involves mainly CH bonding, as does also $4b_2$; charge density contours for these and for the largely O atom, nonbonding "lone-pair" (n) MO $5b_2$ are shown. Contour diagrams for localized MOs are also shown for the O atom lone-pair and C—C and C—O σ bond MOs. The vertical excitation energies of the excited, including Rydberg, states are listed, together with oscillator strengths using both dipole and velocity operators for spectroscopically allowed transitions to them. Comparisons with experiment, including Rydberg transitions, show good agreement. They strongly resemble those for H_2CO. Dipole moments are computed for the ground, $^3(n,\pi^*)$, $^1(n,\pi^*)$, $^3(n,3s)$, and $^1(n,3s)$ states, where n, π^*, and 3s are $5b_2$, $4b_1$, and $9a_1$.

Bruna *et al.* [150], using a double-zeta basis augmented by bond-type polarization, sulfur 3d, and Rydberg-type functions, made a similar study of thioacetone ($(CH_3)_2CS$). A core of 11 MOs (K and L shells and the 3 lowest valence MOs) are kept doubly occupied in the CM. Vertical excitation energies of 12 excited states and of 2 states of the positive ion are given. Comparisons of the energy levels and ground-state dipole moments with those of acetone and of H_2CS are made.

Ha and Keller [151] have used a near double-zeta quality unpolarized Gaussian-lobe basis for SCF calculations plus moderate CM on acetaldehyde (CH_3CHO) and several related molecules (NH_2CHO, HOCHO, and FCHO). Experimental equilibrium geometries were used, with C_s symmetry presumably. The E_h values obtained by CM were -152.8916, -168.8851, -188.7119, and -212.7298 for the respective molecules. Dipole moments were computed (e.g., 3.17 D versus 2.69 D experimental for CH_3CHO), also quadrupole coupling constants for deuterium and diamagnetic contributions to nuclear shielding tensors for protons of the carbonyl group. Using CM, excitation energies and dipole moments for the first singlet and triplet excitations are reported (e.g., 3.94 eV for the triplet and 4.27 eV for the singlet for CH_3HCO); the computed values of the dipole moments considerably exceed the experimental values. Some population analyses are given.

In an SCF computation using an unpolarized basis, Davidson and Allen [152] obtain 1.16 kcal/mole for the increased energy required to rotate the CH_3 group from a conformation in which one methyl H atom eclipses the C=O bond to one in which it eclipses the aldehyde hydrogen. The result is compared with rotational barriers in ethane and propene and their halogenated derivatives and with acetyl halides. The three-fold barrier in acetaldehyde, unlike those in most molecules, is dominated by attractive forces between nonbonded atoms.

Hehre and Lathan [153] made 4-31G calculations that indicate the energy of the keto → enol transformation (12.9 kcal/mole in formaldehyde) is reduced in α substitutions of various groups whereby H of formaldehyde is replaced by CH_3, F, OH, or NH_2.

Vinyl alcohol ($H_2C{=}CHOH$) is the enol isomer of acetaldehyde. Bouma and Radom [154] in a 4-31G calculation optimize the geometry and find the *syn* form to be 2.2 kcal/mole lower in energy than the *anti* form.

The peroxyformyl radical (29) should be capable of exothermal forma-

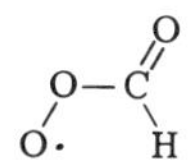

29

tion from O_2 + HCO. Winter *et al.* [155] have made calculations using a rather small basis set on its $^2A''$ ground and $^2A'$ low excited states. Both could be formed exothermally. Other reactions are also considered.

Jensen *et al.* [156] made SCF calculations on three C_s symmetry conformers of glycolaldehyde of which (30) is experimentally the most stable;

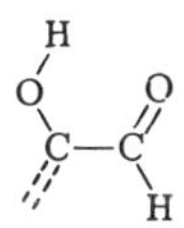

30

they used an unpolarized basis, except for p functions on the hydroxyl hydrogen in the stable conformer. They performed extensive geometrical optimalizations on the three conformers; results are given. They were interested in hydrogen bonding in the first conformer, but the results were not very instructive. One conformer was found to be less stable than the other two by 5 kcal/mole.

4. Glyoxal, Methylglyoxal, and Biacetyl (Dimethylglyoxal)

Glyoxal [$(CHO)_2$] is known experimentally in its stabler trans form (31)

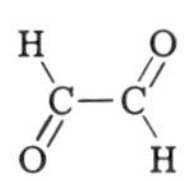

31

of C_{2h} symmetry and its cis form of C_{2v} symmetry. Dykstra and Schaefer [157] have used an unpolarized double-zeta quality basis set for SCF computations on the ground state of both forms and intermediate forms. Starting from the trans form (0°), a barrier of 7.5 kcal/mole is found, peaking at a twist angle of 102°; the *cis* form (180°) is 5.9 kcal/mole up. For the trans form, $E_h = -226.5182$ is found. The dipole moment of the trans form is, of course, zero, while for the cis form, 4.79 D is computed (experiment: 4.8 D). Orbital energies are given (Table IV in [157]) for the last six MOs in the ground state:

$$\ldots (5b_u)^2 (6a_g)^2 (1a_u)^2 (6b_u)^2 (1b_g)^2 (7a_g)^2, \ {}^1A_g$$

The MOs, in the order given, are also designated as σ, σ, π_a, n_b, π_b, and n_a (π^* is $2a_u$). The MOs are schematically pictured. A population analysis shows considerable 2p carbon (antibonding) in n_a but very little in n_b; both are mainly oxygen 2p. The formation of glyoxal from two CO plus H_2 is discussed.

In a second article, Dykstra and Schaefer [158], for both *trans*- and *cis*-glyoxal, compute vertical excitation energies for 10 triplet and 10 singlet states corresponding to all $\pi^* \leftarrow n$, $\pi^* \leftarrow \pi$, and $\sigma^* \leftarrow n$ ($7a_g$) excitations. Dipole moments are computed for the excited states of the cis isomer, assuming vertical excitations. Relations to $(CO)_2$ are discussed.

In a later article, Dykstra *et al.* [159] introduce CM (about 6000 configurations) in calculations on the lowest three excited states. Both vertical and adiabatic excitation energies were computed, using excited geometries previously computed [157] for the $\ldots 7a_g 2a_u$, ${}^3A_u$ ($n_a\pi^*_a$) and $\ldots 1b_g 2a_u$, ${}^3B_u$ ($n_b\pi^*_a$) states, and a newly determined geometry for the $\ldots 6b_u 2a_u$, ${}^3B_g$ ($n_b\pi^*_a$) state. The predicted adiabatic excitation energies T_e are 21,800 (${}^3A_u$), 25,100 (${}^3B_u$), and 29,300 cm^{-1} (${}^3B_g$). Observed states are ${}^3A_u$ at 19,200 and ${}^1A_u$ at 21,900 cm^{-1}. After allowance for d polarization (not included in the basis set) and higher than double CM excitations (unlinked clusters), it is concluded that the excitation energy of the ${}^3B_u$ state is dramatically increased by correlation effects but those for the ${}^3A_u$ and ${}^3B_g$ states are markedly decreased. The four main CM components for the ground and three excited states are listed; in all, there is one main configuration.

In an STO-3G plus modest CM study, Mueller *et al.* [160] again find ${}^3A_u$ to be below ${}^3B_u$.

Using a many-body Green's function method, von Niessen [161] has computed the vertical valence ionization potentials of *trans*- and *cis*-glyoxal. His SCF $E_h = -226.5460$ is somewhat lower than that of Dykstra and Schaefer and there is a slight reversal of the $1b_g$ and $6b_u$ MOs (they are close together in both computations). The author lists all orbital energies for both isomers. Von Niessen uses a polarized basis set unlike Dykstra and Schaefer and a somewhat different geometry. The ionization potentials found are considerably lower (about 1.5 eV) than Koopmans' theorem values: for $7a_g$, $6b_u$, $1b_g$, $1a_u$, $6a_g$, $5b_u$, $5a_g$, and $4b_u$ of *trans*-glyoxal, they are 10.52, 12.39, 14.08, 15.49, 15.79, 17.00, 17.50, and 20.02 eV, respectively. Agreement with experiment is very satisfactory.

Using an unpolarized double-zeta basis set, Nitzche and Davidson [162] calculated vertical transition energies by SCF and CM (single and double substitution) for the first three excited triplet states of glyoxal. They make two sets of calculations: (a) assuming C_s symmetry corresponding to localization of the n MOs on the O atoms and the π MOs in the two C=O bonds (n_a, $n_b = n_1 \pm n_2$; π_a, $\pi_b = \pi_1(CO) \pm \pi_2(CO)$, with a smaller interaction); (b) assuming the overall C_{2h} symmetry of the molecule. In (a), ^3A_u and ^3B_g are degenerate by SCF and also by CM; more complete CM should, however, remove the degeneracy to produce correct wave functions of C_{2h} symmetry. In (b), the CM energies seem to be reasonable, although the SCF energies are not. Note that a similar discussion of localization effects would be applicable to *ionization* from the n and π MOs. However, von Niessen's procedure has perhaps introduced enough CM to take care of this.

In addition to glyoxal, the authors have discussed the $^3n\pi^*$ triplet states of methylglyoxal in a similar manner.

Ha [163], using a moderate unpolarized basis set, has made SCF and CM (up to 387 configurations) calculations on the ground and eight $n\pi^*$ vertically excited, singlet and triplet, states of biacetyl. Ground-state (C_{2h} framework) geometry is used. A table of orbital energies is given. Major CM coefficients are listed in some detail. The computed excitation energies are compared with experimental evidence.

Tyrrell [164] has used an STO-4G basis set in a computation of energies for biacetyl as a function of the OCCO dihedral angle as the molecule is rotated from trans to cis.

5. Acrolein and Oxetene

Acrolein (32) is isoelectronic with glyoxal and butadiene,

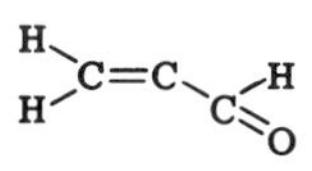

32

and has C_s symmetry. The electron configuration is $\ldots (12a')^2 (1a'')^2 (13a')^2 (2a'')^2 (3a'')^0$; 13a′ is n (on O), 1a″ and 2a″ are π, 3a″ is π^*. Dykstra [165], using an unpolarized double-zeta basis set, made SCF calculations on the ground and $n\pi^*$ and $\pi\pi^*$ singlet and triplet excited states, and gave vertical and optimized geometries and excitation energies, for both cis and trans. The trans isomer is found to be 540 cm^{-1} lower in energy than the cis form in the ground state. The C—O bond lengths are much increased in the excited states. Iwata and Morokuma [166] made somewhat similar calculations. Davidson and Nitzche [167] give a revision of Dykstra's calculation of the vertical energy of the $^1\pi\pi^*$ state, using variationally correct nonorthogonal SCF procedures, making it much higher (7.5 eV excitation energy without Rydberg admixture; observed: 6.5 eV).

With a double-zeta basis set and single- and double-substitution CM (about 6000 determinants), Lucchese *et al.* [168] have calculated vertical and adiabatic excitation energies for the $n\pi^*$, $^3A''$ state (vertical: 27,200, adiabatic: 21,700 cm^{-1}) and the $\pi\pi^*, {}^3A'$ state (vertical: 32,000, adiabatic: 23,100 cm^{-1}). A 1000 cm^{-1} correction for unlinked clusters for the $^3A''$ state gives good agreement with experiment (24,300 cm^{-1}).

van der Meer and Mulder [169] used a 4-31G basis in mapping the electrocyclic reaction connecting acrolein to the oxetene ring system. The transition state involves an 18° torsion around the central bond and an activation energy of 3.4 eV. Oxetene is 43 kcal/mole endothermic to acrolein. For its ring opening, the activation energy is 1.0 eV.

6. Ketene ($H_2C{=}C{=}O$) and Its Isomers

The most accurate calculation on the ground state of ketene and of its higher energy isomers oxirene (33), and ethynal (34) is one by Dykstra

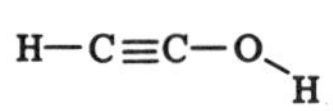

34

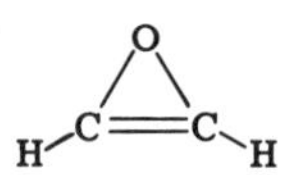

33

[170]. The isomers are possible reaction intermediates. He used a double-zeta plus polarization basis and very extensive CM (keeping only C and O K shells doubly occupied) by the SCEP (self-consistent electron-pairs) method. He tabulates SCF orbital energies and total energies for all three isomers, along with optimized geometries. For ketene, $E_h = -151.6724$ for SCF and -152.1535 for CM. Ethynal lies 35 kcal/mole higher, and oxirene 87 kcal/mole higher than ketene.

Strausz *et al.* [171] have used double-zeta quality SCF calculations to optimize the geometries and compute the geometries and energies of the lowest singlet states of ethynal and of two other isomers, oxiranylidene (36) and formylmethylene (HC̈—CHO) of ketene. They discuss

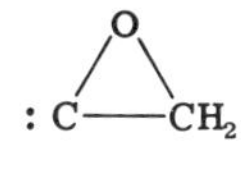

35

reaction paths among the isomers and compute the activation barriers between them. Oxiranylidene is computed to be 75 kcal/mole and formylmethylene 74 kcal/mole above ketene in energy.

Several authors have discussed the excited states of ketene. In SCF calculations using a double-zeta plus polarization (DZP) or plus Rydberg (DZR) basis, Dykstra and Schaefer [172] obtain vertical excitation energies and dipole moments for 18 excited states. The computed dipole moments cover a wide range from -4.24 D (C^+O^-) to $+2.55$ D (C^-O^+); the value for the ground state is -1.81 D. The ground state is $\ldots (7a_1)^2 (1b_2)^2 (1b_1)^2 (2b_2)^2 (2b_1)^2$, $^1A_1$. Optimization of the geometry leads to large decreases in energy and changes in shape. The first three excited singlet and triplet states have minimum energy for C_s symmetry. Lowest is $3b_2 \leftarrow 2b_1$, $^{1,3}A_2$ in C_{2v} symmetry, which become $10a' \leftarrow 2a''$ with geometry (36a) at excitation energies 13,940 and 11,710 cm^{-1}. Next are $3b_1 \leftarrow 2b_1$, $^{1,3}A_1$, which become $10a' \leftarrow 9a'$, $^{1,3}A'$ with geometry (36b) at 18,070 and 15,220 cm^{-1}. Then come $3b_2 \leftarrow 2b_2$, $^{3,1}A_1$, $3a'' \leftarrow 2a''$, $^{3,1}A'$ with conformation (36c) at 24,360 and 23,020 cm^{-1}. Correspondences with

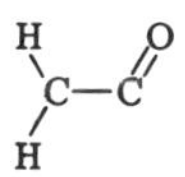

36a

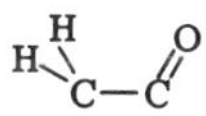

36b

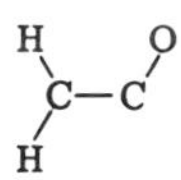

36c

the spectrum are discussed.

Predicted next above the A states just surveyed are the $8a_1 \leftarrow 2b_1$, $^{1,3}B_1$ states. Here the DZR calculation gives much lower excitation energies than the DZP, indicating that these states (near 44,000 cm^{-1} by DZR) have strong Rydberg character, as concluded earlier in a calculation by Basch [173]. This result is to be attributed to the 3s character of the $8a_1$ MO. Basch was interested in the photodecomposition of ketene to CH_2 + CO. As pointed out by Basch and verified by Dykstra and Schaefer, the $^3B_1$ and $^1B_1$ states can dissociate directly to CH_2 + CO without a barrier.

Above $^3B_1$ and $^1B_1$ are five more triplet and five more singlet states, the highest of which involve $8a_1$ excitation and again show Rydberg character. First, however, are $3b_2 \leftarrow 7a_1$, $^{1,3}B_2$, which mainly show a lengthening of the C—O bond from 1.17 to 1.38 Å. Experimentally, the high-frequency part of the spectrum shows a long ns $\leftarrow \pi$ Rydberg series, of which the transition to the $^1B_1$ state discussed above no doubt corresponds to the first member.

In a GVB plus limited CM calculation, Harding and Goddard [174] obtained wave functions, vertical excitation energies, and dipole moments for the ground and three lowest singlet and triplet excited states. The ground-state dipole moment (1.62 D) agrees fairly well with experiment (1.41 D). The vertical excitation energies for $^1A_2$ (3.69 eV) and the lower $^3A_1$ (5.39 eV) agree well with electron impact values (3.8 and 5.35 eV).

7. Carbonic Acid (H_2CO_3)

Using a double-zeta quality unpolarized (except for H atoms) basis followed by CM using all single and double excitations except for K-shell and corresponding virtual MOs, Jönsson *et al.* [175] have calculated the ground state of H_2CO_3, with geometry optimizations. The ground state is slightly more stable in a C_{2v} than in a C_s planar conformation, but with dipole moment 0.01 D in the former and 3.6 D in the latter, so that in water solution the polar form should be preferred. However, as confirmed by thermodynamic calculations, very little H_2CO_3 is present in equilibrium with CO_2 plus H_2O. Computed geometry for the C_{2v} form: $R(\text{C=O}) = 1.211$ Å, $R(\text{C—O}) = 1.338$ Å, $R(\text{O—H}) = 0.943$ Å, O=C—O angle 124.6°, C—O—H angle 108.8°. A population analysis gave a charge distribution $(H^{0.22})_2C^{1.00}(O^{-0.55})_2O^{-0.45}$.

8. Peroxymethylene (H_2COO)

In two articles relevant to the mechanism of ozonolysis, Wadt and Goddard [176] and Harding and Goddard [177] have carried out GVB plus moderate CM calculations on (37). They used a double-zeta basis set, plus

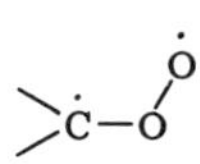

37

d polarization in the case of the second article. The ground state of this Criegee reaction intermediate is found to be a planar singlet biradical. Above the ground state of peroxymethylene is the corresponding triplet state.

Lower than the ground state is a ring state, dioxirane (38). If now the

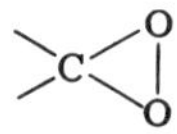

38

O—O bond is broken (dioxymethylene) there are several singlet and triplet states whose energies are computed. Comparisons with the isoelectronic molecule ozone are made. The discussion of reaction mechanisms is in great detail.

9. Formate Anion, and Formic, Protonated Formic, Acetic, and Thioformic Acids

Basch *et al.* [178], using a double-zeta plus Rydberg 3s basis, have made an SCF computation on formic acid (39a) and its anion (39b) ob-

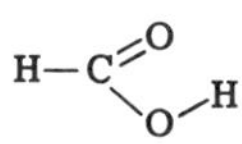

39a

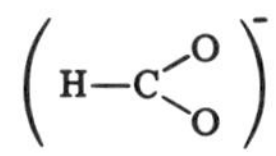

39b

taining $E_h = -188.6877$ for the former and -188.1261 for the latter. They also computed E_h for formyl fluoride. They give orbital energies for all three. In addition, they computed excitation energies to several singlet and triplet excited states: for the lowest of these, 6.9 eV for all three, agreeing fairly well with experimental values at 5.65 eV.

Peyerimhoff [179] has made similar computations on the formate ion ($E_h = -188.1132$, or with very limited CM, -188.1256). She has calculated 14 excited states by SCF plus CM. She makes detailed comparison with states of the isoelectronic molecule O_3 (also NO_2^-), using Mulliken–Walsh diagrams of the MOs. For HCOOH, in a similar computation Peyerimhoff and Buenker [180] obtain $E_h = -188.6552$, and compute 12 excited states and compare with experiment. They also discuss the formate ion in the field of a unit positive charge in various positions and give charge-density contours. In addition, they compute energies, dipole moment, and quadrupole moment tensors for each of several conformations of HCOOH.

In an SCF calculation ($E_h = -188.4962$) Schwartz *et al.* [181] computed the energies for HCOOH in three geometries; the ground state, cis, a trans form 8.1 kcal/mole higher, and an intermediate barrier form with the hydroxyl hydrogen twisted up 90° (13.0 kcal/mole). Atomic and overlap populations and some other details are given. In a minimal-basis calculation, Peterson and Csizmadia [182] give further details on the conformational hypersurface, including a tautomerization barrier of 59 kcal/mole and an in-plane inversion barrier for syn to anti of 54 kcal/mole.

A computation on four valence and very numerous Rydberg states of CHOOH by Demoulin [182a] (ground-state $E_h = -188.4571$) should be mentioned. He used a moderate unpolarized basis set of 83 Gaussian lobes contracted to 34, plus Rydberg functions, and made some empirical adjustments. He gives $\langle X^2 \rangle$, $\langle Y^2 \rangle$, and $\langle Z^2 \rangle$ values and makes extensive comparisons with relatively recent spectroscopic data.

Flood [183] makes SCF calculations on the formic acid monomer ($E_h = -188.4761$) and dimer. This work will be considered in a later chapter in connection with hydrogen bonding.

Hopkinson *et al.* [184] have made SCF calculations on formic acid ($E_h = -188.2635$), the formate ion (-187.7324), and protonated formic acid. For the last-named, the three conformers (cis–trans, cis–cis, and trans–trans, with cis–trans the most stable by 6–7 kcal/mole), with the proton added to the carbonyl oxygen, were found to be more stable by 25 kcal/mole for the last than for the conformer with the proton added to the hydroxyl oxygen. Orbital energies for all the conformers are listed.

With a 7s 3p 1d basis for O and C contracted to 4s 2p 1d and a 10s 6p 1d basis for S contracted to 6s 4p 1d, and 4s contracted to 2s for H, Auberg *et al.* [184a] have made SCF calculations on syn and anti forms of monothioformic acid (O=CH=SH) and dithioformic acid (S=CH—SH). The syn forms are calculated to be the more stable, with barrier heights for syn → anti of 40 and 50 kJ/mole for the mono-and dithio molecules. Computed total dipole moments are 1.78 D for syn and 3.51 D for anti for the monothio acids, 1.44 and 2.99 D for the dithio acids. In the former case, agreement with experiment is fairly good.

Using a double-zeta basis set of Gaussian lobes, Meyer *et al.* [184b] computed the barrier to tunneling of the carboxyl proton in acetic acid (CH_3COOH) as 97 kcal/mole. The cis conformation (of the hydroxyl hydrogen toward the other O) is computed to be 11 kcal/mole lower than the trans, with a barrier 15 kcal/mole near 90° in rotation toward the trans (both COOH conformations are planar). The CH_3 torsional barrier is computed as 0.9 kcal/mole. The orbital energies are tabulated. The infrared frequencies are calculated *in extenso* and compared with experiment, and anharmonicities are interpreted as due largely to Fermi resonances.

10. Protonated Carbon Dioxide

Using a polarized double-zeta basis, Green *et al.* [184c] in an SCF calculation followed by CM (all singles and doubles except for 1s substitutions) showed that in its stable state CO_2 is protonated at the O atom with a probably linear O—C—O arrangement and the H somewhat off the axis. The calculation was made with the hope of finding interstellar CO_2H^+ but preliminary results are negative.

G. CONJUGATED POLYENES

1. 1,3-Butadiene and 1,3,5-Hexatriene

The simplest conjugated polyene is 1,3-butadiene (40) which is stablest

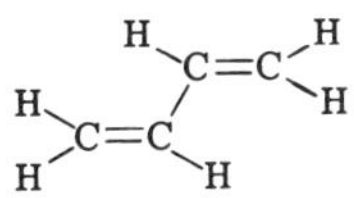

40

in the C_{2h} symmetry trans form. For the four π electrons, the MOs, with χ_1 and χ_2 doubly occupied, are then

$$\chi_1 = c_1\pi_1 + c_2\pi_2 + c_2\pi_3 + c_1\pi_4 \; (c_2 > c_1), \; 1a_u$$

$$\chi_2 = c_1\pi_1 + c_2\pi_2 - c_2\pi_3 - c_1\pi_4 \; (c_1 > c_2), \; 1b_g$$

$$\chi_3 = c_1\pi_1 - c_2\pi_2 - c_2\pi_3 + c_1\pi_4 \; (c_1 > c_2), \; 2a_u$$

$$\chi_4 = c_1\pi_1 - c_2\pi + c_2\pi_3 - c_1\pi_4 \; (c_2 > c_1), \; 2b_g$$

In nearly all the *ab initio* calculations the σ core has been frozen, i.e., with all σ (a_g and b_u) MOs kept doubly occupied. Attention was directed mainly to an understanding of the observed optical and electron impact [185] spectra, which almost exclusively involve excitation of the π electrons. Calculations have been made on both trans and cis (or gauche?) forms. One such calculation, on the trans isomer only, has been made by Hosteny *et al.* [186]. They used a double-zeta basis set plus diffuse s and p functions to deal with Rydberg character. Like others, they first made an all-electron SCF calculation on the ground state ($E_h = -154.8771$) to obtain the σ core. They list orbital energies and give contour maps of the occupied π MOs as obtained by SCF; these are reproduced in Fig. 3.

With the core fixed, they made CM computations with successively single, double, triple, and quadruple π CM substitutions; the final CM result was $E_h = -154.94015$. With single-substitution wave functions, they obtained NOs (natural orbitals) and their occupation numbers. They then proceeded to double substitutions etc. Vertical excitation energies are

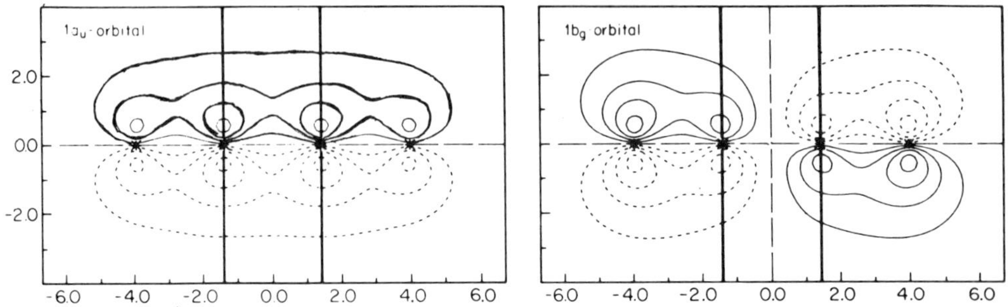

FIG. 3. The SCF π MOs of *trans*-butadiene. The contours are given in the three planes, perpendicular to the plane of the molecule, which contain the three C—C bonds. Distances are in bohrs. From Hosteny *et al.* [186].

given for single excitations to ^3B_g, ^3A_g, ^1B_u, and ^1A_g states with MO excitation from $1a_u$ and from $1b_g$ up to as high as $5a_u$ and $4b_g$ (making use of higher roots of secular equations for the higher states). States 2^1A_g, 3^3B_u, and 1^5A_g involving double-MO excitation were also computed. Values of $\langle x^2\rangle$, $\langle y^2\rangle$, and $\langle z^2\rangle$ for na_u and nb_g MOs of all singly excited states are reported (axes: z perpendicular to the molecule plane, y parallel to the two C=C bonds). Occupation numbers and $\langle z^2\rangle$ values for the $1a_u$, $1b_g$, $2a_u$, and $2b_g$ NOs as obtained at the double-substitution level of the CM computation are also reported.

The lowest excited states are as follows: computed, 1^3B_u at 3.45 eV and 1^3A_g at 5.04 eV; observed by electron impact, 3.22 and 4.91 eV, respectively. Most of the higher states have partial or essentially complete Rydberg character. As seen from the NO$\langle z^2\rangle$ values, only the ground and doubly excited 2^1A_g state, three triplet states (1^3B_u, 1^3A_g, and the doubly excited 3^3B_u), and the doubly excited 1^5A_g have valence-like charge distributions. Comparisons of the butadiene wave functions with those of two ethylene molecules (one or both excited, singlet or triplet) are made. Corresponding to the $\pi\pi^*$, $^1B_{1u}$ or V state of ethylene, four V states of butadiene (V_1, V_2, V_3, V_4) were predicted by Mulliken in 1939, with the optical transition to V_1 very intense, that to V_4 much weaker, those to V_2 and V_3 forbidden in *trans*-butadiene, that to V_1 being observed as a broad intense band with peak at 5.9 eV.

In an earlier calculation with a more restricted AO basis set (E^h = -153.710 for SCF and -154.7723 for CM); Buenker and Whitten [187] discussed the ground and non-Rydberg excited states of *trans*- and *cis*-1,3-butadiene. A $\sigma\pi^*$ excited state ($2a_u \leftarrow 7a_g$), ^1A_u was also considered. Shih *et al.* [188], with E_h = -154.8573 for SCF and -154.9181 for CM for *trans*-butadiene, went further by including diffuse functions (some of σ type) so as to predict Rydberg states, and in many cases using excited state SCF MOs in CM; they made comparisons with experiment. They also made calculations on twisting of the CH_2 groups and considered a  structure for various states.

Buenker *et al.* [189] have given an all-valence-electron treatment of the low-lying valence and Rydberg states of *trans*-butadiene, using extensive CM in their MRD-CI method with an estimated error in the final extrapolations of 0.05–0.10 eV. For CM their final extrapolated $E_h = -155.1900$ (-154.8579 for SCF). Nearly all the computed energies agree closely with those of Shih *et al.*, and moderately well with those of Hosteny *et al.* A weak band at 7.06 eV can be identified with the forbidden transition to the 2^1A_g state. Most of the excited states to which optical transitions are allowed (oscillator strengths are computed) are Rydberg states to which transitions have been observed. One of these, at 6.67 eV computed (6.66 observed), is a ^1B_u state of type $3p\pi \leftarrow \chi_2$ (calculated $f = 0.07$ using the dipole, 0.03 using the velocity operator, observed medium strength). But this is superposed, at 6.72 eV computed, by a transition $3p\sigma \leftarrow \chi_2$ to 2^1A_u (computed $f = 0.05$ and 0.05).

It is puzzling that the computed vertical excitation energy of the $\chi_3 \leftarrow \chi_2$, ^1B_u state,[1] long since identified as the V_1 state, is at 7.67 eV with $F = 0.99$ or 0.41, whereas the observed transition peaks near 5.9 eV, with $f = 0.3$ in solution. The most plausible explanation seems to be that this state has a very different geometry than the ground state. In fact Shih *et al.* had shown that a $\chi_3 \leftarrow \chi_2$ excitation gives a structure that is stabler for  than for ground state  bonding. It seems also that the observed V_1 state must be largely a valence-shell rather than a Rydberg state, although it has some Rydberg character.

Although it is agreed that the trans form of 1,3-butadiene is the most stable, there has been much discussion as to whether the second less stable conformer is cis or gauche. The rotation barrier and energy of the second conformer seem to be rather uncertain. A critical analysis (references but no new *ab initio* calculations) by Bock *et al.* [190] seems to favor a gauche form, as do CM calculations by Dumbacher [191] and a 4-31G calculation by Devaquet *et al.* [192]. The last-named authors also discuss glyoxal, acrolein, and several derivatives. Recent spectroscopic evidence [193] indicates that *cis*-butadiene is planar or nearly so.

Cederbaum *et al.* [194], using a Green's function many-body method, predict the valence-electron photoelectron spectrum. For the outer elec-

[1] Note, however, that the computed state involves considerable CM, and is obtained as a higher root of a secular equation.

trons up to 19 eV, one main line for each, with hardly any weak satellites, is predicted. At higher energies up to 32 eV, (2s AO region of LCAOs), the structure breaks up into very numerous weak lines with only a few stronger lines. Diagrams indicating predicted intensities are shown for ionization of the valence-shell MOs ($3a_g$ to $7a_g$, $3b_u$ to $6b_u$, $1a_u$ and $1b_g$).

The concept of resonance energies in conjugated π-electron hydrocarbons has long been of interest to organic chemists, and many semiempirical calculations have been made. Kollmar [195] has calculated resonance energies for butadiene, benzene, naphthalene, azulene, and several other conjugated molecules, by taking the energy difference between an *ab initio* SCF energy and a model wave function in which the SCF π MOs are replaced by appropriate nonresonating localized π MOs (reference state). He used polarized double-zeta basis sets. Vertical and adiabatic resonance energies were obtained. For butadiene an adiabatic resonance energy of about 9 kcal/mole was obtained.

With a double-zeta basis set Hinchcliffe [196] has made unrestricted SCF calculations with high-spin annihilation, on the positive and negative ions of butadiene and on a triplet state, and calculated ESR proton hyperfine coupling constants. E_h values are also given.

Nascimento and Goddard [197], using double-zeta basis sets, have made SCF and GVB vertical calculations with CM on the valence excited states and the π-ionized positive ion states of *trans*-1,3-butadiene and all *trans*-1,3,5-hexatriene.

2. Cyclopentadiene and Related Molecules and Silacyclopentadiene

Cyclopentadiene (41) is similar to *cis*-1,3-butadiene but with an addi-

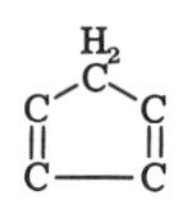

41

tional carbon atom carrying two additional π electrons, and with hyperconjugation in the b_1 MOs of the CH_2 group. The ground state is

$$\ldots (4a_1)^2 (3b_2)^2 (5a_1)^2 (6a_1)^2 (4b_2)^2 (7a_1)^2 (1b_1)^2 (5b_2)^2 (8a_1)^2$$
$$(9a_1)^2 (6b_2)^2 (2b_1)^2 (1a_2)^2, \ {}^1A_1$$

Von Niessen *et al.* [198] have used a Green's function many-electron approach to calculate the ionization potentials ($E_h = -192.6841$ for SCF of cyclopentadiene (C_5H_6) and of sila-cyclopentadiene ($C_4H_4SiH_2$). They list SCF orbital energies for both, also giving population analysis data (atomic populations and overlap populations for bonds). They also list SCF computed one-electron properties (dipole and higher moments, diamagnetic shielding and susceptibility, potential and field gradients at the nuclei, deuteron quadrupole coupling constants). Correlated ionization potentials are listed: for cyclopentadiene; they are 8.53 eV for $1a_2$ (π), 11.35 eV for $2b_1$ (π), 13.60 eV for $6b_2$, 13.90 eV for $9a_1$, 14.53 eV for $8a_1$, 15.45 eV for $5b_2$, 16.22 eV for $1b_1$ (π), 19.12 eV for $7a_1$, 19.76 eV for $4b_2$, 20.49 eV for $6a_1$. There is agreement with experiment of 0.3 eV or better for all but $4b_2$ and $7a_1$. Comparisons are made with several other (aromatic) conjugated dienes (see Chapter XI for these).

Popkie and Kaufman [199] have obtained similar but less accurate SCF orbital energies ($E_h = -192.1535$) using a 6-3G basis, also listing other approximate methods.

Conrad *et al.* [200], using a double-zeta basis set, make computations on the cyclopentadiyl radical (42) or Closs' diradical, of C_{2v} symmetry,

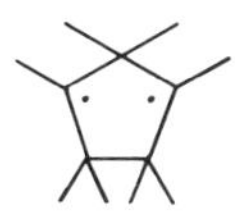

42

which is a higher energy isomer of [2,1,0] pentane. They find for it a ground state ... $2a_2\ 3b_1$, $^3B_2$ with a minimum about 35 kcal/mole above the singlet ground state of [2,1,0] pentane. The singlet state corresponding to this triplet state represents a potential maximum on a potential curve connecting right- and left-handed forms of the pentane. Optimized geometries for the diradical are given. Structural relationships to trimethylenethane are pointed out.

Borden and Davidson [201], using an STO-3G minimal SCF basis followed by CM, have made calculations on the ground and low excited states of the cyclopentadienyl radical and cation. Molecular orbital theory for the radical predicts a π^3, $^2E_1''$ state of D_{5h} symmetry. However, the Jahn–Teller theorem then predicts distortions in an e_2' mode to lower symmetry, but pseudorotation should result. For the cation, three π^2 states are predicted: $^3A_2'$, $^1A_1'$, and $^1E_2'$, with $^3A_2'$ as the ground state, with D_{5h} symmetry. Jahn-Teller distortion of the $^1E_2'$ and pseudorotation as in the radical are predicted, but the $^1A_1'$ state is also involved. In the radical, the energy lowering on Jahn–Teller distortion is calculated as 7 kcal/mole, and in the singlet state of the cation, 13 kcal/mole. The lowest singlet state of the cation is found to be only 7 kcal/mole above the triplet state. Further details are given.

H. MISCELLANEOUS MOLECULES

1. Ethylene Derivatives ($Li_2C_2H_2$, Fluorethylene, Halonium, Cations, Oxirane, Thiorane, and Their Protonation Products, Silylethylene, and Silylsilylene)

Apeloig *et al.* [202] have made STO-3G and unrestricted 4-31G calculations plus very limited CM that show 1,1-dilithioethylene to have remarkable characteristics as compared with ethylene: almost equal energies for planar and 90° twisted molecule (C_{2v} symmetry for both) and for both a distinctly lower energy for a triplet ($^3B_2$ for the planar and $^3B_1$ for the perpendicular form) than for the singlet ($^1A_1$) state. Two-way charge transfer and hyperconjugation with the Li atoms are very strong. The C—C bond has double-bonding length for both planar and perpendicular forms.

Laidig and Schaefer [203] have confirmed the conclusions of Apeloig *et al.* and added further details. They used a basis set of nearly double-zeta quality with d polarization for the carbon atom in simple SCF computations, followed by extended CM (all singles and doubles, except for the 1s shells) and finally a Davidson correction for unlinked clusters. The relative energies found (in kcal/mole) were as follows: twisted $^3B_1$, 0.0; planar $^3B_2$, 1.4; twisted $^1A_1$, 14.0; planar $^1A_1$, 15.5. They optimized the geometries of all these at the SCF level, with bond distances to a precision of 0.001 Å and bond angles to 0.1°. The C—C distances (1.322–1.356 Å) were all close to the normal C=C value. The Li—Li distances in the triplet states were somewhat *less* than in Li_2 (hence the LiCLi angles were only about 75°), suggesting a Li—Li bonded structure complexed to a CH_2 radical. In the singlet states, the Li—Li distances and LiCLi angles were much greater. Orbital energies and population analyses are given for all four states. Dipole moments were computed (about 5.2 D for the singlet; −0.6 and −1.35 for the planar and twisted triplet states, respectively).

Whangbo *et al.* [204] have made STO-3G minimal-basis calculations on the three isomeric difluoroethylenes, also on the fluoropropenes. The 1,1 isomer of difluoroethylene is correctly predicted to be the most stable, but the trans isomer is incorrectly predicted to be more stable than the cis. Similar results have been obtained on the difluoroethylenes by Bock *et al.* [205] using rather good basis sets. They compute total energies, charge populations, and dipole moments also for CH_2=CHF. Various disproportionation reactions are discussed.

Using a small unpolarized Gaussian-lobe basis set followed by rather extensive CM (about 0.17 a.u. energy decrease), Ha [205a] has calculated the energies of seven singlet and four triplet states of nitroethylene. He gives transition energies, and for the lower singlet states, dipole moments and transition oscillator strengths. Three spectroscopically observed transitions are identified: an $n \rightarrow \pi^*$ ($^1A'' \leftarrow {}^1A'$) with energy 4.42 eV (observed: 4.20) and $f = 0.004$ (observed: 0.002); two $\pi \rightarrow \pi^*$ ($^1A' \leftarrow {}^1A'$) at 5.48 and 6.80 eV (observed: 5.12 and 6.12) with $f = 0.005$ and 0.28 (observed: 0.01 and 0.3).

Hopkinson *et al.* [206], using a 4-31G and a double-zeta basis set, have made SCF calculations on the ethylene chloronium ion (43) and its 2-

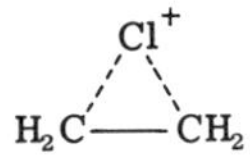

43

chlorethyl isomer ($CH_2X—C^+H_2$). They find the bridged chloronium ion to be the more stable by 9.35 kcal/mole, with a barrier of 6.25 kcal/mole between them. Orbital energies and atomic populations are given. The results contrast with those of their earlier calculations on the corresponding fluorine compounds, in which they found the 2-fluoroethyl cation to be 10 kcal/mole more stable than the bridged ion, with an activation energy of 19 kcal/mole for the latter.

Using good polarized basis sets and the CEPA PNO scheme for correlation, Lischka and Köhler [206a] have computed energies for oxirane C_2H_4O and thiirane C_2H_4S and three isomers of their protonated ions. For $C_2H_4OH^+$, they find protonated acetaldehyde to be the stablest isomer, with the oxiranium ion (44) 29 kcal/mole higher in energy and β-

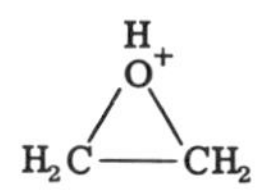

44

hydroxyethyl ion 41 kcal/mole higher. For $C_2H_4SH^+$, they find protonated thioacetaldehyde to be the stablest, with the thiiranium ion 5 kcal/mole higher and the β-mercaptoethyl ion 45 kcal/mole higher.

Ahlrichs and Heinzmann [207], using good polarized basis sets, have made MCSCF calculations (by including $\pi^2 \rightarrow \pi^{*2}$ CM) on the planar singlet ground state and SCF calculations on the "perpendicular" triplet first excited state of silaethylene ($H_2C{=}SiH_2$). They report the triplet, with the two H of SiH_2 first twisted by 90° relative to the CH_2, in addition to being "flapped" by 45° away from the SiC axis. The adiabatic excitation energy is calculated as 28 kcal/mole, as compared with a 90° twisting and 35° flapping barrier of 45 kcal/mole for the singlet state. The vertical energy of the triplet above the planar singlet ground state is calculated as

61 kcal/mole; thus the energy of the triplet state gets lower by 33 kcal/mole on 90° twisting and 45° flapping. Using their improved method of population analysis, the authors find a charge distribution of $(H^{0.05})_2C^{-0.4}Si^{0.5}(H^{-0.1})_2$ in the ground state. The authors also make computations on the cycloaddition dimerization of silaethylene to 1,3-disylacyclobutane; they find a barrier of less than 14 kcal/mole and a reaction energy of about −76 kcal/mole.

With a slightly larger basis set and CM (all singles and doubles except for inner shells) and after including a Davidson correction, Hood and Schaefer [208] place the adiabatic triplet 38 ± 5 kcal/mole above the ground state, with a flap angle of 40° (relative to a perpendicular orientation of the SiH_2). They also give other slightly modified bond lengths and angles.

Other work has shown a need to consider two isomers of silaethylene: methylsilylene $\left(H\ddot{S}i\diagdown CH_3\right)$ and silylmethylene $\left(H_3Si{-}\ddot{C}\diagdown H\right)$. Schaefer [1] reports that in a polarized double-zeta computation plus CM he and John Goddard found methylsilylene to be practically degenerate (computed: 0.08kcal/mole below) with silaethylene, with H_3SiCH 69 kcal/mole higher in energy. With a minimum basis set, Gosavi *et al.* [209] have mapped the reaction surface for the interconversion of the *triplet* states of the three isomers, finding large barriers between them.

In an SCF calculation using a 4-31G basis, Snyder and Wasserman [209a] conclude that the ground state of Si_2H_4 is a singlet state of silylsilylene ($H_3Si{-}\ddot{S}iH$), not disilene ($H_2Si{=}SiH_2$). The authors find a triplet state to be lowest, but estimate that if d functions had been included in the basis, the singlet would be 11 kcal/mole below the triplet. The planar disilene is computed to be 8.6 kcal/mole above the singlet silylsilylene. Electron correlation is not expected to change the relative energy orderings.

2. The Oxolanes and Cyclopentane

Cremer and Pople [209b], using a 4-31G basis with estimates of polarization effects, have studied pseudorotation in cyclopentane, oxolane (45), 1,3-dioxolane, and cyclopentanone. During pseudorotation an out-

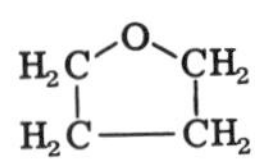

45

of-plane displacement of one atom in the roughly planar molecule circulates around the ring so that the molecule passes alternately through C_s and C_2 symmetry in the case of cyclopentane. Detailed results are presented.

Cremer [210], using basis sets up to a double-zeta plus polarization (d for carbon and p for hydrogen), has made extensive SCF calculations on the conformations of 1,2,3-trioxolane and 1,2,4-trioxolane. He finds the stablest conformations to be

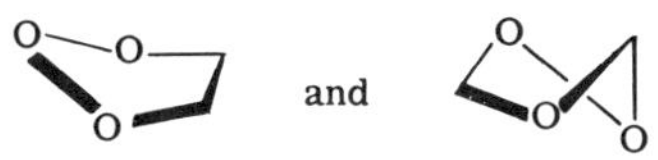

with, respectively, C_s and C_2 symmetry. In following articles he discusses methyl-substituted derivatives.

3. $H_2C{=}Be$, CH_2Li^-, CH_2LiF, CH_2Li_2, and AlC_2H_2

Binkley *et al.* [211] have calculated the energy of the ground ...$(1b_2)^2\ 1b_1\ 5a_1$, ${}^3B_1$ and first excited ... $(1b_2)^2\ (1b_1)^2$, ${}^1A_1$ excited state of H_2CBe, using ordinary 6-31G* for the ${}^1A_1$ and unrestricted 6-31G* for the ${}^3B_1$ state followed in both cases by third-order Moeller–Plesset perturbation theory to obtain considerable amounts (0.09 h and 0.15 h) of correlation energy. Using these results, dissociation energies to CH_2 ${}^3B_1$ plus Be are calculated to be 62 kcal/mole for the ground state and 33 kcal/mole for the $H_2C{=}Be$ ${}^1A_1$ state.

Lamanna and Maestro [212], using a double-zeta basis, made an SCF calculation on the two lowest states of $H_2C{=}Be$ with similar results. They also calculated the corresponding states of H_2CLi^-, which they found unstable to dissociation.

Clark and Schleyer [213], using a 4-31G basis, have made SCF calculations to optimize the geometry of CH_2FLi. Three local minima were found, the lowest a peculiar umbrella-shaped structure with $CH_2Li^+F^-$ ion pair character.

Laidig and Schaefer [214] have made SCF calculations followed by CM (singles and doubles, plus Davidson corrections for linked clusters) on CH_2Li_2, and given SCF-optimized geometries. Lowest in energy is a tetrahedrally arranged singlet (${}^1A_1$) with 120° LiCLi angle, 2.1 kcal/mole higher is a tetrahedral triplet (${}^3B_2$) with 72° angle, 3.9 kcal/mole higher is a planar triplet (${}^3B_1$) with 74° angle, and 8.3 kcal/mole higher is a planar singlet (${}^1A_1$) with 102° angle. The calculated dipole moments are about 5.42 and 4.85 for the two singlets (C^-Li^+), -0.76 and -1.22 for the triplets (C^+Li^-).

Trenary *et al.* [215], using double-zeta SCF plus extended CM, have made calculations on three isomers of AlC_2H_2. They find the vinylidene complex (Al—CH_2) to be the most stable, with a binding energy of 22 kcal/mole versus Al + C_2H_2, and the cis and trans isomers

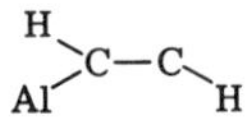

to be less stable by 8 kcal/mole. Experimentally, the vinylidene complex seems to be excluded, but this may be because of a barrier to the necessary 1,2-hydrogen shift.

4. Other Molecules

Using roughly 4-31G basis sets plus d functions for second-row atoms, Oberhammer and Boggs [216] in SCF computations have obtained fully optimized geometries for O=CHCl, O=CCl_2, O=CHF, O=CF_2, O=CFCl, S=CH_2, and S=CF_2, and computed dipole moments and charge densities.

Using basis sets similar to 4-31G in SCF computations, Burdon *et al.* [217] have obtained orbital energies (and compared them with those of benzene) for the cyclopentadienyl anion $C_6H_7^-$ and made some population analyses.

Burdon *et al.* [218] have made a similar study of monofluorocyclohexadienyl anions.

Mostly using an STO-3G minimal basis, Hayes *et al.* [219] have explored the benzene oxide oxepin isomerism and considered related molecules.

Using uncontracted Gaussian basis sets [9s, 5p, up to 1d for P; 5s, 2p for C; 3s for H), Absar and Van Wazer [220] have made SCF calculations for two conformations of the hypothetical phosphonium ylide CH_2PH_2. Electron populations and valence-shell orbital energies are given, along with electron density maps.

Using polarized double-zeta basis sets and the SCEP method for electron correlation, Dykstra *et al.* [221] have obtained geometries and energies for the cumulene carbenes $H_2C{=}(C{=})_nC{:}$ with $n = 0, 1, 2$; and their isomers with a C≡C triple bond. Dipole moments are computed.

With a double-zeta basis plus d orbitals, Gropen and Johansen [222] have made SCF calculations on hydroxyborane (BH_2OH). The molecule is found to be planar with a barrier of 16 kcal/mole for rotation of the H atom of the hydroxyl group. Both σ and π bonding of the OH are involved.

To assist interpretation of the $S(CN)_2$ photoelectron spectrum, Rosmus *et al.* [223] have made SCF computations with a good sp basis plus d on sulfur. They give valence-shell orbital energies and identify them with lone pairs on bonds. The SCN moieties are somewhat nonlinear. The computed dipole moment (3.16 D) agrees well with the experimental (3.02 D). Some orbital contour maps are given.

Using a small basis, Mezey *et al.* [224] have made SCF calculations on a simple sulphilimine (H_2SNH) model. Rotation of the NH hydrogen and pyramidal inversion at the sulphur are considered. The calculated inversion barrier is 24 kcal/mole.

With a minimal basis set, Burdon *et al.* [225] find CH_2F^- to be 13 kcal/mole more stable in the pyramidal than in the planar form, as compared with 1 kcal/mole for CH_3^-.

With an extended 4-31G basis, Olsen [226] has made SCF calculations on various conformations of $F_2C(OF)_2$. A gauche–trans form was found to be the most stable, but a gauche–gauche form is calculated as only 1 kcal/mole less stable. Comparisons with methanediol are made.

Vishveshwara [227] has made 4-31G basis SCF calculations on methanediol ($CH_2(OH)_2$), studying bond length variations in conformers.

Rosen and Fricke [228], assuming C_{4v} symmetry, have made relativistic calculations on UF_5.

Hay [229] has made computations on $Re_2Cl_8^{2-}$ using an effective core potential and a GVB description of the quadruple bond, with two configurations for each bond. Excitation energies for three excited states are computed, using some CM. Also two states of $Re_2Cl_8^{3-}$ are computed.

Rappe and Goddard [230] have made GVB and CM calculations on the low energy states of $NiCH_3$ (see Section VI,F) and $NiCH_2$. Geometries are optimized. The Ni atom is mostly 4s $3d^9$, the 4s combining in a σ bond with the radical. The ground state of $NiCH_2$ is $^1A_1$, with a charge distribution $Ni^{0.56}(CH_2)^{-0.56}$.

Guest *et al.* [231] used a minimal-basis set plus d on Cl to predict C_{4v} but very flexible structure for ClF_4^+ and D_{4h} (square planar) for ClF_5^-. Ungemach and Schaefer [232], using a double-zeta basis but without d on Cl, agree. They also predict D_{4h} for the ClF_4 radical. For the latter, So and Richards [233] with a minimal-basis set find D_{4h} for ClF_4 but C_{4v} on including 3d on Cl. Experiment favors C_{2v} for ClF_4^+, as in SF_4. It would seem that final conclusions should await more accurate calculations.

Otterson and Jensen [234], using a (7s 3p/4s) basis contracted to (4s 2p/2s) have made an SCF calculation on *s*-diformylhydrazine, which is planar; bond lengths are optimized and compared with values, also calculated, for hydrazine and dimethylhydrazine. Atomic populations and stretching force constants are given for diformylhydrazine.

Richartz *et al.* [235] have made a series of MRD-CI calculations on propane (in a C_{2v} conformer) relevant to its electronic spectrum. They list orbital energies, excitation energies, and seven ionization energies, and make comparisons with methane and ethane.

Pacansky and Dupuis [236], using a 4-31G basis, made calculations on propane in C_{2v} conformation, and for three conformations of the *n*-propyl radicals, which were found to be nearly equal in energy.

von Niessen and Diercksen [237] calculate the first 14 ionization potentials of norbornadiene by a many-body Green's function method, and relate these to the photoelectron spectrum. Most easily ionized are two π MOs at 8.56 and 9.58 eV.

Feller *et al.* [238], using a minimal-basis SCF (or two-term MCSCF for the single followed by very extensive CM, have computed potential surfaces for rotations of the OH groups in the lowest singlet and lowest triplet state of dihydroxycarbene ($C(OH)_2$). The ground state is planar and singlet.

Bouma *et al.* [239], using a polarized basis set 6-31G* have calculated the energies of 11 isomers of $C_2H_4O^+$ composition. The stablest is $CH_2{=}CHOH^+$, followed by CH_3CHO^+, and then $CH_3{-}C{=}OH^+$. All 11 are deemed low enough in energy to be good candidates for experimental observation.

REFERENCES

1. H. F. Schaefer, III, *Acc. Chem. Res.* **12,** 288 (1979).
2. J. Weber and A. D. McLean, *J. Am. Chem. Soc.* **98,** 875 (1976); J. Weber, M. Yoshimine, and A. D. McLean, *J. Chem. Phys.* **64,** 4159 (1976).
3. H. Lischka and H.-J. Köhler, *J. Am. Chem. Soc.* **100,** 5297 (1978).
4. P. Millie and G. Berthier, *Int. J. Quantum Chem., Symp.* **2,** 67 (1968).
5. Y. Ellinger, A. Rassat, R. Subra, and G. Berthier, *J. Chem. Phys.* **62,** 1 (1975).
6. J. E. Williams, Jr. and A. Streitwieser, Jr., *J. Am. Chem. Soc.* **97,** 2634 (1975).
7. C. E. Dykstra, A. J. Arduengo, and T. Fukunaga, *J. Am. Chem. Soc.* **100,** 6007 (1978).
8. J. H. Davis, W. A. Goddard, III, and L. R. Harding, *J. Am. Chem. Soc.* **99,** 2919 (1977).

8a. C. E. Dykstra and H. F. Schaefer, III, *J. Am. Chem. Soc.* **100,** 1378 (1978).

8b. M. P. Conrad and H. F. Schaefer, III, *J. Am. Chem. Soc.* **100,** 7820 (1978).

9. I. G. Csizmadia, V. Lucchini, and G. Modena, *Theor. Chim. Acta* **39,** 51 (1975).
10. I. G. Csizmadia, A. J. Duke, V. Lucchini, and G. Modena, *J.C.S. Perkins Trans. II* p. 1808 (1974).
11. D. G. Lister and P. Palmieri, *J. Mol. Struct.* **48,** 133 (1978).

11a. J. Kao, *J. Am. Chem. Soc.* **100,** 4685 (1978).

12. S. D. Peyerimhoff and R. J. Buenker, *J. Chem. Phys.* **51,** 2528 (1969).
13. L. Radom, P. C. Hariharan, J. A. Pople, and P. von R. Schleyer, *J. Am. Chem. Soc.* **95,** 6531 (1973).
14. H.-J. Köhler and H. Lischka, *J. Am. Chem. Soc.* **101,** 3479 (1979).

15. A. Hinchcliffe, *J. Mol. Struct.* **27,** 329 (1975).
16. G. Levin, W. A. Goddard, III, and D. L. Huestis, *Chem. Phys.* **4,** 409 (1974).
17. G. Levin and W. A. Goddard, III, *Theor. Chim. Acta* **37,** 253 (1975); *J. Am. Chem. Soc.* **97,** 1649 (1975).
18. J. Paldus and A. Veillard, *J. Mol. Phys.* **35,** 445 (1978).
19. H. Basch, M. B. Robin, N. A. Kuebler, C. Baker, and D. W. Turner, *J. Chem. Phys.* **51,** 52 (1969).
20. W. von Niessen, L. S. Cederbaum, and W. P. Kraemer, *Theor. Chim. Acta* **44,** 85 (1977).
21. B. Bigot, A. Sevin, and A. Devaquet, *J. Am. Chem. Soc.* **101,** 1095 (1979).
22. B. Bigot, A. Sevin, and A. Devaquet, *J. Am. Chem. Soc.* **101,** 1101 (1979).
23. P. Mezey, R. E. Kari, A. S. Demes, I. G. Csizmadia, R. K. Gosavi, and O. P. Strausz, *Theor. Chim. Acta* **36,** 329 (1975).
24. A. C. Hopkinson, M. H. Lien, I. G. Csizmadia, and K. Yates, *Theor. Chim. Acta* **42,** 97 (1978).
25. E. Oliveros, M. Riviere, J. P. Malrieu, and C. Teichteil, *J. Am. Chem. Soc.* **101,** 318 (1979).
26. B. Bigot, D. Roux, A. Sevin, and A. Devaquet, *J. Am. Chem. Soc.* **101,** 2560 (1979).
27. R. Cimiraglia, M. Persico, and J. Tomasi, *Theor. Chim. Acta* **49,** 13 (1978).
28. A. Skancke, *J. Mol. Struct.* **42,** 235 (1977).
29. A. Skancke and J. E. Boggs, *J. Mol. Struct.* **50,** 173 (1978).
30. A. Skancke and J. E. Boggs, *J. Mol. Struct.* **51,** 263 (1979).
31. A. Skancke and J. E. Boggs, *J. Am. Chem. Soc.* **101,** 4063 (1978).
32. A. Skancke, *J. Mol. Struct.* **30,** 95 (1976).
33. A. Skancke, *Acta Chem. Scand.* **27,** 3243 (1973).
34. W. J. Hehre and J. A. Pople, *J. Am. Chem. Soc.* **97,** 6941 (1975).
35. S. D. Peyerimhoff and R. D. Buender, *Theor. Chim. Acta* **14,** 305 (1969).
36. H. Basch, *Mol. Phys.* **23,** 683 (1972).
37. E. Kochanski and J. M. Lehn, *Theor. Chim. Acta* **14,** 281 (1969).
38. K. Vasudevan and W. E. Kammer, *Chem. Phys.* **15,** 103 (1976).
39. B. Bigot, R. Ponec, A. Sevin, and A. Devaquet, *J. Am. Chem. Soc.* **100,** 6575 (1978).
40. B. Bigot, A. Sevin, and A. Devaquet, *J. Am. Chem. Soc.* **100,** 6924 (1978).
41. M. B. Robin, H. Basch, N. A. Kuebler, K. B. Wiberg, and G. B. Ellison, *J. Chem. Phys.* **51,** 45 (1969).
42. L. Radom, W. A. Lathan, W. J. Hehre, and J. A. Pople, *J. Am. Chem. Soc.* **93,** 5339 (1971).
43. W. J. Hehre and J. A. Pople, *J. Am. Chem. Soc.* **97,** 6941 (1975).
44. K. B. Wiberg, G. B. Ellison, J. J. Wendoloski, W. E. Pratt, and M. D. Harmony, *J. Am. Chem. Soc.* **100,** 7837 (1978).
45. J. Kao and L. Radom, *J. Am. Chem. Soc.* **100,** 379 (1978).
46. J. H. Davis, W. A. Goddard, III, and R. C. Bergman, *J. Am. Chem. Soc.* **98,** 4015 (1976); **99,** 2427 (1977).
47. H.-L. Hase, C. Mühler, and A. Schweig, *Tetrahedron* **34,** 2983 (1978).
47a. R. Shepard, A. Banerjee, and J. Simons, *J. Am. Chem. Soc.* **101,** 6174 (1979).
47b. B. A. Levi, E. S. Blurock, and W. J. Hehre, *J. Am. Chem. Soc.* **101,** 5537 (1979).
48. K. Krogh-Jesperson, D. Cremer, D. Poppinger, J. A. Pople, P. von R. Schleyer, and J. Chandrasekhar, *J. Am. Chem. Soc.* **101,** 4843 (1979).
49. E. R. Davidson, *J. Chem. Phys.* **57,** 1999 (1972); S. T. Elbert, S. R. Langhoff, and E. R. Davidson, *J. Chem. Phys.* **57,** 2005 (1972).

50. S. T. Elbert and E. R. Davidson, *Int. J. Quantum Chem.* **8,** 857 (1974).
51. A. Kapur, R. P. Steer, and P. G. Mezey, *J. Chem. Phys.* **71,** 588 (1979).
52. N. R. Carlsen, L. Radom, N. V. Riggs, and W. R. Rodwell, *J. Am. Chem. Soc.* **101,** 2233 (1979).
53. M. B. Robin, "Higher Exicted States of Polyatomic Molecules," Vol. 2, pp. 122–159. Academic Press, New York, 1975.
54. R. H. Staley, L. B. Harding, W. A. Goddard, III, and J. L. Beauchamp, *Chem. Phys. Lett.* **36,** 589 (1975).
55. L. B. Harding and W. A. Goddard, III, *J. Am. Chem. Soc.* **97,** 6300 (1975).
56. L. Z. Stenkamp and E. R. Davidson, *Theor. Chim. Acta* **44,** 405 (1977).
57. L. E. Nitzsche and E. R. Davidson, *J. Chem. Phys.* **68,** 3103 (1978).
58. E. Oliveros, M. Riviere, C. Teichteil, and J.-P. Malrieu, *Chem. Phys. Lett.* **57,** 220 (1978).
59. N. C. Baird and H. B. Kathpal, *J. Am. Chem. Soc.* **98,** 7532 (1976).
60. L. E. Nitzsche and E. R. Davidson, *J. Am. Chem. Soc.* **100,** 720 (1978).
61. G. Fogarasi, P. Pulay, F. Török, and J. E. Boggs, *J. Mol. Struct.* **57,** 259 (1979).
62. S. Skaarup and J. E. Boggs, *J. Mol. Struct.* **30,** 389 (1976).
63. D. Poppinger, L. Radom, and J. A. Pople, *J. Am. Chem. Soc.* **99,** 7806 (1977).
64. J. F. Harrison and G. Shalhoub, *J. Am. Chem. Soc.* **97,** 4172 (1975).
65. D. Poppinger and L. Radom, *J. Am. Chem. Soc.* **100,** 3675 (1978).
66. R. R. Lucchese and H. F. Schaefer, III, *J. Am. Chem. Soc.* **99,** 13 (1977).
67. J. F. Harrison, A. Dendramis, and R. E. Lord, *J. Am. Chem. Soc.* **100,** 4352 (1978).
68. M. E. Zandler, J. D. Goddard, and H. F. Schaefer, III, *J. Am. Chem. Soc.* **101,** 1072 (1979).
69. N. C. Baird and K. F. Taylor, *J. Am. Chem. Soc.* **100,** 1333 (1978).
70. P. G. Mezey, M. A. Robb, K. Yates, and I. G. Csizmadia, *Theor. Chim. Acta* **49,** 277 (1978).
71. S. Skaarup, P. J. Skancke, and J. E. Boggs, *J. Am. Chem. Soc.* **98,** 6106 (1976).
72. K. Vasudevan, S. D. Peyerimhoff, R. J. Buenker, W. E. Kammer, and H.-L. Hsu, *Chem. Phys.* **6,** 187 (1975).
73. M. Peric, R. J. Buenker, and S. D. Peyerimhoff, *Can. J. Chem.* **55,** 1533 (1977).
73a. R. A. Back, C. Willis, and D. A. Ramsay, *Can. J. Chem.* **56,** 1575 (1978).
74. M. Peric, R. J. Buenker, and S. D. Peyerimhoff, *Mol. Phys.* **35,** 1495 (1978).
75. N. C. Craig, and I. W. Levin, *J. Chem. Phys.* **71,** 400 (1979).
76. J. A. Pople, R. Krishnan, H. B. Schlegel, and J. S. Binkley, *Int. J. Quantum Chem.* **14,** 545 (1978).
77. R. Ahlrichs and V. Staemmler, *Chem. Phys. Lett.* **37,** 77 (1976).
78. J.-H. Davis and W. A. Goddard, III, *J. Am. Chem. Soc.* **99,** 7111 (1977).
79. W. von Niessen, W. Domcke, L. S. Cederbaum, and W. P. Kraemer, *J. Chem. Phys.* **67,** 44 (1977).
80. C. A. Parsons and C. E. Dykstra, *J. Chem. Phys.* **71,** 3025 (1979).
81. D. J. Pasto, *J. Am. Chem. Soc.* **101,** 6852 (1979).
82. S. P. Walsh and W. A. Goddard, III, *J. Am. Chem. Soc.* **97,** 5319 (1975).
83. J. Lievin and G. Verhaegen, *Theor. Chim. Acta* **42,** 47 (1976); **45,** 269 (1977).
84. T.-K. Ha and U. P. Wild, *Chem. Phys.* **4,** 300 (1974).
85. N. P. Ernsting, J. Pfab, and J. Römelt, *J.C.S. Faraday Trans. II* **74,** 2286 (1978).
86. B. J. Duke, *J. Mol. Struct.* **50,** 109 (1978).
87. J. A. Ryan and J. L. Whitten, *J. Am. Chem. Soc.* **94,** 2396 (1972).
88. J. Lievin, J. Braulet, and G. Verhaegen, *Theor. Chim. Acta* **52,** 75 (1979).
89. A. D. McLean, G. H. Loew, and D. S. Berkowitz, *J. Mol. Spectrosc.* **64,** 184 (1977).

90. K. Kitaura, Y. Osamura, K. Nishimoto, and S. Yamabe, *Chem. Phys. Lett.* **55,** 531 (1978).
91. Y. Osamura, K. Kitaura, K. Nishimoto, and S. Yamabe, *Chem. Phys. Lett.* **63,** 406 (1979).
92. A. C. Hopkinson and I. G. Csizmadia, *Theor. Chim. Acta* **34,** 93 (1974).
93. K. Yamaguchi, S. Yabushita, and T. Fueno, *J. Chem. Phys.* **71,** 2321 (1979).
94. J. Almlöf and A. A. Ischenko, *Chem. Phys. Lett.* **61,** 79 (1979).
95. Y. Ozias and A. Julg, *J. Mol. Struct.* **57,** 183 (1979).
96. T.-K. Ha, *J. Mol. Struct.* **21,** 311 (1974).
97. J. M. Howell, A. R. Rossi, and R. Bissell, *Chem. Phys. Lett.* **39,** 312 (1976).
98. A. Hinchcliffe and J. C. Dobson, *Theor. Chim. Acta* **39,** 17 (1975).
99. J. F. Olsen, *J. Mol. Struct.* **57,** 245 (1979).
100. M. Barber, J. A. Connor, M. F. Guest, I. H. Hillier, M. Schwarz, and M. Stacey, *J.C.S. Faraday Trans. II* **69,** 551 (1973).
101. O. Gropen and H. M. Seip, *Chem. Phys. Lett.* **25,** 206 (1974).
101a. S. P. So, *J.C.S. Faraday Trans. II* **75,** 1047 (1979).
101b. A. Hinchcliffe, J. Mol. Struct. **53,** 147 (1979).
101c. P. A. Kollman, W. F. Trager, S. Rothenberg, and J. E. Williams, *J. Am. Chem. Soc.* **95,** 458 (1973).
101d. G. Merényi, G. Wettermark, and B. Roos, *Chem. Phys.* **1,** 340 (1973).
101e. J. F. Olsen, *J. Mol. Struct.* **57,** 245 (1979).
101f. G. Fogarasi, P. Pulay, K. Molt, and W. Sawodny, *Mol. Phys.* **33,** 1565 (1977).
101g. Y. Ozias and A. Julg, *J. Am. Chem. Soc.* **101,** 3487 (1979).
101h. W. J. Bouma and L. Radom, *J. Am. Chem. Soc.* **101,** 3487 (1979).
102. See G. Herzberg, "Molecular Spectra and Molecular Structure, II, Electronic Spectra and Structure of Polyatomic Molecules," pp. 518–522. Van Nostrand, Princeton, New Jersey, 1966.
103. B. J. Garrison, H. F. Schaefer, III, and W. A. Lester, Jr., *J. Chem. Phys.* **61,** 3039 (1974).
104. T. H. Dunning, Jr. and N. W. Winter, *J. Chem. Phys.* **55,** 3360 (1971).
105. R. J. Buenker and S. D. Peyerimhoff, *J. Chem. Phys.* **53,** 1368 (1970).
106. W. Meyer and P. Pulay, *Theor. Chim. Acta* **32,** 253 (1974).
107. M. D. Newton, E. Switkes, and W. N. Lipscomb, *J. Chem. Phys.* **53,** 2645 (1970).
108. D. L. Wilhite and J. L. Whitten, *J. Chem. Phys.* **58,** 948 (1973).
109. S. R. Langhoff, S. T. Elbert, and E. R. Davidson, *Int. J. Quantum Chem.* **7,** 999 (1973).
110. S. Bell, *Mol. Phys.* **37,** 255 (1979).
111. R. A. Tawil and S. R. Langhoff, *J. Chem. Phys.* **63,** 1572 (1975).
112. T. Tokuhiro, B. R. Appleman, G. Fraenkel, P. K. Pearson, and C. W. Kern, *J. Chem. Phys.* **57,** 20 (1972).
113. D. B. Neumann and J. W. Moskowitz, *J. Chem. Phys.* **50,** 2216 (1969).
114. D. L. Yeager and V. McCoy, *J. Chem. Phys.* **60,** 2714 (1974).
115. L. B. Harding and W. A. Goddard, III, *J. Am. Chem. Soc.* **97,** 6293 (1975).
116. I. Özkan, S. Y. Chu, and L. Goodman, *J. Chem. Phys.* **63,** 3195 (1975).
117. S. R. Langhoff and E. R. Davidson, *J. Chem. Phys.* **64,** 4699 (1976).
118. J. M. F. van Dijk, M. J. H. Kemper, J. H. M. Kerp, and H. M. Buck, *J. Chem. Phys.* **69,** 2453 (1978).
119. J. M. F. van Dijk, M. J. H. Kemper, J. H. M. Kerp, and H. M. Buck, *J. Chem. Phys.* **69,** 2462 (1978).
120. M. J. H. Kemper, J. M. F. van Dijk, and H. M. Buck, *J. Chem. Phys.* **70,** 2854 (1979).

121. R. R. Lucchese and H. F. Schaefer, III, *J. Am. Chem. Soc.* **100,** 298 (1978).
122. M. J. H. Kemper, J. M. F. van Dijk, and H. M. Buck, *J. Am. Chem. Soc.* **100,** 7841 (1978).
123. J. D. Goddard and H. F. Schaefer, III, *J. Chem. Phys.* **70,** 5117 (1979).
124. J. L. Whitten and M. Hackmeyer, *J. Chem. Phys.* **51,** 5584 (1969).
125. S. D. Peyerimhoff, R. J. Buenker, W. E. Kammer, and H.-L. Hsu, *Chem. Phys. Lett.* **8,** 129 (1971).
126. J. L. Whitten, *J. Chem. Phys.* **56,** 5458 (1972).
127. S. R. Langhoff, S. T. Elbert, C. F. Jackels, and E. R. Davidson, *Chem. Phys. Lett.* **29,** 247 (1974).
128. J. E. Mentall, E. P. Gentieu, M. Krauss, and D. Neumann, *J. Chem. Phys.* **55,** 5471 (1971).
129. S. D. Peyerimhoff and R. J. Buenker, *Adv. Quantum Chem.* **9,** 86 (1975).
130. J. F. Gouyet and M. T. Prat, *J. Chem. Phys.* **64,** 946 (1976).
131. L. B. Harding and W. A. Goddard, III, *J. Am. Chem. Soc.* **99,** 677 (1977).
132. L. S. Cederbaum, W. Domcke, and W. von Niessen, *Chem. Phys. Lett.* **34,** 60 (1975).
133. W. Domcke and L. S. Cederbaum, *J. Chem. Phys.* **64,** 612 (1976).
134. P. W. Langhoff, S. R. Langhoff, and C. T. Corcoran, *J. Chem. Phys.* **67,** 1722 (1977).
135. P. M. Guyon, W. A. Chupka, and J. Berkowitz, *J. Chem. Phys.* **64,** 1419 (1976).
136. M. V. Pires, G. Galloy, and J. C. Lorquet, *J. Chem. Phys.* **69,** 3242 (1978).
137. S. Saebø, *Chem. Phys. Lett.* **40,** 462 (1976).
138. O. P. Strausz, C. Kozmutza, E. Kapuy, M. A. Robb, G. G. Theodorakopoulos, and I. G. Csizmadia, *Theor. Chim. Acta* **48,** 215 (1978).
139. P. Cremaschi and M. Simonetta, *Theor. Chim. Acta* **43,** 351 (1977).
140. R. Johansen and E. W. Nilssen, *J. Mol. Struct.* **49,** 131 (1978).
141. J. D. Dill, C. L. Fischer, and F. W. McLafferty, *J. Am. Chem. Soc.* **101,** 6531 (1979).
142. J. Del Bene and S. Radovich, *J. Am. Chem. Soc.* **100,** 6936 (1978).
143. J. E. Del Bene and A. Vaccaro, *J. Am. Chem. Soc.* **98,** 7526 (1976).
144. P. E. M. Siegbahn, *J. Chem. Phys.* **70,** 5391 (1979).
145. P. J. Bruna, S. D. Peyerimhoff, R. J. Buenker, and P. Rosmus, *Chem. Phys.* **3,** 35 (1974).
146. A. Kapur, R. P. Steer, and P. G. Mezey, *J. Chem. Phys.* **69,** 968 (1978); **70,** 745 (1979).
147. P. Rosmus and H. Bock, *J.C.S. Chem. Commun.* **7,** 334 (1979).
148. T. Yamabe, K. Yamashita, K. Fukui, and K. Morokuma, *Chem. Phys. Lett.* **63,** 433 (1979).
149. B. Hess, P. J. Bruna, R. J. Buenker, and S. D. Peyerimhoff, *Chem. Phys.* **18,** 267 (1976).
150. P. J. Bruna, R. J. Buenker, and S. D. Peyerimhoff, *Chem. Phys.* **22,** 375 (1977).
151. T.-K. Ha and L. Keller, *J. Mol. Struct.* **27,** 225 (1975).
152. R. B. Davidson and L. C. Allen, *J. Chem. Phys.* **54,** 2828 (1971).
153. W. J. Hehre and W. A. Lathan, *J.C.S. Chem. Commun.* p. 771 (1972).
154. W. J. Bouma and L. Radom, *J. Mol. Struct.* **43,** 267 (1978).
155. N. W. Winter, W. A. Goddard, and C. F. Bender, *Chem. Phys. Lett.* **33,** 25 (1975).
156. H. H. Jensen, H. Møllendal, and E. W. Nielssen, *J. Mol. Struct.* **30,** 145 (1976).
157. C. E. Dykstra and H. F. Schaefer, III, *J. Am. Chem. Soc.* **97,** 7210 (1975).
158. C. E. Dykstra and H. F. Schaefer, III, *J. Am. Chem. Soc.* **98,** 401 (1976).
159. C. E. Dykstra, R. R. Lucchese, and H. F. Schaefer, III, *J. Chem. Phys.* **67,** 2422 (1977).
160. W. B. Mueller, J. F. Harrison, and P. J. Wagner, *J. Am. Chem. Soc.* **100,** 33 (1978).

161. W. von Niessen, *J. Am. Chem. Soc.* **99,** 7151 (1977).
162. L. E. Nitzche and E. R. Davidson, *Chem. Phys. Lett.* **58,** 171 (1978).
163. T.-K. Ha, *Chem. Phys. Lett,* **57,** 64 (1978).
164. J. Tyrrell, *J. Am. Chem. Soc.* **101,** 3766 (1979).
165. C. E. Dykstra, *J. Am. Chem. Soc.* **98,** 7182 (1976).
166. S. Iwata and K. Morokuma, *J. Am. Chem. Soc.* **97,** 966 (1975).
167. E. R. Davidson and L. E. Nitzche, *J. Am. Chem. Soc.* **101,** 6524 (1979).
168. R. R. Lucchese, H. F. Schaefer, III, and C. E. Dykstra, *Chem. Phys. Lett.* **51,** 600 (1977).
169. K. van der Meer and J. J. C. Mulder, *Theor. Chim. Acta* **37,** 159 (1975).
170. C. E. Dykstra, *J. Chem. Phys.* **68,** 4244 (1978).
171. O. P. Strausz, R. K. Gosavi, and H. E. Gunning, *Chem. Phys. Lett.* **54,** 510 (1978).
172. C. E. Dykstra and H. F. Schaefer, III, *J. Am. Chem. Soc.* **98,** 2689 (1976).
173. H. Basch, *Theor. Chim. Acta* **28,** 151 (1973).
174. L. B. Harding and W. A. Goddard, III, *J. Am. Chem. Soc.* **98,** 6093 (1976).
175. B. Jönsson, C. Karlström, H. Wennerström, and B. Roos, *Chem. Phys. Lett.* **41,** 317 (1976).
176. W. R. Wadt and W. A. Goddard, III, *J. Am. Chem. Soc.* **97,** 3004 (1975).
177. L. B. Harding and W. A. Goddard, III, *J. Am. Chem. Soc.* **100,** 7180 (1978).
178. H. Basch, M. B. Robin, and N. A. Kuebler, *J. Chem. Phys.* **49,** 5007 (1968).
179. S. D. Peyerimhoff, *J. Chem. Phys.* **47,** 349 (1967).
180. S. D. Peyerimhoff and R. J. Buenker, *J. Chem. Phys.* **50,** 1846 (1969).
181. M. E. Schwartz, E. F. Hayes, and S. Rothenberg, *J. Chem. Phys.* **52,** 2011 (1970).
182. M. R. Peterson and I. G. Csizmadia, *J. Am. Chem. Soc.* **101,** 1076 (1979).
182a. D. Demoulin, *Chem. Phys.* **17,** 471 (1976).
183. E. Flood, *J. Mol. Struct.* **21,** 221 (1974).
184. A. C. Hopkinson, K. Yates, and I. G. Csizmadia, *J. Chem. Phys.* **52,** 1784 (1970).
184a. E. Auberg, S. Sandal, and H. M. Seip, *J. Mol. Struct.* **57,** 95 (1979).
184b. R. Meyer, T.-K. Ha, H. Frei, and H. H. Gunthard, *Chem. Phys.* **9,** 393 (1975).
184c. S. Green, H. Schor, P. Siegbahn, and P. Thaddeus, *Chem. Phys.* **17,** 479 (1976).
185. O. A. Mosher, W. M. Flicker, and A Kuppermann, *J. Chem. Phys.* **59,** 6502 (1973).
186. R. P. Hosteny, T. H. Dunning, Jr., R. R. Gilman, A. Pipano, and I. Shavitt, *J. Chem. Phys.* **62,** 4764 (1975).
187. R. J. Buenker and J. L. Whitten, *J. Chem. Phys.* **49,** 5381 (1968).
188. S. Shih, R. J. Buenker, and S. D. Peyerimhoff, *Chem. Phys. Lett.* **16,** 244 (1972).
189. R. J. Buenker, S. Shih, and S. D. Peyerimhoff, *Chem. Phys. Lett.* **44,** 385 (1976).
190. C. W. Bock, P. George, M. Trachtman, and M. Zanger, *J.C.S. Perkin Trans. II* p. 26 (1979).
191. B. Dumbacher, *Theor. Chim. Acta* **23,** 346 (1972).
192. A. J. P. Devaquet, R. E. Townshend, and W. J. Hehre, *J. Am. Chem. Soc.* **98,** 4068 (1976).
193. M. E. Squillacote, R. S. Sheridan. O. L. Chapman, and R. A. L. Anet, *J. Am. Chem. Soc.* **101,** 3657 (1979).
194. L. S. Cederbaum, W. Domcke, J. Schirmer, W. von Niessen, G. H. F. Diercksen, and W. P. Kraemer, *J. Chem. Phys.* **69,** 1591 (1978).
195. H. Kollmar, *J. Am. Chem. Soc.* **101,** 4832 (1979).
196. A. Hinchcliffe, *J. Mol. Struct.* **27,** 329 (1975).
197. N. A. C. Nascimento and W. A. Goddard, III, *Chem. Phys.* **36,** 147 (1979); *Chem. Phys. Lett.* **60,** 197 (1979).
198. W. von Niessen, W. P. Kraemer, and L. S. Cederbaum, *Chem. Phys.* **11,** 385 (1975).

199. H. E. Popkie and J. J. Kaufman, *J. Chem. Phys.* **66,** 4827 (1977).
200. M. P. Conrad, R. M. Pitzer, and H. F. Schaefer, III, *J. Am. Chem. Soc.* **101,** 2245 (1979).
201. W. T. Borden and E. R. Davidson, *J. Am. Chem. Soc.* **101,** 3771 (1979).
202. Y. Apeloig, P. von R. Schleyer, J. S. Binkley, and J. A. Pople, *J. Am. Chem. Soc.* **98,** 4332 (1976).
203. W. D. Laidig and H. F. Schaefer, III, *J. Am. Chem. Soc.* **101,** 7184 (1979).
204. M.-H. Whangbo, D. J. Mitchell, and S. Wolfe, *J. Am. Chem. Soc.* **100,** 3698 (1978).
205. C. W. Bock, P. George, G. J. Mains, and M. Trachtman, *J.C.S. Perkins Trans. II* p. 814 (1979).
205a. T.-K. Ha, *Mol. Phys.* **27,** 793 (1974).
206. A. C. Hopkinson, M. H. Lien, K. Yates, and I. G. Csizmadia, *Theor. Chim. Acta* **44,** 385 (1977).
206a. H. Lischka and H.-J. Köhler, *Chem. Phys. Lett.* **63,** 326 (1979).
207. R. Ahlrichs and R. Heinzmann, *J. Am. Chem. Soc.* **99,** 7452 (1977).
208. D. M. Hood and H. F. Schaefer, III, *J. Chem. Phys.* **68,** 2985 (1978).
209. R. K. Gosavi, H. E. Gunning, and O. P. Strausz, *Chem. Phys. Lett.* **59,** 321 (1978).
209a. L. C. Snyder and Z. R. Wasserman, *J. Am. Chem. Soc.* **101,** 5222 (1979).
209b. D. Cremer and J. A. Pople, *J. Am. Chem. Soc.* **97,** 1358 (1975).
210. D. Cremer, *J. Chem. Phys.* **70,** 1898 (1979).
211. J. S. Binkley, R. Seeger, J. A. Pople, J. D. Dill, and P. von R. Schleyer, *Theor. Chim. Acta* **45,** 69 (1977).
212. U. Lamanna and M. Maestro, *Theor. Chim. Acta* **36,** 103 (1974).
213. T. Clark and P. von R. Schleyer, *J.C.S. Chem. Commun.* p. 883 (1979).
214. W. D. Laidig and H. F. Schaefer, III, *J. Am. Chem. Soc.* **100,** 5972 (1978).
215. M. Trenary, M. E. Casida, B. R. Brooks, and H. F. Schaefer, III, *J. Am. Chem. Soc.* **101,** 1638 (1979).
216. H. Oberhammer and J. E. Boggs, *J. Mol. Struct.* **55,** 283 (1979).
217. J. Burdon, I. W. Parsons, and E. J. Avramides, *J.C.S. Perkins Trans. I* p. 1268 (1979).
218. J. Burdon, I. W. Parsons, and E. J. Avramides, *J.C.S. Perkins Trans. II* p. 1201 (1979).
219. D. M. Hayes, S. D. Nelson, W. A. Garland, and P. A. Kollman, *J. Am. Chem. Soc.* **102,** 1255 (1980).
220. I. Absar and J. R. Van Wazer, *J. Am. Chem. Soc.* **94,** 2382 (1972).
221. C. E. Dykstra, C. A. Parsons, and C. L. Oates, *J. Am. Chem. Soc.* **101,** 1962 (1979).
222. O. Gropen and R. Johansen, *J. Mol. Struct.* **25,** 161 (1975).
223. P. Rosmus, H. Stafast, and H. Bock. *Chem. Phys. Lett.* **34,** 275 (1975).
224. P. Mezey, A. Kucsman, G. Theodorakopolous, and I. G. Csizmadia, *Theor. Chim. Acta* **38,** 115 (1975).
225. J. Burdon, D. W. Davies, and G. del Conde, *J.C.S. Perkins Trans. II* p. 1205 (1979).
226. J. F. Olsen, *J. Mol. Struct.* **49,** 361 (1978).
227. S. Vishveshwara, *Chem. Phys. Lett.* **59,** 30 (1978).
228. A. Rosen and B. Fricke, *Chem. Phys. Lett.* **61,** 75 (1979).
229. P. J. Hay, *J. Am. Chem. Soc.* **100,** 2897 (1978).
230. A. K. Rappe and W. A. Goddard, III, *J. Am. Chem. Soc.* **99,** 3966 (1977).
231. M. F. Guest, M. B. Hall, and I. H. Hillier, *J.C.S. Faraday Trans. II* p. 1829 (1973).
232. S. R. Ungemach and H. F. Schaefer, III, *J. Am. Chem. Soc.* **98,** 1658 (1976).
233. S. P. So and W. G. Richards, *J.C.S. Faraday Trans. II* p. 55 (1979).
234. T. Otterson and H. H. Jensen, *J. Mol. Struct.* **26,** 355 (1975).

235. A. Richartz, R. J. Buenker, and S. D. Peyerimhoff, *Chem. Phys.* **31,** 187 (1978).
236. J. Pacansky and M. Dupuis, *J. Chem. Phys.* **71,** 2095 (1979).
237. W. von Niessen and G. H. F. Diercksen, *J. Electron Spectrosc.* **16,** 351 (1979).
238. D. Feller, W. T. Borden, and E. R. Davidson, *J. Chem. Phys.* **71,** 4987 (1979).
239. W. J. Bouma, J. K. MacLeod, and L. Radom, *J. Am. Chem. Soc.* **101,** 5540 (1979).

CHAPTER X

SOME MOLECULES OF LOW SYMMETRY

A. MOLECULES WITH "FREE ROTATION"

1. General Considerations

Molecules containing such groups as CH_3, NH_2, or OH change only a little in energy when the group in question is rotated about its bond to another atom. Thus at room temperature, thermal energy is sufficient to permit "free rotation" through two or more conformations differing by one or at most a few kilocalories in energy. In such cases, more than one conformer and/or intermediate forms corresponding to rotational excitation exist in equilibrium at room temperature. For the barriers between different conformers, SCF calculations are adequate to give essentially correct barrier heights.

Many of these molecules have at least one conformer possessing a plane of symmetry, which facilitates computations, and for several such cases, the results of computations for a particular conformer have been reported in Chapter IX (see Section IX,D,2 for *N*-methylacetamide; E,7 for nitrosomethane; F,3 for acetone, acetaldehyde, etc.).

In many other molecules reported in Chapter IX, rotational isomers (cis and trans, syn and anti) exist separated by a barrier too high to be readily surmounted at room temperature.

Many writers have sought to analyze the factors that contribute to the small barriers existing for "free rotation." However, for a full understanding, one needs to use the energy decomposition methods initiated by Morokuma [1]. Morokuma and Umeyama have applied them to the cases of CH_3OH and CH_3NH_2 [2]. Although the methods were originally developed for the interaction of closed-shell molecules to form molecular complexes, they can be applied here by treating the molecules as combinations of two radicals. The result is that of the various contributing terms (electrostatic, polarization, exchange, charge transfer, and mixed items), exchange makes the largest contribution by far to the rotational barriers (computed: 2.7 kcal for CH_3NH_2 and 1.5 kcal for OH). The same authors also discuss C_2H_6 (barrier 2.9 kcal), with the same conclusion (see Section VI,G). This conclusion is in agreement with the judgement of other authors.

Radom *et al.* [3], using 4-31G SCF computations, have discussed the Fourier component analysis of internal rotation potentials in saturated molecules. In the simplest cases, such as C_2H_6, CH_3NH_2, and CH_3OH, internal rotation is well described by a threefold potential function

$$V(\Phi) = \tfrac{1}{2}V_3(1 - \cos 3\Phi)$$

The potential functions for hydrazine, hydroxylamine, and hydrogen peroxide are more complicated, involving terms in $\cos \Phi$, $\cos 2\Phi$, and $\cos 3\Phi$. In still more complicated cases, terms in $\sin \Phi$ and $\sin 2\Phi$ must be added. The paper of Radom *et al.* lists energy minima and maxima and corresponding rotation angles as computed for a large number of molecules, and as obtained from functions $V(\Phi)$, which they deduce to match the *ab initio* computations.

Cremer *et al.* [4] in 4-31G SCF calculations have computed rotation barriers for double methyl rotors of the form $(CH_3)_2X$ and compared these with the barriers for corresponding single rotors CH_3X. The double rotors considered are propane, dimethylamine, dimethyl ether, singlet dimethylcarbene, isobutene, acetone, dimethyborane, and isopropyl cation—too numerous to list under separate molecules. Qualitatively, the changes in going from a single to a double rotor are well explained by a quasi-aromatic stabilization involving bonding between methyl groups if X is a π donor and a destabilization if it is a π acceptor.

In this chapter, we shall not attempt to review even all the best of the very numerous *ab initio* computations that have been made on rotational barriers or other aspects of molecules of low symmetry, but rather we have selected examples or special features that illustrate what has been done. On the other hand, we must emphasize that Pople and collaborators have made very many relevant calculations, of which we have touched on only a few. For these we refer readers to their new book (see Chapter 1). A useful general review on rotation barriers has been given by Payne and Allen [4a].

2. CH_3OH, CH_3SH, CH_3BH_2, CH_3NH_2, $(CH_3)_2O$, $CH_3—C≡C—CH_3$, and Related Molecules; *n*-Butane

Tel *et al.* [5], using a double-zeta basis, made SCF computations on CH_3OH in four geometries, Lowest in energy is a tilted variation (see below) of the staggered form

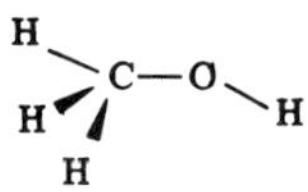

with the eclipsed form computed to be 1.44 kcal higher in energy (1.07 kcal experimentally). For in-planar (H_3—C—O—H) inversion the computed barrier was 32 kcal/mole. Proton affinities were computed for CH_3O^- and CH_3OH and compared with those for OH^- and H_2O. Valence-shell orbital energies are listed. Excitation energies (singlet and triplet) and oscillator strengths are computed for 16 electronic transitions of CH_3OH in both staggered and eclipsed geometry. Comparison is made with the experimental absorption spectrum. The dipole moment (2.3–2.4 D for differing computations, 1.69 D for experiment) and its components are computed and displayed.

Flood *et al.* [6] with a double-zeta basis set made SCF computations "on the tilt and asymmetry of methyl groups in asymmetric environments," namely, in CH_3OH, CH_3NH_2, CH_3BH_2, and CH_3NO. In these molecules the threefold axis of the CH_3 group, treated as rigid, is tilted by an angle α relative to the direction of the C—O, C—N, or C—B bond, and at the same time one C—H bond is lengthened by 0.006–0.008 Å relative to the other two. There is experimental evidence which confirms the computed effects. Computed are $\alpha = 3.9°$ in CH_3OH (staggered), 2.6 or 2.9° in CH_3BH_2, 3.4° in CH_3NH_2 (staggered), 2.2° in CH_3NO (eclipsed). These effects are attributed to repulsions between the unique CH_3 hydrogen atom and lone pairs or bonds in the attached group. A detailed discussion is given.

Kern *et al.* [7] have also discussed barrier and tilt in CH_3OH. They find $\alpha = 3.4$ and $-1.8°$ for the staggered and eclipsed forms, respectively.

They note that inclusion of tilt raises the barrier about 25%.

Pappas [8], using a double-zeta basis set, has made an SCF computation of the rotation barrier (calculated: 1.37, experimental: 1.26 kcal/mole) and the dipole moment (1.63 D). Hey also calculated the SH barrier in cysteine [$H_2C(SH)CNH_3{}^+O_2{}^-$] as 3.1 kcal/mole.

Schlegel *et al.* [9], using a 4-31G basis, have computed anharmonic force fields and vibrational frequencies for CH_3NH_2, CH_3OH, and CH_3SH. Agreement with experiment is good.

Ha *et al.* [10] have made double-zeta SCF calculations on four different conformations of the methyl groups in dimethyl ether $(CH_3)_2O$. Optimized geometries and barriers to internal rotation are given; the latter are compared with experimental evidence. Dipole moment parameters for various conformations are reported.

Blom *et al.* [11], using the results of 4-31G calculations, have determined the complete general valence force field for dimethyl ether, and calculated infrared and Raman frequencies and intensities. The infrared but not the Raman intensities agree well with experiment.

Radom *et al.* [12] have made mainly 4-31G calculations to determine constants in a Fourier expansion of the torsional potential for rotations of 1,2-disubstituted ethanes $RCH_2{\cdot}CH_2R'$, where R, R′ are CH_3, Li; Li, Li; and Li, F.

Kollman *et al.* [13] have calculated rotational barriers in R—C≡C—R′, where R, R′ are CH_3, CH_3; NH_2, NH_2; OH, OH; CF_3, CH_3; SiH_3, CH_3; also in propene, *cis*- and *trans*-2-butene. In dimethylacetylene and its CF_3 and SiH_3 analogs, the eclipsed structure is found to be most stable.

Peterson and Csizmadia [14], using a minimal Gaussian basis set, have constructed an analytic conformational hypersurface for *n*-butane by fitting 63 SCF energy points. Two minima and two maxima were determined, also four saddle and four supersaddle points.

3. Protonated Methanol ($H_3C—OH_2{}^+$)

Tel *et al.* [5] made computations on $H_3C—OH_2{}^+$, in addition to CH_3OH and concluded that it shows no barrier to either rotation or inversion. However, Flood and Nilssen [15], in an SCF computation using a larger basis set including d orbitals, found the $C—OH_2$ fragment to be nonplanar with an inversion barrier of 3.4 kcal/mole. They examined the potential surface for the reaction $CH_3OH_2 \rightarrow CH_3{}^+ + H_2O$. They present a graph displaying orbital energies as a function of reaction coordinate during this dissociation, give overlap populations, and discuss changes in bonding.

4. 1,2-Dihydroxyethylene

With a 4-31G basis, Larson *et al.* [16] have made SCF computations on the energies of various conformers of the cis and trans forms of HOHC=CHOH. The results are explained in terms of hyperconjugative interaction. Reference is also made to previous studies on CH_2=CHOH and CH_2=CHSH in the same connection. All these studies are relevant to conformational isomerism in ester and vinyl ethers due to rotation about the C—O single bond.

B. HYDROCARBON MOLECULES, RADICALS, IONS, AND SOME DERIVATIVES

1. Ethyl Radical and Ions and Haloethyl Ions

Pacansky and Dupuis [17] made restricted and unrestricted SCF calculations on the ethyl radical in staggered and eclipsed conformations (both C_s). They found minimum energy for a variety of the staggered configuration in which the H_2 of the CH_2 group are bent upward by 6°, toward the unique CH_3 hydrogen atom, relative to a planar CCH_2 arrangement (inversion barrier 0.5 kcal for a change from +6° to −6°).

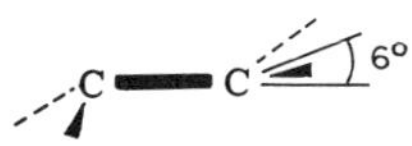

The planar staggered and the eclipsed conformations are both found to be 0.2 kcal higher in energy. The C—C bond (1.498 Å) is slightly shorter than in ethane. Bond lengths and angles are given.

Claxton and Overill [18], using a small basis set, have made an unrestricted SCF calculation followed by spin annihilation on C_2H_5, followed by calculations of isotropic coupling constants and anisotropic coupling tensors for the nuclei. Computations were made for both staggered and eclipsed conformations. For comparison with experiment, the results were averaged since there should be essentially free rotation (of CH_2 as well as of CH_3). Agreements with experiment are rather good.

Williams and Streitwieser [19] have made SCF calculations using a double-zeta plus carbon d basis on the ethyl anion, and given ionization potential, electron affinity (−2.1 eV for the radical, expected to be smaller in magnitude if allowance were made for correlation), proton affinities, and electronic density diagrams.

Lischka and Köhler [20] have made PNO-CI and CEPA-PNO calculations on $C_2H_5^+$ using rather good basis sets including carbon d polarization. They find the nonclassical bridged structure

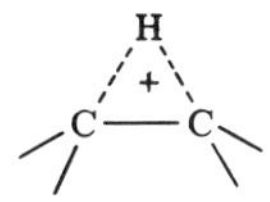

to be more stable by 6.4 kcal/mole than the classical structure. (In solution, solvation probably stabilizes the classical structure.)

They have also made similar computations on the haloethyl ions. In $C_2H_4F^+$, structure $H_3C—C^+ FH$ is the most stable, being 33 kcal more stable than the bridged structure, which in turn is 10 kcal more stable than the classical structure $CH_2F—C^+H_2$. Corresponding figures for $C_2H_4Cl^+$ are 4 and 16 kcal/mole.

2. $C_3H_7^+$ Isomers

Radom *et al.* [21] optimized geometries for nine $C_3H_7^+$ isomers in minimal-basis (STO-3G) SCF computations, then used a 4-31G basis for computing energies. Lischka and Köhler (see above), using the geometries of Radom *et al.*, show that the most stable $C_3H_7^+$ isomer is the isopropyl cation. They also show that the stablest form of protonated cyclopropane has the added H atom in a bridged position at one edge of the ring. They find this structure to be about 8 kcal/mole and the *n*-propyl cation 16.5 kcal/mole less stable than the isopropyl cation. The edge-protonated cyclopropane is found to be about 5 kcal/mole more stable than a corner-protonated form. Much less stable is a face-protonated form. In general, their work shows that correlation is important for the correct estimation of energy differences between isomeric ions; their conclusions differ considerably from those of Radom *et al.*

3. CH_5^+ and $C_2H_7^+$

The ion CH_5^+ is well known in mass spectroscopy. Dyczmons *et al.* [22] have made good SCF calculations, including mid-C—H-bond functions for polarization, and found either of two C_s configurations $H_3CH_2^+$ to be the most stable, with essentially free rotation between the two variations. [Originally a trigonal bipyramidal structure (D_{3h}) had been assumed.] As they note, equilibrium geometries of small molecules are rather insensitive to correlation, so their conclusions can probably be accepted. The C_{2v}, C_{4v}, and D_{3h} models showed increasing energy, all being unstable with respect to deformation into C_s.

In a somewhat more accurate SCF calculation using d functions on C atoms and p on the H, Hariharan *et al.* [23] have confirmed the results of Dyczmons *et al.*, finding the C_s to be 16 kcal/mole more stable than the D_{3h} model.

Köhler and Lischka [24], using polarized basis sets in an SCF calculation, followed by valence-shell correlation (CEPA-PNO) using localized SCF MOs, have calculated the energies for several conformations of $C_2H_7^+$. They find the stablest conformation to be one in which the added proton is attached to the C—C bond of C_2H_6 in a bridged structure (two variations similar in energy).

4. 2-Norbornyl Cation

In a 4-31G SCF calculation, Goetz *et al.* [25] find almost no difference in energy between the classical (1a) and nonclassical (1b) forms of the 2-norbornyl cation

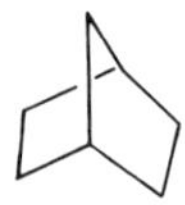

1a

1b

but estimate that polarization and correlation would favor the nonclassical form.

Köhler and Lischka [26] in similar calculations concur but, after also examining other structures, find an edge-protonated nontricyclene form to be about equally as stable as the nonclassical form.

Using a 4-31G basis set, Tyrell *et al.* [27] have made SCF calculations on cyclopropyl and α-halocyclopropyl anions.

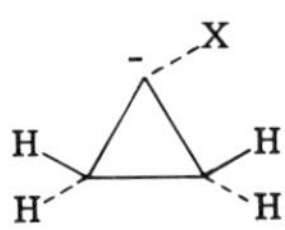

The X atom is found to be bent out of the cyclopropane ring plane by 58°, 64°, and 61° for X = H, F, and Cl, with respective barriers for passage through the ring plane of 17, 42, and 40 kcal/mole.

5. Spin Densities in Hydrocarbon Radicals

Chipman [28] has calculated isotropic hyperfine coupling constants arising from contact spin density distributions in vinyl and ethynyl, methyl and ethyl, allyl and cyclohexadienyl radicals, using the unrestricted SCF method with spin projection and a variety of basis sets. Good results are obtained with an STO-3G basis set.

Jemmis *et al.* [28a] have made 4-31G calculations to determine the optical configurations of lithiated C_3H_4 molecules (allene, propyne, and cyclopropene). Stablest for monolithiation is lithiopropyne Li—C≡CH_3, of C_{3v} symmetry. For $C_3H_2Li_2$, a C_s bridged structure (2a) is found to be the most stable. For C_3HLi_2, (2b) was found to be stablest, and for C_3Li_4, (2c).

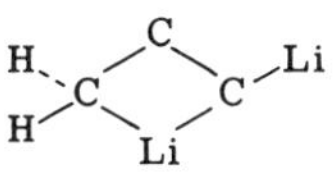

2a

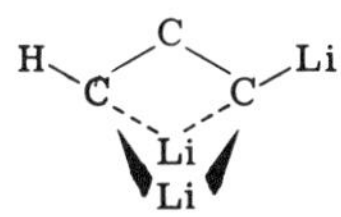

2b

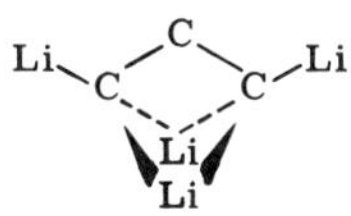

2c

C. SOME MOLECULES CONTAINING O ATOMS

1. Hydrogen Peroxide (H_2O_2) and Hydrogen Trioxide (H_2O_3)

With SCF calculations on H_2O_2 it has proved unusually difficult to obtain good agreement with experiment. The problem seems to have been largely solved in an SCF calculation by Cremer [29] using a polarized basis set including d on O and p on H, and supplemented by correlation corrections obtained by Møller–Plesset perturbation theory. The equilibrium geometry is calculated to be skewed from the cis position toward the trans position by a dihedral angle of 119.3°, with barrier heights 8.13 and 0.94 kcal/mole for cis and trans respectively; also R(OO) = 1.451 Å, R(OH) = 0.967 Å, and the OOH angle 99.3°. Agreement with experiment is very good. Rescaling of basis functions accompanied by complete geometry optimization, at all rotational angles, inclusion of polarization functions, and correction for correlation errors are of particular importance for the final results.

Rodwell *et al.* [30] have obtained very similar results using a good polarized basis set but for electron correlation using strongly orthogonal geminal procedures. Results for the equilibrium skew geometry are as follows: dihedral angle 117.3°, barrier heights 3.49 and 3.2 kJ/mole for cis and trans, respectively; R(OO) = 1.49 Å, R(OH) = 0.965 Å, OOH angle 99.3°. The authors give a table of calculated values for several one-electron properties including the dipole moment (calculated: 1.90 D, experimental : 2.26 D) and the quadrupole moment.

Hinchcliffe [31] has made SCF calculations on HSSH, H_2SS, FSSH, and F_2SS using ample polarized basis sets, and obtained optimized geometries. Two stable isomers of F_2S_2 are known experimentally: FSSF analogous to H_2O_2 with gauche symmetry and pyramidal F_2SS. The calculations give good agreement with experiment; calculated energies ($E_h \simeq -993.785$ for FSSF and -993.734 for F_2SS) and complete tables of orbital energies are given. Corresponding calculated results for H_2S_2 ($E_h \cong -796.230$ for HSSH and -796.180 for H_2SS) are given; only HSSH is known experimentally. Some population analyses, and computed dipole and quadrupole moments are given.

In double-zeta calculations on $(CH_3)_2S_2$ Pappas [8] calculated a dihedral angle of 86°, a cis barrier of 16°, and a trans barrier of 9°, also E(eclipsed) − E(staggered) = 3.10 kcal/−mole (experimental: 3.2), and a dipole moment of 2.06 D (experimental: 1.99 D). Comparisons with H_2S_2 are made, and plots of the electrostatic potential of H_2S_2 are given.

Olsen [32], using a 4-31G basis, has made SCF calculations on CF_3OOH, CF_3OOF, and $CF_2(OF)_2$. For the first two, torsional potential curves and barriers have been computed. O—O bonds in CF_3OOH are similar to those in H_2O_2, while those in CF_3OOF are much longer. The three-fold barrier to rotation of the CF_3 group in CF_3OOF is predicted to be 4.4 kcal/mole.

The simplest molecular double rotor is H_2O_3, which is known experimentally as a molecule so reactive that experimental data on its structure are not available. Cremer [33] has made a very intensive study of this molecule using the same procedures as for H_2O. Minimum energy is found for a C_2 structure

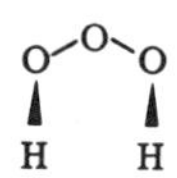

six other saddle points or maximum energy structures are identified, and geometrical parameters are optimized. Contour maps for various parameters as a function of two dihedral angles are shown, and a Fourier analysis is given. Comparisons are made with H_2O_2 and with the pseudorotation of five membered rings, in particular, between the conformational tendencies of H_2O_3 and 1,2,3-trioxolane (cf. Section IX,H,2). H_2O_3 is discussed as a prototype for general double rotors $X(OY)_2$, with X = CH_2, Y = H or CH_3, and X = PO_2^-, Y = H or CH_3, with citations of recent literature.

2. Glycine, Glycolic Acid, Glycolaldehyde, and $M_2C(OCH_3)_2$

Vishveshwara and Pople [34] in a 4-31G SCF calculation have calculated the rotational potential surfaces of glycine (α-aminoacetic acid) (3a)

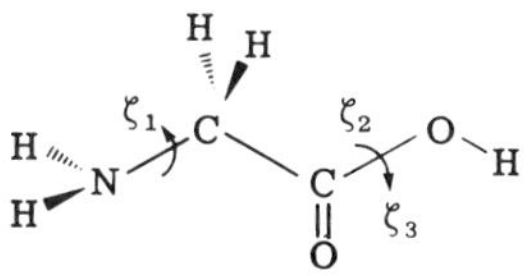

$\Phi_1 = \Phi_2 = \Phi_3 = 0°$

3a

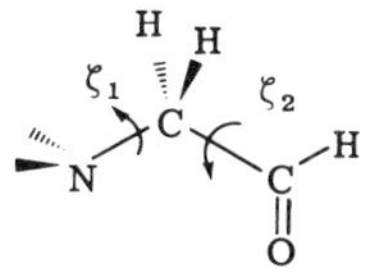

$\Phi_1 = \Phi_2 = 0°$

3b

and the corresponding aldehyde (3b). They show conformational potential energy maps as functions of the angles Φ_1 and Φ_2. These show shallow local minima in addition to a global minimum at $\Phi_1 = 180°$, $\Phi_2 = 0°$ (planar NCCOH). Maxima occur at about 7 kcal/mole higher. Comparisons are made with other aldehydes (for these, see Section IX,F,7).

Although the isolated molecule of glycine is as just described, in polar solvents it takes the zwitterion form (4). Tse *et al.* [35] have made 4-31G

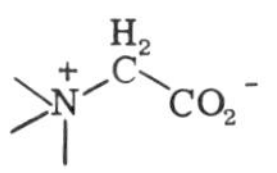

4

computations on various conformations of this. The unsolvated zwitterion is calculated to be 29 kcal/mole higher in energy than the nonzwitterionic structure; solution in water then involves a hydration energy of about 50 kcal for the zwitterion. Other details are given.

In an earlier article, Ryan and Whitten [36], assuming zwitterion forms in both cases, used a more-than-minimal basis set to compute total and orbital energies for glycine and glycylglycine.

Newton and Jeffrey [37], using a 4-31G basis, have studied the conformational energetics of glycolic acid ($HOCH_2COOH$), the glycolate anion, and glycolaldehyde ($HOCH_2CHO$). In the stablest conformer, there is hydrogen bonding between the H of the first hydroxyl group and the O of the C=O bond. Further details are given.

Jeffrey *et al.* [38], using a 4-31G basis, have made SCF calculations on $(CH_3O)_2CH_2$ as a model for the acetyl moieties in methyl pyranosides ("structural moieties of carbohydrates"). Comparisons with previous calculations on methane diol and methoxymethanol are made. Relations to geometries in the pyranosides are discussed.

3. CH_2OH

In an approximately double-zeta SCF calcualtion Ha [39] finds CH_2OH to be nonplanar with an out-of-plane angle of 25° for the CH_2 group with a barrier of 0.4 kcal/mole. The rotational barriers found for the OH group are 1.8 (trans) and 3.5 kcal/mole (cis).

4. Oxazolidine

Hotokka and Pyykkö [40], using a split-valence basis set, have made a conformational analysis of oxazolidine (5). The ring atoms are displaced by varying amounts from a planar arrangement.

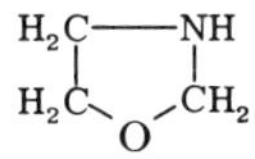

5

5. Oxaziriidine and Nitrone

Bigot *et al.* [41] have used minimal-basis plus limited CM to obtain potential curves for reaction paths in the transformation of oxaziridine (6) (see Section IX,C,1 for related molecules) to nitrone ($H_2C{=}N^+HO^-$) and to formamide ($O{=}CH{\cdot}NH_2$). Use is made of the lowest excited states of oxaziridine in some of the transformations.

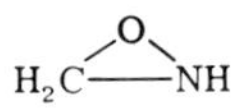

Oliveros *et al.* [42] have used double-zeta plus diffuse AOs followed by extensive CM calculations to study the mechanism of thermal and photochemical rearrangements of oxaziridines into amides. A table of orbital energies is given. The lowest singlet excited state is predicted at 5.59 eV, as well as two triplet states at 4.61 and 4.96 eV. Excitation is partially Rydberg; CM coefficients are given. Also the ground state requires a CM description. The transformation of oxaziridine to formamide is calculated to involve an energy drop of 3.39 eV (78.1 kcal/mole). Migration pathways of the atoms are predicted.

Wallmeier and Kutzelnigg [42a], using a polarized basis set of double-zeta quality followed in some cases by PNO-CI for CM, have made a comparative study of H_3NO, H_2NOH, H_3PO, H_3POH, H_2POF, H_2SO, HSOH, HClO, ArO, and related molecules in an endeavor to understand the nature of the X—O semipolar bond. They conclude that although d orbitals are often important, it is as polarization rather than valence orbitals.

D. MISCELLANEOUS SYSTEMS

1. Adsorption of H Atoms on Metal Clusters

Several authors have made SCF calculations on metal atom clusters with an H atom attached. Thus, Bauschlicher *et al.* [43] have made such computations for Be clusters using a double-zeta basis up to Be_{10}, and a minimal basis for up to Be_{22}. Four different high-symmetry sites for the approach of the H atom were considered simulating approach to the (0001) surface of a Be crystal. Two sites are at the center of triangles of Be atoms, one is an edge site, one is an overhead site (over one Be atom). Binding energies of about 50 kcal/atom and an approach distance of about 0.1 Å were found for the first three sites, and about 30 kcal/atom and 0.14 Å for the overhead site. Thus dissociative adsorption of H_2 on Be (0001) should probably be exothermic. A calculation indicates that diffusion of H into the bulk is unfavorable for an ideal (0001) surface. Vibration frequencies for motions normal to the surface were computed. The overhead site gave the largest frequency ($\omega_e = 2060\ cm^{-1}$ for Be_{22}; others were 1360, 1230, and 1220 cm^{-1}).

2. CH_3PH_2 and PH_3CH_2

For CH_3PH_2, Absar and Van Wazer [44], using uncontracted moderate basis sets with and without d on P, made SCF calculations for three relative orientations of the CH_3 and PH_2 groups. They found a barrier of 1.83 kcal/mole for the eclipsed versus the stabler staggered rotamer, as compared with the experimental value of 1.96. The authors tabulate orbital energies and population analyses and show overall and valence-shell-MO density maps.

The same authors [45] made a similar calculation for the ylide PH_3CH_2, which is often described in terms of resonance between $P^+H_3—CH_2^-$ and $PH_3═CH_2$. Here there is practically no difference in energy between eclipsed and staggered forms. A population analysis and electron density maps are given. Effects of including d on phosphorus on total energy, orbital energies, and charge on P are compared for PH_3, P_2H_4, OPH_3, OPF_3, HCP, CH_3PH_2, and PH_3CH_2. In PH_3CH_2 with d inclusion, total energy is lowered by 0.10 a.u. and charge on P is +0.40. Both effects are larger for OPH_3 and especially OPF_3. There is evidently considerable $d\pi—p\pi$ bonding in PH_3CH_2.

With a polarized double-zeta or triple-zeta basis and valence-shell electron correlation using CEPA and other methods, Lischka [46] has made improved calculations on PH_3CH_2, also on its protonated adduct

$PH_3CH_3^+$. He finds PH_3CH_2 to be slightly stabler with a 10° rotation of the CH_2 plane from an eclipsed orientation, but orientation is almost free. Calculated bond lengths and P—C force constant are given, also the dipole moment (3.18 D). Contour diagrams of the highest occupied MO (10a′) and orbital energies are given. Population analyses are discussed. The calculated proton affinity is given.

Trinquier and Malrieu [47], using pseudopotential calculations with a fairly good basis set, give potential curves for the formation of PH_3CH_2 from PH_3 plus CH_2. For minimum energy approach (no barrier), the CH_2 plane is at first at a considerable angle (CH_2 pyramidalization) to a HPC line, but becomes practically coplanar in PH_3CH_2 (3.7° versus 10° by Lischka).

3. Silylamine (H_3SiNH_2)

In SCF calculations using a good basis set with d functions on N and Si, Lehn and Munsch [48] find a weakly nonplanar structure (inversion barrier 0.65 kcal/mole, pyramidality angle 31°) for silyamine. There is some $d\pi$—$p\pi$ bonding with charge transfer from Si to N.

4. Difluoramine and Dimethylamine

Using a moderate basis set plus d functions on the N atom, Skaarup *et al.* [49] have made geometrically optimized SCF computations on the pyramidal equilibrium and the planar transition states for F_2NH and $(CH_3)_2NH$. The calculated inversion barriers are 42 kcal/mole for the former and 2.6 (experimental: 4.4 ± 1) for the latter. Tilt and asymmetry as found here in the methyl groups of $(CH_3)_2NH$ have been discussed in Section A,2.

5. Cyanide–Isocyanide Isomerization

Using an STO-3G minimal basis, Moffat [50] has calculated optimum geometries and energies for the cyanides and isocyanides of the isomers (H C≡N and H—C≡N═O) of HNCO, and their hypothetical
  \ /
   O

transition states. The energies of isomerization from NCNCO, NCCNO, NCOCN, and NCONC to the corresponding isocyanides were calculated to be 42, 30, 45, and 41 kcal/mole, respectively, while the respective energy barriers were found to be 84, 68, 108, and 107 kcal/mole. Net charges and overlap populations were examined.

6. The Boranes and Carboranes

The B_2H_6 molecule has been discussed in Section VIII,C,2. Concerning the boranes and carboranes, there are many articles by Lipscomb and collaborators, and by others. A 1975 article by Lipscomb [51] surveys much of this work. Among others, Switkes *et al.* [52] have made minimal-basis SCF computations on B_4H_{10}, B_5H_9, and B_5H_{11}, and given orbital energies and forms of MOs, charge distributions, overlap populations, electron density contour maps, and dipole moments (nearly twice the experimental values). Magnetic properties are computed. Localized MOs are obtained in a second article. Various detailed results are given.

Laws *et al.* [53] have made a minimal-basis SCF study of $B_{10}H_{14}$. Electron density and density difference maps are given. Atomization energy and ionization potential agree well with experiment. Dipole moment, diamagnetic susceptibility, and shielding constants are computed.

Camp *et al.* [54] have discussed the relative energies of three isomers each of $C_2B_3H_5$, B-fluoro-$C_2H_2B_3F_3$, $C_2B_4H_6$, B-fluoro-$C_2B_4F_4$, $C_4B_2H_6$, and B-fluoro-$C_4H_4B_2F_2$. One of each of these has a classical configuration.

Using STO-3G methods, Brown and Lipscomb [55] have discussed adducts of triborane (B_3H_7) with the Lewis bases NH_3, H_2O, $(CH_3)_2O$, and CO. Six possible structures were studied. In all, a C_s structure was preferred. Also $B_3H_8^-$ and derivatives are discussed.

For B_4Cl_4, B_4F_4, and B_4H_4, see Section VII,C.

Pepperburg *et al.* [56] have made 4-31G and other calculations (including some CM) on the transient boron hydrides) B_2H_4, B_3H_7, B_3H_9, B_4H_8, and B_4H_{12}. Preferred symmetries are found to be as follows: D_{2d} for B_2H_4, C_{2v} or possibly C_s for B_3H_7, D_{3h} for B_3H_9, C_s or possibly D_{2h} for B_4H_8, and D_{4h} for B_4H_{12}. Topological rules and other features are discussed.

Possibilities of flexional behavior in $B_3H_8^-$ (4-31G calculations) and B_5H_{11} are studied by Pepperburg *et al.* [57].

Using up to 4-31G and polarized double-zeta basis sets, Pepperburg *et al.* [58] have studied the role of BH_5 in the hydrolysis of BH_4^-. The results indicate that BH_5 is probably a barely stable intermediate (versus BH_3 + H_2) of C_s symmetry.

7. Diphosphine, Hydrazine, Aminophosphine, and Hydrazyl

Ahlrichs *et al.* [59] have used a (10s 7p 1d) basis contracted to (6s 4p 1d) for P and (4s 1p) contracted to (2s 1p) for H in an SCF calculation on P_2H_4, followed by PNO-CI and CEPA calculations for CM. They obtain energy and dipole moment curves as a function of the rotation angle Φ from cis (0°) to trans (180°). The minimum energy is at about 80° (gauche form, as in N_2H_4, where Φ is near 90°), and there is a secondary minimum for trans about 2.5 kcal/mole higher (in N_2H_4 there is no such secondary minimum). The gauche to cis barrier is about 20 kJ/mole, while the gauche to trans barrier is about 3.5 kJ/mole. The energy curve is very flat from 50° to 310°.

Crowley *et al.* [60] have made 4-31G calculations for N_2H_4, P_2H_4, and H_2NPH_2 on the rotational and pyramidal inversion processes. For H_2NPH_2 they find $\Phi = 90°$, with a rotational barrier of 12 kcal/mole (presumably to cis). They subject their results to PMO analysis.

Chandler and McLean [61] made polarized double-zeta SCF calculations followed by first-order CM (1771 functions) on the lowest two electronic states of the hydrazyl ($H_2N—NH$) radical. The ground state is nonplanar with the NH_2 group depressed about 28° from a planar configuration. From one to the other of the two equivalent ground-state configurations, the barrier for rotation of the NH hydrogen is 25 kcal/mole, whereas for passage through a structure with a linear NNH arrangement it is 51 kcal/mole. Dipole moment components are given. The excited state is of C_s symmetry, with vertical and adiabatic excitation energies of 4.32 and 1.37 eV (cf. 0.85 and 0.75 eV for HO_2); computed oscillator strength for excitation is 0.0066. Further details are given.

8. Azoalkanes

Kao and Huang [62] have made minimal-basis STO-3G calculations to investigate the structures, energies, and conformations of *trans*- and *cis*-methyldiazene, *trans*-azomethane, diazetine, pyrazoline, 1,2,4,5-tetrazacyclohexa-1,4-diene, and *trans,trans*-1,2,4,5-tetraaza-1,4-pentadiene. A force field is developed and applied to a study of more than 50 molecules.

REFERENCES

1. K. Morokuma, *J. Chem. Phys.* **55,** 1236 (1971); K. Kitaura and K. Morokuma, *Int. J. Quantum. Chem.* **10,** 325 (1976).
2. K. Morokuma and H. Umeyama, *Chem. Phys. Lett.* **49,** 333 (1977).
3. L. Radom, W. J. Hehre, and J. A. Pople, *J. Am. Chem. Soc.* **94,** 2371 (1972).
4. D. Cremer, J. S. Binkley, J. A. Pople, and W. J. Hehre, *J. Am. Chem. Soc.* **96,** 6900 (1974).

4a. P. W. Payne and L. C. Allen, "Modern Theoretical Chemistry," Vol. 4. Plenum, New York, 1977.

5. L. M. Tel, S. Wolfe, and I. G. Csizmadia, *J. Chem. Phys.* **59,** 4047 (1973).
6. E. Flood, P. Pulay, and J. E. Boggs, *J. Am. Chem. Soc.* **99,** 5570 (1977).
7. C. W. Kern, R. M. Pitzer, and O. J. Sovers, *J. Chem. Phys.* **60,** 3583 (1974).
8. J. A. Pappas, *Chem. Phys.* **12,** 397 (1976).
9. H. B. Schlegel, S. Wolfe, and F. Bernardi, *J. Chem. Phys.* **67,** 4181 (1977).
10. T.-K. Ha, P. Groner, A. Bauder, and H. H. Günthard, *Chem. Phys.* **33,** 27 (1978).
11. C. E. Blom, C. Altona, and A. Oskam, *Mol. Phys.* **34,** 557 (1977).
12. L. Radom, P. J. Stiles, and M. Vincent, *J. Mol. Struct.* **48,** 431 (1978).
13. P. A. Kollman, C. F. Bender, and J. McKelvey, *Chem. Phys. Lett.* **28,** 407 (1974).
14. M. R. Peterson and I. G. Csizamida, *J. Am. Chem. Soc.* **100,** 6911 (1978).
15. E. Flood and E. W. Nilssen, *J. Mol. Struct.* **30,** 129 (1976).
16. J. R. Larson, N. D. Epiotis, and F. Bernardi, *J. Am. Chem. Soc.* **100,** 5713 (1978).
17. J. Pacansky and M. Dupuis, *J. Chem. Phys.* **68,** 4276 (1978).
18. T. A. Claxton and R. E. Overill, *J. C. S. Faraday Trans. II* **70,** 1005 (1974).
19. J. E. Williams, Jr. and A. Streitwieser, Jr., *J. Am. Chem. Soc.* **97,** 2634 (1975).
20. H. Lischka and H.-J. Köhler, *J. Am. Chem. Soc.* **100,** 5297 (1978).
21. L. Radom, J. A. Pople, V. Buss, and P. von R. Schleyer, *J. Am. Chem. Soc.* **94,** 311 (1972).
22. V. Dyczmons, V. Staemmler, and W. Kutzelnigg, *Chem. Phys. Lett.* **5,** 361 (1970).
23. P. C. Hariharan, W. A. Lathan, and J. A. Pople, *Chem. Phys. Lett.* **14,** 385 (1972).
24. H.-J. Köhler and H. Lischka, *Chem. Phys. Lett.* **58,** 175 (1978).
25. D. W. Goetz, H. B. Schlegel, and L. C. Allen, *J. Am. Chem. Soc.* **99,** 8118 (1977).
26. H.-J. Köhler and H. Lischka, *J. Am. Chem. Soc.* **101,** 3479 (1979).
27. J. Tyrell, V. M. Kolb, and C. Y. Meyers, *J. Am. Chem. Soc.* **101,** 3497 (1979).
28. D. M. Chipman, *J. Chem. Phys.* **71,** 761 (1979).
28a. E. D. Jemmis, J. Chandrasekhar, and P. von R. Schleyer, *J. Am. Chem. Soc.* **101,** 2848 (1979).
29. D. Cremer, *J. Chem. Phys.* **69,** 4440 (1978).
30. W. R. Rodwell, N. R. Carlsen, and L. Radom, *Chem. Phys.* **31,** 177 (1978).
31. A. Hinchcliffe, *J. Mol. Struct.* **55,** 127 (1979).
32. J. F. Olsen, *J. Mol. Struct.* **49,** 361 (1978).
33. D. Cremer, *J. Chem. Phys.* **69,** 4456 (1978).
34. S. Vishveshwara and J. A. Pople, *J. Am. Chem. Soc.* **99,** 2422 (1977).
35. Y.-C. Tse, M. D. Newton, S. Vishveshwara, and J. A. Pople, *J. Am. Chem. Soc.* **100,** 4329 (1978).
36. J. A. Ryan and J. L. Whitten, *J. Am. Chem. Soc.* **94,** 2396 (1972).
37. M. D. Newton and G. A. Jeffrey, *J. Am. Chem. Soc.* **99,** 2413 (1977).
38. G. A. Jeffrey, J. A. Pople, J. S. Binkley, and S. Vishveshwara, *J. Am. Chem. Soc.* **100,** 373 (1978).
39. T.-K. Ha, *Chem. Phys. Lett.* **30,** 379 (1975).
40. M. Hotokka and P. Pyykkö, *J. Mol. Struct.* **51,** 133 (1979).
41. B. Bigot, D. Roux, A. Senin, and A. Devaquet, *J. Am. Chem. Soc.* **101,** 2560 (1979).
42. E. Oliveros, M. Riviere, J. P. Malrieu, and G. Teichteil, *J. Am. Chem. Soc.* **101,** 318 (1979).
42a. H. Wallmeier and W. Kutzelnigg, *J. Am. Chem. Soc.* **101,** 2804 (1979).
43. C. W. Bauschlicher, Jr., D. H. Liskow, C. F. Bender, and H. F. Schaefer, III, *J. Chem. Phys.* **62,** 4815 (1975); C. W. Bauschlicher, Jr., P. S. Bagus, and H. F. Schaefer, III, *IBM J. Res. Dev.* **22,** 213 (1978).
44. I. Absar and J. R. Van Wazer, *J. Chem. Phys.* **56,** 1284 (1972).
45. I. Absar and J. R. Van Wazer, *J. Am. Chem. Soc.* **94,** 2382 (1972).

46. H. Lischka, *J. Am. Chem. Soc.* **99,** 353 (1977); see also A. Strich, *Nouv. J. Chim.* **3,** 105 (1979).
47. G. Trinquier and J.-P. Malrieu, *J. Am. Chem. Soc.* **101,** 7169 (1979).
48. J. M. Lehn and B. Munsch, *Chem. Commun.* p. 994 (1970).
49. S. Skaarup, L. L. Griffin, and J. E. Boggs, *J. Am. Chem. Soc.* **98,** 3140 (1976).
50. J. B. Moffat, *Int. J. Quantum Chem.* **15,** 547 (1979).
51. W. N. Lipscomb, *in* "Boron Hydride Chemistry" (E. L. Muetterties, ed.), p. 39. Academic Press, New York, 1975.
52. E. Switkes, I. R. Epstein, J. A. Tassell, R. M. Stevens, and W. N. Lipscomb, *J. Am. Chem. Soc.* **92,** 3837 (1970); E. Switkes, W. N. Lipscomb, and M. D. Newton, *J. Am. Chem. Soc.* **92,** 3847 (1970), localized MO's.
53. E. A. Laws, R. M. Stevens, and W. N. Lipscomb, *J. Am. Chem. Soc.* **94,** 4467 (1972).
54. R. N. Camp, D. S. Marynick, G. D. Graham, and W. N. Lipscomb, *J. Am. Chem. Soc.* **100,** 6781 (1978).
55. L. D. Brown and W. N. Lipscomb, *Inorg. Chem.* **16,** 1 (1977).
56. I. M. Peppenburg, T. A. Halgren, and W. N. Lipscomb, *Inorg. Chem.* **16,** 363 (1977).
57. I. M. Pepperburg, D. A. Dixon, W. N. Lipscomb, and T. A. Halgren, *Inorg. Chem.* **17,** 587 (1978).
58. I. M. Pepperburg, T. A. Halgren, and W. N. Lipscomb, *J. Am. Chem. SOC.* **98,** 3442 (1976).
59. R. Ahlrichs, R. Heinzmann, and C. Zirz, *Theor. Chim. Acta* **43,** 29 (1976).
60. A. H. Crowley, D. J. Mitchell, M.-H. Whangbo, and S. Wolfe, *J. Am. Chem. Soc.* **101,** 5224 (1979).
61. G. S. Chandler and A. D. McLean, *J. Am. Chem. Soc.* **71,** 2175 (1797).
62. J. Kao and T.-N. Huang, *J. Am. Chem. Soc.* **101,** 5546 (1979).

CHAPTER XI

AROMATIC MOLECULES

A. GENERAL CONSIDERATIONS

Chapters II–X have dealt with inorganic molecules and with aliphatic organic molecules, arranged largely in terms of symmetry and size. Aromatic molecules are a rather distinctive class of mostly large molecules and ions that follow Hückel's well-known rule of possessing $4n + 2$ π electrons. Simplest is the cyclopropanyl cation

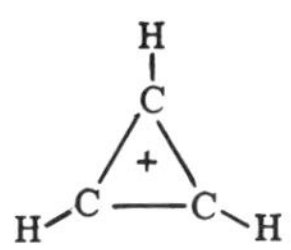

with $n = 0$. More typical are benzene (C_6H_6), the cyclopentadienyl anion ($C_5H_5^-$), the cycloheptanyl cation ($C_7H_7^+$), pyrrole (C_5H_5NH), furan (C_5H_5O), and thiophene (C_5H_5S), all with $n = 2$. These are all monocyclic. Then there are very numerous polycyclic compounds containing one or more fused or unfused rings similar to those just mentioned, and/or containing heteroatoms or aliphatic or inorganic substituent atoms or

TABLE 1

Symmetry Operations and Characters for the D_{6h} Group

D_{6h}	E	$2C_6(x)$	$2C_6^2 \equiv 2C_3$	$C_6^3 \equiv C_2''$	$3C_2$	$3C_2'$	σ_h	$3\sigma_v$	$3\sigma_d$	$2S_6$	$2S_3$	$S_6^3 \equiv S_2 \equiv i$	Coordinates
A_{1g}	1	1	1	1	1	1	1	1	1	1	1	1	
A_{1u}	1	1	1	1	1	1	−1	−1	−1	−1	−1	−1	
A_{2g}	1	1	1	1	−1	−1	1	−1	−1	1	1	1	
A_{2u}	1	1	1	1	−1	−1	−1	1	1	−1	−1	−1	x
B_{1g}	1	−1	1	−1	1	−1	−1	−1	1	1	−1	1	
B_{1u}	1	−1	1	−1	1	−1	1	1	−1	−1	1	−1	
B_{2g}	1	−1	1	−1	−1	1	−1	1	−1	1	−1	1	
B_{2u}	1	−1	1	−1	−1	1	1	−1	1	−1	1	−1	
E_{1g}	2	1	−1	−2	0	0	−2	0	0	−1	1	2	
E_{1u}	2	1	−1	−2	0	0	2	0	0	1	−1	−2	y, z
E_{2g}	2	−1	−1	2	0	0	2	0	0	−1	−1	2	
E_{2u}	2	−1	−1	2	0	0	−2	0	0	1	1	−2	

groups. The rings in aromatic molecules are normally planar, π MOs being those MOs that have a nodal surface in the ring plane. Aromatic molecules are of varied symmetry.

The most symmetrical is benzene (symmetry D_{6h}) with group theoretical species and characters as given in Table 1. The $C_5H_5^-$ and $C_7H_7^+$ molecules have symmetries C_{5h} and C_{7h}, respectively. For cyclic or fused-ring aromatic molecules we shall assume the x axis to be perpendicular to the molecule plane, along or parallel to the symmetry axes of the π MOs.

Because of the large numbers of electrons present, most of the calculations on aromatic molecules have been made at a lower level of approximation than is feasible for smaller molecules.

B. SIX-MEMBERED RING COMPOUNDS AND RELATED MOLECULES

1. Benzene

A very good SCF computation on ground-state benzene was made by Ermler and Kern [1] using a (9s 5p 1d/4s 1p) polarized Gaussian basis set contracted to (4s 2p 1d/2s 1p) with differently scaled s and p exponents. They found $E_h = -230.74938$ and estimated the SCF limit as -230.82 ± 0.02. They give orbital energies and several computed one-electron properties. Ermler *et al.* [2] made a somewhat improved computation using an (11s 7p 1d/7s 1p) contracted to (5s 3p 1d/3s 1p) basis, with resulting $E_h = -230.7771$. They give orbital energies, MOs in terms of coefficients, and a table of one-electron properties in atomic units (electric field gradient components, charge densities at nuclei, and nuclear-electron potentials, each property also broken down into nuclear, σ, and π contributions), along with expectation values of powers of R and of X^2, Y^2, and Z^2 with origin at ring center, and overlap populations between carbon atoms.

We make brief reference to the earlier-favored procedure of describing aromatic and other wave functions in terms of resonance among a set of valence-bond (VB) π-electron structures. Buenker *et al.* [3] made an SCF MO calculation ($E_h = -230.3745$) followed by extensive CM for the π electrons only ($E_h = -230.4543$). Norbeck and Gallup [4] obtained very nearly the same E_h values in an SCF MO and a CM calculation, but used a different approach in the CM calculation. Namely, they combined the SCF MO σ-electron core with each of the 22 types of VB resonance structures in turn, which can be reported by Rumer diagrams such as

etc., and computed the ${}^1A_{1g}$ energy corresponding to the particular diagram. They then combined them and analyzed the final CM function into components corresponding to the different diagrams. Their main conclusion is that the polar and multipolar resonance structures are much more important than was thought in the early days. A greatly improved "resonance energy" for benzene is one of the results. Tantardini *et al.* [5], using a minimal-basis set, presented similar calculations on the resonance structures and their CM ($E_h = -230.3160$).

For the ionization potentials and one-electron properties of benzene, the best results by far are those obtained by von Niessen *et al.* [6] in their many-body approach, which takes good account of electron correlation and reorganization. They used an unpolarized Gaussian basis set, (9s 5p/4s) contracted to (4s 2p/2s), yielding an SCF E_h of -230.5806. Their final results are given in Table 2. Of special interest is the ordering of the second (σ) and third (π) ionization potentials, a question which has

TABLE 2
Orbital and Ionization Energies of Benzene

MO	$-\epsilon$ (eV)	Ionization potential (eV)	
		Theory	Experiment[a]
$1a_{1g}$	305.8		
$1e_{1u}$	305.8		
$1e_{2g}$	305.7		
$1b_{1u}$	305.7		
$2a_{1g}$	31.35		
$2e_{1u}$	27.65		
$2e_{2g}$	22.45	20.01	19.3
$3a_{1g}$	19.32	17.48	17.0
$2b_{1u}$	17.52	15.75	15.5
$1b_{2u}$	16.90	14.83	14.7
$3e_{1u}$	16.03	14.46	14.1
$1a_{2u}$ (π)	13.63	12.26	12.2
$3e_{2g}$	13.50	11.95	11.7
$1e_{1g}$ (π)	9.17	9.10	9.3

[a] Photoelectron (from centroids) and Rydberg spectroscopy.

long been disputed. From Rydberg series, Potts *et al.* [7] identified the 11.49 eV ionization potential as $1a_{2u}$ (π), but others (e.g., [8]) identified it as σ. It is of some interest that in an early semiempirical calculation [9] Mulliken obtained 14.33 eV for $1a_{2u}$ and 9.45 eV for $1e_{1g}$, thus favoring the larger value for $1a_{2u}$. To confirm the identifications of the $3e_{2g}$ and $1a_{2u}$ potentials, von Niessen *et al.* computed vibrational profiles; agreement with experiment is good. They also give tables of SCF-computed one-electron properties.

The most accurate computations on the π-electron excited states of benzene appear to be those of Hay and Shavitt [8]. They used a double-zeta basis of contracted Gaussians, and first made an SCF computation on the ground state ($E_h = -230.6410$). Then for the excited states they held the ground-state σ core fixed while they varied the π configurations ("frozen σ core method") in fairly extended CM computations. Before doing so, they added two diffuse π functions on each carbon atom to the two main π functions. Although single and double substitutions account for nearly all the correlation energy in the ground state, it is necessary to include triple substitutions for excited states. Since there are too many (sometimes over 7000) of these for practical computation, natural orbitals (NOs) were obtained from a ground-state CM wave function including all 182 single and double substitutions. By a selection procedure among the NOs guided by their occupation numbers, the total numbers of configurations included for various excited states were restricted to from 469 to 2636 (most near the middle of this range).

The computed excitation energies of singlet and triplet valence-type and Rydberg states from various orbital excitations are shown in Figs. 1 and 2. Valence-type states $^5A_{1g}$ from the double excitation $1e_{1g}{}^2 \rightarrow 1e_{2u}{}^2$ and $^5E_{1u}$ from $1a_{2u}\,1e_{1g} \rightarrow 1e_{2u}{}^2$ mixed with $1e_{1g}{}^2 \rightarrow 1e_{2u}\,1b_{2g}$ were also computed. In the computations, three to six roots were recorded for the secular equations of each symmetry. Many of the computed states are of valence type, including most of the $\pi \rightarrow \pi^*$ and the quintet states; others are of Rydberg type (the MOs $2a_{2u}$, $2e_{1g}$, and $1e_{2u}$ correspond to united-atom $3p\pi$, $3d\pi$, and $4f\pi$ atomic orbitals, respectively). The $1e_{1g} \rightarrow 1e_{2u}$ excitations are the lowest of the known excitations. Of these, only the $^1E_{1u} \leftarrow {}^1A_{1g}$ is allowed; it is a V $\leftarrow$ N transition of high intensity. In the $^1E_{1u}$ state, the π^* MO differs from that in the other π^* states in being larger ("diffuse"). In this regard, and also in the fact that the wave function in terms of VB structures has strong polar character, the $^1E_{1u}$ state here resembles the V state of ethylene. The $\pi\pi^*$ states $^1B_{1u}$ and $^3B_{2u}$ also have strong polar character, and share with the $^1E_{1u}$ state the fact that their computed excitation energies are about 1.4 eV too high as compared with experiment. For the other states, including the Rydberg states with

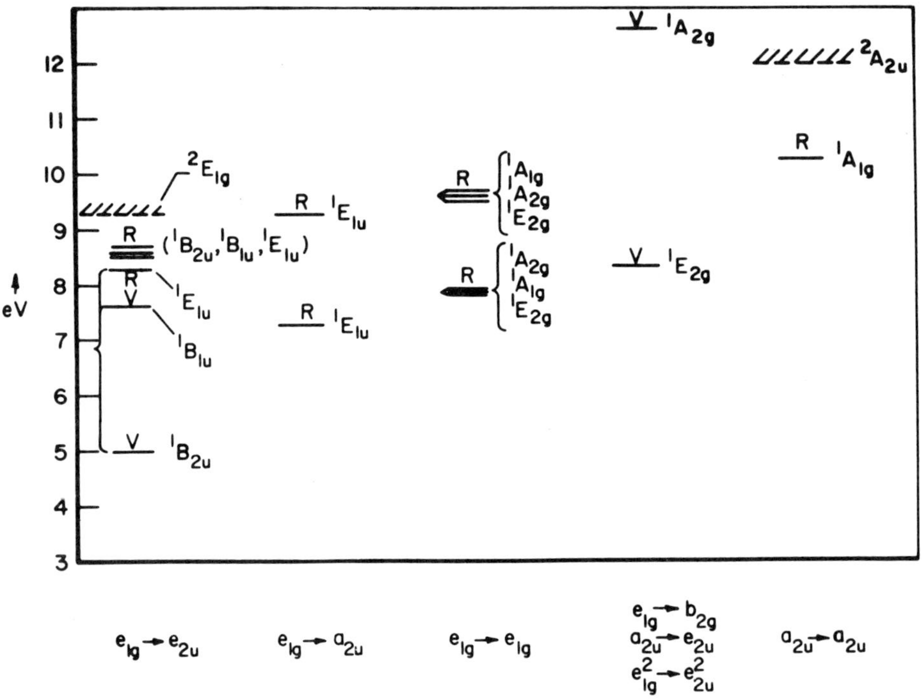

FIG. 1. Energy levels of π-electron singlet states of benzene from CM calculation, arranged by dominant orbital excitation as shown beneath each column. R and V refer to Rydberg or valence excitation. The two π ionization limits are aso shown. From Hay and Shavitt [8].

united-atom $n = 3$, agreement with experiment is fairly good. The $^1E_{2g}$ states (in part doubly excited) are of some special interest; their relation to the experimental evidence is discussed. An NO analysis is given for occupation numbers and $\langle z^2 \rangle$ values of the main occupied NOs.

Rose *et al.* [10] used the equations of motion method to calculate singlet and triplet excitation energies and intensities for transfers out of the $3a_{1g}$, $1b_{2u}$, $2b_{1u}$, $3e_{1u}$, and $3e_{2g}$ σ MOs as well as the $1a_{2u}$ and $1e_{1g}$ π MOIs into the excited $1e_{2u}$ (π^*) and $1b_{2g}$ MOs to give B_{2u}, B_{1u}, E_{1u}, and E_{2g} states. They began with an SCF computation using a rather small basis—(7s 3p/3s) contracted to (3s 2p/1s)—giving $E_h = -230.2184$. They used in part double excitations and obtained results mostly about 1 eV above experiment for the eight states considered. For the $^1E_{1u} \leftarrow {}^1A_{1g}$ allowed transition, they obtained a transition moment of 1.74 a.u., as compared with an experimental value of 1.61. a.u. ($f = 0.88$). Additionally, they obtained a $\pi^* \leftarrow \sigma$ transition to a $^1A_{2u}$ state with f about 0.03. Electron spin resonance spectra ($D = 0.1580$, $E = -0.0064\ \text{cm}^{-1}$) for the $\pi\pi^*$ $^3B_1$ lowest excited state show that, at least in C_6D_6 host crystals, it has less than D_{6h} symmetry. This experimental evidence and hyperfine experiments indicate that the hexagon is elongated (two long bonds). Theoretical calculations have been made, but for a more conclusive test Langhoff *et al.* [11] made double-zeta level SCF computations with geometry optimization, followed by some CM studies, on the electron spin dipole–dipole contribution D to the zero-field splitting. A large CM wave function gave $D = 0.1676\ \text{cm}^{-1}$ for D_{6h} symmetry, but both elongated and squeezed-ring forms with only 2–3% distortions gave zero-field splittings that differ qualitatively from experiment. The results are very sensitive to certain types of CM ($\pi^2 \rightarrow \pi'^2$ or $\pi \rightarrow \pi'$ and $\sigma\pi \rightarrow \sigma'\pi'$ increase D, $\sigma^2 \rightarrow \sigma'^2$ or $\sigma \rightarrow \sigma'$ decrease D, whereas $\sigma^2 \rightarrow \pi^2$ and $\pi^2 \rightarrow \sigma^2$ have little effect). The indicated conclusion is that the actual distortion is very small, and it is not unlikely that it is due to environmental effects and disappears in the free molecule. However, there is a need for better calculations, with a bigger basis set and more complete electron correlation.

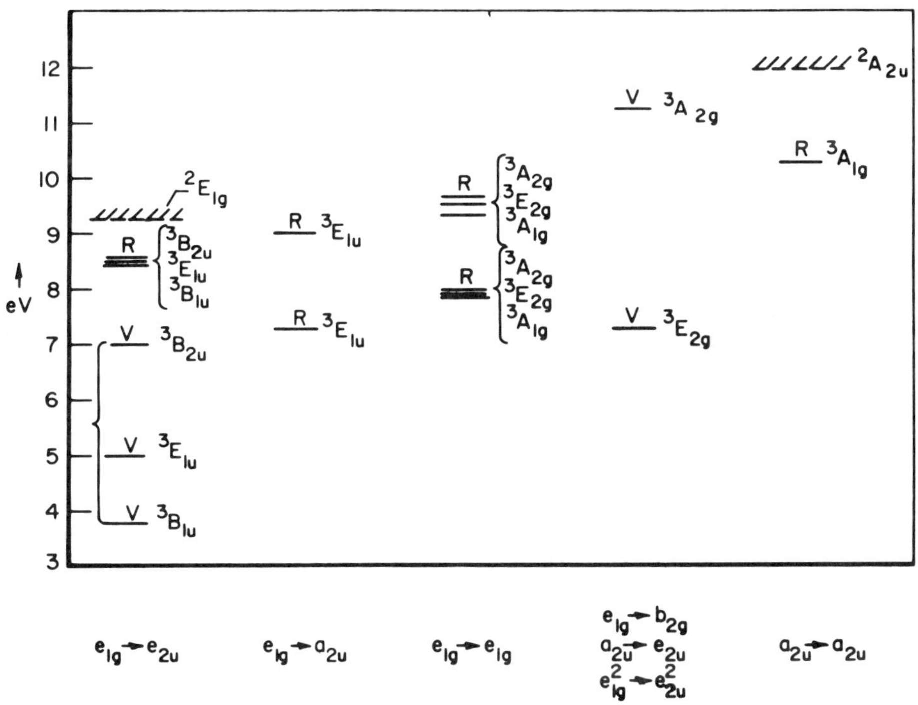

FIG. 2. Energy levels of π-electron triplet states of benzene. (See Fig. 1 for explanation.) From Hay and Shavitt [8].

2. Borazine

Earlier Peyerimhoff and Buenker [12] made SCF and CM computations on benzene and the isoelectronic molecule borazine ($B_3N_3H_6$). They used a (10s 5p/5s) Gaussian-lobe basis contracted to (3s 1p/1s), afterward somewhat re-expanded to include two pπ contracted functions. Although the π electrons are not as uniformly distributed in borazine as in benzene, they conform to Hückel's $4n + 2$ rule for an aromatic molecule, with polar B^-N^+ π bonds; the situation is similar to that in the aromatic molecules furan, thiophene, and pyrrole; however, the symmetry (D_{3h}) is higher here. The authors give a list of orbital energies in borazine and a detailed discussion of the character of the bonding. In the BN σ framework there is very strong charge transfer from B to N, strongly outweighing the strong N $\rightarrow$ B π charge transfers. The overall effect may be represented qualitatively by

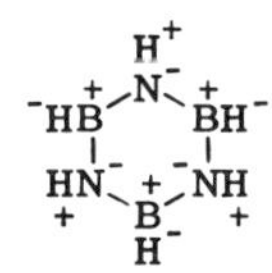

Of the total π population in each BN pair, a population analysis indicates 0.2 e on B, 0.4 e in the π bond, and 0.4 e on the N. As compared with $H_3N{\cdot}BH_3$ (cf. Section VII,F,2), there is more N $\rightarrow$ B π transfer here then N $\rightarrow$ B pσ transfer there. Valence excited states of benzene and borazine are computed.

3. The Benzenium Ion (Protonated Benzene)

Minimal-basis (STO-3G) calculations by Hehre and Pople [13] suffice to show that the proton is attached to one carbon atom to form a CH_2 group with its plane perpendicular to that of the ring. (Whether the ring may perhaps become slightly nonplanar is not established, but if not, the symmetry is C_{2v}.) The form in which the proton bridges a C—C bond is found to be less favored, and structures corresponding to edge and face protonation are found to be poorer still.

Along with their benzene computations (see above), Ermler *et al.* [2] made detailed SCF computations on the benzenium ion using a large basis set. We assumed a C_{2v} symmetry ion made up of a C_5H_5 part with the same CC and CH distances (1.395 and 1.085 Å) as in benzene, and a CH_2 part for which, by computing a none-point potential energy surface, we optimized the HCH angle and the distances from the C atom to the neighboring C atoms. We assumed a CH distance of 1.0975 Å in the CH_2 part.

The resulting HCH angle was 103.5° and the CC distance was 1.45 Å, giving a CCC angle of 113.5° at the CH_2 group. We numbered the atoms as follows:

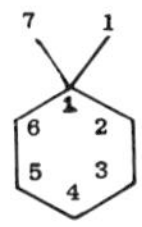

From a comparison of the total SCF energy with that of benzene, we obtained a proton affinity of 189.1 kcal/mole, fairly close to the experimental value of 183 ± 3.

A comparison of the σ and π MOs of benzene with the σ, π and quasi-π MOs of benzenium is of interest; some of their orbital energies (we give a complete list) are shown in Table 3.
As compared with benzene, the MOs which are degenerate in benzene are of course split, and *all* ϵ become more negative. For most MOs, the shifts are surprisingly uniform (about 0.21 a.u.), but for the b_1 MOs, they are larger (about 0.3 a.u.) as a result of bonding to the quasi-π H_2 orbitals of the CH_2 group (hyperconjugation). In the splitting of $1e_{1g}$ of benzene into $2b_1$ and $1a_2$ of benzenium, the form of the $1a_2$ MOs remains almost exactly the same as in benzene and its ϵ is shifted only -0.21 a.u. The form and bonding characteristics of the π and quasi-π MOs (a_2 and b_1) are displayed in a table in our article. An interesting consequence of the downward ϵ shift in benzenium is that the π^* MO of benzene, which there is unbound,

TABLE 3
Orbital Energies ϵ of MOs in Benzene and Benzenium

Benzene		Benzenium		
MO	ϵ	MO	ϵ	shift
$1e_{2u}$	+0.119	$2a_2$	−0.068	−0.187
		$3b_1$	−0.183	−0.302
$1e_{1g}$	−0.337	$1a_2$	−0.546	−0.209
		$2b_1$	−0.640	−0.303
$3e_{2g}$	−0.496	$7b_2$	−0.699	−0.203
		$11a_1$	−0.715	−0.219
$1a_{2u}$	−0.501	$1b_1$	−0.811	−0.310
$3e_{1u}$	−0.589	$6b_2$	−0.794	−0.205
		$10a_1$	−0.800	−0.211
$1b_{2u}$	−0.621	$5b_2$	−0.842	−0.221
$2b_{1u}$	−0.644	$9a_1$	−0.863	−0.219
$1a_{1g}$	−11.238	$1a_1$	−11.495	−0.257

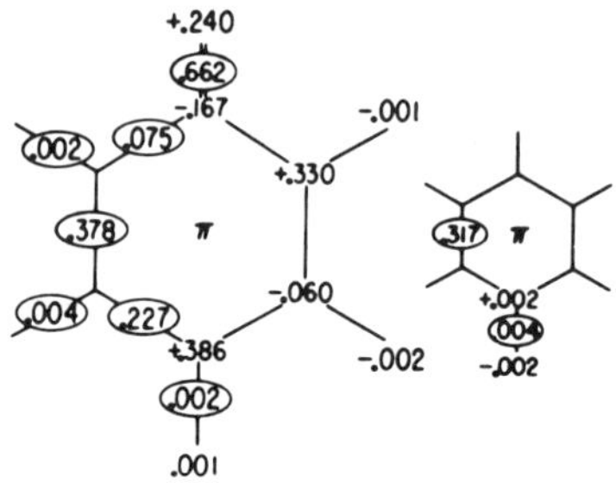

FIG. 3. π and quasi-π charges and overlap populations (encircled) in benzenium ion and benzene, omitting C_1C_3, C_1C_4, H_1H_7, etc. overlap terms. From Ermler *et al.* [2].

becomes bound (and split) in benzenium, suggesting that the radical C_6H_7 and perhaps even $C_6H_7^-$ may be stable.

Various one-electron properties are computed (electric field gradients at nuclei, charge densities at nuclei, nuclear-electron potentials, powers of R and X^2, Y^2, Z^2 with origin at the center of the ring). The values of the field gradients at the nuclei are of particular interest since large changes from benzene are found in the C_1 and C_4 positions. These results, especially for the π electrons, are connected with a population analysis, results of which for the π and quasi-π electrons are shown in Fig. 3, taken from our article, where other similar diagrams for the σ and total charge distributions are also shown. As is seen from the overlap populations in Fig. 3, bonding (C_2C_3 and C_5C_6) is enhanced in the central positions as compared with benzene (partial quinoid structure). From the changes, one can estimate that, as compared with the *assumed* 1.395 Å bond lengths, the C_2—C_3 and C_5—C_6 bonds are shortened in benzenium to about 1.38 Å and the C_3—C_4 and C_4—C_5 bonds are lengthened to about 1.415 Å. Further, the C_2—C_6 distance is predicted to increase by 0.006 Å. These variations in geometry, and the further possibility of a slightly nonplanar ring, should not affect the accuracy of our numerical results very much. One additional result is an estimate of a hyperconjugation energy of about 6.5 kcal/mole for the benzenium ion.

4. Toluene and the *p*-Toluenium Ion

Ermler and Mulliken [14] have made SCF computations on toluene and the toluenium ion similar to those of Ermler *et al.* [2] on benzene and the benzenium ion. For the toluenium ion, four low-energy positions of protonation of toluene can be considered: ipso (at the methyl group), and ortho, meta, and para to the methyl group. According to STO-3G calculations by Devlin *et al.* [15], the respective proton affinities for these four positions *relative* to those of benzene are 0.3, 6.5, 2.0, and 8.5 kcal/mole. To simplify the calculations we have restricted them to the stablest form *p*-toluenium, with the CH_3 group oriented with two H atoms on a line parallel to the ring plane, giving C_s symmetry. For the methyl group we assumed tetrahedral symmetry, and for the rest of the molecule we assumed the same geometry that we used or found in our calculations on the benzenium ion.

For toluene and toluenium we obtain $E_h = -269.8245$ and -270.1399, respectively, and we give a comparative tabulation of the orbital energies of the MOs of benzene, toluene, toluenium, and benzenium. Toluene and toluenium, of course, have some extra MOs without analogs in the other two. Three of the MOs (12a′, 14a′, and 16a′) are π and quasi-π and involve hyperconjugation at the CH_2 and CH_3 ends of the molecule or ion. A fourth (9a″) is a π MO essentially localized on the ring and practically the same as a benzene MO. The nature of the weak (sacrificial) hyperconjugation in toluene and the much stronger hyperconjugation in toluenium are discussed.

A table of electric field gradients at the nuclei for toluene and toluenium is given; also nuclear electron potentials and charge densities at the nuclei, and values of R^{-1}, Z, X, R^2, X^2, Y^2, Z^2, and XZ are listed. A comparative diagram for benzene, benzenium, toluene, and toluenium showing the field gradients at the nuclei is reproduced here as Fig. 4. A figure based on population analyses, showing π-electron charges and overlap populations for toluene and toluenium, is reproduced here as Fig. 5. Diagrams of the σ and total populations are also given in our article. Protonation of toluene greatly strengthens C_2C_3 and C_5C_6 bonding and greatly weakens C_1C_2, C_1C_6, C_4C_3, and C_4C_5 bonding, for both σ and π bonds.

Our computed dipole moment for toluene is -0.130 a.u. (-0.330 D) in the plane, together with a small component (0.031 D) perpendicular to the plane. The direction corresponds to $CH_3{}^+C_6H_5{}^-$, as is usually expected. The magnitude agrees with the observed 0.36 D. On the other hand, the population analysis (though not to be entirely trusted, of course) makes the CH_3 group total population $-0.102\ e$ even though the π part is $+0.018$ e (see Fig. 5). This result would indicate that the direction of the dipole moment is due not to electron transfer from the CH_3 but instead to ring polarization of the σ electrons in a direction away from the CH_3 group.

5. The Ethylenebenzenium Cation

In an STO-3G minimal-basis calculation, Hehre [16] has shown that the stablest structure of the $C_8H_9{}^+$ ethylenebenzenium cation is one in which

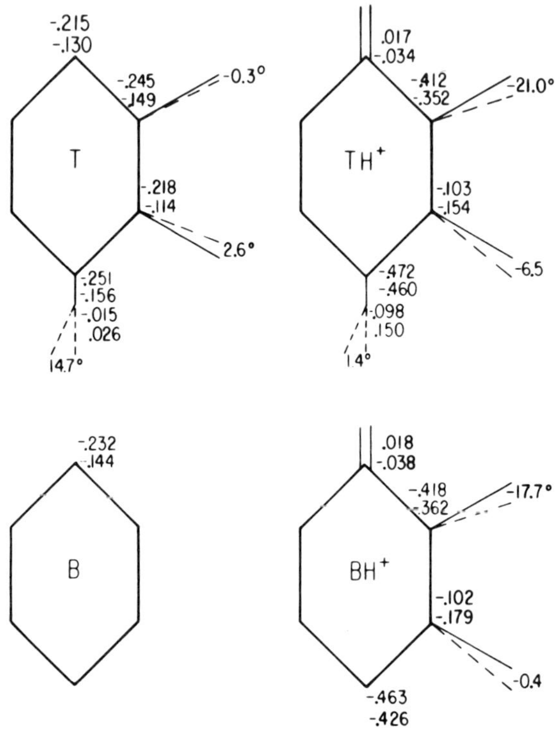

FIG. 4. Electric field gradients in atomic units at unique carbon atoms for toluene (T), toluenium (TH^+), benzene (B), and benzenium (BH^+). The upper value at each center is q_{bb} and the lower q_{cc}. The angles indicate the deviation of the principal axis system from the C—H bond direction in the plane of the ring (C_2 and C_3) and perpendicular to this plane (C_{4M}). From Ermler and Mulliken [14].

the two H atoms of the CH_2 group of the benzenium ion are replaced by two CH_2 groups with a single bond between them.

6. Valence Isomers of Benzene

Newton *et al.* [17] have made SCF calculations on the benzene isomers Dewar benzene (1), benzvalene (2), and prismane (3), using STO-3G for

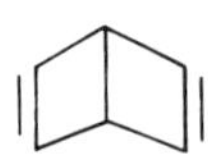  

1 **3** **2**

geometries and 4-31G for energies. They give geometries, orbital energies, dipole moments, and some nuclear spin–spin coupling constants.

7. The Benzynes

Three isomers of benzyne (bisdehydrobenzene) should occur, designated ortho, meta, and para, with the following structures, the first two of

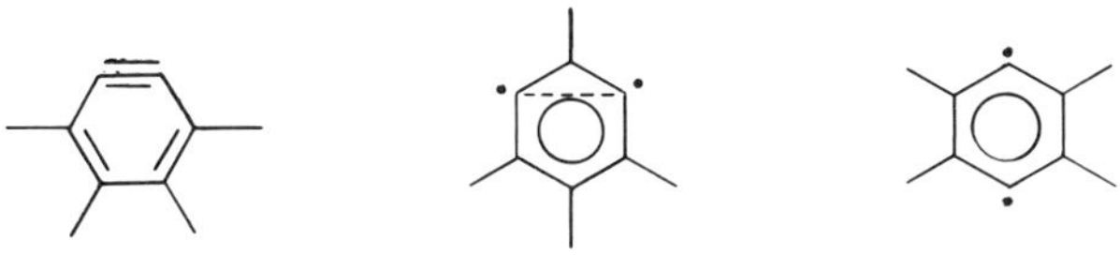

ortho meta para

symmetry C_{2v} and the third D_{2h}. Removal of two H atoms leaves two odd electrons which in *o*-benzyne are paired in a σ bond so as to produce a triple bond. The odd-electron σ pair in the other two isomers may be regarded as a biradical structure, or in *p*-benzyne perhaps as a weak cross-bond. In both these isomers the six π electrons should interact almost in the same way as in benzene. The odd-electron σ pair in all three isomers should interact not only to give the lowest singlet state, say Φ_s^2, ${}^1A_1$, but also a low-energy triplet $\Phi_s\Phi_a$, ${}^3B_2$ and two higher singlet states $\Phi_s\Phi_a$, ${}^1B_2$ and Φ_a^2, ${}^1A_1$, a set like the N, T, V, and Z states of H_2. Here Φ_s and Φ_a are the symmetric and antisymmetric combination of the two odd-electron orbitals.

For accuracy, as Wilhite and Whitten [18] have pointed out in an SCF plus CM calculation, one should introduce two-determinant CM between

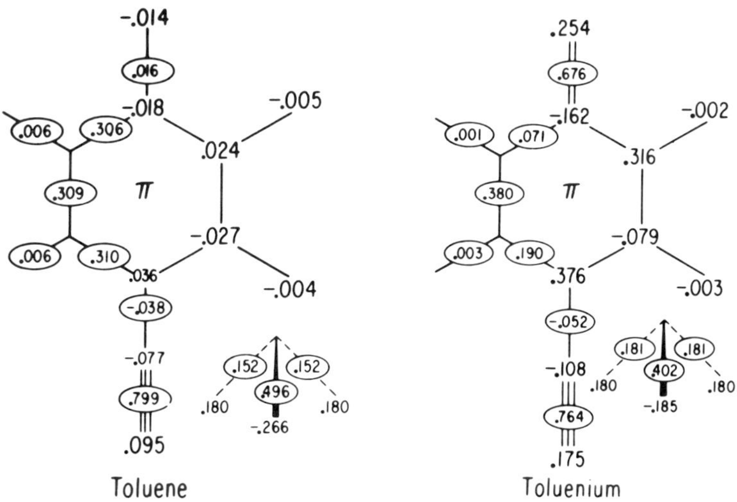

FIG. 5. Computed π and quasi-π charges and overlap populations (encircled) in toluene and toluenium. From Ermler and Mulliken [14].

the two ${}^1A_1$ functions. They have done so. They have then introduced large-scale CM in the computation of excitation energies of various singlet and triplet states of *o*-benzyne (including $\sigma \rightarrow \sigma^*$, $\pi \rightarrow \sigma^*$, $\pi \rightarrow \pi^*$ types).

Wilhite and Whitten assumed benzene geometry for the benzyne rings. Noell and Newton [19] have made new calculations, using a 4-31G basis and two-determinant wave functions to determine the equilibrium geometries and energies for the lower-lying singlet and triplet states. Their results, which differ considerably from those of Wilhite and Whitten, are as follows. For the singlet states, *o*-benzyne is the most stable (and has been observed experimentally), having $E_h = -229.1080$, with *m*-benzyne and *p*-benzyne computed to be, respectively, 14.5 and 23.3 kcal/mole higher. The triplets are computed to be higher than the singlets by 28.1, 12.2, and 1.4 kcal/mole for *o*-, *m*-, and *p*-benzyne, respectively. Overlap populations indicate much stronger π bonding in the triple than in the double bonds in *o*-benzyne, but nearly uniform π bonding in all the π bonds, as in benzene, in the other two isomers. There are also interesting variations in the σ bonds.

8. The Phenyl Cation ($C_6H_5^+$)

Dill *et al.* [20], optimizing geometries at the STO-3G level, calculating energies at 4-31G, and making corrections for correlation, discuss the ${}^1A_1$ singlet ground state and two excited states ${}^3B_1$, estimated to be 20 kcal/mole higher, and ${}^3A_2$ higher still. In the singlet state, the angle α is opened to 145°, but in the triplet states the angles are more benzene-like.

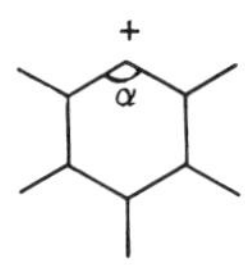

9. Azabenzenes and Phosphorin

Almlöf *et al.* [21] have made comparative SCF computations on benzene and the azabenzenes pyridine (4), pyridazine (5), pyrimidine (6), pyrazine (7), *s*-triazine (8), and *s*-tetrazine (9). They used (7s 3p/4s) basis

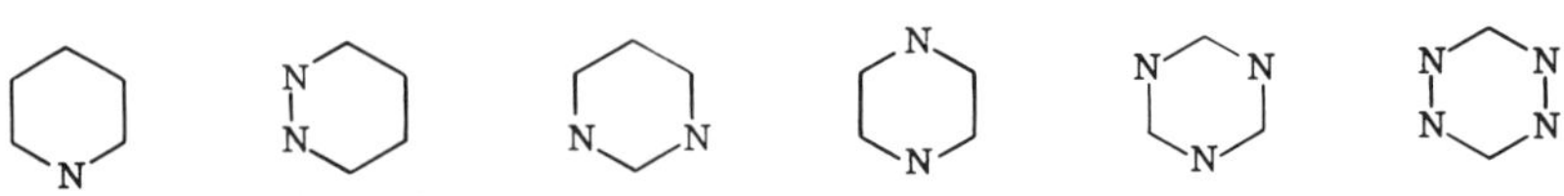

4 5 6 7 8 9

sets contracted to (4s 2p/2s), plus 2pσ polarization functions on the H atoms. The results are presented in tables of orbital and total energies. By means of corrections increasing regularly from 0 to 5.7 eV with orbital energy, estimated ionization potentials are obtained for the valence-shell electrons that agree very well with those from photoelectron spectroscopy. The same correction is used for *corresponding* MOs of benzene and all the azabenzenes. For pyridine (E_h = -246.417) the lowest ionization potentials agree well with the observed values, except that it is uncertain whether the π MO $1a_2$ or the lone-pair σ MO $11a_1$ comes first. For all the azabenzenes, a number of molecular properties (dipole moments, quadrupole moments, diamagnetic susceptibilities, potential and field gradients at the nitrogen nuclei, quadrupole coupling constants for N^{14}, and also $\langle x^2 \rangle$, $\langle y^2 \rangle$, $\langle z^2 \rangle$, and $\langle r^2 \rangle$ are reported. Contour maps are shown of the electrostatic potentials around the molecules.

Palmer *et al.* [22] report similar but much less accurate calculations including additional isomers, along with pentazine and hexazine. The forms of the MOs, especially the lone-pair orbitals, are extensively discussed.

Ha [22a], using a moderate basis set followed by moderate CM, has calculated the energies of numerous singlet and triplet states of *s*-tetrazine, including many nπ^* and $\pi\pi^*$ states. Orbital energies are listed. Charge-density contours of some of the MOs are given. The first nπ^* transition is calculated at 2.80 eV (observed about 2.5 eV). The singlet–triplet split of the first nπ^* states ($^1B_{3u}$ and $^3B_{3u}$) is calculated as 0.76 eV. The first $\pi\pi^*$ excited state is calculated at 5.99 eV, slightly above the observed range of absorption.

The lowest reported SCF energy values for the azabenzenes from pyridine to tetrazine, except for pyrazine, are those of Mulder *et al.* [23], who used an unpolarized (9,5/42) basis for C and N, with H (4/2). For pyrazine others [20, 26] obtained lower values (E_h = -262.624 versus -262.584 by Mulder *et al.*). Mulder *et al.* have used their SCF wave functions in very extensive calculations of permanent multipole moments, polarizabilities, and long-range interaction coefficients of the azabenzenes.

Del Bene [24] has used minimal-basis SCF calculations to investigate the structural, energetic, and electronic properties of the 2- and 4-R-substituted pyridines, where R is CH_3, NH_2, OH, or F, or C_2H_3 (vinyl), CHO, or CN.

von Niessen *et al.* [25], using a basis set of double-zeta quality for the heavy atoms, have used their many-body approach, which takes account of correlation, to compute good vertical ionization energies for the 12 most easily ionized MOs of pyridine (C_6H_5N) and phosphoridine (C_6H_5P). For pyridine, the lowest computed ionization potentials are $1a_2$ (π), 9.57 eV; $11a_1$ (n), 9.59; $2b_1$ (π), 10.24; $7b_2$, 12.87; and $1b_1$ (π), 13.43 eV; agreement with experiment and with the estimates of Almlöf is good. In phosphoridine, analogous MOs are in a somewhat different order: $2b_1$, 8.74 eV; $1a_2$, 9.20; a_1 (n), 9.44; b_2, 11.72; $1b_1$, 12.15 eV; again, agreement with experiment is good. Tables of one-electron properties are given in detail for both molecules. Self-consistent field E_h values are -246.5491 for pyridine, -532.7372 for phosphoridine.

In an early SCF computation on pyridine ($E_h = -245.6219$), Clementi [26] reported an N atom population 1.41 e for the "lone pair" (n) MO, the rest of its population being on the C and H atoms.

Hackmeyer and Whitten [27] made SCF (ground-state $E_h = -262.2681$) and moderate CM (ground-state $E_h = -262.3579$) computations on pyrazine (symmetry D_{2h}) and its electronic spectrum.[1] With two N atoms, there are two lone pairs that interact with each other and the other σ electrons to give $n_+ = n_1 + n_2 + \lambda\sigma$ and $n_- = n_1 - n_2 + \mu\sigma$; they report $\lambda \gg \mu$. They give a table of orbital energies, the highest six of which are $1b_{3u}$ (π) with $\epsilon = -0.64$; $3b_{3g}$ (σ) at -0.59; $5b_{1u}$ (n_-) at -0.53; $1b_{2g}$ (π) at -0.48; $6a_g$ (n_+) at -0.438; $1b_{1g}$ (π) at -0.435; then, unoccupied, $2b_{3u}$ (π^*) at 0.05; $1a_u$ (π^*) at 0.06; and $2b_{2g}$ (π^*) at 0.27. All these and three more are used in their CM, in which 23 singlet and triplet excited states and two quintet states were computed, using single and double substitutions. The major CM coefficients are listed. The resulting computed excitation energies are compared with spectroscopic data.

Hinchcliffe [29] reports $E_h = -262.6236$ for an SCF calculation on pyrazine using an (11s 6p/5s) basis contracted to (5s 4p/3s). He obtains ϵ values of the same order as Hackmeyer and Whitten, but somewhat different in magnitude. He uses unrestricted SCF, with one spin annihilation, to compute the energies of the cation (^2A_g, n_+ ionization) and the anion ($^2B_{3u}$, π^* attachment), and the $^3B_{1u}$ first excited state. He gives population analyses and a table of one-electron properties for all four states.

With a double-zeta basis Wadt and Goddard [30] use the GVB method plus CM to compute the vertical excitation energies of 15 singlet and 11 triplet states of pyrazine, along with 4 states of the cation: ^2A_g (n) at 9.13 eV (experimental: 9.63), $^1B_{2g}$ (π) at 10.01 (10.18), $^2B_{1u}$ (n) at 10.48 (11.35), and $^2B_{2g}$ (π) at 11.32 (11.77).

Canuto *et al.* [31] make SCF calculations using C_{2v} symmetry on the $n\pi^*$ and $\pi\pi^*$ triplet and singlet vertical excitation energies for the excited states of pyrazine. They first point out the necessity in SCF computation of using C_{2v} rather than D_{2h} symmetry with localized MOs, unless adequate CM is used (cf. with discussion of O_2 in [31a], pp. 116–117). Note

[1] Strangely, they give $E_h = -472.7986$, which is impossible, but Richards *et al.* [28] quote -262.3579.

that [27] and [30] *do* use CM. They compare the results of using C_{2v} and D_{2h} symmetry and compare their results with those of [27, 30] and experiments. Further, they calculate the N atom 1s core hole states of pyrazine, and compare them with those of N_2 in localized symmetry.

Using a minimal basis, Bendazzoli *et al.* [32] used SCF computations to estimate singlet–triplet matrix elements and obtain a computed radiative lifetime (10 msec) for the short-lived component of the $^3B_1$ lowest excited state of pyrazine, in good agreement with experiment. Phosphorescent lifetimes for three other triplet states and rates for several nonradiative triplet → singlet transitions are also computed.

Del Bene [33], using an up to 4-31G basis, computed relative proton affinities of pyrazine and the diazines, finding the order pyridine > 1,2-diazine > 1,3-diazine > 1,4-diazine to be in agreement with experiment.

10. Pyridinium, Pyrylium, and Thiopyrylium Cations, and Phosphorin

Palmer *et al.* [34] have made a minimal-basis SCF calculation of several molecules of the C_{2v} symmetry type,

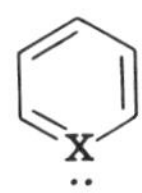

where X is S^+, O^+, CH_2^+, NH^+, or P. They give tables of orbital energies, population analyses, and a correlation diagram. They discuss aromatic character and other features, and give references to their earlier articles. They compute 0.99 D, phosphorin end positive, for the dipole moment of phosphorin (experiment: 1.54 D).

C. OTHER COMPOUNDS CONTAINING NONCONDENSED SIX-MEMBERED RINGS

1. Phenol, Anisole, Thiophenol, and Thioanisole

Using an STO-3G minimal-basis set, including d functions on S, Matsushita *et al.* [35] have made ground-state SCF calculations on C_6H_5OH, $C_6H_5OCH_3$, C_6H_5SH, and $C_6H_5SCH_3$. They have reported atomic populations and ^{13}C-NMR chemical shifts, and the energy curves as a function of the angle of rotation of the substituent group.

Using a double-zeta basis, Palmer *et al.* [36] give orbital energies and total energies for phenol.

2. Styrene

Bendazzoli *et al.* [37] have made a minimal-basis SCF calculation followed by limited CM ($E_h = -303.959$) for the ground state and the lowest four singlet states of styrene (10) They use C_{2v} pseudosymmetry and show

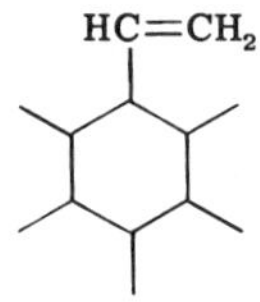

10

a qualitative diagram of the π MOs. They compute oscillator strengths: 0.0083 for $S_1 \leftarrow S_0$ with long axis polarization (observed $f = 0.002$), 0.0005 for $S_2 \leftarrow S_0$ (not observed), and 0.61 for $S_3 \leftarrow S_0$ (observed 0.24, but the discrepancy may be due to a computed excitation energy that is too high). The excitation energies were computed as a function of the angle of twist and CC distance in the ethylenic CH_2 group. Double substitutions in the CM are important. The S_1 excited state is planar. Relévance to cis–trans isomerization is discussed.

A minimal-basis set SCF calculation by Almlöf *et al.* [38] indicates that the molecule is probably planar but with a CCC angle at the vinyl group of about 124°, and a barrier of 4 kcal/mole for rotation of the vinyl group about an orientation perpendicular to the ring plane.

3. Doubly Substituted Benzenes

Using a double-zeta basis set in SCF calculations, Palmer *et al.* [36, 39] have investigated a variety of doubly substituted benzenes, C_6H_4XY, where X and Y are H, Me, OH, F, and have used their results in the identification of photoelectron spectra. They give tables of orbital energies of all the isomers in each case.

Domcke *et al.* [40] have made STO-3G minimal-basis plus partial CM to explain the 1s ESCA spectrum of paranitroaniline ($H_2N \cdot C_6H_4NO_2$). Configuration mixing is used between the single-hole configurations and the manifold of two-hole–one-particle excited configurations. While the $N(1s)NH_2$ hole shows one main peak, the $N(1s)NO_2$ hole is split into three main peaks involving a negative $\pi\pi^*$ excitation. The calculated energy separation of the $N1s(NO_2)$ and $N1s(NH_2)$ main peaks is about 7 eV (experimental: 6.3 eV with the NO_2 peak distinctly split). The negative shake-up energy of the NO_2 peak is interpreted.

4. Benzyl, Anilino, and Phenoxyl Radicals

Using a basis set of approximately double-zeta quality in an unrestricted SCF calculation followed by spin annihilation, Hinchcliffe [41] has calculated the energies and ESR hyperfine coupling constants of these radicals, the coupling constants being in fair agreement with experiment.

5. Stilbene

By a minimal basis STO/3G SCF calculation plus some CM, Orlandi *et al.* [42] have obtained energies of the ground and singlet and triplet $\pi\pi^*$ excited states for the cis and trans isomers of stilbene (C_6H_5—CH═CH—C_6H_5) and for twisted geometry. The calculated barrier for ground-state cis–trans conversion is about 70 kcal/mole (experimental: about 50), with trans the more stable. The results for the excited states are discussed in connection with cis–trans photoisomerization. Two of the excited states have minima at 90° twist.

Wolf and Schmidtke [43], using a minimal basis set with 82 contracted Gaussians, find minimal energy for *cis*-stilbene when the phenyl rings are rotated 52° out of a planar arrangement of the molecule. For *trans*-stilbene they find a rotation of 19°. The C—C═C angles were found to be 126° for *cis*-stilbene and 127° for *trans*-stilbene. The trans isomer is found to be more stable by 6 kcal/mole (experimental: about 3). It is not clear why these results differ so much from those of Orlandi *et al.*, who found both cis and trans planar in the ground state.

6. Biphenyl

Using a basis set of double-zeta quality, in an SCF computation, Almlöf [44] has found an equilibrium twist angle of 32° between the phenyl rings in biphenyl (C_6H_5—C_6H_5), with rotation barriers of 1.2 and 4.5 kcal/mole, respectively, for the planar and perpendicualr orientations. A table of orbital energies for the occupied valence orbitals is given and compared with those for the corresponding MOs of benzene.

D. FIVE-MEMBERED HETEROCYCLES AND RELATED MOLECULES

The five-numbered ring compounds furane (11), thiophene (12), and pyrrole (13), all of C_{2v} symmetry, satisfy the Hückel $4n + 2$ rule and

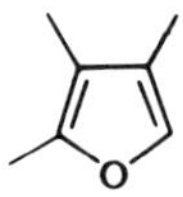

11

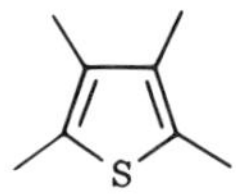

12

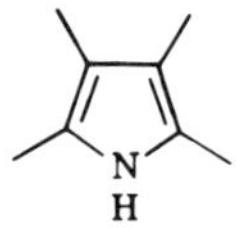

13

thereby qualify as being aromatic. The cyclopentadienyl anion (C_{5v} symmetry) also qualifies, but cyclopentadiene (see Section IX,G,2) does not.

1. Furane, Thiophene, and Related Molecules

For furane, using an unpolarized basis for C and O (9s 5p) contracted to (4s 2p) and a polarized basis (4s 1p) contracted to (2s 1p) for the H atom, Siegbahn [45] made an SCF calculation. He lists orbital energies and total energy ($E_h = -228.5738$) and gives a population analysis.

For thiophene, Gelius *et al.* [46] use a (10s 6p 1d) basis for S, (7s 3p) for C, and (4s 1p) for H, contracted to double-zeta, in an SCF calculation. They give orbital energies, a population analysis, and some computed molecular properties and find $E_h = -550.999$. Relevant to an understanding of electrophilic substitution in thiophene, they give some contour maps of molecular potentials near the molecule.

von Niessen *et al.* [47] have computed the ionization potentials of furane and thiophene with their many-body approach. For furane they used the same basis set as Siegbahn except for omission of polarization on the H atoms. For thiophene they used a basis set of double-zeta quality (12s 9p 1d) for S and (9s 5p/4s) for C and H, both contracted. In their SCF computations, they obtained $E_h = -228.4869$ for furane and -551.0771 for thiophene. They give SCF orbital energies, population analyses, and an extensive table of computed properties (dipole moments, $\langle x^2 \rangle$, $\langle y^2 \rangle$, $\langle z^2 \rangle$, $\langle r^2 \rangle$, total second moments, quadrupole, third and octopole moments, charge densities at nuclei, diamagnetic shieldings at nuclei, diamagnetic susceptibilities, potentials at nuclei, electric fields and field gradients at nuclei, deuteron and ^{17}O or ^{33}S quadrupole coupling constants). Agreements with experiment where available are rather good. The computed dipole moments have their negative end on the heteroatom.

The computed valence ionization potentials are listed. Agreements with photoelectron spectra are for the most part very good. For furane, the lowest computed IPs are $1a_2$ (π) at 8.87 eV, $2b_1$ (π) at 10.36, $9a_1$ at 13.30, $8a_1$ at 14.10, $6b_2$ at 14.69, $5b_2$ at 15.17, and $1b_1$ (π) at 15.47. For thiophene, they are $1a_2$ (π) at 8.77 eV, $3b_1$ (π) at 8.98, $11a_1$ at 12.01, $2b_1$ (π) at 12.88, etc. The computed pole strength (fractional strength of main line versus satellites) is given for each IP.

The carbon 1s IPs of benzene, furane, pyrrole, and thiophene have been discussed in comparison with photoelectron spectra (ESCA) data by Gelius *et al.* [48]. For the heteromolecules these show potential or overt splittings of about 0.34 eV for thiophene, 0.98 eV for pyrrole, and 1.20 eV for furane. These splittings may be attributed to the differential effects of the heteroatom on the nearer as compared with the more distant pairs of carbon atoms, increasing with the electronegativity of the heteroatom. Roughly in agreement are splittings of various authors predicted from the *ab initio* computed orbital energies: 0.59 [46], 1.20 [49] (see below), and 1.50 eV [45]. For each molecule, there are two pairs (one $2a_1$ and $2b_2$, the other $3a_1$ and $2b_2$) of carbon 1s ϵ's, with almost equal energies within each pair (e.g., for furan: 307.79 a.u. for $2a_1$ and the same for $1b_2$, -306.30 for $3a_1$ and -308.28 for $2b_2$); each orbital pair evidently belongs to the pair of C atoms nearer to, or further from, the heteroatom. One would wish for IPs instead of ϵ's but there are complications in computing these when equivalent atoms are present [50], and von Niessen *et al.* [47] do not give them.

For thiophene, Bendazzoli *et al.* [51], using Gelius' basis set have made SCF plus CM computations on the six lowest excited states (three singlet, three triplet) of symmetry A_1, and the eight lowest of each of the symmetries B_1, B_2, and A_2. They have computed f values for the allowed transitions. Comparison with the observed vapor and single crystal UV spectra leads to the unambiguous identification of the 41,595 cm^{-1} vapor spectrum ($\tilde{A} \leftrightarrow \tilde{X}$) as $5b_1 \leftarrow 3b_1$, mixed with some $2a_2 \leftarrow 1a_2$ ($\tilde{A} = {}^1A_1$); the weak 48,330 cm^{-1} spectrum ($\tilde{B} \leftarrow \tilde{X}$), which disappears in the crystal, is identified as a Rydberg transition or transitions ($\tilde{B} = {}^1B_1$); and the 53,270 cm^{-1} spectrum ($\tilde{C} \leftarrow \tilde{X}$) as mainly $5b_1 \leftarrow 1a_2$ ($\tilde{C} = {}^1B_2$). Two states at 30,250 and 37,270 cm^{-1}, found by electron impact, are identified as the ${}^3A_1$ and ${}^3B_2$ states corresponding to the ${}^1A_1$ and ${}^1B_2$ states just mentioned; all these are $\pi^* \leftarrow \pi$ transitions. MCD data are also cited in support.

Azaanalogs of furan are isoxazole (14), 1,2,5-oxadiazole (15), and 1,3,4-oxadiazole (16). SCF calculations on these and furan have been

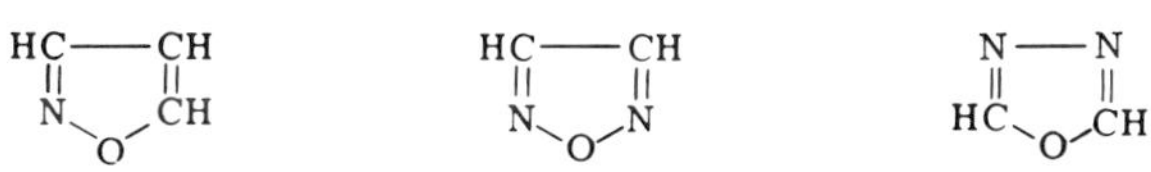

14 15 16

made by Ha [52] using a basis set of double-zeta quality. Orbital energies are given for the valence and core MOs. The following molecular properties were computed and then compared with experiment when possible: dipole moment, quadrupole moment, average diamagnetic shielding, diamagnetic susceptibility, electric field gradient, and nuclear quadrupole coupling constant.

Salmona *et al.* [53] have made STO-3G minimal basis computations on thiazole (17)

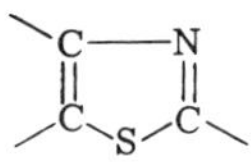

17

and isothiazole (18)

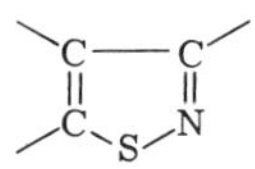

18

and give their minimum vertical orbital energies.

John and Radom [54] have made STO-3G minimal-basis calculations on conformations, stabilities, and charge distributions in numerous 2- and 3-monosubstituted furans. Dipole moments are computed. Conclusions are drawn about a number of characteristics, including donor or acceptor capabilities.

Kao and Radom [55] have made similar STO-3G calculations on 2- and 3-monosubstituted thiophenes. The discussion emphasizes donor and acceptor properties.

2. Pyrrole, Pyrazole, and Imidazole

Preston and Kaufman [56] made SCF calculations on pyrrole and pyrazole (19) using a (9s 5p/4s 1p) basis set contracted to (4s 2p/3s 1p). They

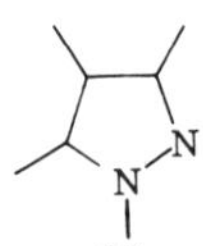

19

give orbital energies for all MOs and overlap populations and obtain $E_h = -208.7764$ for pyrrole, -224.7401 for pyrazole. They compute dipole moments (1.98 D for pyrrole, experimental: 1.74 D; 2.65 D for pyrazole, experimental: 2.21 D), quadrupole moments (in moderately good agreement with experiment for pyrrole), and $\langle R^2 \rangle$. Relations of the calculated orbital energies, which are tabulated, to the photoelectron IPs are discussed. For the most loosely bound MOs, the ϵ values for pyrrole in atomic units are $1a_2$ (π), -0.299; $2b_1$ (π), -0.349; $1b_1$ (π), -0.571; $6b_2$, -0.547; $5b_2$, -0.583, $9a_1$, -0.528; $8a_1$, -0.597. For the first two, there is good agreement with the observed IPs: $1a_2$, 8.13 eV (observed IP: 8.21) and $2b_1$, 9.50 eV (observed IP: 9.20). There are similar agreements for pyrazole.

Tanaka *et al.* [56] made a slightly less accurate SCF calculation ($E_h = -208.7416$) for pyrrole, but they then introduced some π-electron CM to give $E_h = -208.8107$, and thereafter calculated vertical energies of several excited and three ionic states. They used a (9s 5p/4s) basis contracted to (4s 2p/2s) plus two diffuse $2p\pi$ orbitals on C and N to permit calculation of Rydberg states. In the CM, they used only single and double π substitutions on groups of two or three reference states. They began with iterative NO calculations on the ground state, and used the resulting NOs for NO iteration of the other states.

The observed states at 4.35, 5.9, 6.8, 7.2, and 7.4 eV are tentatively identified with calculated $^3B_2$ ($3b_1 \leftarrow 1a_2$, $\pi^* \leftarrow \pi$) at 4.27 eV, $^1B_2$ ($3b_1 \leftarrow 1a_2$) at 6.02 eV, $^1A_1$ ($2a_2 \leftarrow 1a_2$ and $3b_1 \leftarrow 2b_1$, $\pi^* \leftarrow \pi$) at 6.51 eV, $^1B_2$ ($4b_1 \leftarrow 1a_2$) at 7.07 eV, and $^1A_1$ ($3a_2 \leftarrow 1a_2$) at 7.30 eV. The $^3B_2$, a $^3A_1$ computed at 5.45 eV, and the first $^1A_1$ are valence-type states (small $\langle x^2 \rangle$). The $^1B_2$ states and the second $^1A_1$ state are Rydberg states, as is shown by large computed $\langle x^2 \rangle$ values. Dipole moments are computed for all the states. For the positive ion, the CM IPs are 8.10 eV for $^2A_2$ (observed: 8.21) and 8.94 eV for $^2B_1$ (observed: 9.20). For a second $^2B_1$ ($\infty \leftarrow 1b_1$ with considerable CM), 13.33 eV is found (observed: 13.0, weak, attributable to depletion by satellites, as indicated by the fact that the Koopmans' theorem value is $-\epsilon = 15.65$ eV).

Butscher and Thunemann [57] have made an SCF plus, in part, CM, calculation of similar SCF accuracy ($E_h = -208.7692$ for the ground state) for pyrrole, including seven singlet (among them three Rydberg) and six triplet excited states, and the ion. Identifications with experimentally observed excited states are made, with good agreements.

Using a basis of double-zeta quality, Ha [52] has made SCF calculations on pyrazole and imidazole (20) and given valence-shell and core orbital

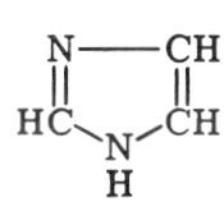

20

energies for these. He made calculations of several molecular properties. In a separate article [58] he gives calculated values of nuclear quadrupole coupling constants.

Keijzers *et al.* [59] have made SCF calculations, with up to triple-zeta basis, on the ground and low-lying excited states of the pyromellitic acid amhydride (PMDA) molecule (21) Results for the ground state of the

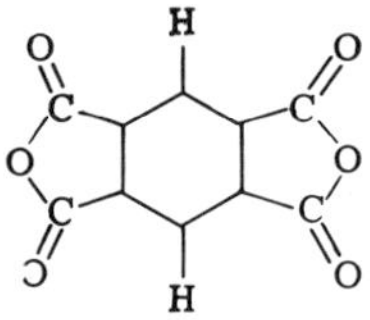

21

anion and cation are also included. The ground state of the molecule has D_{2h} symmetry, but for the nπ^* states (n mostly lone pairs on C=O and ether oxygens) much lower energies are obtained in SCF approximation by using MOs localized on one of the oxygen atoms, a sort of effect first found for 1s ionization in O_2 (cf. pp. 116–117 of [80]). The same localization effect occurs in the ions, but hardly at all in the $\pi\pi^*$ excitations.

3. Porphine and Metal–Porphine Complexes

Kashiwagi *et al.* [60] have made SCF computations on the assumed-planar cobalt–porphine complex, using a (12s 6p 4d) basis set contracted to (6s 5p 3d) for cobalt, a (7s 3p) set contracted to (3s 2p) for the N atoms, (7s 3p) and (3s) for C and H contracted to a minimal basis. For the $^2A_{1g}$ ground state (symmetry D_{4h}) a population analysis gives a charge $+1.78\ e$ on the Co atom and $-0.57\ e$ on each of the four N atoms, with very little covalent bonding. The computations give a $^4B_{2g}$, a $^4A_{2g}$, and two ^4E_g states below the $^2A_{1g}$ ground state, but this discrepancy must be attributed to lack of electron correlation. Three low-energy doublet states and two ionic states (at 6.51 and 6.77 eV) are computed (observed IPs: 6.44 and 6.62 eV).

An earlier SCF calculation on porphine by Almlöf [61] used similar basis functions but with somewhat different orbital exponents and with the N atom bases contracted to minimal, but with two contracted functions for the two central H atoms. The energy was found to be minimized by the following D_{2h} arrangement of the central atoms:

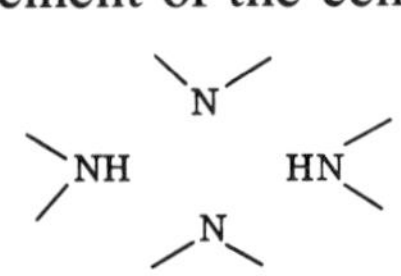

The NH vibrational frequencies were computed and compared to experimentally measured quantities. Separate spin-restricted calculations were made on 10 low ionized states; the lowest ($^2B_{1u}$) corresponds to an ionization potential of 8.0 eV. These and the others are 0.4–0.9 eV lower than corresponding Koopmans' theorem ($-\epsilon$) values.

An article by Ohno [62] surveys *ab initio* computations on metal–porphine complexes, including unpublished work from his laboratory. Electronic structures of Co-, Cu-, Fe-, as well as metal-free porphine are studied by SCF and limited CM. Rather good basis sets were used. Calculated ionization potentials, electron affinites, excitation energies, and related oscillator strengths are given. Charge distributions are studied by population analysis.

E. CONDENSED AROMATIC RING SYSTEMS

1. Aromatic Hydrocarbons

Using a double-zeta basis set (10s 5p/4s) contracted to (4s 2p/2s), Hinchcliffe [63] has made an unrestricted SCF calculation on naphthalene ($C_{10}H_8$), $E_h = -383.2142$, its lowest π triplet, its cation, and its anion. The orbital energies are listed; the D_{2h} notation differs from ours. In their notation, after correcting two misprints, the most easily ionized MOs are, first, four π MOs: $1a_u$ with $\epsilon = -0.2815$ a.u.; $2b_{1u}$ with $\epsilon = -0.3288$; $1b_{3g}$ with $\epsilon = -0.3910$; $1b_{2g}$ with $\epsilon = -0.4478$; then three σ MOs: $6b_{1g}$ with $\epsilon = -0.4836$ a.u.; $9a_g$ with $\epsilon = -0.4875$; $7b_{3u}$ with $\epsilon = -0.5327$; then the first π MO, $1b_{1u}$ with $\epsilon = -0.5439$ a.u.; etc. Koopmans' theorem gives a first ionization energy of 0.282 a.u., in excellent agreement with experiment.

For the four states considered, the charge densities and potentials at the nuclei, and the total and π portions of the second moments $\langle x^2 \rangle$, $\langle y^2 \rangle$, and $\langle z^2 \rangle$ are given. Also, after one spin annihilation ESR coupling constants are given for the cation, anion, and triplet state. For the anion, agreement with experiment is fairly good except for the β protons. Most of the computed properties of the σ core are fairly constant for all four states considered.

In an earlier article Buenker and Peyerimhoff [64], using a similar primitive basis but contracted to (3s 1p/1s), made an SCF calculation on naphthalene ($E_h = -382.7883$) and on its isomer azulene ($E_h = -382.7082$). For each, they also computed two doubly excited states. In agreement with chemical evidence, the seven-ring carbon atoms in azulene are relatively strongly positive and the five-ring atoms negative; this effect shows up markedly among the K-shell orbital energies.

In earlier days, many calculations of resonance energies were made by semiempirical methods for π-electron systems. By the "resonance energy" is meant the lowering of energy from that of an idealized system in which all the π electrons are in localized two-atom bonds (or in pairs etc.) and the actual conjugated system. Resonance energies can now be sensibly computed by *ab initio* methods. Kollmar [65] has done this for several molecules, including naphthalene, for which, using an SCF calculation, he obtains 43 kcal/mole, with only partial conjugation and alternating bond lengths.

2. Carbazole and Trinitrofluorenone

Batra *et al.* [66] have made minimal-basis SCF calculations on carbazole (22) and 2:4:7 trinitro-9-fluorenone (22) They give orbital and total

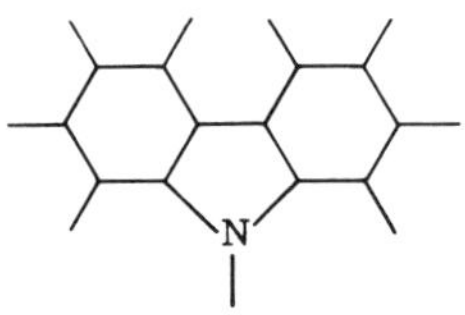

22

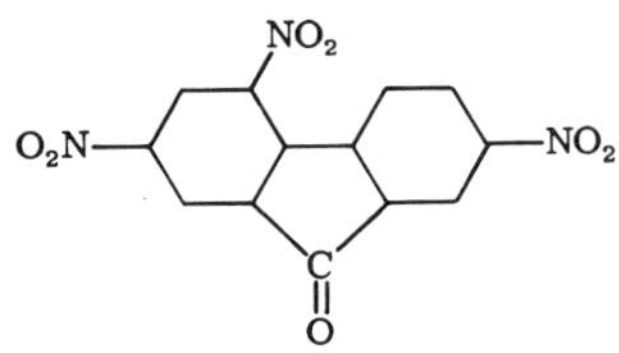

23

energies and gross populations for both, and energies of four singlet and four triplet excited states for carbazole. The two molecules play an important role in the IBM photocopier. Carbazole is an electron donor toward trinitrofluorenone as acceptor.

3. Promazine and Chlorpromazine

Chlorpromazine, a widely used tranquilizer, and promazine are derivatives of phenothiazene (24), with a $CH_2—CH_2—CH_2—N(CH_3)_2$ group

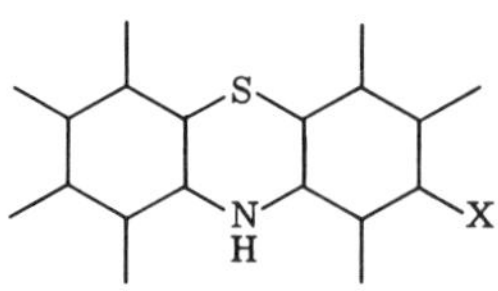

24

substituted for H at N, and in chlorpromazine a Cl substituted for H in the position marked X. Popkie and Kaufman [67] have made a minimal-basis

SCF calculation on these two molecules (7s 3p/3s contracted to 2s 1p/1s on the C, N, and H; 11s 6p contracted to 3s 2p for the S). They give orbital and total energies and overlap populations.

Petrongolo *et al.* [68] have used these wave functions to calculate the electrostatic potential in the neighborhood of these two molecules. The results are discussed in connection with probable sites of attack and reaction paths of electrophilic reagents in interaction with the molecules.

F. THE NUCLEIC ACID BASES AND THE NUCLEOSIDES

There are four bases, or nucleotides, which are used in the genetic code and are nucleic acid building blocks. Two are derivatives of pyramidine (25) and two are derivatives of purine (26). The former are cytosine (27) and thymine (methyluracile) (28). The latter are adenine (29) and guanine (30). It may be argued as to whether some of these bases are properly

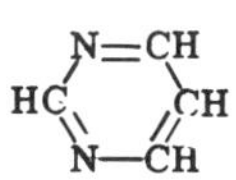

25

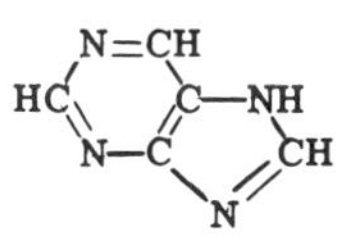

26

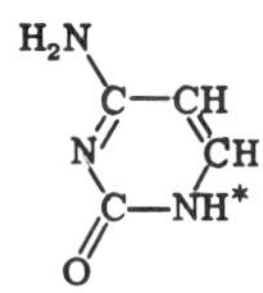

27

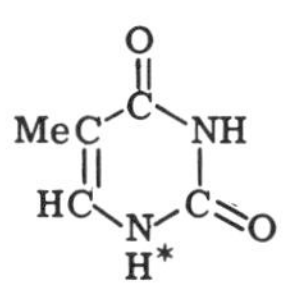

28

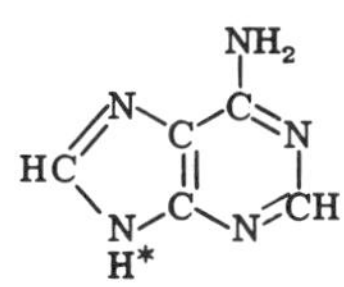

29

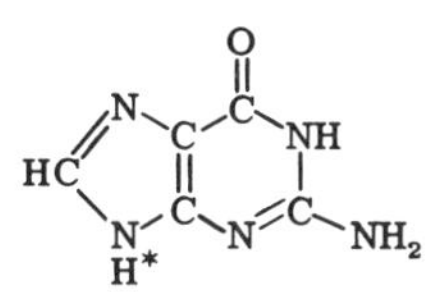

30

classified as aromatic.

Clementi *et al.* [69], using a minimal-basis set contracted from 7s 3p on C, N, and O and 3s on H made SCF calculations on the four nucleotides just mentioned. They give orbital energies and compare them with atomic orbital energies, and give total energies and population analyses. They also discuss electron flow (σ and π charge transfer).

On thymine, Snyder *et al.* [70] have made a minimal (or subminimal?) basis set SCF calculation ($E_h = -446.790$ versus -449.591 for Clementi *et al.*). They have also calculated the two anions obtained by removing a proton at either of the two NH positions. For thymine and the two deprotonated anions, they have computed four singlet and four triplet states each of the types $\pi^* \leftarrow \sigma$ and $\pi^* \leftarrow \pi$. For the triplet states, they have computed the ESR parameters D and E.

Pullman *et al.* [71] with a minimal-basis set have given excellent contour maps of the σ and π charge density distributions in adenine, cytosine, and thymine. They are based on previous minimal-basis calculations by Mély and Pullman [72].

With a minimal-basis set, Mezey and Ladik [73] have computed the energies of 8 tautomers of adenine, with energies 12–48 kcal/mole above that of the accepted stablest tautomer. They also compute 17 tautomers of guanine with energies 5, 12, 16, 22, ... 60 kcal/mole above that of the tautomer calculated to be the stablest, but which differs from the accepted stablest tautomer, perhaps because of solution effects.

Bonaccorsi *et al.* [73a, 73b], using a small basis set, have computed molecular electrostatic potentials in theneighborhood of adenine, thymine, cytosine, and adenine. These are relevant to the identification of protonation sites. In a later article (73c) Pullman and Berthod give a general discussion of electrostatic molecular potentials in hydrogen-bonded systems.

When the NH group H atom [with an asterisk in formulas (27)–(30)] is replaced by a ribose side group (31):

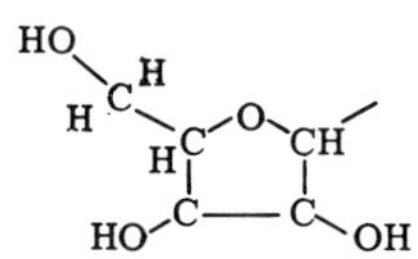

31

one gets a nucleoside. Corresponding to uracil is uridine. Using essentially the same basis as Clementi *et al.*, Pullman and Berthod [74] have made SCF calculations for the known anti form of uridine (attachment of the ribose CH carbon as indicated, with the ribose turned toward the left for anti, or toward the right for syn). They tabulate orbital energies for ribose, uracil (recall that thymine is methyluracile) and uridine. They discuss the relative positions of the orbital energies in the separate and fused systems. They also display contour diagrams of the molecular electrostatic potential in the neighborhood of uridine; these show greater changes on fusion than do the molecular orbitals.

Kwiatkowski and Pullman [75] give similar calculations on both syn and anti forms of adenosine, which is the purine nucleoside corresponding to adenine. Computed dipole moments are 1 D for the anti and 4.3 D for the syn form, as compared with 2–3 D predicted and observed for adenine.

The orbital energies for both syn and anti forms and for adenine and ribose are tabulated. Comparison of orbital energies with the observed photoelectron spectrum of adenine is made.

In the DNA helix, appropriate pairs of the four bases shown above are connected internally by hydrogen bonding, and externally by ribose and phosphate side groups. Pullman and collaborators have made minimal-basis SCF calculations on the molecular electrostatic potentials around base pairs and around the other constituents of the nucleic acids. These potentials are important in assessing the reactivities of various regions of the DNA. No attempt will be made here to give detailed reviews of these articles, but for references and a summary we refer to three articles that give a picture of the integrated results for the molecular potential in the neighborhoods of various base pairs.

In the first paper, Pullman *et al.* [76] present results of a computation on a minihelix

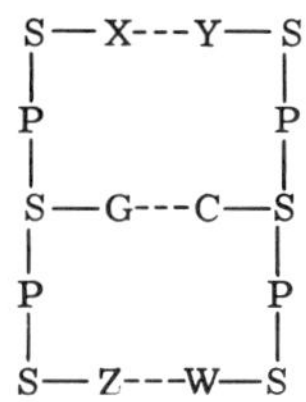

surrounding a guanine–cytosine (G–C) base pair (S = sugar, i.e., ribose, P = phosphate groups; X–Y and Z–W refer to other base pairs). The sugar–phosphate backbone within the DNA strongly enhances the minima in the G–C pair above. This matter is discussed in detail. Perahia and Pullman [77] similarly discuss the region of the adenine–thymine base pairs, and Perahia *et al.* [78] discuss the potential due to the sugar–phosphate backbone. In a related article Clementi *et al.* [79] made minimal SCF computations for the $(PO_4—CH_2)^-$ group, which characterizes part of the repeating unit of nucleic acids.

REFERENCES

1. W. C. Ermler and C. W. Kern, *J. Chem. Phys.* **58,** 3458 (1973).
2. W. C. Ermler, R. S. Mulliken, and E. Clementi, *J. Am. Chem. Soc.* **98,** 388 (1976).
3. R. J. Buenker, J. L. Whitten, and J. D. Petke, *J. Chem. Phys.* **49,** 2261 (1968).
4. J. M. Norbeck and G. A. Gallup. *J. Am. Chem. Soc.* **96,** 3386 (1974).
5. G. F. Tantardini, M. Raimondi, and M. Simonetta, *J. Am. Chem. Soc.* **99,** 2913 (1977).
6. W. von Niessen, L. S. Cederbaum, and W. P. Kraemer, *J. Chem. Phys.* **65,** 1378 (1976).

7. D. W. Potts, W. C. Price, D. G. Strets, and T. A. Williams, *Discuss. Faraday Soc.* **54,** 168 (1972).
8. P. J. Hay and I. Shavitt, *J. Chem. Phys.* **60,** 2865 (1974).
9. R. S. Mulliken, *J. Chim. Phys.* **46,** 704 (1949).
10. J. B. Rose, T.-I. Shibuya, and V. McCoy, *J. Chem. Phys.* **60,** 2700 (1974).
11. S. R. Langhoff, E. R. Davidson, and C. W. Kern, *J. Chem. Phys.* **63,** 4800 (1975).
12. S. D. Peyerimhoff and R. J. Buenker, *Theor. Chim. Acta* **19,** 1 (1970).
13. W. J. Hehre and J. A. Pople, *J. Am. Chem. Soc.* **94,** 6901 (1972).
14. W. C. Ermler and R. S. Mulliken, *J. Am. Chem. Soc.* **100,** 1647 (1978).
15. J. L. Devlin, III, J. F. Wolf, R. W. Taft, and W. J. Hehre, *J. Am. Chem. Soc.* **98,** 1990 (1976).
16. W. J. Hehre, *J. Am. Chem. Soc.* **94,** 5919 (1972).
17. M. D. Newton, J. M. Schulman, and M. M. Manus, *J. Am. Chem. Soc.* **96,** 17 (1974).
18. D. L. Wilhite and J. L. Whitten, *J. Am. Chem. Soc.* **93,** 2858 (1972).
19. J. O. Noell and M. D. Newton, *J. Am. Chem. Soc.* **101,** 51 (1979).
20. J. D. Dill, P. von R. Schleyer, J. S. Binkley, R. Seeger, J. A. Pople, and E. Haselbach, *J. Am. Chem. Soc.* **98,** 5428 (1976).
21. J. Almlöf, B. Roos, U. Wahlgren, and H. Johansen, *J. Electron Spectrosc.* **2,** 51 (1973).
22. M. H. Palmer, A. J. Gaskell, and R. H. Finlay, *J. C. S. Perkin Trans. II* p. 778 (1974).
22a. T.-K. Ha, *Mol. Phys.* **29,** 1829 (1975).
23. F. Mulder, G. van Dijk, and C. Huiszoon, *Mol. Phys.* **38,** 577 (1979).
24. J. Del Bene, *J. Am. Chem. Soc.* **101,** 6184 (1979).
25. W. von Niessen, G. H. F. Diercksen, and L. S. Cederbaum, *Chem. Phys.* **10,** 345 (1975).
26. E. Clementi, *J. Chem. Phys.* **46,** 4731 (1967); also on the positive ion, *J. Chem. Phys.* **47,** 4485 (1967).
27. M. Hackmeyer and J. L. Whitten, *J. Chem. Phys.* **54,** 3739 (1971).
28. W. G. Richards, T. E. H. Walker, L. Farrell, and P. R. Scott, "Bibliography of *Ab Initio* Molecular Wave Functions, Supplement for 1970–73," p. 291. Oxford Univ. Press (Clarendon), London and New York, 1974.
29. A. Hinchcliffe, *Int. J. Quantum Chem.* **11,** 767 (1977).
30. W. R. Wadt and W. A. Goddard, III, *J. Am. Chem. Soc.* **97,** 2034 (1975); W. R. Wadt, W. A. Goddard, III, and T. H. Dunning, Jr., *J. Chem. Phys.* **65,** 438 (1976).
31. S. Canuto, O. Goscinski, and M. Zerner, *Chem. Phys. Lett.* **68,** 232 (1979).
31a. R. S. Mulliken and W. C. Ermler, "Diatomic Molecules: Results of Ab Initio Calculations." Academic Press, New York, 1977.
32. G. L. Bendazzoli, G. Orlandi, and P. Palmieri, *J. Chem. Phys.* **67,** 1948 (1977).
33. J. E. Del Bene, *J. Am. Chem. Soc.* **99,** 3617 (1977).
34. M. H. Palmer, R. H. Findlay, W. Moyes, and A. J. Gaskell, *J. C. S. Perkin Trans. II* p. 841 (1975).
35. T. Matsushita, Y. Osamura, N. Misana, K. Nishimoto, and Y. Tsuno, *Bull. Chem. Soc. Jpn.* **52,** 2521 (1979).
36. M. H. Palmer, W. Noyes, M. Speirs, and J. N. A. Rudyard, *J. Mol. Struct.* **52,** 293 (1979).
37. G. L. Bendazzoli, G. Orlandi, P. Palmieri, and G. Poggi, *J. Am. Chem. Soc.* **100,** 392 (1978).
38. J. E. Almlöf, P. V. Isacsson, P. J. Mjöberg, and W. M. Ralomski, *Chem. Phys. Lett.* **26,** 215 (1974).

39. M. H. Palmer, W. Noyes, M. Speirs, and J. N. A. Rudyard, *J. Mol. Phys.* **49,** 105 (1978).
40. W. Domcke, L. S. Cederbaum, J. Schirmer, and W. von Niessen, *Chem. Phys.* **39,** 149 (1979).
41. A. Hinchcliffe, *Chem. Phys. Lett.* **27,** 454 (1974).
42. G. Orlandi, P. Palmieri, and G. Poggi, *J. Am. Chem. Soc.* **101,** 3492 (1979).
43. A. Wolf and H.-H. Schmidtke, *Theor. Chim. Acta* **48,** 37 (1978).
44. J. Almlöf, *Chem. Phys.* **6,** 135 (1974).
45. P. Siegbahn, *Chem. Phys. Lett.* **8,** 245 (1971).
46. U. Gelius, B. Roos, and P. Siegbahn, *Theor. Chim. Acta* **27,** 171 (1972).
47. W. von Niessen, W. P. Kraemer, and L. S. Cederbaum, *J. Electron Spectrosc.* **8,** 179 (1976).
48. U. Gelius, C. J. Allan, G. Johansson, H. Siegbahn, D. A. Allison, and K. Siegbahn, *Phys. Scr.* **3,** 237 (1971).
49. H. J. T. Preston and J. J. Kaufman, *Int. J. Quantum Chem., Symp.* **7,** 207 (1973).
50. Compare our review on O_2, *in* "Diatomic Molecules: Results of Ab Initio Calculations," p. 116. Academic Press, New York, 1977.
51. G. L. Bendazzoli, F. Bertinelli, P. Palmieri, A. Brillante, and C. Taliani, *J. Chem. Phys.* **69,** 5077 (1978).
52. T.-K. Ha, *J. Mol. Struct.* **51,** 87 (1979).
53. G. Salmona, R. Faure, E. J. Vincent, C. Guimon, and G. Pfister-Guillouzo, *J. Mol. Struct.* **48,** 205 (1978).
54. I. G. John and L. Radom, *J. Am. Chem. Soc.* **100,** 3981 (1978).
55. J. Kao and L. Radom, *J. Am. Chem. Soc.* **101,** 311 (1979).
56. K. Tanaka, T. Nomura, and T. Noro, *J. Chem. Phys.* **67,** 5738 (1977).
57. W. Butscher and K.-H. Thunemann, *Chem. Phys. Lett.* **57,** 224 (1978).
58. T.-K. Ha, *Chem. Phys. Lett.* **37,** 315 (1976).
59. C. P. Keijzers, P. S. Bagus, and J. P. Worth, *J. Chem. Phys.* **69,** 4032 (1978).
60. H. Kashiwagi, T. Tadaka, S. Obara, E. Miyoshi, and K. Ohno, *Int. J. Quantum Chem.* **14,** 13 (1978).

62. K. Ohno in "Horizons of Quantum Chemistry," K. Fukui and B. Pullman, eds., D. Reidel, Dordrecht, Holland, 1980.
63. A. Hinchcliffe, *J. C. S. Faraday Trans. II* **73,** 1627 (1977).
64. R. J. Buenker and S. D. Peyerimhoff, *Chem. Phys. Lett.* **3,** 37 (1969).
65. H. Kollmar, *J. Am. Chem. Soc.* **101,** 4832 (1979).
66. I. P. Batra, P. S. Bagus, E. Clementi, and H. Seki, *Theor. Chim. Acta* **32,** 279 (1974).
67. H. E. Popkie and J. J. Kaufman, *Int. J. Quantum. Chem.* **10,** 569 (1976).
68. C. Petrongolo, H. J. T. Preston, and J. J. Kaufman, *Int. J. Quantum Chem.* **13,** 457 (1978).
69. E. Clementi, J. M. André, M. C. André, D. Klint, and D. Hahn, *Acta Phys. Acad. Sci. Hung.* **27,** 493 (1969).
70. L. C. Snyder, R. G. Shulman, and D. B. Neumann, *J. Chem. Phys.* **53,** 256 (1970).
71. A. Pullman, M. Dreyfus, and B. Mély, *Theor. Chim. Acta* **16,** 85 (1976).
72. B. Mély and A. Pullman, *Theor. Chim. Acta* **13,** 278 (1968).
73. P. G. Mezey and J. J. Ladik, *Theor. Chim. Acta* **52,** 129 (1979).
73a. R. Bonaccorsi, A. Pullman, E. Scrocci, and J. Tomasi, *Theor. Chim. Acta* **24,** 51 (1972).
73b. R. Bonaccorsi, E. Scrocco, J. Tomasi, and A. Pullman, *Theor. Chim. Acta* **36,** 339 (1975).

73c. A. Pullman and H. Berthod, *Theor. Chim. Acta* **48,** 269 (1978).
74. A. Pullman and H. Berthod, *Int. J. Quantum Chem., Quantum Biol. Symp.* **4,** 327 (1977).
75. J. S. Kwiatkowski and B. Pullman, *Int. J. Quantum Chem.* **15,** 499 (1979).
76. A. Pullman, C. Zakrewska, and D. Perahia, *Int. J. Quantum Chem.* **16,** 395 (1979).
77. D. Perahia and A. Pullman, *Theor. Chim. Acta* **50,** 351 (1979).
78. D. Perahia, A. Pullman, and B. Pullman, *Theor. Chim. Acta* **51,** 349 (1979).
79. E. Clementi, G. Corongiu, and F. Lelj, *J. Chem. Phys.* **70,** 3726 (1979).

CHAPTER XII

MOLECULAR COMPLEXES AND COMPOUNDS, HYDROGEN BONDING, HYDRATION OF IONS, AND PROTON AFFINITIES

A. GENERAL CONSIDERATIONS

This chapter contains a survey of *ab initio* calculations on molecular complexes: first, miscellaneous complexes not included in Chapter VII, then organometallic compounds, followed by hydrogen bonding in neutral systems, and finally, hydrogen bonding in ionic systems and in hydrated ions. Section VI,B contains a survey of the general nature of molecular complexes and compounds and of Morokuma's analysis of their binding energies into components. Reference should be made to that survey. Sections VI,H–J contain detailed discussions of calculations on systems of trigonal symmetry, in particular, NH_3BH_3, NH_3BF_3, NH_3 + HCl, $NH_3{\cdot}X_2$ (X = halogen), and similar systems, along with $CO{\cdot}BH_3$. *Ab initio* computations are especially helpful in determining the experimentally inaccessible geometry of complexes. Note here that most of the experimental studies of complexes have been in solution. The computations, of course, refer to gas-phase complexes. Three reviews may be mentioned here, one by A. and B. Pullman [1] on biomolecules in solution; lectures by Clementi [2] on water polymers and liquid water, and on ions and biomolecules in water solution as studied by *ab initio* methods; and reviews by Kollman [3] on hydrogen bonding and donor–acceptor interactions.

B. MISCELLANEOUS COMPLEXES AND COMPOUNDS

1. Li, Na, and Al Complexes

Trenary *et al.* [4], using polarized basis sets of better than double-zeta quality, made SCF calculations on MD, where M = Li or Na, and D = NH_3, H_2O, HF, PH_3, H_2S, or HCl; also FNH_3 is discussed. (Such complexes are not known experimentally.) Formation energy, geometry (mostly C_{3v} or C_{2v}), dipole moments, and charge transfer (M^-D^+) are predicted. Predicted to be the most stable are $LiNH_3$ (14.5 kcal/mole) and LiH_2O (12.8 kcal/mole) with $LiPH_3$ and $LiSH_2$ only 0.5 and 1.8 kcal/mole; $NaOH_2$ is 5.2 kcal/mole.

In a second article [5], $LiOH_2$ was studied with a larger basis set, reducing the SCF dissociation energy to 10.4 kcal/mole, but extensive CM increased it again to 12.2 kcal/mole: conclusion, the earlier SCF dissociation energies are not far wrong. In this article LiC_2H_4 and $Li(^2P){\cdot}OH_2$ were also studied; in the latter, binding is increased to 27 kcal/mole for $^2B_2$ and 22.5 for $^2B_1$. Further, $Al(^2P$ ground state$){\cdot}OH_2$ was studied; the predicted dissociation energy for the $^2B_2$ state using the basis set of the first article was 4.4 kcal/mole. The AlC_2H_4 molecule was found to be not strongly bonded.

2. Van der Waals Complexes ($H_2CO{\cdot}He$, $C_2H_4{\cdot}X_2$)

In connection with collisions of H_2CO at low temperatures in interstellar space, Garrison *et al.* [6] have made SCF calculations using a basis set better than triple zeta plus polarization on the system $H_2CO{\cdot}He$ with the H_2CO held rigid, and with various directions of approach of the He to the H_2CO. In a second calculation of Garrison *et al.* [7], CM is introduced in order to include the dispersion energy, which SCF does not consider but which by far dominates the well depth of the van der Waals minimum. The maximum depth is attained for in-plane approaches of He toward the C atom at a center of mass separation of about 7.5 a.u. and corresponding to a temperature in the range 35–40°K (1°K = 3.1668×10^{-6} a.u.).

In a somewhat similar discussion of the systems $C_2H_4{\cdot}X_2$ (X = Cl, Br, or I), Prissette *et al.* [8] find that the energy is a minimum for Cl_2 for an axial approach of the X_2 along a line perpendicular to the C_2H_4 plane at its midpoint. However, they computed the dispersion energy by perturbation theory. Earrlier, Jaszúnski and Kochanski [9] made an SCF calculation on $C_2H_4 \cdot Cl_2$ using a double-zeta basis with a d function on the Cl. They found a stabilization energy of only 0.63 kcal/mole at a distance of 3.7 Å from the nearer Cl to the midpoint of the C_2H_4. For the dispersion energy at 3.7 Å, 1.44 kcal/mole is found. Hence the complex should probably be called a van der Waals rather than an electron donor–acceptor complex.

In Section C,3, van der Waals complexes $M(CO)_4R$, where R is a rare gas atom, are discussed.

Kollman [10], using a 4-31G basis, has made SCF calculations on Ne···HF, $(H_2)_2$, $(N_2)_2$, (H_2) (N_2), and $(Cl_2)_2$. For the formation energy $-\Delta E$ he finds 0.23 kcal/mole at 3.3 Å, to which should be added an estimated dispersion energy of 0.14 kcal/mole. For Ne···FH, the SCF calculation gives $-\Delta E = 0.06$ kcal/mole. For $(H_2)_2$, $(N_2)_2$, and $(Cl_2)_2$, he has computed $-\Delta E$ for several structures, but finds T structures to be the most stable for $(H_2)_2$, with $-\Delta E = 0.013$ (to be augmented by about 0.025 for dispersion energy), and for $(N_2)_2$, with $-\Delta E = 0.25$ kcal/mole (estimated dispersion energy about 0.1 kcal/mole). On the other hand, for $(N_2)(H_2)$, with $-\Delta E = 0.08$, a collinear arrangement was found to be the most stable. For $(Cl_2)_2$, with $\Delta E = 0.19$ kcal/mole by a SCF calculation, the dispersion force would favor a collinear form, but the other forces a T shape; the result is a compromise with $\theta = 75°$, near an L shape ($\theta = 90°$ for the L shape). Umeyama *et al.* [11] find a similar shape for $(F_2)_2$.

Using double-zeta plus polarization functions, Lischka [12] made SCF computations, followed by CM calculations using the pair-natural-orbitals (IEPA PNO) method to approximate the dispersion energy, for the He···HF and He···H_2O van der Waals complexes. He tried only two geometric arrangements in each case. He concluded that He···H—F is more stable than He···F —H, and He ··· — O(H)(H) [H's pointing away from He] more stable than He ··· O(H)(H) [H's pointing toward He], the former with $-\Delta E = 0.00024$ a.u. at $R_0 = 6.39$ a.u., the latter with $-\Delta E = 0.00023$ a.u. at $R_0 = 6.39$ a.u. (R_0 approximately equals the HeF or HeO distance). In another article [13], Lischka made similar computations on the systems H_2···HF and H_2···H_2O. Comparing configurations H—H···F—H and H|H ··· H—F, the latter was strongly preferred with $-\Delta E = 0.0024$ and $R_0 = 5.22$. Comparing 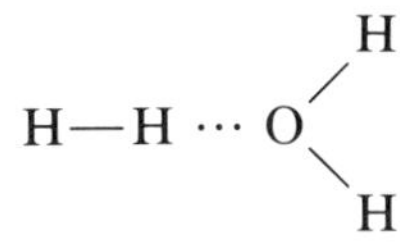 and 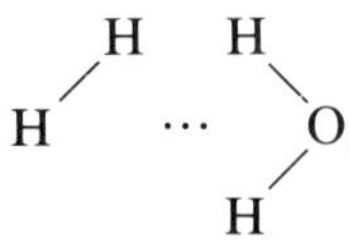, the former was preferred with $-\Delta E = 0.0017$ and $R_0 = 6.07$. Detailed discussions are given.

3. ROR and Benzene Complexes of Carbonyl Cyanide and ROR Complexes of Tetracyanoethylene, and $H_2CO \cdot C_2H_4$

Lathan and Morokuma have made minimal-basis (STO-3G), and some 4–31G. For the charge-transfer, excited, state $H_2O^+ \cdot [CO(CN)_2]^-$ the H_2O $CO(CN)_2$ and $C_2(CN)_4$ complexes of ROR (ether) molecules using H_2O as a stand-in for ROR [14]. These weak nπ electron donor–acceptor complexes (n donor ROR, π acceptor $CO(CN)_2$ or TCNE) are apparently too strong to be rated as van der Waals complexes. Binding energies of the complexes are analyzed by the Morokuma procedure (see Section VI,H); it is concluded that the electrostatic energy is the predominant stabilizing term. Gas-phase $CO(CN)_2$ complexes are known experimentally. Geometries are determined for both ground and charge-transfer states.

For $CO(CN)_2 \cdot H_2O$ the optimum ground-state location is for the O of the H_2O approximately 2.7 Å directly above the center of the $CO(CN)_2$. The water hydrogens are then assumed to be directed away from the carbonyl cyanide, since the electron donor action comes from the O; their orientation around the intermolecular line makes little difference. A stabilization energy of 3.5 kcal/mole was found by STO-3G but 8.0 kcal/mole by 4-13G. For the charge-transfer, excited state $H_2O^+ \cdot [CO(CN)_2]^-$ the H_2O lies 2.5 Å above the C=O oxygen of $CO(CN)_2$ with one H pointed toward one CN nitrogen and with a binding energy of 96 kcal/mole between H_2O^+ and $CO(CN)_2^-$. For both ground and CT states, individual atomic charges from a population analysis are shown in a figure.

For TCNE·H_2O it is concluded that the approach of the H_2O is directly from above the midpoint of the C=C bond, with a formation energy by STO-3G of 3 kcal/mole in the ground state at about 3 Å distance. In the CT state, the computed binding energy of H_2O^+ to $TCNE^-$ is 80 kcal/mole at 2.7 Å.

For both acceptor molecules and their anions, E_h is computed: -293.435 by STO-3G and -296.845 by 4-31G for $(CN)_2CO$, -293.307

and −296.857 for $(CN)_2CO^-$; −439.250 for TCNE and −439.189 for $TCNE^-$, by STO-3G. Dipole moments of $(CN)_2CO$ are also computed: 1.00 D by STO-3G and 0.19 D by 4-31G. In determining the geometries of the complexes, maps are made of the interaction energy as a function of the distance of the H_2O above the acceptor molecule plane. Also, an electrostatic potential map for $(CN)_2CO$ is given depicting energy contours as a function of geometry in a plane 2.7 Å above the $(CN)_2CO$ plane.

In another article Lathan *et al.* [15] make a dispersion energy correction for the ground state of $CO(CN)_2 \cdot H_2O$ and in a perturbation energy calculation they find a dispersion energy contribution of 0.43 kcal out of a total 3.96 kcal using STO-3G, 1.15 kcal dispersion energy out of a total 9.17 kcal using 4-31G, with both at 2.7 Å. They also make SCF and dispersion energy calculations on the complex $CH_2O \cdot C_2H_4$ with the planes of the two partners assumed parallel, and with a 4-31G basis they find a formation energy of 1.06 kcal/mole at 3.75 Å, 0.38 kcal of which is dispersion energy.

For the $CO(CN)_2 \cdot$benzene $\pi\pi$ complex with the planes of the two molecules assumed parallel, they find a total binding energy of 1.54 kcal/mole at 3.6 Å using STO-3G, 0.89 kcal/mole of which is dispersion energy. However, an experimental estimate of the stabilization energy is about 5 kcal/mole. In a discussion, they note that 4-31G tends to overestimate charge effects but that inclusion of d functions should improve the results.

4. SO_2 and O_3 Complexes

Lucchese *et al.* [16] made minimal-basis SCF calculations on the $n\pi$ complexes $H_3N \cdot O_3$ and $(CH_3)_3N \cdot SO_2$, and double-zeta (also plus sulfur d) calculations on $H_3N \cdot SO_2$. The complex $(CH_3)_3N \cdot SO_2$ is the strongest yet studied in the gas phase (still stronger in solution). The calculations disclose a geometry in which the acceptor molecule is oriented nearly perpendicular to the R_3N C_{3v} symmetry axis, with its central atom (S in SO_2) approximately on the axis. With the minimal-basis set, the predicted binding energies are 2.24, 4.06, and 5.00 kcal/mole, but for $H_3N \cdot SO_2$ with sulfur d orbitals, the calculated value is increased from 5 to 9.3 kcal/mole. Dipole moments are computed (4.30 D for $H_3N \cdot SO_2$). The dipole interactions of the two partners are adjudged to be important contributions to the binding (experimental: μ = 1.47 D for NH_3, 1.63 D for SO_2). The $H_3N \cdot SO_2$ complex is computed to be a stronger complex than $(CH_3)_3N \cdot SO_2$, whereas experimentally, in *solution,* the reverse is the case. Similarly, the $H_3N \cdot X_2$ complexes (see Section VI,I) were calculated to be stronger than the $(CH_3)_3N \cdot X_2$ complexes (X = halogen atoms).

5. Two Weak Complexes ($H_2CO \cdot F_2$ and $FCl \cdot FH$)

Umeyama *et al.* [11] in a study of several complexes have concluded that the very weak complex (1a) is parimarily a weak electrostatic–charge transfer complex. The weak complex of experimental structure (1b) is calculated to be only a little ($\Delta E = -3.23$ kcal/mole versus -2.54) more stable than an H-bonded form (1c) but only by using a 4-31G** basis that

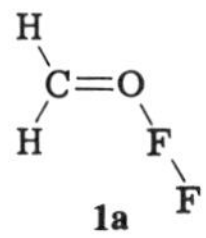

1a

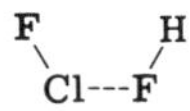

1b

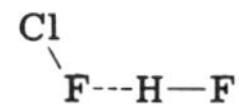

1c

includes polarizations; a 4-31G basis gives a reverse relation. Other details are given.

6. Some Complexes of Boranes

Brown and Lipscomb [17], using a 4-31G basis, made SCF calculations on the complexes of NH_3 and H_2O with triborane (B_3H_7). They studied six different structures, the stablest structure being of C_s symmetry with one BHB bridge bond.

Using STO-3G minimal-basis sets, Fitzpatrick and Fanning [18] made geometry optimization calculations on complexes of borane and trifluoroborane with cyclopropenone and dimethylcyclopropenone. The BH_3 proved to be stronger than the BF_3 complexes, and dimethylation increases the basicity of the cyclopropenone system.

C. ORGANOMETALLIC COMPOUNDS AND COMPLEXES

There is a great variety of complexes or often very stable compounds in which one or more metal atoms or ions are linked to organic compounds or radicals. These fall into several types, which we discuss below.

1. Metal–Ethylene and Metal–Acetylene Complexes

Many complexes and complex ions are familiar in which an ethylene or acetylene molecule, in general with other ligands, is attached to a transition-metal atom or ion that contains d electrons in its outer shell. An understanding of these complexes is relevant to the understanding of catalysis by such metals. Let us consider the ethylene complexes according to the well-known Dewar–Chatt–Duncanson model, in which the binding is attributed largely to a two-way charge-transfer interaction, with a σ bond to the metal from ethylene using its π electrons, and a π bond from the metal into the π^* MO of ethylene (back-bonding). It will be useful for the following to describe this model more specifically. Let axes be chosen as shown in the following diagram

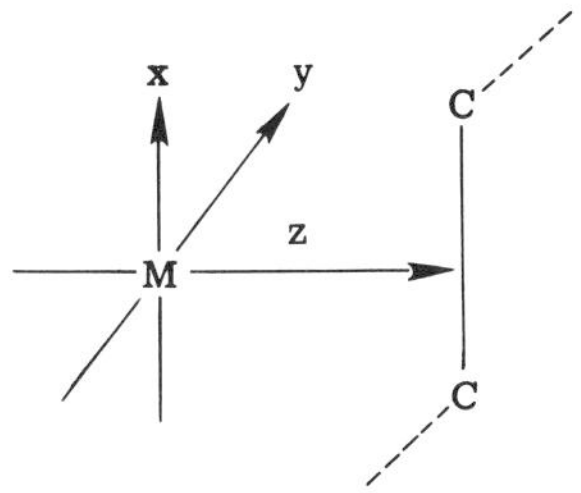

with the H atoms of C_2H_4 either in a plane parallel to the xy plane (the usual "face" or "top" model) or, as an unlikely alternative, in the xz plane ("side" model). In either case, the symmetry is C_{2v}. The π MO of ethylene is now an a_1 MO and can mix with (partially charge transfer into) metal AOs of symmetry a_1.

For example, taking Ni as the M atom, the ground state is ... $3s^2\,3p^6\,3d^8\,4s^2$, $^3F_4$. Slightly higher (205 cm^{-1} up) is ... $3d^9\,4s$, $^3D_3$. The closed-shell state ... $3d^{10}$, ^1S at 14,729 cm^{-1} is less favorable.

For the Ni AOs, three types of description are appropriate. We may use symbols σ, π, δ appropriate to $C_{\infty v}$ symmetry around the z axis, although the π and δ AOs are no longer degenerate as they are for $C_{\infty v}$.[1] We may then further designate these AOs according to their angular behavior with respect to rotations around the z axis (ϕ) or with reference to inclination to the z axis (θ), or we may designate the d AOs in customary ways such as d_{xz}, $d_{x^2-y^2}$, d_{z^2}. Or we may label the AOs as a_1, a_2, b_1, b_2 in accordance with the actual C_{2v} symmetry.

To match the a_1 MO $\pi + \pi$ of ethylene we have available from Ni the 4s and $4p\sigma$ ($\cos\theta$) AOs or a hybrid of these; $3d\sigma$ ($\cos 2\theta$) or $3d_{z^2}$; and $3d\delta$ ($\cos 2\phi$), briefly $3d\delta_{\cos}$ or $3d_{x^2-y^2}$. Of the last two, $3d\sigma$ is much more useful because of better overlap. Assuming that $3d\sigma$ and a $4s$–$4p\sigma$ hybrid are used for the binding, there remain eight 3d electrons of Ni, as follows: two each in $d\pi_x$ ($\sin\theta\cos\theta\cos\phi$) or d_{xz} or b_1; in $d\pi_y$ ($\sin\theta\cos\theta\sin\phi$) or d_{yz} or b_2; in $d\delta_{\sin}$ ($\sin 2\phi$) or d_{xy} or a_2; and in $d\delta_{\cos}$ ($\cos 2\phi$) or $d_{x^2-y^2}$ or a_1. To

[1] The symbol π was originally devised for certain twofold degenerate MOs of $C_{\infty v}$ or $D_{\infty h}$ symmetry. It is unfortunate that it is now used also for nondegenerate MOs (half-a-π MOs) in planar molecules like C_2H_4 and C_6H_6.

match the $\pi-\pi$ or π^* MO of ethylene, the $d\pi_x^2$ shell of Ni, which may now[1] be called π^2 of Ni, is now available to give back bonding.

With these preliminaries, we refer to the SCF (plus a little CM) calculations of Basch *et al.* [19] on $Ni \cdot C_2H_4$ and $Ni_2 \cdot C_2H_4$. These authors used an *ab initio*-based effective core potential or pseudopotential for the inner electrons of Ni and C and a single-zeta contracted basis for Ni 3d and 4p and a double-zeta basis for 4s. They computed the $^3A_1$ ground state of NiC_2H_4 derived from Ni (... $3d^9\,4s$, ^3D) plus ground state C_2H_4. They found that a $4s-4p\sigma$ hybrid pointing away from the C_2H_4 was needed. For the C_2H_4 they made the computations for two geometries: (a) the normal planar geometry and (b) the C—C bond stretched 0.02 Å and the two CH_2 groups bent back 20°, changes that are indicated qualitatively by other work on complexes. They give the following electron configuration description, where core MOs are omitted, and the outer MOs are characterized by approximate descriptions:

$$[1a_1(\mathrm{C{-}H})]^2\,[2a_1(\mathrm{C_2H_4}\sigma)]^2\,[3a_1(\mathrm{C_2H_4}\pi)]^2\,[4a_1(3d_{x^2-z^2})]^2$$
$$[5a_1(3d_{2y^2-x^2-z^2})]\,[6a_1(4s-\lambda 4p\sigma)]\,[1b_1(\mathrm{C{-}H})]^2\,[2b_1(3d_{xz})]^2\,[1b_2(\mathrm{C{-}H})]^2$$
$$[2b_2(3d_{yz})]^2\,[1a_2(\mathrm{C{-}H})]^2\,[2a_2(3d_{xy})]^2$$

They give a population analysis but no table of orbital energies. They show potential energy curves against R(NiC) distance for bent and planar C_2H_4 geometries with and without CM (this being confined to $\pi \to \pi^*$ and $\sigma \to \sigma^*$ excitations in the C_2H_4). The lowest energy is for a stable complex with $R(\mathrm{NiC}) = 2.10$ Å with CM, in the bent form. Low-lying $^3A_1$, $^3A_2$, $^3B_1$, and $^3B_2$ states (up about 0.5–1 eV) are mentioned. Higher, among others, are charge-transfer states.

For $Ni_2C_2H_4$ two configurations, π-$Ni_2C_2H_4$ with one Ni atom behind the other and di-σ-$Ni_2C_2H_4$ with a σ bond from each C to an Ni atom, are considered. It is concluded that the former is stable in a $^3A_1$ state, but the latter is not.

Upton and Goddard [20] have used GVB plus CM methods to evaluate the geometry of NiC_2H_4 and NiC_2H_2. Their result for NiC_2H_4 is that R(NiC) is 2.07 Å, R(CC) is lengthened by 0.02 Å, and the H atoms are bent back by only 2°, much less than considered by Basch *et al.* For NiC_2H_2 the values are 2.01 Å for R(NiC) and a bending of 5°. Binding energies of 14 and 17 kcal/mole are found for nickel–ethylene and nickel–acetylene complexes, respectively. Di-σ-complexes for $Ni_2C_2H_2$ and $Ni_2C_2H_4$ and di-π-$Ni_2C_2H_2$ complexes are briefly considered. It is concluded (in harmony with the results of Basch *et al.*) that delocalization of the Ni and C_2H_4 or C_2H_2 orbitals in the complex is relatively small (less

[1] See the footnote on p. 367.

than in the Dewar–Chatt–Duncanson model). The bonding is predicted as follows: the Ni with configuration hy3d^9 with a 4s–4pσ hybrid hy directed away from the C_2H_4 or C_2H_2 (Pauli effect) behaves much like a bare 3d^9 Ni^+ in its effect on the C_2H_4 or C_2H_2 (σ bond), while for back-bonding there is only a slight delocalization of the 3dπ_{yz} of Ni into the π^* of the ligand. The ultraviolet-visible spectrum shows an intense broad band near 2800 Å that can be identified with an in part charge-transfer transition into π^* of C_2H_4.

Pitzer and Schaefer [21] have made SCF computations on the ground states of $Ni(C_2H_4)_2$ and $Ni(C_2H_4)_3$ and eleven excited states of $Ni(C_2H_4)_2$. They used large basis sets (14s 9p 6d) contracted to (10s 6p 4d) for Ni, (10s 6p) contracted to (6s 6p) for C, (5s) contracted to (2s) for H, plus three additional p functions and one more s function for Ni in the case of $Ni(C_2H_4)_2$. For the $^3B_{3u}$ ground states of "planar" (D_{2h}) and twisted (D_{2d}) $Ni(C_2H_4)_2$ energies were found to be equal within 0.1 kcal/mole. For $Ni(C_2H_4)_3$ an "upright" geometry was found to be 24 kcal/mole more stable than a "planar" one. The ionization potentials of both molecules were predicted and very large deviations from Koopmans' theorem predictions were found.

Åkermark *et al.* [22] have discussed chemical bonding and reactivity in nickel–ethylene complexes, not just NiC_2H_4 but also $NiL_2C_2H_4$, where L is one of the ligands H_2O, NH_3, PH_3 for Ni(0) or Ni^{2+}, or F^-, Cl^-, CH_3^-, CN^-, or NH_2^- for Ni^{2+}. In such complexes the Ni usually has a square-planar coordination geometry, i.e., the metal is approximately quadratically coordinated by the central atom of the ligand and the carbon atoms of ethylene. The basis sets were approximately as follows: (12s 7p 5d) contracted to (4s 4p 2d) for Ni; (7s 3p) for C, N, O, F and (10s 6p) for P and S, all contracted to minimal except that for C_2H_4 a double-zeta contraction was used. They do not say with what configuration of the Ni or Ni^{2+} they are dealing. However, they list π and π^* populations for C_2H_4 which in NiC_2H_4 (1.81 *e* in π and 0.52 *e* for π^*) seem to be radically at variance with the conclusions of Upton and Goddard, and are hard to believe. Still bigger effects are reported for $NiL_2C_2H_4$, e.g., 1.83 *e* in π and 0.98 *e* in π^* for L = H_2O, but here, of course, part of the charge transfer would come from the H_2O. Even for Ni^{2+} C_2H_4, the reported populations are 1.37 for π, 0.02 for π^*. In NiC_2H_4, the reported charges are $Ni^{0.33}C_2H_4^{-0.33}$, in Ni^{2+} C_2H_4, $Ni^{1.39}C_2H_4^{0.61}$. Gross charges on Ni and C_2H_4 are listed for the various complexes. Orbital energies for highest occupied and lowest unoccupied MOs are listed.

As a simpler model situation with respect to the metal atom, Swope and Schaefer [23] have considered the interaction of $Be^+(^2S)$, $Be(^1S)$, and $Be(^3P)$ with ethylene and with acetylene in double-zeta basis SCF calculations. The results for the first two are not surprising: $1s^2\,2s^2$, ^1S gives a repulsion; $1s^2\,2s$, ^2S an attraction. Following Basch [24] (see below) the authors have compared the attraction energies in the Be^+ complexes with those computed for the electrostatic plus polarization[2] effect of a bare proton (with no 1s basis orbital), and found all but 1.5 kcal of the binding in $Be^+C_2H_2$ and 1.0 kcal in BeC_2H_4 were accounted for. Presumably addition of 1s for the proton to the basis set would give additional charge-transfer delocalization stabilization in the formation of the bridged $C_2H_5^+$ ion (see Section B,1). For the $1s^2\,2p$, ^3P beryllium, $1s\,2p\sigma$, $^3A_1$ is repulsive, $1s\,2p\pi$, $^3B_1$ gives an attraction (relative to the $^3P_2$ asymptote), while $1s\,2p\pi_y$, $^2B_2$ gives a flat curve.[3] These results are understandable as follows: In $^3A_1$, there is a Pauli repulsion between the $2p\sigma$ electron and the ethylene or acetylene closed shells. (In the above NiC_2H_4, a similar repulsion for 4s induced a $4s-4p\sigma$ hybridization that decreased the overlap with the C_2H_4 closed shell; nevertheless, some stabilization occurred by forward charge transfer from C_2H_4, mostly from the π_x MO, into this hybrid and the available Ni orbitals of a_1 symmetry.) In $^3B_1$, there is stabilization of the $2p\pi_x$ MO by back charge transfer into the π_x^* MO of ethylene or acetylene.

Basch [24], using a (15s 10p 7d) basis contracted to (5s 4p 3d) for Ag^+, (8s 4p) contracted to (2s 2p) for C, and (4s) contracted to (1s) for H, made an SCF calculation on the $Ag^+C_2H_4$ complex. The model and axis choice is of a face complex like that given above for NiC_2H_4. However, we recall that in Ag^+ there is a complete 4d shell, giving a $4d^{10}$, ^1S state. The C_2H_4 molecule was assumed to be planar but the C=C bond distance was taken as 1.37 Å; the Ag^+—C bond distance was taken as 2.49 Å. The orbital energies of Ag^+, C_2H_4, and $Ag^+C_2H_4$ are listed in a comparative table, and the fractional amounts of 5s, 5p, and 4d Ag^+, π and π^* of C_2H_4, and other LCAO components of the $Ag^+C_2H_4$ MOs are listed. The electron configuration is

$$\ldots[12a_1(5\%4d)]^2\,[13a_1(87\%d)]^2\,[2a_2(92\%d)]^2\,[14a_1(99\%d)]^2\,[6b_2(98\%d)]^2$$

$$[7b_1(98\%d \text{ and } 2\%\pi^*)]^2\,[3a_2(8\%4d)]^2\,[15a_1(8\%d,\ 6.5\%5s,\ 3.5\%p,\ 82\%\pi)]^2$$

Thus we see extensive σ charge transfer from C_2H_4 π into 3d, 5s, and 5p of Ag^+ and only slight back charge transfer into π^* of C_2H_4 from Ag^+. The last-mentioned result is less than has been predicted from the Dewar–Chatt model, but this is not too surprising when one considers the energy

[2] Electrostatic: interaction with the multipole field of the C_2H_4; polarization: of all the C_2H_4 MOs.

[3] Here we have reversed the ^3B_i subscripts as compared with Swope and Schaefer, because their x and y axes are switched as compared with those we used above for NiC_2H_4.

required to ionize Ag^+. The result is in accord with most of the work on NiC_2H_4. Various population analyses are given. A calculation on the side complex showed it to be less stable than the top complex.

For comparison with those of $Ag^+C_2H_4$, Basch computed orbital energies for what he called PC_2H_4, where P indicates a proton but presumably without including a corresponding 1s AO in the basis set (see the discussion above of this model by Swope and Schaefer).

Less accurate calculations were made on $Ag^+C_2H_4$ and also on PdC_2H_4 and on $Pd(C_2H_4)_n$ with $n = 3$ and 4 by Murrell and Scollary [25]. The $Ag^+C_2H_4$ results are consistent with those of Basch. In PdC_2H_4 (orbital energies of outer electrons are listed) there is a gain of 0.05 *e* in the Pd a_1 orbitals (versus 0.17 *e* for Ag^+) and a loss to ethylene π^* of 0.24 *e* in the b_1 orbitals (much greater than the 0.02 *e* in Ag^+). Thus with the neutral metal, back bonding is important. This conclusion is more nearly in accord with that of Åkermark *et al.* on NiC_2H_4 and is contrary to that of Upton and Goddard (see above).

Using a better than double-zeta basis set (no d functions on carbon) Swope and Schaefer [26] have made SCF calculations on C_2H_2 complexes $^7A_1$ from Mn^+ 4s $3d^5$, ^7S; $^6B_1$, $^6A_2$, and $^6B_2$ from Mn 4s $3d^6$, ^6D; $^8A_1$, $^8B_1$, and $^8B_2$ from Mn 4s 4p $3d^5$, ^8P; and $^6A_1$ from Mn $4s^2$ $3d^4$, ^6S. They compare their results with those [23] on the interaction of acetylene with Be^+ $1s^2$ 2s, ^2S, Be $1s^2$ 2s 2p, ^3P, and Be $1s^2$ $2s^2$, ^1S. Orbital energies are listed and a population analysis is given for the $^6B_2$ state, with configuration ... $7a_1^2$ $4b_2^2$ $8a_1^2$ $9a_1^2$ $3b_1^2$ $5b_2^2$ $10a_1$ $1a_2$ $11a_1$ $4b_1$ $12a_1$. The last doubly occupied MO $5b_2$ shows a back transfer of 0.22 *e* from Mn into π^* of C_2H_2, while the $9a_1$ and $3b_1$ Mos show forward transfer from π of C_2H_2 to Mn. There is some 4s 4p hybridization on Mn. The $^6A_1$ state from $4s^2$ $3d^5$, ^6S Mn is repulsive like that from Be ^1S. The curves of the three states $^8B_2$, $^8B_1$ and $^8A_1$ from 4s 4p $3d^5$, ^8P Mn resemble those from $1s^2$ 2s 2p, ^3P Be but are all less attractive. Difficulties arise with the basis set for d-shell orbitals in the SCF calculation, but a hopefully good compromise is used. Results of computations on the complexes of ethylene with the Mn states referred to above are mentioned; they run parallel to those with the acetylene complexes.

2. Sandwich Compounds

Many relatively stable compounds are now known in which a metal atom is interposed between two organic radicals or molecules. One such compound is dibenzene chromium $Cr(C_6H_6)_2$. Best known is ferrocene, in which an iron atom is located symmetrically between two cyclopentadienyl radicals. The structure can be described roughly as $Fe^{2+}(C_5H_5^-)_2$, where $C_5H_5^-$ has an aromatic ($4n + 2\pi$ electron) structure. However, the simple ionic structure is modified by some covalency or polarization. The vapor molecule has D_{5h} (eclipsed) symmetry. In the solid it is distorted and was at first thought to have D_{5d} (staggered) symmetry. The ground state [27], using D_{5d} symbols, is as follows:

$$\ldots a_{1g}^2(\pi)\, e_{2u}^2(\sigma)\, e_{2g}^4(\sigma)\, a_{1g}^2(3d\sigma)\, e_{2g}^4(\sigma)\, a_{2u}^2(\pi)\, e_{2u}^4(\sigma)\, e_{2g}^4(3d\delta)$$

$$e_{1g}^4(\pi) e_{1u}^4(\pi) e_{2g}(\pi)^0 e_{1g}^0(3d\pi) e_{2u}^0(\pi),\ {}^1A_{1g}$$

Here the MOs called π and σ are largely of dicyclopentadienyl form. Those called $3d\sigma$, π, or δ are largely Fe orbitals that are approximately σ, π, or δ with reference to the symmetry axis through the Fe atom and the centers of the two rings; this may be taken as the z axis. The two rings are not quite planar. Following an article by Coutière *et al.* [29], in an SCF calculation Veillard and Demuynck [28] give a table of orbital energies for the valence-shell MOs and show pictures illustrating their forms. The basis set used by Coutière *et al.* was minimal except for the 3d orbitals of iron, which are split.

The second a_{1g} MO involves bonding with $3d\sigma_{Fe}$, the a_{2u} with $4p\sigma_{Fe}$, the e_{1g} with $3d\pi_{Fe}$, and the second e_{2g} with $3d\delta_{Fe}$. Other MOs involve some bonding or antibonding with iron. A population analysis gives a net charge distribution $Fe^{1.23}(C_5H_5)_2^{-1.23}$. When ferrocene is ionized to give the ferricinium ion, some remarkable shifts of charge, mostly toward the Fe, take place in some of the MOs, so that the ionization potentials are not at all what one would expect by Koopmans' theorem from the orbital energies of the neutral molecule. Coutière *et al.* report a shift of 0.86 e in the e_{1g} MO from ring $2p\pi$ to $3d\pi_{Fe}$, a shift of 0.24 e in the e_{1u} from ring $2p\pi$ to $4p\pi_{Fe}$, and a reverse shift of 0.35 e in the a_{1g} and e_{2g} MOs, making a net shift of 0.78 e from rings to Fe. The net result is that the most easily ionized MOs are e_{2g} and then a_{1g}, followed by e_{1u} and then e_{1g}, and the charge distribution in the ion becomes $Fe^{+1.45}(C_5H_5)_2^{-0.45}$. The order of the predicted ionization potentials is in good agreement with the results of photoelectron spectroscopy.

Rohmer *et al.* [30] have made SCF computations on several excited states of ferrocene, using the same near minimal-basis set and geometry as for the ground-state calculations [29], with resulting plausible comparisons of excitation energies as shown in Table 1 with observed spectra. In addition, the very weak absorption observed at 14,000–15,100 cm^{-1} is attributed to $e_{2g} \rightarrow e_{1g}$ transitions to the ${}^3E_{1g}$ and ${}^3E_{2g}$ states. Note that only the transition identified with band VI is a dipole-allowed transition; it is relatively intense.

Bagus *et al.* [31] have made a considerably more accurate SCF calculation on ferrocene and its ion than Coutière *et al.*, but with essentially the

TABLE 1
Excitation Energies of Ferrocene, According to SCF Computations[a]

Excitation	Excited state	Excitation energy,[b] cm^{-1}	Nature of the excitation	Band position,[c] cm^{-1}
$e_{2g} \rightarrow e_{1g}$	$^1E_{2g}$	13,300	$d \rightarrow d$	21,800 (IIa)
	$^1E_{1g}$	14,200	$d \rightarrow d$	24,000 (IIb)
$a_{1g} \rightarrow e_{1g}$	$^1E_{1g}$	21,800	$d \rightarrow d$	30,800 (III)
$e_{1u} \rightarrow e_{1g}$	$^1A_{1u}$	44,900	L → M CT[d] $\pi \rightarrow d$	37,700 (IV)–41,700 (V)
	$^1A_{2u}$	47,700	L → M CT $\pi \rightarrow d$	50,000 (VI)
	$^1E_{2u}$	46,300	L → M CT $\pi \rightarrow d$	—
$e_{2g} \rightarrow e_{2u}$	$^1A_{1u}$	60,700	M → L CT $d \rightarrow \pi^*$	—
	$^1A_{2u}$	62,000	M → L CT $d \rightarrow \pi^*$	—
	$^1E_{1u}$	61,400	M → L CT $d \rightarrow \pi^*$	—
$a_{1g} \rightarrow e_{2u}$	$^1E_{2u}$	75,500	M → L CT $d \rightarrow \pi^*$	—
$e_{1g} \rightarrow e_{2g}$	$^1E_{1g}$	85,800	Intraligand $\pi \rightarrow \pi^*$	—
	$^1E_{2g}$	84,700	Intraligand $\pi \rightarrow \pi^*$	—
$a_{2u} \rightarrow e_{2g}$	$^1E_{2u}$	117,000	Intraligand $\pi \rightarrow \pi^*$	—

[a] From Veillard and Demuynck [28].
[b] Computed.
[c] In the experimental spectrum.
[d] L → MCT refers to ligand to metal charge transfer.

same conclusions. They used a basis set of better than double-zeta quality in that it included three d-type functions on Fe and a p-type polarization function on the H atoms. They made calculations for both the eclipsed (D_{5h}) vapor phase and staggered (D_{5d}) solid-phase conformations, but found the energies to be almost identical. They assumed planar cyclopentadienyl rings, even though electron diffraction results for the vapor indicate that the H atoms may be bent slightly out of the plane toward the Fe. They computed the following ionization energies in electron volts (vertical experimental values are given in parentheses) by subtracting SCF energies of the ion and the ground state: e_{2g}, 5.69 (6.86); a_{1g}, 7.46 (7.23); e_{1g}, 8.79 (8.72); e_{1u}, 8.85 (9.38); a_{2u}, 13.03 (12.2); e_{2u}, 13.60 (13.6). Their computed values are considerably smaller than those of Coutière *et al.* In the MOs e_{2g}, e_{1g}, and a_{1g}, they report the following populations on the Fe atoms in ferrocene and (in parentheses) in ferricinium: e_{2g}, 3.77 (2.88), 3d; e_{1g}, 0.79 (1.39), 3d; a_{1g}, 0.23 (0.33), 4s; 1.95 (1.96), 3d. In a_{2u}, bonding is with $4p\sigma_{Fe}$.

Bagus *et al.* also discuss Mössbauer isomer shifts. They note that the difference between the shifts in ferrocene and in ferricinium salts is quite small, and correlate this, using computations of electron density near the nucleus for ferrocene and ferricinium, with the small differences in the positive charge on the iron in ferrocene and ferricinium as obtained from population analysis. For ferrocene, they find $Fe^{+1.39}$, for ferricinium +1.46 for $^2E_{2g}$, 1.47 for $^2A_{1g}$, 1.38 for $^2E_{1g}$, 1.27 for $^2E_{1u}$, 1.26 for $^2A_{2u}$, and 1.36 for $^2E_{2u}$; these results are similar to those of Coutière *et al.*

Besides sandwich molecules, there are also half-sandwich molecules MC_5H_5. Alexandratos *et al.* [32] have made an SCF computation on the simplest of these, LiC_5H_5, with minimal and double-zeta basis sets. The molecule is experimentally known in solution. Geometry optimization showed C_{5v} symmetry with the hydrogens bent out of the C_5H_5 plane, *away* from the Li, by 3.56°. The structure is found to be near to $Li^+C_5H_5^-$. Optimized dimensions and E_h values, but no orbital energies, are given.

Jemmis *et al.* [33] have made a similar SCF (STO-3G and 4-31G) study of C_5H_5BeH, along with a minimal (STO-3G) study of $Be(C_5H_5)_2$. Several geometries were explored. Beryllocene and a number of molecules of the type C_5H_5BeY are known experimentally. For C_5H_5BeH, the lowest energy was found for a C_{5v} (pentahapto) geometry. The H atoms were computed to be bent *toward* the Be atoms by 0.8°, with the positive end of the molecule, surprisingly, on the C_5H_5. The bond structure is ... $a_1^2\,a_1^2\,e_1^4$, $^1A_1$, with the a_1 MOs formed from the a_1 π MOs of C_5H_5 and the hybrid $2s{-}2p\sigma$ orbital of Be (and $1s_H$ of BeH), and the e_1 MO formed from the e_1 MO of C_5H_5 and the $2p\pi$ AO of Be. For $(C_5H_5)_2Be$, an unsymmetrical structure with one C_5H_5 ring bound symmetrically (pentahapto) but the other bound only at one corner (monohapto) is found (possibly $C_5H_5^-Be^+{—}C_5H_5$); however, without a more accurate calculation this result is adjudged to be somewhat inconclusive. By electron diffraction a C_{5v} structure (the two C_5H_5 rings parallel but not equidistant from the Be) is found, but this is not supported by the calculations.

A minimal-basis "NOCOR" calculation by Ewig *et al.* [34] on TlC_5H_5 (known experimentally) shows a C_{5v} structure that can be represented by ... $a_1^2\,a_1^2\,e_1^4$, $^1A_1$ with MOs similar to those in C_5H_5BeH, but with a net charge of $+0.1\ e$ on the Tl. Other details on the MOs are given.

Hillier and Saunders [35] have made SCF computations on $Ni(C_5H_5)NO$, with (assumed?) C_{5v} symmetry. They used a minimal STF basis set expanded economically into GTFs, except that for the Ni 3d AOs they used two STFs. They list MO orbital energies and configurations of the MOs in terms of AOs on the basis of a population analysis. The most easily ionized MO is mostly carbon $2p\pi$ of C_5H_5 but contains some N and O 2p and some Ni 3d and 4p, representing probably bonding from Ni to NO with some charge transfer into the $2\pi(\pi^*)$ MO of NO. There is some bonding to C_5H_5 but mostly charge transfer. The net charges are $-0.8\ e$ on C_5H_5, $+0.82\ e$ on Ni, with the NO nearly neutral. Thus the overall structure is approximately $Ni^+(C_5H_5)^-NO$.

Guest *et al.* [36] have made SCF computations, with a minimal-basis set, or somewhat better for Cr 3d, on dibenzene chromium (symmetry D_{6h}) and benzene chromium tricarbonyl (symmetry C_{3v}), and for the two lowest valence states of their positive ions. Tables of orbital energies are given for the valence electrons of both molecules. Koopmans' theorem holds for all the MOs except the $8a_{1g}$ in $(C_6H_6)_2Cr$ and $17a_1$ in $(C_6H_6)Cr(CO)_3$, which have the lowest ionization potentials (see Table 2)

TABLE 2

Calculated and Experimental Vertical IPs of Dibenzene Chromium and Benzene Chromium Tricarbonyl[a]

	Orbital	Experiment IP (eV)	Koopmans' theorem IP (eV)	ΔSCF IP (eV)
$Cr(C_6H_6)_2$	$8a_{1g}$	5·5	11·3	5·1
	$4e_{2g}$	6·5	7·5	4·9
$C_6H_6Cr(CO)_3$	17e	7·5	8·4	6·1
	$17a_1$		11·1	6·1

[a] From Guest *et al.* [36].

experimentally and by ΔSCF calculations (subtraction of SCF energies of ion and neutral molecule).

In $Cr(C_6H_6)_2$, the first ionization ($8a_{1g}$) is nearly all Cr 3d. The second, $4e_{2g}$, which shows a smaller deviation from Koopmans' theorem, is a major source of bonding, by back bonding from Cr 3d into the $1e_u$ unoccupied antibonding π MO of benzene. According to the population analysis, there is a net transfer from Cr of 0.12 *e* to 2pπ and 0.08 *e* to 2s, *per carbon atom,* making an overall charge distribution of $Cr^{+1.5}(C_6H_6)_2^{-1.5}$. In benzene chromium tricarbonyl, the transfers from C_6H_6 are similar, and there are transfers from Cr to CO also. Both the first and second ionizations are mostly 3d. A detailed discussion is given.

Rohmer *et al.* [37] have reported a double-zeta (for the valence orbitals) SCF computation on the experimentally known *bis*(π-allyl)nickel, and for two states of its positive ion. The molecule is a typical sandwich compound with symmetry C_{2h}. For the allyl radical and ions (see Section IX,B,2), the electron configuration is ... $b_1^2\ a_2^n$, with $n = 0$, 1, and 2 for the cation, radical, and anion, respectively. The bonding b_1 MO is of the form $\pi_1 + \pi_2 + \pi_3$, the nonbonding a_2 MO of the form $\pi_1 - \pi_3$.

The allyl groups are in parallel planes above and below the Ni atom. The orbital energies and the forms of the valence MOs and their relations to the NI orbitals and the bonding and nonbonding allyl MOs are given. The charge distributions in terms of individual atomic orbitals as given by a population analysis are listed; overlap populations are also given.

In this molecule, as in ferrocene (see above) Koopmans' theorem cannot be used in predicting the ionization potentials. The ground state of the ion is a ^2A_g state resulting from removal of an electron from an MO that may be identified with $9a_g$ of the neutral molecule. Slightly higher according to an earlier less accurate (minimal basis) computation [38] are a ^2B_g state, then two ^2A_g states, corresponding to removal from $3b_g$, $11a_g$, and $10a_g$, respectively; finally, above these is a ^2A_u state corresponding to removal from the $7a_u$, which by Koopmans' theorem should give the ground state of the ion. Higher states obey Koopmans' theorem. In the ion states that disobey Koopmans' theorem, the MO is reorganized during ionization to increase its metal character. Energies (E_h) and ionization potentials for the ^2A_g and ^2A_u ions from the more accurate computation [36] are listed.

Bénard [39], using double-zeta SCF computations, has obtained deformation density plots on t-$[C_5H_5Fe(CO)_2]_2$ and $(\eta^5C_5H_5Ni)_2C_2H_2$. These systems are not typical sandwich compounds, but are conveniently included here. In the first of these, the two Fe atoms are linked by CO bridges; in the second there may be a direct Ni—Ni bond. Contour maps are given for sections of the deformation density distributions. A detailed discussion is not feasible.

3. The Carbonyls and Related Complexes

The carbonyls $M(CO)_n$ are well-known compounds. In related compounds, N_2, CO, CN^-, C_2H_4 may be substituted for one or more CO groups. SCF calculations, although near minimal, have been made on some of these. For $Ni(CO)_4$, with symmetry T_d, and $Cr(CO)_6$, with symmetry O_h (cf. Section VII,A), Hillier and Saunders [40] have made SCF calculations using basis sets of 65 and 85 contracted Gaussian-type orbitals for the respective molecules (including two 3d orbitals for Ni and Cr). Thus the basis sets were probably near minimal except for the metal 3d orbitals. A complete roster of MO orbital energies is given for each molecule, along with a population analysis in terms of total and percentage populations of each atomic orbital, and overlap populations. The orbital energies are compared with experimental values for the most easily ionized MOs for $Ni(CO)_4$: 11.7 (8.9) for $9t_2$, 13.5 (9.8) for 2e; for $Cr(CO)_6$, 10.7 (8.4) for $2t_{2g}$, 17.5 and 17.9 for $8t_{1u}$ and $5e_g$ (13.3), respectively; the numbers in parentheses are the experimental values. Shifts of 1s orbital energies for C and O atoms as compared with those in CO are listed: for

$Cr(CO)_6$, C_{1s} shift -0.3 (-2.5), O_{1s} -0.5 (-1.7); for $Ni(CO)_4$, C_{1s} -0.3 (-1.2), O_{1s} -0.4 (-1.4).

The two most easily ionized MOs of $Ni(CO)_4$ and the most easily ionized MO of $Cr(CO)_6$ are predominantly metal in character; all the other valence-shell MOs are mostly ligand. The sequence of ionization energies is in accord with Koopmans' theorem, although as is shown by SCF calculations on the ions in a later article [41], the two most easily ionized MOs of $Ni(CO)_4$ are considerably reorganized in the ion. The population analyses indicate a $+0.47$ *e* charge on the Ni in $Ni(CO)_4$, and $+0.70$ *e* on the Cr in $Cr(CO)_6$. There are σ transfers from the 5σ MO of CO (largely 2s) and lesser π transfers into the $2\pi(\pi^*)$ MO of CO. The authors do not give E_h values.

Demuynck and Veillard [42] have reported a similar SCF computation on $Ni(CO)_4$ but apparently with fewer [51] contracted Gaussian-type functions. They report $E_h = -1953.95$. They give a complete roster of MO orbital energies that differs appreciably from that of Hillier and Saunders, covering a somewhat narrower spread. From a population analysis, they find a charge of $+0.24$ *e* on the Ni. They report σ transfer from each CO to Ni of about 0.16 *e*, and π transfer from Ni to each CO of about 0.11 *e*. As compared with Ni $3d^{10}$, there is computed a loss of 0.57 *e* in $3dt_2$ and 0.20 *e* in 3de, and a gain of 0.57 *e* in the $4pt_2$ MO, mostly through the $8t_2$ and $9t_2$ MOs. However, all these figures should not be taken too seriously in view of the limited size of the basis set, the lack of CM, and the shortcomings of population analysis. In this article, the authors also compute localized Ni—C MOs and give a contour diagram. They also extensively discuss $Ni(CN)_4^{2-}$.

Mitcheson and Hillier [43] made a partial CM calculation on a model of $Ni(CO)_4$ with one CO not equivalent to the others, and used it for a calculation of satellite structure in the photoelectron spectrum. This spectrum is interpreted in terms of states arising from single and double excitations from the highest filled (metal 3d) MO to the first virtual MO (CO π^*) of the ion. They propose that the satellite structures in hexacarbonyls can be similarly explained.

Jansen and Ros [44] have made comparative calculations on $Ni(CO)_4$ and the isoelectronic molecule $Ni(N_2)_4$. They use a relatively small basis set: $E_h = -1941.14$ for $Ni(CO)_4$. For the charge on Ni, they find -0.32 *e* for $Ni(CO)_4$ and -0.49 *e* for $Ni(N_2)_4$. In detail, they find populations $3d^{8.46}4s^{0.46}4p^{1.40}$ for $Ni(CO)_4$ and $3d^{8.69}4s^{0.46}4p^{1.34}$ for $Ni(N_2)_4$. However, the results of the two articles mentioned above, which find essentially no 4s population on Ni, may be more reliable in this regard. The σ donation is found to be slightly larger but the π back donation is considerably less in $Ni(N_2)_4$ than in $Ni(CO)_4$.

With a basis set similar to that of Hillier and Saunders [40], Hillier *et al.* [41] made SCF calculations on $Fe(CO)_2(NO)_2$, symmetry C_{2v}, and $Co(CO)_3NO$, symmetry C_{3v}, which they compare with $Ni(CO)_4$. They also made SCF calculations on the five lowest states of the positive ion of these two molecules, and found that Koopmans' theorem is not a satisfactory guide. They do not tabulate orbital energies, but give some population analyses. The most easily ionized MOs in $Co(CO)_3NO$ are 8e with 57% Co character and $8a_1$ with 78%, together giving the first photoelectron band, followed by 7e with 86% Co character, giving the second band. The two bands are only 1.1 eV apart as compared with 5.4 eV by Koopmans' theorem. Similar effects occur for $Fe(CO)_2(NO)_2$. In both these nitrosyls, the NO ligand bears a significant net negative charge, greater than for the CO.

Veillard and Demuynck [45] discuss $Fe(CO)_5$, with trigonal bipyramid symmetry D_{3h}, giving a population analysis and charge transfers. They also discuss $Fe(CO)_4C_2H_4$, where a C_2H_4 substitutes for an equatorial CO. Self-consistent field calculations on these molecules and on other $M(CO)_5$, where M = V, Cr, Mn, and also on $Mn(CO)_5^-$, $Co(CO)_5^-$, $Co(CO)_4CH_3$, $Cu(H_2O)_5^{2+}$, $CuCl_5^{3-}$, $MnCl_5^{2-}$, VF_5, and VF_5^- are contained in a significant article by Demuynck *et al.* [46]. The Gaussian basis sets are minimal for the inner shells and metal 4s and 4p, but double-zeta for the valence shells. There is competition between the trigonal bipyramid (TBP) and the square pyramid (SP) structures in these molecules (respectively, D_{3d} and D_{4h}). In $V(CO)_5$, $Cr(CO)_5$, and $Mn(CO)_5$ the D_{4h} structure is predicted to be the more stable, but in the other molecules D_{3h} is preferred. The transformation between the two symmetries by the Berry mechanism [47] is explored, and interligand exchange barriers are computed, these barriers (or at least a lower limit to them) being merely the difference between the TBP and SP energies. Table 3 shows the (metal-3d) configurations, E_h values, and the SP minus TBP barriers for most of the molecules mentioned above. Lists of orbital energies are not given. Conclusions, including the assumption of the Berry mechanism for intramolecular exchanges, appear to be in harmony with experimental evidence where available.

In the same article Demuynck *et al.* made a variety of calculations with differing geometries on $Fe(CO)_4C_2H_4$. They found the most stable form to have the C_2H_4 in an equatorial position facing the metal atom, as in NiC_2H_4 (see Section C,1). Testing varied geometries, they found the most stable conformation of $Co(CO)_4CH_3$ to have C_{3v} symmetry with the CH_3 in an axial location.

Using a near-minimal basis set, except for two 3d orbitals on the Fe atom, Connor *et al.* [48], Hall *et al.* [49], and Connor *et al.* [50] have made SCF calculations on the respective compounds iron tricarbonyl buta-

TABLE 3

Total and Relative Energies of the TBP and SP Conformations[a,b]

	TBP conformation			SP conformation			
	Electronic state and configuration[c]		Total energy (a.u.)	Electronic state and configuration[c]		Total energy (a.u.)	Relative energy (kcal)
VF_5	$^1A_1'$	d^0	−1439.661	$^1A_1$	d^0	−1439.658	1.9
VF_2^-	$^2E''$	$(e'')^1$	−1439.890	$^2B_2$	$(b_2)^1$	−1439.884	3.8
$MnCl_5^{2-}$	$^5A_1'$	$(e'')^2\ (e')^2$	−3441.378	$^5B_1$	$(b_2)^1\ (e)^2\ (a_1)^1$	−3441.370	5.0
$CuCl_5^{3-}$	$^2A_1'$	$(e'')^4\ (e')^4\ (a_1')^1$	−3929.233	$^2B_1$	$(b_2)^2\ (e)^4\ (a_1)^2\ (b_1)^1$	−3929.229	2.5
$Cu\ (H_2O)_5^{2+}$	$^2A_1$	$(b_1)^2\ (b_2)^2\ (a_1)^2\ (a_2)^2\ (a_1)^1$	−2014.747	$^2A_1$	$(a_1)^2\ (b_1)^2\ (a_2)^2\ (b_2)^2\ (a_1)^1$	−2014.747	0.
$V\ (CO)_5$	$^2E'$	$(e'')^4\ (e')^1$	−1503.818	^2E	$(b_2)^2\ (e)^3$	−1503.828	−6.3
$Cr\ (CO)_5$	$^3A_2'$	$(e'')^4\ (e')^2$	−1603.980	$^1A_1$	$(b_2)^2\ (e)^4$	−1603.996	< -10
$Mn\ (CO)_5$	$^2E'$	$(e'')^4\ (e')^3$	−1710.173	$^2A_1$	$(b_2)^2\ (e)^4\ (a_1)^1$	−1710.189	−10
$Mn\ (CO)_5^-$	$^1A_1'$	$(e'')^4\ (e')^4$	−1710.200	$^1A_1$	$(b_2)^2\ (e)^4\ (a_1)^2$	−1710.199	0.6
$Fe\ (CO)_5$	$^1A_1'$	$(e'')^4\ (e')^4$	−1822.5624	$^1A_1$	$(b_2)\ (e)^4\ (a_1)^2$	−1822.5614	0.6
$Co\ (CO)_5^+$	$^1A_1'$	$(e'')^4\ (e')^4$	−1941.038	$^1A_1$	$(b_2)^2\ (e)^4\ (a_1)^2$	−1941.032	3.8

[a] From Demuynck *et al.* [46].

[b] A positive value for the relative energy stands for the TBP more stable than the SP.

[c] The electronic configuration refers to the metal d orbitals.

diene, iron tricarbonyl cyclobutadiene, and iron tricarbonyl trimethylene-methane. In all, population analysis shows a substantial net negative charge on the organic ligand. The forms of the MOs are presented in terms of detailed population analyses. Photoelectron spectra are presented and their peaks interpreted using the ΔSCF method (Koopmans' theorem fails badly).

Using a much smaller basis set than Demuynck *et al.* used (minimal except for the 3d and 4p functions, where the basis for CO is roughly comparable to that of Demuynck *et al.*), but supplementing it by CM (about 700 single and double substitutions), Hay [51] has made SCF plus CM computations in SP and TBP geometries of the ground state and nine singlet and six triplet excited states of $Cr(CO)_5$. As ground state he finds ... $b_2{}^2\, e^4$, $^1A_1$ of C_{4v} and as first excited state ... $(e'')^4\,(e')^2$, $^3A_2$ of D_{3h}, in agreement with Demuynck *et al.* Hay's E_h for the ground state is -1599.23 (SCF) or -1599.37 (CM), as compared to that obtained by Demuynck *et al.* -1603.996 (SCF). He gives computed excitation energies and plots of energy as a function of equatorial bending angle for the states investigated. The visible absorption band is identified as a $^1E \leftarrow {}^1A_1$ transition.

Demuynck *et al.* [52] considered the possibility that $M(CO)_5$, where M = Cr or Mo, might form a bond with a rare-gas atom Ar, Kr, or Xe. For this purpose, they made SCF calculations up to double-zeta quality for $Cr(CO)_5Ar$, $Cr(CO)_5Kr$, $Mo(CO)_5Kr$, and $Mo(CO)_5Xe$, but found a repulsive interaction when the larger basis set was used. However, inclusion of the dispersion energy should, of course, give an attraction.

Using a near-minimal basis (but two 3d orbitals for Mn 3d), Guest *et al.* [53] have reported SCF computations on $Mn(CO)_5X$, where X = H, CH_3, Cl, CN. The symmetry is C_{3v} in every case. Orbital energies for the valence MOs of each are tabulated, together with population analyses. In $Mn(CO)_5H$ and $Mn(CO)_5CH_3$, the most easily ionized MO is that of X, in $Mn(CO)_5Cl$ and $Mn(CO)_5CN$, the first two are those of X, while in each case the next two MOs are predominantly Mn 3d. All this is true provided Koopmans' theorem is valid, but experience in other cases (see above) indicates that this is probably not true, and the predominantly metal MOs may be the most easily ionized. The population analysis indicates that the Mn configuration is essentially $3d^6$. Bonding of Mn to X is essentially σ. The C—O bonding is somewhat affected by X with respect to the degree of π donation from Mn.

With a minimal basis except for two 3d orbitals on the Ni, Hillier and Saunders [35] made an SCF computation on the ground states of $Ni(CN)_4{}^{2-}$, symmetry D_{4h}. They give a table of MO orbital energies together with the atomic compositions of the MOs from a population analysis. In this, the six highest filled MOs are nearly all CN π, while a number of lower ones contain a considerable or sometimes predominant percent of Ni (mostly 3d) and more or less σ CN. The metal–ligand σ bonds are formed by Ni $3d_{x^2-y^2}$ and Ni $3d_{z^2}$.

Demuynck *et al.* [54, 42] have made a more intensive study of $Ni(CN)_4^{2-}$ and shown through SCF calculations on $Ni(CN)_4^-$ states that, as in $Ni(CO)_4$ and other cases (see above), the energy order is not that given by Koopmans' theorem, but some of the predominantly metal MOs are more easily ionized. They give a table of orbital energies [54] not very different from that of [35]; herein, they give details of every MO as a linear combination of C, N (s, pσ, pπ), and Ni ($3d_i$ and $4p_x$) orbitals. A positive charge of about 0.5 *e* is found on the Ni. They also compute excitation energies of many excited singlet and triplet states, six of them by subtracting SCF computed energies of ground and excited states. As with the ionization energies, excitation of predominantly metal orbitals gives the lowest energies. Comparisons are made with the absorption spectrum. First at 22,000–27,000 cm^{-1} there are three weak bands that are interpreted as mainly 3d → 3d. Then at 32,000–37,600 cm^{-1} there are three stronger bands that agree fairly well with computed excitations $4a_{2u} \leftarrow$ 3d b_{2g}, a_{1g}, and e_g (computed 33,900–46,900 cm^{-1}). A localized MO is computed for (NI—C) and a contour map for it is shown.

In an SCF computation already discussed in Section VII,E (ground-state symmetry D_{2d}), Demuynck *et al.* [55] have dealt with $CuCl_4^{2-}$, mostly for symmetry D_{4h}. They tabulate MO orbital energies [56, 55] and give LCAO coefficients for the MOs for the $^2B_{1g}$ ground state (for D_{4h}). As in ferrocene, $Ni(CN)_4^{2-}$, and other cases, SCF calculations on the once-ionized molecule show that the predominately metal MOs are more easily ionized than Koopmans' theorem would predict. Also, calculations on 11 excited states show that the 3 lowest energy excitations are metal to metal; next are 5 ligand to metal transitions.

Smit *et al.* [57] have made more accurate (essentially double-zeta) SCF calculations on $CuCl_4^{2-}$, and also on $NiCl_4^{2-}$, in symmetry T_d. The respective ground states are ... $9t_2^5\,3e^4$, $^2T_2$ and ... $9t_2^4\,3e^4$, $^2T_1$. Orbital energies are listed for the valence MOs in these states. Ligand field splittings and Racah parameters for $NiCl_4^{2-}$ are discussed. Also SCF energies of three excited states of $NiCl_4^{2-}$ are given.

4. Dioxygen Complexes

Dedieu *et al.* [58] have made minimal-basis (except for a split 3d on the Co) calculations on a simplified model of the Co (acacen) LO_2 complexes (L = none, H_2O, CO, CN^-, or imidazole). The structure of Co (acacen) L is as follows:

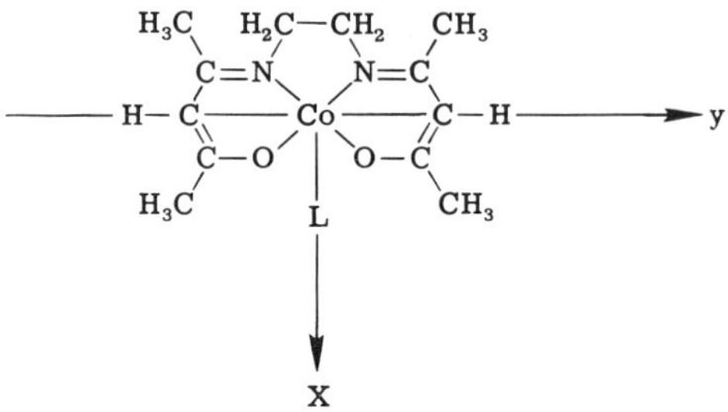

and the simplification consisted in replacing the CH_3 and the bridging CH_2 groups by H atoms. With the axes as shown, the O_2 is located on the z axis above the Co atom. The electron spin in the observed complex is $\frac{1}{2}$. Three positions for the O_2 are considered: linear (on the z axis); the nearer O atom on the z axis and the other bent to the one side (CoOO angle 126°); and the O—O axis perpendicular to the z axis so that the CoOO forms an equilateral triangle. The relative stabilities are found to depend on the ligand, with the bent structure the most stable by 4–26 kcal/mole depending on the nature of the ligand (4 kcal with no ligand). This structure has C_s symmetry. Calculations made on the model molecule without the O_2 show a structure $(d_{xz})^2\,(d_{yz})^2\,(d_{x^2-y^2})^2\,(d_{z^2})^1$. In free O_2 there are two π_g electrons whose MOs in the bent complex could be either symmetric ($\pi_g{}^a$) or antisymmetric ($\pi_g{}^b$) to a plane through the z axis and the O_2. The calculations indicate that $\pi_g{}^a$ interacts with d_{z^2} of Co and forms an occupied pair, leaving one electron in the $\pi_g{}^b$ to give the spin $S = \frac{1}{2}$. The structure would then involve roughly Co(III) and O_2^-. (Population analysis gives a charge of about $-0.5\ e$ on the O_2.) The interaction between $3d_{z^2}$ and $\pi_g{}^a$ decreases in the series L = none, H_2O, Im, CN^- as a result of destabilization of $3d_{z^2}$ by σ donor interaction of L. In all cases, the linear conformation is found to be the second most stable and, of course, the calculations are not accurate enough to exclude its being the most stable. There is much experimental evidence, but none of it quite conclusive as to geometry. However, it perhaps favors an O_2^- structure.

Rather similar SCF calculations with a similar basis set were made by Dedieu *et al.* [59] on $FePO_2$, $FePO_2NH_3$, $Fe(PO_2)(NH_3)_2$, and FeP(CO), where P denotes porphyrin, as models for the hemoglobins. The results indicated a bent structure for the O_2 relative to the z axis. A diamagnetic structure with $(\pi_g{}^a)^2$ as in the Pauling model [60], corresponding to a neutral O_2 and Fe(II), was found. (Oxyhemoglobin is diamagnetic [60].) For FePCO a linear structure was found to be the most stable. Further details are given in a longer article [61]. For the simplified model indicated, Dedieu *et al.* [62] in a double-zeta type of calculation confirm their previous calculations. They also made a "rather qualitative" calculation of the stretching frequency of the O—O bond, bearing on a controversy that the observed frequency favors a $(FeP)^{+}O_2^{-}$ model. They also make calculations with an imidazole ligand, whose interaction with the dioxygen ligand in oxyhemoglobin and oxymyoglobin should be strongly influenced by the charge on the O_2 and an NO distance.

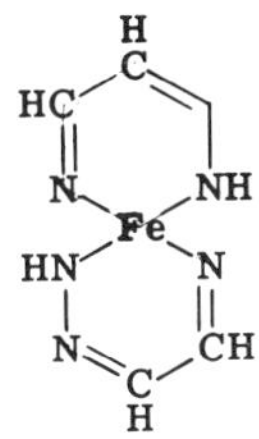

Olafson and Goddard [63] report on "*ab initio* quality calculations on model systems" (four NH_2 groups in a plane in D_{4h} symmetry to simulate porphyrin) to represent hemoglobin and myoglobin. Their results are "consistent with Pauling's model."

D. HYDROGEN BONDING BETWEEN NEUTRAL MOLECULES

1. Survey

Hydrogen bonding is a special case, of particular importance, of the formation of molecular complexes (cf. Section VI,H). This is made clear by Morokuma in a review [64] in which he emphasizes his procedure for analyzing the binding energy of a complex into the sum of several components (ES, POL, EX, CT, and MIX—see Section VI,H,1). Kollman has also published a useful review [65] on hydrogen bonding and donor–acceptor interactions.

Hydrogen bonding may be divided into two main categories: (1) normal hydrogen bonding between neutral molecules, discussed in this section, (2) much stronger hydrogen bonding in or with cations and anions, discussed in Section E. In both cases, one often speaks of proton donors and proton acceptors (or, respectively, electron acceptors and electron donors). Most often considered in *ab initio* computations are 1 : 1 complexes (dimers), but there are also n : 1 and m : n complexes. Proton affinities are the binding energies of a special case of category two in which the proton donor is simply an H^{+} nucleus.

Morokuma in his review has summarized the geometry and his computed analysis into components for nine molecule pairs in category 1 (reproduced in Table 4). Each of these complexes is strongly ES in nature. All are "linear" (see diagram that follows); for water, with the two H atoms of the acceptor molecule in a plane that makes an angle θ with the OHO axis. For the water dimer $(H_2O)_2$, comparisons are made with the

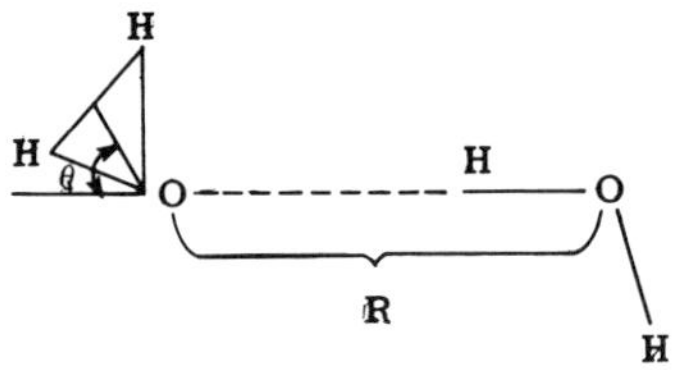

less stable bifurcated form (2a) and the cyclic form (2b) (see Table 5).

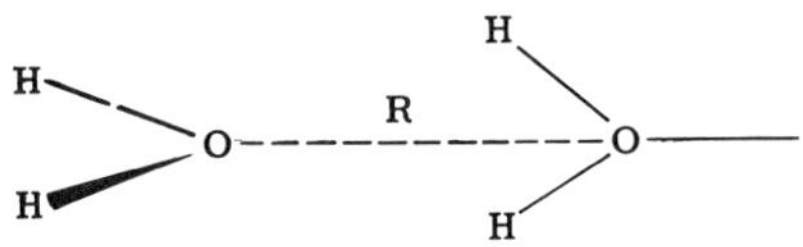

2a

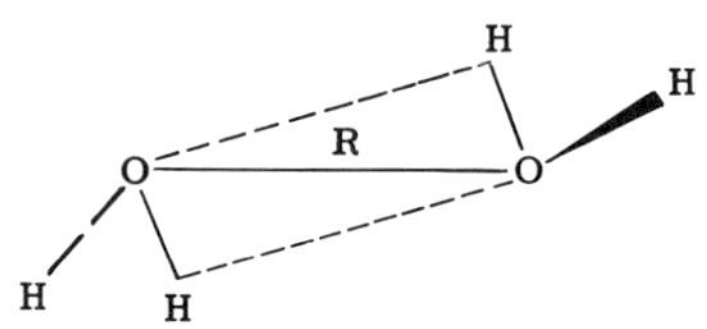

2b

Morokuma gives a figure (see Fig. 1) showing how the angle θ in the proton acceptor molecule in $(H_2O)_2$ varies with each of the energy components and with the total energy. He contrasts normal hydrogen complexes with salt dimers, discussed in Section VIII,C,5. In these, the ES is so large that the typical hydrogen-bonded linear structure is replaced by a cyclic (D_{2h}) structure.

It should be noted that Table 4, although giving a good picture of the energy component analysis, comes from an article [66] in which a relatively small (4-31G) basis set is used. Hence the binding energies (too large) and geometries are not very accurate (see below for some specific cases). The same remarks apply to articles of Kollman *et al.* and of Topp and Allen [67] in which, using a 4-13G basis set the authors report the results of SCF calculations on a great variety of hydrogen-bonded dimers, including complexes of H_2O, HCl, H_2S, PH_3, HCN, HNC, HCP, CH_2 NH, H_2CO,

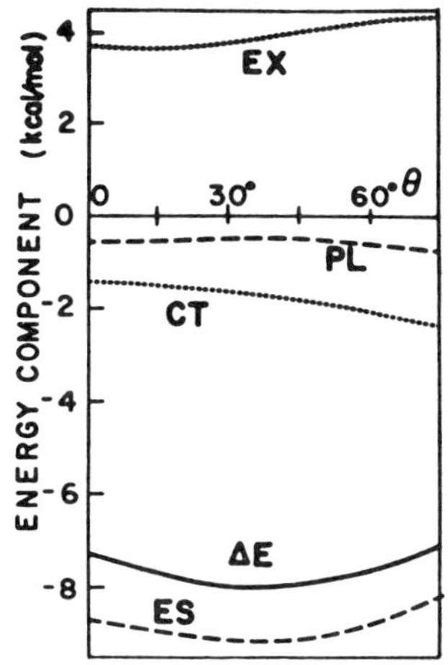

FIG. 1. Energy components for linear $(H_2O)_2$ as functions of θ at R = 2.98 Å. From Morokuma [64].

H_2CS, HCF_3, C_2H_2, C_2H_4, C_6H_6, F^-, and H_3O^+. In another article using a 4-31G basis, Kerns and Allen [68] discuss cyclic hydrogen bonds for $(H_2O)_2$, $(HF)_2$, $(H_2S)_2$, and $(HCl)_2$, and bifurcated bonds for (H_3N) (H_2O), $(H_2O)_2$, $(HF)(H_2O)$, and $(H_2S)_2$. These correspond to saddle points on the potential energy surface. Geometries, total and binding energies, charge transfers, and charge density difference plots are reported.

TABLE 4
Energy Components[a] at the Optimized Intermolecular Geometries[b,c]

Proton acceptor	Proton donor	R_e, Å	θ, deg	ΔE	ES	EX	PL	CT	MIX
H_3N	HF	2.68	0	−16.3	−25.6	16.0	−2.0	−4.1	−0.7
H_2O	HF	2.62	6	−13.4	−18.9	10.5	−1.6	−3.1	−0.4
HF	HF	2.71	60	−7.6	−8.2	4.5	−0.4	−3.2	−0.3
H_3N	HOH	2.93	0[d]	−9.0	−14.0	9.0	−1.1	−2.4	−0.4
H_2O	HOH	2.88	60[d]	−7.8	−10.5	6.2	−0.6	−2.4	−0.5
H_3N	HNH_2	3.30	0[d]	−4.1	−5.7	3.6	−0.6	−1.3	−0.2
H_2O	HNH_2	3.22	60[d]	−4.1	−4.6	2.5	−0.3	−1.5	−0.2
H_3N	HCH_3	4.02	0[d]	−1.1	−0.6	0.5	−0.3	−0.7	−0.0
H_2O	HCH_3	3.80	60[d]	−1.1	−0.5	0.5	−0.1	−0.9	−0.0

[a] In kilocalories per mole.
[b] From Morokuma [64].
[c] All the complexes are linear, having a linear X . . . H—Y bond. θ is the angle between this bond and the molecular axis of the proton acceptor.
[d] Assumed.

TABLE 5

Comparison of Energy Components[a] at the Calculated R_e for Three Structures of $(H_2O)_2$[b]

Geometry: R_e (Å):	Linear 2.88	Bifurcated 2.90	Cyclic 2.85
ΔE	−7.8	−6.4	−6.1
ES	−10.5	−7.4	−7.4
EX	6.2	2.3	4.7
PL	−0.6	−0.3	−0.3
CT	−2.4	−1.0	−2.8
MIX	−0.5	−0.1	−0.3

[a] In kilocalories per mole.
[b] From Morokuma [64].

In addition to Morokuma's energy component analysis, Yamabe and Morokuma [69] have made a corresponding charge-density analysis and applied it to the dimers $(H_2O)_2$, $H_2CO \cdot H_2O$, and cyclopropenone·H_2O. For these they show contour diagrams for the difference-densities between the dimers and the sum of the densities of the monomers. The results are similar for all these cases. The exchange component decreases the density markedly along the O···H—O line and increases it locally on the O and H atoms involved in the hydrogen bond. The polarization component, although small for the energy, produces extensive and widespread changes in the charge density. The charge-transfer component produces most notably a strong covalent buildup between the H and O atoms of the hydrogen bond.

For the water dimer, Yamabe and Morokuma also report a component analysis of the computed dipole moments.

In the following reports, unless otherwise stated, the geometry of the partners in each hydrogen-bonded system has been *assumed* to be the same as in the monomers.

2. Water Dimers and Polymers

The water dimer $(H_2O)_2$ has been the subject of many *ab initio* computations [70]. All agree that the linear conformation is the most stable. A charge-density component analysis of the water dimer has already been discussed in Section D,1, along with a component analysis of the computed dipole moment. The most accurate SCF calculation, estimated to be within 0.0002 a.u. of the SCF limit by use of a very extensive basis set, made by Popkie *et al.* [71], yields $E_h = -152.1378$; however, this is for

$R_{OO} = 3.00$ Å and $\theta = 0°$, whereas the best values are about 2.92 Å and $\theta = 42°$ (see [67] below).

Using a smaller but still rather large (double-zeta plus polarization) basis set, Popkie *et al.* also computed the SCF interaction energy of two water molecules placed at 216 different relative locations varying in R_{OO} and in orientation. From these data plus more accurate values from 26 points near the energy minimum, they derived an analytical potential, in terms of interatomic distances, for the energy of interaction of the water molecules. For the attractive regions of this potential, the fit to the data is good to a standard deviation of 0.0002 a.u. Using this potential, in a Monte Carlo simulation of water at 277°, 298°, and 348°K, they obtained preliminary results for pair correlation functions in agreement with experimental data. Note that in the foregoing procedure, the three conformations of $(H_2O)_2$ illustrated above are included with due weight among much more varied configurations.

In a later, more elaborate article, Kistenmacher *et al.* [72] obtained a revised pair potential for the water dimer. They used this potential to compute energies for water polymers $(H_2O)_n$, with $n = 3–8$, in thousands of possible configurations and concluded that, at least if three- and four-body effects are minor, the trimer and tetramer (in partial disagreement with previous workers) are cyclic. Diagrams for the geometry of these are given. A separate SCF calculation on the trimer at the optimal cyclic configuration gave an energy about 10% lower, and a similar result was obtained for the tetramer, indicating that the analytical expression for the pair or two-body potential should be fairly good in statistical computations on an extended system such as water. In applying this potential to water, Kistenmacher *et al.* [73] used semiempirical corrections for the effect of electron correlation.

Next, Matsuoka *et al.* [74] made a CM study to obtain a reasonably good value of the correlation energy function for the potential surface of the water dimer. They also obtained dimerization energies of −5.6, −4.9, and −4.2 kcal/mole corresponding to the potential minima, respectively, of the linear, cyclic, and bifurcated forms of the dimer. They used essentially the same double-zeta plus polarization basis set as in [70–73]. They transformed the SCF MOs (10 occupied and 48 virtual) to localize MOs by the method of Edmiston and Ruedenberg, then transformed the virtual MOs to approximate NOs, and used the first 15 of these in configuration state functions (CSFs) using single and double substitutions. They then calculated the CM energies for the dimers of each of a considerable number of carefully selected geometrical configurations, and obtained analytical formulas to express the resulting potential. Additional details, including a division into *inter* and *intra* wave functions, are given.

Finally, Lie *et al.* [75] used the CM interaction pair potential just obtained in a Monte Carlo simulation of water at 25°C. From the results they computed radial distribution functions and X-ray and neutron scattering intensities, which were found to be in satisfactory agreement with experiment, the agreement being extremely good for the radial distribution functions. Allowing for zero-point energy, the total interaction energy is −6.8 kcal/mole as compared with the experimental value −8.1. It is suggested that the discrepancy may be attributed to lack of allowance for three- and even four-body interactions (as to this, compare the second preceding paragraph). Lie and Clementi [76], using a somewhat earlier *ab initio* potential, calculated second virial coefficients for steam. Agreements with experiments are only moderate.

Aside from the articles of Clementi and collaborators just discussed, many other *ab initio* calculations have been published [70], mostly SCF. With apologies to the other authors, we shall refer here only to a computation by Diercksen *et al.* [77] on the water dimer. They used a double-zeta plus polarization basis set, as did Matsouka *et al.*, but limited themselves to a small portion of the linear-model potential surface. (It is not unlikely that there is a small departure from the linear model for minimum energy.) However, they used much more extensive CM (56,268 CSFs, including all single and double substitutions except for the 1s orbitals), and thereby obtained more accurate results for the correlation energy, though in a limited region of configuration space. They estimated that higher than double substitutions would contribute less than 0.1 kcal/mole to the correlation energy. Their correlation energy (−0.01388 a.u.) is considerably less than that of Matsouka *et al.* (−0.001656 a.u.) for the linear model. For the equilibrium geometry of $(H_2O)_2$, they find d(OO) = 2.919 Å, θ = 42.4°. For the binding energy they find a value of 6.05 kcal/mole without allowing for zero-point energy. Making an estimated correction of 1 kcal/mole for the latter, the dimerization energy of water turns out to be 5.1 ± 0.3 kcal/mole, close to the experimental value of 5.2 ± 1.5.

3. Hydrogen Fluoride Dimers

The principal constituents of HF vapor are a monomer, a dimer, and a cyclic hexamer. Several *ab initio* computations have been made on the dimer, but perhaps the most accurate of the earlier SCF equilibrium results are those of Diercksen and Kraemer [78], who used an (11s 7p 1d/6s 1p) basis set contracted to (5s 4p 1d/3s 1p). They examined the linear and a cyclic dimer for a number of interatomic distances and θ values, and found the former to be more stable with R_{FF} = 2.85 Å, θ = 40° (experimental: 40 ± 5°), and $-\Delta E$ = 4.50 kcal/mole (experimental: 6.0 ± 1.5).

Yarkony *et al.* [79], using a double-zeta plus polarization basis, made a massive exploration of the interaction potential between two rigid HF molecules. They made computations of the energy for 294 geometric configurations both planar and nonplanar, and report attempts at an analytical fit to these points. They show contour maps of parts of the potential surface. For the predicted equilibrium geometry they find $R = 5.29$ a.u., $\theta = 52°$ with a departure of about 5° from linearity in a plane. For the dissociation energy they find 4.63 kcal/mole.

Lischka [80] has studied the effect of electron correlation, using a strictly linear model ($\theta = 0°$). He used an (11s 7p 2d/6s 1p) basis contracted to (7s 4p 2d/4s 1p), and obtained an SCF $E_h = -200.130735$ at $R_{FF} = 5.50$ a.u., $\theta = 0°$, as compared with Diercksen and Kramer's $E_h = -200.11974$ at $R_{FF} = 5.50$ a.u. and $\theta = 40°$ (zero-point corrections not included in either case). Using a localization approach, Lischka computed the valence-shell correlation energy by IEPA (which is nonvariational), by CEPA (but with no single substitutions), and by PNO-CI (which furnishes a variational upper bound including double substitutions). More than double substitutions are approximately included in the IEPA and CEPA. The final conclusion is that correlation makes a relatively small net effect on the computed interaction energy $-\Delta E$, the F·F distance, and the intermolecular vibrations. Force constants for the latter are given. For PNO-CI at $R_{FF} = 5.50$ a.u., $E_h = -200.544615$; for CEPA, $E_h = -200.577605$. For various technical details reference must be made to Lischka's paper.

4. Mixed Dimers of H_2O and of HF

Results of 4-31G basis SCF computations on mixed dimers of H_2O are reported in [66–69] and in Table 4. Diercksen *et al.* [81] have made a more accurate calculation on $NH_3{\cdot}H_2O$ using an (11, 7, 1/6, 1) Gaussian basis contracted to (5, 4, 1/3, 1). Two sets of structures $H_3N{\cdots}HOH$ and $H_2O{\cdots}HNH_2$ were investigated, with the former found to be the more stable, with a binding energy $-\Delta E$ of 6.28 kcal/mole. For the latter, $-\Delta E = 4.83$ kcal/mole. Calculations were also made with the NH_3 assumed planar; here $-\Delta E = 3.92$ and 2.35 in the former and latter cases. The $H_3N{\cdots}HOH$ calculation was also made with the NH_3 pyramid inverted: $-\Delta E = 0.23$; from the data, two curves are shown for the dimer with the NH_3 normally H-bonded or inverted.

Morokuma [82], using a minimal SCF basis set, studied the $H_2CO{\cdot}H_2O$ dimer for several relative positions of the monomers. His optimal geometry involved a linear $O{\cdots}HOH$ bond from the carbonyl oxygen in a planar structure with an inclination $\theta = 64°$. The computed binding energy $-\Delta E$

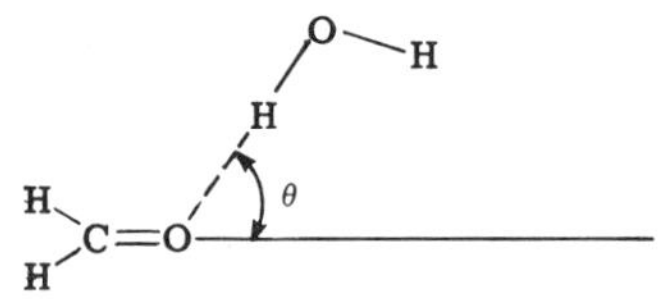

was 3.5 kcal/mole at an O···H distance of 1.89 Å. A Morokuma analysis of $-\Delta E$ was made and showed charge transfer to be important, the amount of charge transferred ($H_2CO \rightarrow H_2O$) being about 0.03 e [at R(O···H) = 1.796 Å]. In formaldehyde (see Section IX,F,1) with electron configuration ··· $5a_1^2\, 1b_1^2\, 2b_2^2$, the last three MOs are, respectively, σ, π, and n, in which the $2b_2$ MO is roughly a nonbinding O atom 2p AO with symmetry axis in the plane but perpendicular to the C=O axis. In the dimer, it is clearly the $2b_2$ MO that is most active in the hydrogen bonding, with some charge transfer to the H-bonded H. Morokuma also examined other configurations, including one with the HOH plane perpendicular to the H_2CO plane and with H bonding again to the O atom (or alternatively to the C atom or an intermediate point on the C=O axis); here one would speak of π bonding, since the $1b_1$ MO of H_2CO would be involved. But the computed binding energy is only 0.4 kcal/mole at a distance of 2.5 Å. The possibility of a hydrogen bond H_2O···HCHO was also considered, but with the best orientation $-\Delta E$ was only 0.6 kcal/mole. In addition to his computation on the planar complex, Morokuma made extensive calculations on the complex with the two H atoms of H_2CO bent down as they are in familiar excited states ("bent complex").

The trimer H_2CO···$2H_2O$ was also considered. In this, the second H_2O attaches itself to the first as in the H_2O dimer, with a computed *additional* binding energy of 2.5 kcal/mole.

Using a 6-31-G basis, Swaminathan *et al.* [83] made SCF calculations for $H_2CO \cdot H_2O$ at 200 different configurations, and to these data fitted an analytical potential $V_{12-1-3} = \sum_{i<j} a_i a_j / r_{ij}^{12} + b_i b_j / r_{ij} + c_i c_j / r_{ij}^3$, along with a less successful V_{12-1} potential. For the V_{12-1-3} potential, the fitting resulted in a standard deviation of 0.23 kcal/mole. Contour maps of orientationally optimized geometries are shown and discussed. The minimum energy contour is at $-\Delta E = 5.427$ kcal/mole, which is larger than Morokuma's value, but with about the same orientation.

Iwata and Morokuma [84], using minimal (STO-3G), minimal plus diffuse p, and 4-31G basis sets, made SCF computations using the "electron–hole potential method," on the (n, π^*) and (π, π^*) singlet and triplet excited states, and also for $H_2CO \cdot 2H_2O$. It is well known experimentally that with aldehydes and ketones, the (n, π^*) transition generally shows a rather large shift toward higher frequencies ("blue shift") on hydrogen bonding, while the (π, π^*) transition in conjugated molecules makes a small red shift. These shifts should correspond to vertical transitions. The calculations on $H_2CO \cdot H_2O$ resulted in predictions of blue shifts approaching 2000 cm^{-1} for the ^{3}(n, π^*) and 1600 cm^{-1} for the ^{1}(n, π^*) vertical transitions, and smaller blue shifts for the (π, π^*) transitions, the exact results varying somewhat with the basis set used. Charge transfer from H_2CO to H_2O is much reduced in the (n, π^*) states. This is understandable because the charge on the carbonyl oxygen is much less in the (n, π^*) transition. Vertical excitation energies are listed, and population analyses and energy decomposition (Morokuma) analyses are given.

Similar calculations on the (n, π^*) transition for the trimer $H_2CO \cdot 2H_2O$ predicted a blue shift of 2150 cm^{-1}, nearly the same as the corresponding shift (1900 cm^{-1}) for acetone in water solution (minimal basis?). The additional $-\Delta E$ on adding the second water molecule was 2.50, 1.06, and 2.64 kcal/mole in the ground, ^{3}(n, π^*), and ^{1}(π, π^*) states, respectively. The (n, π^*) states relative to H_2CO are unstabilized and are more unstable than for the dimer.

Iwata and Morokuma [85] made similar computations on the *trans*-acrolein–water complexes $C_3H_4O \cdot H_2O$ and $C_3H_4O \cdot 2H_2O$. For the ground states, a planar (C_s) geometry was assumed. With a 4-31G basis [which

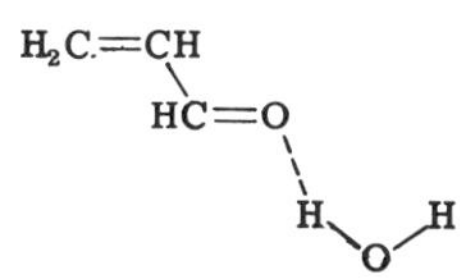

(see Section D,1) gives too large values], $-\Delta E$ = 7.57 kcal/mole for the dimer and 14.21 for the trimer. Excitation energies for (n, π^*) and (π, π^*) singlet and triplet transitions are given. Blue shifts are found for the (n, π^*) transitions: for the dimer, 2800 cm^{-1} for the triplet and 2400 cm^{-1} for the singlet; for the trimer, 5500 cm^{-1} for the triplet and 4700 cm^{-1} for the singlet. Smaller red shifts are found for (π, π^*) transitions: for the dimer, -60 cm^{-1} for the triplet and -700 cm^{-1} for the singlet; for the trimer, -300 cm^{-1} for the triplet and -2000 cm^{-1} for the singlet. The (π, π^*) red shifts here contrast with small blue shifts in $H_2CO \cdot H_2O$; the difference is evidently a result of conjugation in the present case; a detailed explanation is given.

As already mentioned in Section D,1, Yamabe and Morokuma [69] have made charge-density-change analyses on the cyclopropenone–water complex. First they made geometry and energy studies on the complex using a 4-31G basis. They found a planar geometry with an O···H distance

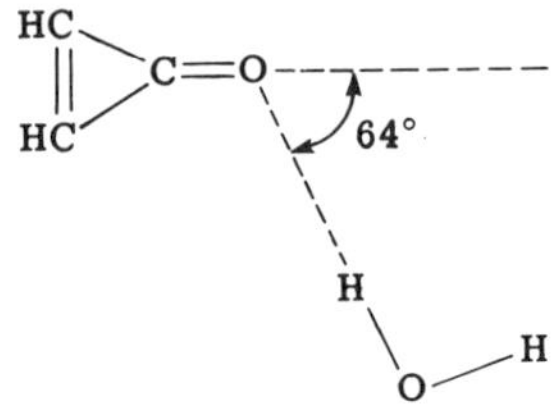

of 1.79 Å, shorter than that (1.89 Å) in $H_2CO{\cdot}H_2O$, and a larger $-\Delta E$ (5.75 kcal/mole as against 3.50). These differences they explain as a result of the stronger C^+O^- character of the C=O bond in cyclopropenone. The charge-transfer component of the charge-density change is larger in the cyclopropenone than in the formaldehyde complex because of the smaller O···H distance. They examined the possibility of a π complex in both cases and found a calculated binding energy of 2.36 kcal/mole for the cyclopropenone as against 0.40 kcal/mole for the formaldehyde complex (but with a minimal basis for the latter—see above).

Using a double-zeta basis for carbon, Ungemach and Schaefer [86] have made SCF computations on $CH_4{\cdot}H_2O$. They arrive at a linear (O···H—C) C_s geometry with an O–C distance of 3.85 Å, and a binding

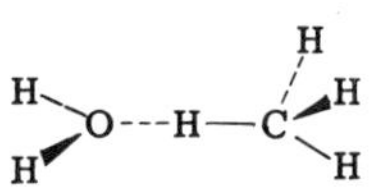

energy of 0.5 kcal/mole plus a guessed dispersion energy of 0.1 kcal/mole. On adding a d polarization function, R(OC) becomes 4.01 Å and $-\Delta E$ becomes 0.34 kcal/mole. However, they express doubt as to whether this result is more nearly correct. They find that on bringing up a second water molecule, there is a repulsion.

Harrison *et al.* [87] have used a 6-31G basis set to make 225 SCF computations of energy for a variety of relative locations and orientations of a CH_4 and an H_2O molecule. They fit these data with an analytical potential in inverse powers of interatomic distances (see $V_{12,1,3}$ under the account of $H_2CO{\cdot}H_2O$ in [72]). They give contour maps of the computed potential as a function of water location and orientation. Finally, they use Møller–Plesset perturbation theory to estimate the dispersion energy contribution, and find electron correlation to stabilize the system by 0.4 kcal/mole—at the configuration of Ungemach and Schaefer. However, in further calculations using the Møller–Plesset correction they arrive at a different equilibrium configuration (contour map shown) with $-\Delta E =$

1.17 kcal/mole. They also use their potential energy formula in obtaining a Monte Carlo simulation of a dilute aqueous methane solution.

From the above, it appears uncertain whether we should speak of H-bonding between water and methane.

Del Bene [88] in STO-3G and 4-31G calculations has made calculations on hydrogen bonding of HF and H_2O with the π electrons of C_2H_2, C_2H_4, HCN, and H_2CO. With C_2H_2 and C_2H_4, planar complexes of C_{2v} symmetry were formed with the HF lying perpendicular to the middle of the π bonds. The structures of HOH·C_2H_2 and HOH·C_2H_4 were similar except that the O—H···π bonds differed somewhat from linearity. In the cases of H_2CO (cf. with Morokuma above [82]) and HCN, the π-bonded complexes were less stable than the usual σ-bonded complexes.

del Conde and Novaro [89], using a double-zeta basis, have calculated H bonding of HOH to SH_2 and of HSH to OH_2. They find the former to be stabilized for $R(\mathrm{OS}) = 3.82$ Å by 2.08 kcal/mole and the latter by 5.1 kcal/mole for $R(\mathrm{OS}) = 3.44$ Å.

Alagona *et al.* [90], using the 4-31G basis, have made calculations on the ten most stable complexes involving H_2O, HF, HCl, F^-, and Cl^- in an inquiry as to whether water acts as a catalyst for some nucleophilic reactions in inert solvents with hydrogen halides as nucleophilic agents. Their answer is distinctively positive.

Using up to double-zeta basis sets, Clark and Cromaty [91] investigated electronic reorganizations accompanying core and valence ionization in H_3N···HOH, H_2O···HOH, HF···HOH, H_2O···HNH_2, H_2O···HF, and $(H_2O)_3$. Creation of a 1s hole causes substantial changes in hydrogen bond energies. Many details are given.

5. Mixed Dimers with Biomolecules

Clementi and associates have made *ab initio* computations with a rather small basis set on the interaction energy between a water molecule in a great variety of locations and orientations and each of 21 naturally occurring amino acids. From the results, Clementi *et al.* [92] have constructed analytical potentials of the form

$$\Delta E = \sum_{i} \sum_{i \neq j} (-A_{ij}{}^{ab}/r_{ij}{}^{6} + B_{ij}{}^{ab}/r_{ij}{}^{12} + C_{ij}{}^{ab}/r_{ij})$$

where i and j refer to one of the atoms in H_2O and to one of the atoms in the amino acid, a is an index that distinguishes different types of amino acid atoms, and b distinguishes between H and O atoms in H_2O. The fitting constants A, B, and C are obtained by assigning charges, from a population analysis, to each atom and assigning the atoms to 21 classes.

Based on 1690 individual SCF computations, the standard deviation of the fit is 0.6 kcal/mole. Note that the particular mode of obtaining the fitting formula should be judged by the degree of fit independent of the somewhat arbitrary manner in which it was obtained. For each amino acid a contour diagram in a suitable plane is given showing how the amino acids are "seen" energetically by water. Hydrophobic regions are especially evident from these diagrams.

In another article on "reliability and transferability of the pair potentials," Bolis and Clementi [93] have computed the interaction energy of a water molecule with phenylalanine at 75 relative locations, and fitted these to a pair potential of the form of ΔE as given above. They find that the 75 computed energies can be duplicated to 1 kcal/mole by making use of the pair potentials reported in the previous article.

In another article, Scordamaglia *et al.* [94] determine the coefficients in the above ΔE expression for pair potentials in the case of a water molecule interacting with each of the DNA bases adenine, guanine, cytosine, and thyamine. Contour diagrams for each case are given for ΔE in the plane of the DNA base and in planes 4.0 and 6.0 a.u. above this plane.

The results of the above calculations are reviewed in a lecture by Clementi [2].

In an article that would be out of place here since it involves more than H bonding, Pullman *et al.* (see Section XI,E) discuss the electrostatic potential around a segment of a DNA helix in the neighborhood of a guanine–cytosine base pair. More directly relevant to H bonding (again see Section XI,E), they present and discuss the electrostatic potentials in the neighborhood of each of the four DNA bases.

Pullman and collaborators have published articles which, though largely relevant to Section D,4, are motivated by the idea of using simpler systems as models for peptides and proteins. Using a minimal basis set, Alagona *et al.* [95] have investigated sites for the hydration of formamide and Pullman *et al.* [96] sites for *N*-methylacetamide (for the monomer, see Section IX,D,2). The main results are shown in the accompanying Fig. 2 taken from [97]. The H bonds are for the most part nearly linear (O···H—O or N—H···O), with small energy maxima ($\theta = 0°$) in the O···H—O case for an orientation along the C=O axis between the two maxima shown in Fig. 2.

Port and Pullman [97] report on several model studies relevant to the hydration of peptides and proteins. As a model representation of the H-bonded peptide linkage in proteins they disucss a noncyclic "linear" dimer as in Fig. 3 [97], of formamide. For the stable cyclic dimer, see Section D,6. Actually, the calculation shows that the O···H—N bond would be in one of two sites rotated by 88° in opposite directions (see Fig. 3) so

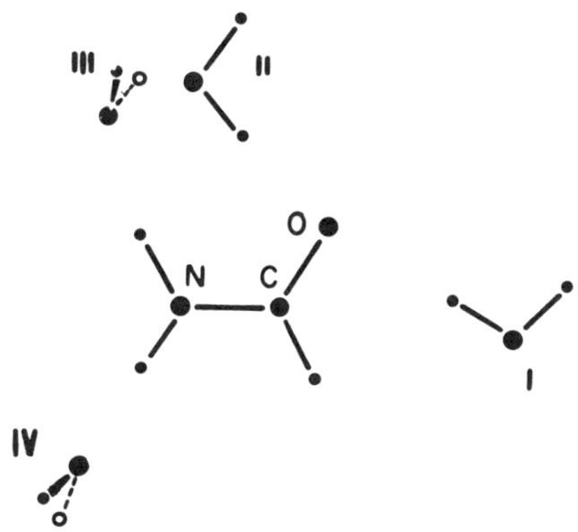

FIG. 2. The stable hydration sites of formamide. From Alagona *et al.* [95].

as to be about perpendicular to the proton acceptor --C═O. Further, in a protein α helix one peptide monomer plane would be rotated 60° with respect to the other (twisted dimer). Disregarding these qualifications for the moment, the authors inquire how an added water molecule might H bond itself in the linear model as shown in Fig. 3. Their calculations indicate two locations as shown in the diagram, with ΔE values in kilocalories per mole as given by the numerical inserts. These are compared with those for the monomer. For the *twisted* dimer, $-\Delta E$ for the lower ($\Delta E = -7.1$) location is slightly increased, while the upper location is excluded; this model and two others with hydration at the second C═O or as N—H···OH_2 appear relevant with respect to the hydration of an α helix in a protein.

Port and Pullman as further models discuss the hydration of formic acid, formate ion, ethylamine, ethylammonium, ethanol, and *p*-cresol, with results as shown in Fig. 4 taken from their article.

Ottersen *et al.* [98], using a (7s 3p 4s) basis contracted to (4s 2p/2s),

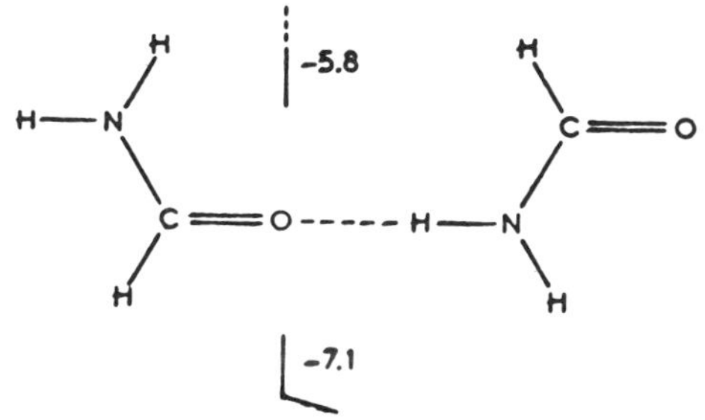

FIG. 3. Linear antiparallel dimer of formamide showing the most favorable hydration sites with their calculated interaction energies (kcal/mole). From Port and Pullman [97].

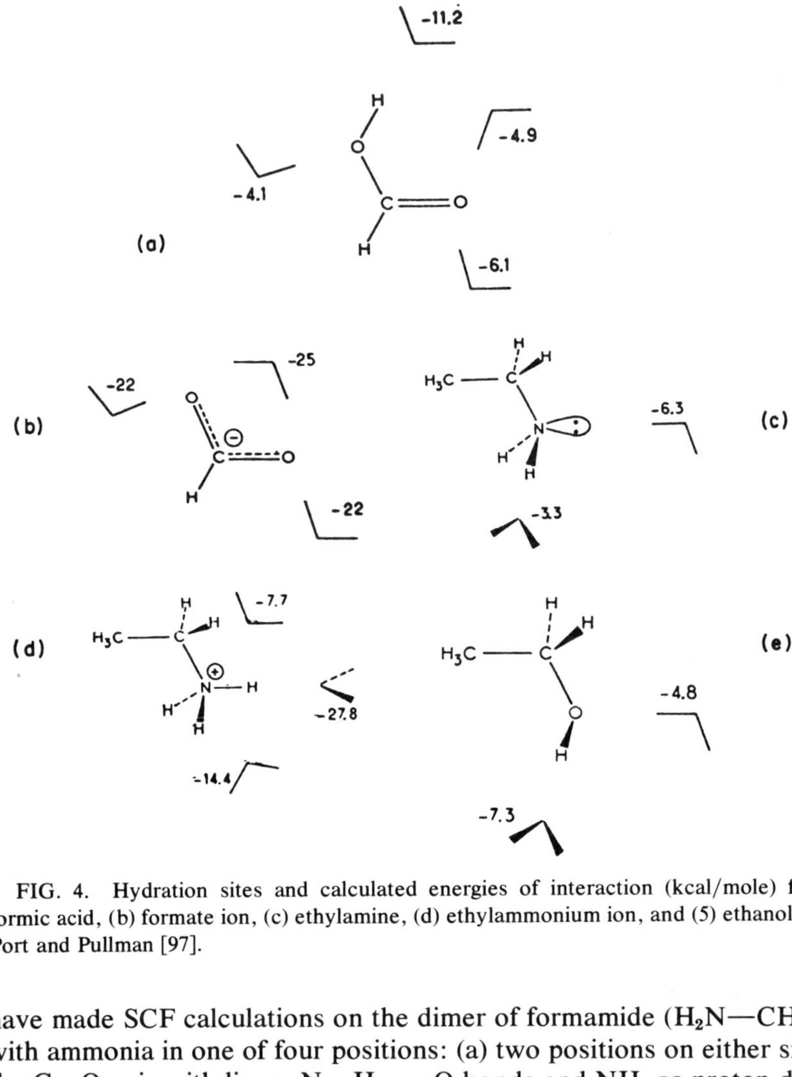

FIG. 4. Hydration sites and calculated energies of interaction (kcal/mole) for: (a) formic acid, (b) formate ion, (c) ethylamine, (d) ethylammonium ion, and (5) ethanol. From Port and Pullman [97].

have made SCF calculations on the dimer of formamide ($H_2N{-}CH{=}O$) with ammonia in one of four positions: (a) two positions on either side of the C=O axis with linear N—H——O bonds and NH_3 as proton donor, and a $\mathrm{C{=}O{\diagdown}H}$ angle less than but approaching 90°; (b) two linear N—H--N bonds with NH_3 as proton acceptor. In each case, the H bonds are cis or trans to the C=O direction. In all cases, the C—O bond lengths were found to be lengthened about 0.005 Å and the C—N bond lengths shortened about 0.008 Å. Computed formation energies were about 6–7 kcal/mole for NH_3 as proton donor and 10–11 for NH_3 as proton acceptor. The NO bond lengths found in N—H . . . O were 3.06 Å (cis) and 3.16 Å (trans), and the NN bond lengths found in N—H . . . N were 3.01 Å (cis) and 3.04 (trans). A population analysis and some force constants are given. The results indicate that the dimer formation has a marked effect on the conjugated peptide linkage.

Orita and Pullman [99] determined the preferred sites of water fixation around urea, finding that up to five water molecules can be bound in the first hydration shell.

Glyoxal and guanine add to form a composite molecule. Demoulin *et al.* [100] have studied the mechanism of the reaction, as being a two-step reaction.

Peters and Peters [101], using a minimal basis set, have obtained a conformational energy map for the simplest possible dipeptides.

6. The Formic Acid and Formamide Dimers

The carboxylic acids form stable dimers. The simplest of these is the formic acid dimer $(HCOOH)_2$. The diagram shows the experimental geometry

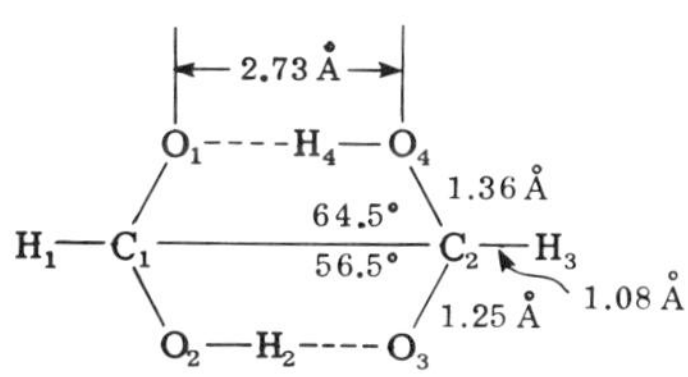

The symmetry is C_{2h} as compared with C_s for the monomer (see Section IX,F,8). Several SCF calculations have been made on this dimer. Using the experimental geometry, Clementi *et al.* [102] used contracted (9s 5p/4s) basis sets with and without polarization functions (p on H_2, H_4 and $3d\sigma$ on C and O) to compute potential curves for the transfer of, e.g., H_2 from O_2 to O_3, leaving all other features of the geometry unchanged. This transfer of an H atom would create an entity built from two radicals.[4] An ascending single-minimum energy curve, including a shoulder, resulted.

Then they computed energy curves for the simultaneous transfer of H_2 to O_3 and H_4 to O_1. A double-minimum curve resulted, with its maximum when the two H atoms are both roughly halfway across. The second min-

[4] The O_2 and O_3 would now have an odd electron, but on O_3 the odd electron would have to go into an antibonding O_3H MO that would be polarized toward O_2; or more accurately, the C_2O_3 σ bond would share the antibonding. A literal *proton* transfer, not comtemplated here, would result in a (polarized) $O_2^+\cdots HO_3^-$ structure.

imum is very considerably higher than the first because allowance has not been made for a simultaneous readjustment of the two C—O distances to a new equilibrium. Clementi *et al.* give a diagram showing valence-electron orbital energies of the dimer for each of three basis sets.

Ady and Brickmann [103] made SCF calculations using an (8s 3p/4s) Gaussian-lobe basis contracted to (3s 1p/2s) for the two-electron integrals but uncontracted for the one-electron integrals. They made calculations like those of Clementi *et al.*, but now also allowed the C—O (C=O and C—OH) distances to readjust during the H-atom exchange process so as to minimize the energy at all stages of the transfer. This yields a double-minimum curve that is symmetrical. For the equilibrium dimer, they assume the experimental geometry (in particular, 2.73 Å for the OO distance in O—H···O). They show energy contours for a process in which the sum of the C=O and C—OH distances for each monomer is held constant, and the O—H···O segment is kept linear, at all stages. A maximum energy occurs at a saddle point where both OH distances are equal, and where CO distances are equal. They note that for an adiabatic process there is a curved path. They list all occupied orbital energies for the dimer.

Del Bene and Kochenour [104] made an STO-3G minimal-basis SCF calculation on $(HCOOH)_2$, and calculated an energy curve for the inversion of the dimer structure (simultaneous and symmetrical exchange of the H-bonded hydrogens, keeping the CO distance involved always optimized). Their curve has a maximum (D_{2h} symmetry) height of 56 kcal if the experimental 2.73 Å O—··O distance is used (however, with their computed O–O distance of 2.54 Å in O—H···O, the height of the maximum is only 23 kcal).

Iwata and Morokuma [105], using their electron–hole potential method for excited states and an STO-3G basis plus diffuse p on C and O, made computations on the ground states and $^3(n, \pi^*)$, $^1(n, \pi^*)$, and $^3(\pi, \pi^*)$ excited states on HCOOH and its dimer. They show computed curves of the energies involved in gradually transferring O_2 to O_3 in $(HCOOH)_2$, and simultaneously H_4 from O_4 to O_1, with all CO lengths held fixed, for all the states mentioned. In all, the result is an unsymmetrical and double-minimum curve. They note that the $O_1 \cdots H_4$ distance of 1.75 Å in $(HCOOH)_2$ found experimentally is shorter than that (1.89 Å) in $H_2CO \cdot H_2O$, and correspondingly the binding energy for the H bond is greatter. They give a table in which dimer binding energies and blue shifts on dimerization for the excited states are compared for $(HCOOH)_2$, H_2CO—HOH, and $H_2CHCHCO$—HOH (for the latter two, see Section D,4). A Morokuma energy decomposition analysis for the several states is given.

Scheiner and Kern [106] with a 4-31G basis made energy computations on $(HCOOH)_2$ in which the geometry was held fixed, with the OO distance in O—H···O kept at 2.73 Å, except that the two H-bonded H atoms were allowed to move independently, on the O—H···O and O···H—O lines. They give a contour diagram for the potential energy as a function of the two H-bonded OH distances, r_1 and r_2, each varying from 1.01 to 1.72 Å. The diagonal ($r_1 = r_2$) has a maximum 46 kcal/mole up. A curved path ($r_1 > r_2$) shows slightly lower energies. When $r_1 = r_2 = 1.72$, there has been a proton exchange but with C=O and C—OH distances artificially constrained to their initial values. When r_1 has its equilibrium value 1.01 Å but $r_2 = 1.72$ Å, there has been a transfer of a single atom (see footnote on p. 397) but of course under geometry constraints.

Carnovale *et al.* [107] have made an improved photoelectron spectrum of $(HCOOH)_2$ and compared it with their 4-31G calculations on the MOs. Their 12 ionization potentials can be matched fairly well against the computed Koopmans' theorem orbital energies if the latter are multiplied by a (somewhat sliding) factor (average about 0.85).

Hashimoto *et al.* [108] with a mininal basis set have calculated what happens in β^- decay of the cyclic dimer of formamide when it is monotritiated at one of the external hydrogens. Here ^3H becomes He^+, which they conclude is repelled, leaving a positively charged dimer ion.

Dreyfus *et al.* [109] have made SCF calculations on the cyclic formamide dimer with a (7s 3p/3s) basis set contracted to (2s 1p/1s) and with a (4s 2p/3s) set contracted to (2s 1p/2s). They obtain dimerization energies of 14 kcal/mole with the first and 19 kcal/mole with the second basis. They tabulate orbital energies for the occupied MOs and give a population analysis. They also give contour maps for the charge density changes from monomers to dimers, along with a map of charge transfers from one monomer to the other in hydrogen bonding. Both the proton donor and acceptor show a global gain in electrons at the expense of the bridge hydrogen and the carbon atoms. The changes are reflected in the 1s orbital energies. Dreyfus and Pullman [110] have made detailed calculations on the linear dimer of formamide (C=O···H—N, only one H-bond), which is, of course, less stable than the cyclic dimer, but is instructive as to the H bond between peptide units (see Section D,5).

Ottersen and Jensen [111] have made SCF calculations on the cyclic dimer of formamide, using a basis set (7s 3p/4s) contracted to (4s 2p/2s), plus a p function on the bridge hydrogens. In this way they obtained $E_h = -337.4112$, as compared with $E_h = -328.9292$ by Dreyfus *et ao*. They also made computations on the monomer, but Stenkamp and Davidson (see Section IX,D,2) have done much better. Unlike Dreyfus *et al.*, they optimized the CO and CN distances, obtaining 1.233 and 1.354 Å, respectively, as compared with 1.216 and 1.354 Å in the monomer. Some force constants and populations are listed.

7. Intramolecular Hydrogen Bonding in the Enol Form of Malonaldehyde

Isaacson and Morokuma [112] used an STO-3G minimal basis set to investigate intramolecular hydrogen bonding in the enol form of malonaldehyde (3). They conclude that a symmetrical hydrogen bond is not quite at-

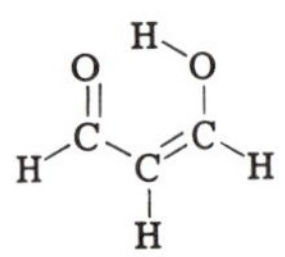

3

tained. With calculations also on FHF, they propose two criteria for such attainment: (1) $R_{X-Y} < 2.3$ Å; (2) the existence of a conjugated backbone structure. Here X and Y are the C═O and the O—H oxygens. Their computed OO distance is 2.38 Å, involving a symmetrical double minimum with the H atom located roughly 0.14 Å off the C_{2v} axis of a symmetrical structure and 0.23 Å above the OO line, with R_{OH} about 1.08 Å.

Del Bene and Kochenour [104] made similar calculations with similar results. They find the actual asymmetric C_s structure to be 6.6 kcal/mole more stable than the C_{2v} symmetric structure. They give a symmetric potential curve for the transfer of a proton from one O to the other. This has a double maximum with a slight dimple (local minimum at C_{2v} symmetry) between the two maxima.

Catalán *et al.* [113] made a similar calculation with similar results, confirming the dimple in the potential curve at C_{2v}. They report in the abstract a barrier of 10.6 kcal/mole, but only 6.6 in the text and in a figure. They made a full geometry optimization, and among other results find $R_{OH} = 1.01$ Å.

Meanwhile, with some assumptions about the geometry (everyone has assumed planarity) Karlström *et al.* [114] made calculations on the C_s and C_{2v} forms using a larger basis set: (7s 3p/4s) contracted to (4s 2p/2s), plus p functions on the H-bond hydrogen. They find the C_s form to be more stable than the C_{2v} form by 48 kJ/mole. Their potential curve for H-bond exchange shows no dimple. Strangely, they report that a minimal-basis calculation shows no significant energy difference between the C_s and C_{2v} geometries.

In another article [115] they repeat the calculation using CM (selected

The ground state is a triplet, with the lowest singlet 2.0 eV higher in $Ni(CH_3)_2$, but in the dihydrate only 0.4 eV higher. Further details and possible applications are given.

Hinchcliffe [125], using STO-3G, 4-31G, and a more extended basis, and assuming a planar C_{2v} structure of the heavy atoms, made SCF calculations on $(CH_3)_2O \cdot HCl$ in which the calculated $R(OH)$ increased from 1.46 to 1.97 Å from the smallest to the largest basis set, while ΔE decreased from 49 to 22 kcal/mole. Carnovale *et al.* [126], using a minimum plus somewhat extended Cl basis set, found evidence that a C_s structure with the OHCl axis bent by 34° out of the C_2OHCl plane is about 1 kJ/mole more stable than the planar structure. They used Hinchcliffe's orbital energies (minus 0.9 eV) in satisfactory interpretations of their photoelectron data on the complex.

Del Bene [127], using an STO-3G minimal basis, determined structure and H-bond energies for complexes of water with 2-substituted pyridines RC_5H_4N, where R is CH_3, NH_2, OH, F, C_2H_3, CHO, or CN.

E. HYDROGEN BONDING AND ION BINDING IN CHARGED SYSTEMS

1. The Interaction of Ions with Water or Ammonia

Clementi and Popkie [128] initiated a series of SCF calculations on the energy surface of a water molecule in the fields of positive and negative ions, beginning with $Li^+ \cdot H_2O$. Adequate basis sets were used: for Li, (7s, 1p) contracted to (3s, 1p); for O, (11s, 7p, 1d) contracted to (4s, 3p, 1d); H, (6s, 1p) contracted to (2s, 1p). Kistenmacher *et al.* [129] extended the work to complexes of Na^+, K^+, F^-, and Cl^- with H_2O, using as basis sets: for Na^+ and F^-, (13s, 8p, 1d) contracted to (7s, 4p, 1d); for K^+, (17s, 11p, 1d) contracted to (9s, 6p, 1d); for Cl^-, (17s, 10p, 1d) contracted to (9s, 6p, 1d). Calculations were made for a number of planar and nonplanar configurations, but minimal energies were found for planar configurations. These were of C_{2v} symmetry for the hydrated cations, with the H atoms directed away from the cations; these complexes *cannot be called hydrogen bonded.* For the hydrated anions, hydrogen bonding of one H to the anion is found, departing somewhat from linearity in the $A^- \cdots H{-}O{\diagdown}H$ bond (by about 4.5° for $F^- \cdot H_2O$ and 14.6° for $Cl^- \cdot H_2O$).

The equilibrium $C^+ \cdots O$ or $A^- \cdots O$ distances found were 1.84 Å for Li^+; 3.58, 4.25, and 5.08 a.u. for Li^+, Na^+, and K^+, respectively; 4.75 a.u. for

F^-; and 6.26 a.u. for Cl^-. The binding energies $-\Delta E$ found (in kcal/mole) were 35.2, 25.2, and 17.5 for Li^+, Na^+, and K^+, respectively; 23.54 for F^-; and 11.86 for Cl^-. In a subsequent article, Kistenmacher *et al.* [130] obtained corrections to the binding energies (zero-point energies from computed vibration frequencies and estimated correlation energy corrections) to bring them to heats of formation ΔH; the corrections are small.

Diercksen and Kraemer [131] made similar studies on the Li^+H_2O and Na^+H_2O complexes, also on H_3O^+ and [132] on F^-H_2O. They used (11, 2) sets contracted to (5, 2) for Li, (14, 8, 1) contracted to (8, 6, 1) for Na, (11, 7, 1) contracted to (5, 4, 1) for O, and (6, 1) contracted to (3, 1) for H. They found $-\Delta E = 36.0$ kcal/mole for Li^+H_2O, 25.2 for Na^+H_2O, and $C^+\cdots O$ distances of 3.57 a.u. for Li^+ and 4.23 a.u. for Na^+. In a later article [133] they studied correlation effects in the cases of Li^+H_2O and F^-H_2O [also H_3O^+ and $(H_2O)_2$]. By extensive CM (all single and double substitutions except for 1s), they found correlation energy contributions to $-\Delta E$ of the order of ± 1 kcal/mole.

With some somewhat less exact calculations, David [134] questions whether $(LiH_2O)^+$ is planar.

Woodin *et al.* [135], using a polarized double-zeta basis set in SCF and GVB computations (the latter with some CM that made scarcely any difference), found Li^+ affinities of 40 kcal/mole for NH_3 (experimental: 41 $\pm$ 2) and 38 kcal/mole for H_2O (experimental: 36 $\pm$ 3).

Using up to 6-31G** on polarized double-zeta basis sets, Berthod and Pullman [136] have computed the Li^+ and K^+ affinities of NH_3 and NH_2CH_3. Only the polarized double-zeta basis gives agreement with the experimental order of the gas-phase affinities for Li^+. The K^+ affinities for H_2O and NH_3 agree with experiment. Up-to-date references to other recent work are given.

Kraemer and Diercksen studied the systems $F^-\cdot 2H_2O$ [137] and $Li^+\cdot 2H_2O$ [138] in SCF calculations using basis sets similar to those given above. For $-\Delta E$ they found 24.1 kcal/mole for the first H_2O in $F^-\cdot(H_2O)_2$ and 20.8 for the second (experiment: 23.3 and 16.6). In this calculation they made the small simplification of assuming "linear" H bonds, and found as the most stable configurations a C_{2h} geometry in which F^- lies between two H-bonded H atoms, one belonging to one O—H bond of each of the two H_2O molecules. Other less stable geometries are discussed. For $Li^+\cdot 2H_2O$ the stablest structure is found to be of D_{2d} symmetry with the Li^+ on a line between the two H_2O molecules in mutually perpendicular planes with the H atoms directed away from the Li^+.

In another article, Kistenmacher *et al.* [139] discuss small clusters of water molecules surrounding Li^+, Na^+, K^+, F^-, or Cl^- ions at 0°K. They used their previously obtained two-body potentials to compute the energies of various configurations of $C^+(H_2O)_n$ and $A^-(H_2O)_n$ with $n = 2 \ldots 10$. For $n = 2$, they agree with Kraemer and Diercksen, except for not quite linear H bonds in $F^-(H_2O)_2$. Similar geometries to those for Li^+ occur for Na^+ and K^+, and to those for F^- for Cl^-. They give diagrams showing the appearance of the clusters for various n values. For $n < 5$, symmetrical configurations are the lowest energy configurations. For higher n, some of the water molecules tend to form a second solvation shell. The coordination numbers for the inner shells are computed to be about 4 for Li^+, between 5 and 6 for Na^+, 5–7 for K^+, 4–6 for F^-, and 6–7 for Cl^-. The average distances from Li^+, Na^+, and K^+ to the oxygen nuclei of the water molecules in the first coordination sphere are, respectively, 1.9–2.0, 2.3–2.4, and 2.8–2.9 Å. For F^- and Cl^- the distances are 2.7–2.8 and 3.4–3.5 Å. Tables of stabilization energies and ion–oxygen distances are given. Many-body effects are tested, and it is concluded that they tend to reduce somewhat the stabilization energies obtained using the pair potentials. Some calculations were made at 298°K, and from these, density distributions for the various atoms are calculated. These are compared with Monte Carlo calculations for $F^-(H_2O)_{27}$. The conclusions reached are believed to remain valid for large n values.

Watts *et al.* [140] have made Monte Carlo calculations on the distributions of the atoms of 50 water molecules in a cubic box surrounding a Li^+F^- pair held fixed in the center. They used pair potentials previously derived (see above). Calculations were made at 298°K for Li^+ to F^- distances of 3.0, 3.5, and 4.0 A, and at 500°K for a 3.5 Å Li^+F^- distance. Graphs and diagrams are shown for the x, y, and z distributions of the O and H atoms in the box. They cluster strongly about the center at 298°K, but scatter considerably toward the boundaries at 500°K. Both ions have four nearest neighbors: O atoms for Li^+, H atoms for F^-. Many details are given. In a second article by Fromm *et al.* [141] a similar but slightly more refined Monte Carlo calculation is made, this time with 200 water molecules and with Li^+F^- distances of 2.0, 6.0, and 10.0 Å at $T = 298$°K and also $T = 500$°K in the case of $R(Li^+F^-) = 6.0$ Å. The Monte Carlo simulations provide reproducible information on the cluster shape and cluster structure of the O and H atoms about the Li^+F^- pair, and on the corresponding coordination numbers for the Li^+ and the F^-.

For the system $H_2O{-}Li^+{-}F^-$, Kress *et al.* [142] have made SCF calculations for about 600 spatial configurations (250 different energies, many configurations having equal energy because of symmetry), compared their results with the sums of computed two-body energies (Li^+H_2O, F^-H_2O, and Li^+F^-), and fitted the computed $H_2O{—}Li^+{—}F^-$ energies to a simple analytical formula. The Li basis set was (11s, 3p) contracted to (7s, 3p); the basis sets for F^-, O and H were as in [103, 104]. The results, given in detail, show for many of the 250 configurations nonnegligible, both positive and negative, three-body corrections to the sums of the two-body energies. However, rather more than half the corrections are small (1 kcal/mole or less). A number of contour diagrams are given showing the total interaction energy for the Li^+ ion at various locations around each of several fixed positions of the H_2O and F^- entities. An analytical formula to fit the computed energies of the sampling of the many-dimensional potential surface is then given. Three minima occur, with stabilization energies of about 201 kcal/mole for $H_2O{\cdot}Li^+F^-$ (C_{2v}), 191 kcal/mole for $F^-{\cdot}HOH{\cdot}Li^+$ (C_s), and 186 kcal/mole for $F^-H_2O{\cdot}Li^+$ (C_{2v}).

Using the results of [114], Chan *et al.* [143] have computed dipole moments of the $H_2O{\cdot}Li^+F^-$ system in 252 configurations. The results are fitted with an analytical expression. Contour maps and three-dimensional representations are shown.

With the results of previous CM calculations on water–water interactions, Clementi and Barsotti [144] used Monte Carlo techniques to study the distribution of water molecules surrounding a single ion. The computed coordination numbers in the first shell are 4.0, 4.3, 5.1, 3.85, and 4.3 for Li^+, Na^+, K^+, F^-, and Cl^-, respectively, with corresponding shell radii of 2.28, 2.59, 3.27, 1.99, and 2.85 Å. The second and third shells are discussed.

Ikuta [145], using mostly a double-zeta basis, has calculated geometry and energies for $Br^-(H_2O)_n$, $n = 1$–4. In contrast to the linear H-bonded structures of $F^-(H_2O)$ and $Cl^-(H_2O)$, $Br^-(H_2O)$ is found to be 1.3 kcal/mole more stable in the ion–dipole C_{2v} geometry than in that for H bonding. The hydration energy increases linearly with n from 14 kcal/mole for $n = 1$ to 51 kcal/mole for $n = 4$; there is rather good agreement with the experimental increments as n increases. For $Br^-(H_2O)_4$ a regular tetrahedral ion–dipole structure is the most stable.

Using STO-3G and 4-31G basis sets, Pullman and Armbruster [146] made calculations on H bonding of one to four NH_3 molecules or an H_2O molecule to an NH_4^+ ion, for various directions of approach. Much the most stable were approaches with the NH_3 or H_2O symmetry axis lined up with an N^+—H bond, and the N of NH_3 or O of H_2O approaching the H of NH_4^+. The results of a population analysis to obtain charges on the atoms are given. For a single NH_3 or H_2O, the calculated $-\Delta E$ values (STO-3G) are 42 and 37 kcal/mole at distances of 2.5 and 2.4 Å respectively. Use of 4,31G gives smaller $-\Delta E$ values (31 and 28 kcal/mole) and larger distances (2.75 and 2.65 Å). Rotation of the NH_3 or H_2O around its symmetry axis makes very little difference in ΔE. Successive additions of water molecules to NH_4^+ yield successive $-\Delta E$ increments of 37, 28, 22, and 18 kcal/mole. With four water molecules, a first solvation shell is formed.

Further calculations indicate that for an NH_3 first solvation shell, H_2O would be preferred for a second shell. [Thus, in $NH_4^+\cdots(NH_3)_4$, $-\Delta E$ is greater for the attachment of H_2O than of NH_3 to the already bound NH_3]. The results are explained in terms of changes in charge distribution.

In an earlier article on H bonding by $N(CH_3)_4^+$ and $N(C_2H_5)_4^+$, using an STO-3G basis after making calculations for several geometries, Port and Pullman [147] found that H_2O molecules place themselves preferentially in four symmetrical positions bisecting three N—C bonds (i.e., about the axis of the fourth N—C bond, opposite to the methyl group), with values of $-\Delta E$ for stepwise hydration of 10.3, 8.9, 8.4, and 7.7 kcal/mole. In [146] H bonding by $NH_3CH_3^+$ is also considered.

Corongiu and Clementi [148] have made SCF and CM calculations on the interaction between a Be^{2+} ion and an H_2O molecule. They first note that the asymptote $Be^{2+} + H_2O$ is higher (experimental: 85.4 kcal/mole) than the asymptote $Be^+(^2S) + H_2O^+(^2A_1)$ of a configuration that must mix in the formation of the ground state of $(BeH_2O)^{2+}$. (Note also that $H_2O^+(^2A_1)$ is higher than the ground state $1a_1^2\,2a_1^2\,1b_2^2\,3a_1^2\,1b_1$, $^2B_1$ of H_2O^+.) They then made SCF calculations, both for in-plane (stable) and out-of-plane approaches, with each of two basis sets for Be: (a) 1s only, giving a strictly $Be^{2+} + H_2O$ potential curve, with $E_h = -89.849$ at the minimum; and (b) a basis (11s, 3p) contracted to (4s, 2p) permitting charge transfer H_2O to Be^{2+}, giving $E_h = -89.881$. Finally a CM calculation, using all single and double substitutions except for 1s(O), brings E_h to -90.101 at R(BeO) = 2.90 a.u., involving a notably large molecular extra correlation energy. They then discuss $[Be—(H_2O)_n]^{2+}$.

Noell and Morokuma [149] have proposed a simple model of solvation whereby the effect of solvent molecules is simulated by the inclusion of fractional point charges at the solvent atomic centers, while other molecules are dealt with by the usual *ab initio* computation. A wave function for the entire system is obtained. Tests on the hydration of Li^+ and of F^- ions, including both inner and outer solvation shells, agreed well with complete *ab initio* computations. In another test, the effect of hydration on the neutral $H_3N{\cdot}HF$ complex was to enhance the stability of the NH_4^+---F^- form of the complex.

Rode and Reibnegger [150] have used a modified less empirical approach for the calculation of larger ion–ligand systems ("MESQUAC"). They have tested this on Li^+, Na^+, K^+, Mg^{2+} and Ca^{2+} ions with coordination numbers from 2 to 6.

Delpuech *et al.* [151], using polarized double-zeta basis sets, have made SCF calculations to study reaction paths for proton transfer to or from NH_4^+ ions in H-bonded reactions with NH_3 and with H_2O. They examined the following systems: $[NH_4^+\cdots NH_3]$, $[NH_4^+\cdots OH_2]$, $[NH_4^+\cdots OH_2\cdots OH_2]$,

$[NH_4^+\cdots OH_2\cdots NH_3]$, and $[H_2O\cdots NH_4^+\cdots OH_2\cdots OH_2]$. They optimized distances of R(NH) of NH_4^+ as a function of N–N or N–O distance, and found corresponding energies and energy barriers (mostly small).

Veillard *et al.* [152], using a double-zeta basis for most of the valence shells and minimal for the inner shells made SCF calculations on $Al^{3+}(H_2O)_n$ and $Cu^{2+}(H_2O)_n$ with n up to 7.

Pullman *et al.* [153], using a minimal basis set, have made a detailed study of the first (up to six water molecules), second, and possibly third hydration shells of the dimethylphosphate anion in its gg, gt, and tt forms.

Pullman *et al.* [154] have made SCF calculations to study "through-water" versus "direct" binding of hydrated Na^+ and MG^{2+} ions with the $PO_4H_2^-$ anion.

2. Other Ion–Molecule Interactions

Kollman [155] used a 4-31G basis, except double-zeta for K, in calculations on the K^+ affinities for NH_3, CH_3NH_2, $(CH_3)_2NH$, H_2O, CH_3OH, $(CH_3)_2O$, HCN, CH_3CN, and $OCHNH_2$, and compared them with Li^+ affinities.

Del Bene [156], using up to a 6-31G basis, computed Li^+ affinities of R_2CO bases, finding an energy order $NH_2 > CH_3 > OH > H > F$. The symmetries found for the complexes are C_{2v}, except C_s for $(OH)_2C{=}O{\cdot}Li^+$, the Li^+ being near the O. It is indicated that the interaction is predominantly an ion–dipole attraction. The changes in geometry of R_2CO on complexation are described.

Pullman *et al.* [157], using STO-3G calculations, investigated the binding of Li^+, Na^+, and K^+ to $(CH_3)_2O$ and to two types of conformations of the 12-ring, 4-oxygen crown ether (cyclic —O—CH_2—O— structure). The cation is poised above the crumpled ring.

Rode and Kraft [158] with moderate basis sets made calculations on the complexes of glyoxal with Li^+, Na^+, K^+, Be^{2+}, Mg^{2+}, and Ca^{2+} ions. Stablest are chelate configurations, except for Li^+ where cis and trans complexes are about equally as stable.

On the interaction of Be^+ with C_2H_2 and C_2H_4, see Section XII,C,1 and [23].

Demoulin *et al.* [159], using a moderately good basis set, have investigated the complexes of Zn^{2+} with H_2O, CO_2, ImH (imidazole 6), the imi-

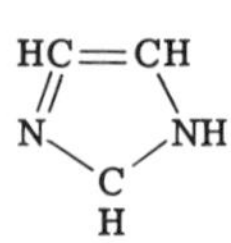

6

dazolate anion Im^-, and OH^-. Correlation corrections are believed to be small. The complexes are found to be planar with bond strengths in the order $CO_2 < H_2O < ImH < Im^- < OH^-$. Population changes on complexing and other details are given.

Using pseudopotentials, Pullman *et al.* [160] have studied the affinities of the DNA bases for Zn^{2+}, finding the order guanine $\simeq$ cytosine $>$ adenine $>$ uracil.

Kothekar *et al.* [161] with an ample basis set made an SCF calculation on the Zn^{2+} complexes of H_2S and HS^-. $Zn^{2+}SH_2$ is found to be nonplanar (C_s symmetry) with a 60° angle, $Zn^{2+}SH^-$ nonlinear with a 75° angle. Binding energies in the order $OH^- > SH^- > NH_3 > OH_2 > SH_2$ are found. Charge transfers to Zn^{2+} from a population analysis are in the order $SH^- > SH_2 > NH_3 > OH^- > OH_2$.

Perahia *et al.* [162] with a small basis find that in Mg^{2+}·uracil, binding is essentially to the O atom.

Perahia *et al.* [163] with a minimal basis calculated binding energies of Na^+ to the four DNA bases and compared them with the proton affinities of the bases.

Depsipeptides contain alternating ester and amide groups and assist the passage of alkali ions through membranes. Pullman [164] has made SCF computations on the binding of Na^+ and K^+ to amide and ester carbonyl groups, finding the affinities to be greater for Na^+ than for K^+, and greater for the amide than for the ester, and discusses the geometry.

Marynick and Schaefer [165] in minimal-basis SCF calculations studied model phosphate–metal ion solvation complexes, namely $RO_2PO_2^-$ (R = H or CH_3) with Li^+, Na^+, K^+, Be^{2+}, Mg^{2+}, Ca^{2+}, H_2O, and Cl^-. Binding energies calculated vary in the order $Be^{2+} > Mg^{2+} > Ca^{2+} > Li^+ > Na^+ > K^+$.

Gresh and Pullman [166] explored six modes of interaction of the NH_4^+ ion with the dimethylphosphate anion.

Pullman and Armbruster [167] have studied the intrinsic capacities of $C_2H_5^+$ and CH_3^+ ions to bind the anionic oxygens of the phosphate groups of the nucleic acid; they are found to be larger than their affinity for the carbonyl oxygens or for the nitrogens of cytosine or guanine.

3. Internally Hydrogen Bonded Ions

The hydrated proton, or H_3O^+, has been discussed in Section VI,B,2. On further hydration, distinctive strongly H-bonded structures are indicated at $H_5O_2^+$, $H_7O_3^+$, and $H_9O_4^+$. After that, further hydration by weaker H bonding should occur in water solutions. Crystalline salts of $H_5O_2^+$ and $H_7O_3^+$ are known, and $H_5O_2^+$ is known in the gas phase.

Kraemer and Diercksen [168] made SCF calculations on $H_5O_2^+$ using an (11s, 7p/6s) basis contracted to (5s, 4p/3s), plus polarization functions (d on O and p on H). Minimal energy was obtained with $R(OO) = 2.39$ Å in a D_{2d} or C_{2v} structure. A potential curve for the energy of the hydrogen-

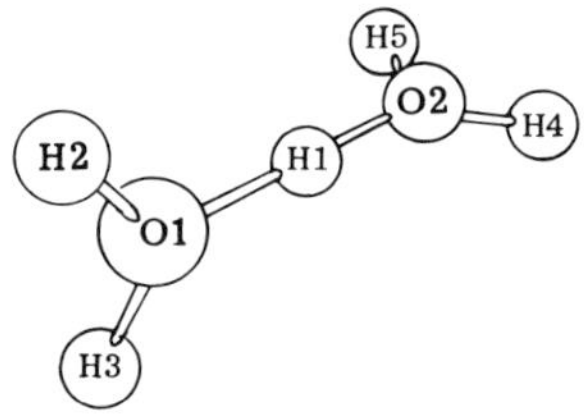

bonding H atom as a function of its distance between the two O atoms shows a very broad flat curve with perhaps a single minimum or probably a widely separated double minimum with an extremely low barrier between the minima. Other configurations showed higher energy. A binding energy relative to H_2O plus H_3O^+ of 32.2 kcal/mole was computed.

Newton and Ehrenson [169] in a less accurate (4-31G) calculation concurred (D_{2d} symmetry and $R_{OO} = 2.36$ Å). They also investigated the higher hydrattes, with configurations as shown in Fig. 5, taken from their article. As the hydration increases, the O—H--OH atom potential curves clearly take on double-minimum form with increasing R_{OO}: 2.46 Å in $H_7O_3^+$, symmetry C_{2v}, $R_{OH} = 1.04$ Å to one minimum, 1.42 Å to the other, (cf. 0.95 Å in H_2O, 0.964 Å in H_3O^+, assumed); 2.54 Å in $H_9O_4^+$, $R_{OH} = 1.00$ and 1.54 Å, symmetry D_{3h} (Eigen type structure). Variations of these structures with somewhat higher energies are computed. Additional hydration leads to attachment of H_2O molecules in the same linear mode as in $(H_2O)_2$ but with larger binding energies than there. The low barriers for proton tunneling account for facile proton exchange. Extensive discussion is given.

Meyer *et al.* [170] discuss correlation effects for proton transfer in $H_5O_2^+$, using a double minimum H bond with $R_{OO} = 2.74$ Å. The potential barrier is lowered 2.6 kcal/mole by correlation, and R_{OH} is decreased 0.02 Å. The CM and CEPA calculations are used on a (9s 5p 1d/4s) basis contracted to (5s 3p 1d/3s), except (4s 1p $1d_\sigma$) contracted to (3s 1p $1d_\sigma$) for the H-bonded proton.

Janoscek *et al.* [171] emphasize the extremely high polarizability of the hydrogen bonds in $H_5O_2^+$ and $H_3O_2^-$ groupings, with resulting facile proton exchange. They used Diercksen's machine program [121].

The examples of the H_3O^+ hydrates discussed above show that for a bond X--H--X there is a critical XX length above which the potential of the H atom in the bond goes from single minimum to double minimum in

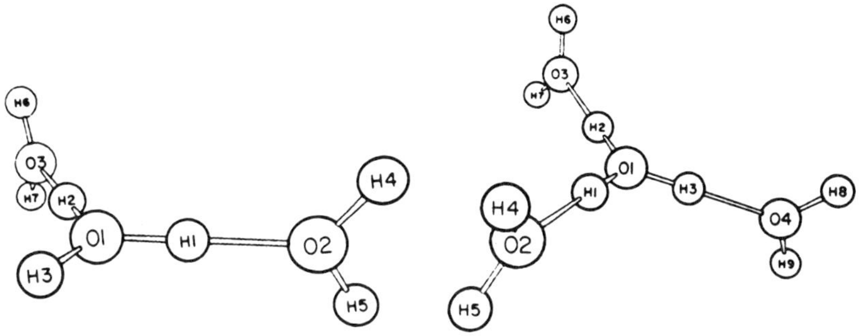

A B

FIG. 5. (A) optimized chain structure of $H_7O_3^+$; (B) optimized branched (Eigen-type) structure of $H_9O_4^-$.

form. Another example has been discussed in Section D,7. An example of the single minimum case is seen in HF_2^- (see Section III,H). Kollman and Allen [172] have discussed HF_2^- and $H_5O_2^+$ using an essentially double-zeta-quality basis, and have shown difference density plots (HF_2^- versus H plus $2F^{-1/2}$ and $H_5O_2^+$ in D_{2d} symmetry versus H^+ plus $2H_2O$).

Emsley *et al.* [173], using an extended (5s 3p/3s 1p) basis set have computed the energy of formation of a linear hydrogen bond HCOO—H--F⁻ and find 245 kJ/mole for HCOOH + F⁻ as compared with 212 kJ/mole for F—H--F⁻ using the same basis set. They refer to it as "the strongest hydrogen bond?" but give no details except to show that the linear bond is much more stable than a bifurcated bond. However, Bouma and Radom [174] point out that the proper comparison with F⁻ + HF is not for HCOOH + F⁻ but for HCOO⁻ + HF. The formation energy in their 4-31G calculation is only 139 kJ/mole for the latter as compared with 311 kJ/mole for the former, and one should not speak of an HCOOC—F⁻ hydrogen bond.

With a (4s 2p/2s 1p) basis set, Emsley *et al.* [175] find another very strong hydrogen bond in the biformate anion in the configuration

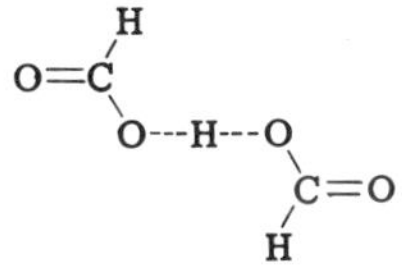

They examine also four other configurations found to be of higher energy.

For the $H_3O_2^-$ ion, Kraemer and Diercksen [176], using an (11·7·1/6·1)

basis contracted to (5·4·1/3·1), found an asymmetrical C_s structure

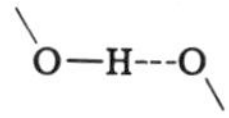

with R(OO) = 2.51 Å, R(OH) = 1.11 and 1.40 Å and with a barrier of about 1.5 kcal/mole in the H atom double-minimum potential, and a binding energy between H_2O and OH^- of 24.3 kcal/mole, in agreement with Kebarle's 1971 vapor-phase experimental value of 24.0 kcal/mole. It should be noted that optimizations were not quite complete.

For $H_3O_2^-$ using the same basis set, Roos *et al.* [177] have made a CM calculation including all single and double substitutions except for 1s correlation. Only a linear O—H--O bond was considered, and the other OH distances and the HOH angle were constrained to their values for H_2O. They now find for SCF, R(OO) = 2.55 Å and R(OH) = 1.02 (and 1.54) Å, and for CM, R(OO) = 2.465 Å, R(OH) = 1.09 (and 1.37) Å, but conclude that with a fully optimized geometry the H bond would probably be symmetrical; for the dissociation to $H_2O + OH^-$ they found 28 kcal/mole.

Earlier, with a less accurate basis set but more extensive geometry variation in an SCF calculation on $H_3O_2^-$, Newton and Ehrenson [169] found R(OO) = 2.45 Å, R(OH) = 1.11 Å, and only 0.1 kcal/mole for the energy barrier to proton tunneling.

Lukovits *et al.* [178], with basis sets up to (8s 4p 1d/3s) contracted to (5s 3p/2s), have made SCF calculations on the $HCO_2^- \cdot H_2O$ (formate ion hydratte) complex in 10 different configurations, and found a C_{2v} structure involving two symmetrically arranged nonlinear H bonds

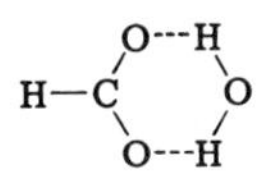

to be the most stable, with R(CO) = 3.23 Å, $\Delta E = -19$ kcal/mole.

4. Proton Affinities

Although a number of *ab initio* articles are relevant to the subject of proton affinities, we shall mention only two. Kollman and Rothenberg [179] have used 4-31G SCF calculations in a study of the basicities of a large number of molecules, for which they have calculated dipole moments, threshold and 1s orbital energies, proton affinities, Li^+ affinities, H bond potential surfaces, etc.

With the help of a Morokuma component analysis, Umeyama and Morokuma [180], using 4-31G calculations, have studied the origin of alkyl substituents effects in the proton affinity of amines, alcohols, and ethers.

REFERENCES

1. A. Pullman and B. Pullman, *Q. Rev. Biophys.* **7,** 4 (1975).
2. E. Clementi, "Lecture Notes in Chemistry," Vol. 2. Springer-Verlag, Berlin and New York, 1976.
3. P. A. Kollman, *in* "Modern Theoretical Chemistry, Vol. 4, Applications of Electronic Structure Theory," (H. F. Schaefer III, ed.) Chap. 3. Plenum, New York, 1977; P. A. Kollman, *J. Am. Chem. Soc.* **99,** 4875 (1977).
4. M. Trenary, H. F. Schaefer, III, and P. A. Kollman, *J. Am. Chem. Soc.* **99,** 3885 (1977).
5. M. Trenary, H. F. Schaefer, III, and P. A. Kollman, *J. Chem. Phys.* **68,** 4047 (1978).
6. B. J. Garrison, W. A. Lester, Jr., and H. F. Schaefer, III, *J. Chem. Phys.* **63,** 1449 (1975).
7. B. J. Garrison, W. A. Lester, Jr., P. Siegbahn, and H. F. Schaefer, III, *J. Chem. Phys.* **63,** 4167 (1975).
8. J. Prissette, G. Seger, and E. Kochanski, *J. Am. Chem. Soc.* **100,** 6941 (1978); J. Prissette and E. Kochanski, *Nouv. J. Chim.* **2,** 107 (1978).
9. M. Jaszúnski and E. Kochanski, *J. Am. Chem. Soc.* **99,** 4624 (1977).
10. P. Kollman, *J. Am. Chem. Soc.* **99,** 4875 (1977).
11. H. Umeyama, K. Morokuma, and S. Yamabe, *J. Am. Chem. Soc.* **99,** 330 (1977).
12. H. Lischka, *Chem. Phys. Lett.* **20,** 448 (1973).
13. H. Lischka, *Chem. Phys.* **2,** 191 (1973).
14. W. A. Lathan and K. Morokuma, *J. Am. Chem. Soc.* **97,** 3615 (1975).
15. W. A. Lathan, G. R. Pack, and K. Morokuma, *J. Am. Chem. Soc.* **97,** 6624 (1975).
16. R. R. Lucchese, K. Haber, and H. F. Schaefer, III, *J. Am. Chem. Soc.* **98,** 7617 (1976).
17. L. D. Brown and W. N. Lipscomb, *Inorg. Chem.* **16,** 1 (1977).
18. N. J. Fitzpatrick and M. O. Fanning, *J. Mol. Struct.* **50,** 127 (1978).
19. H. Basch, M. D. Newton, and J. W. Moskowitz, *J. Chem. Phys.* **69,** 584 (1978).
20. T. H. Upton and W. A. Goddard, III, *J. Am. Chem. Soc.* **100,** 321 (1978).
21. R. M. Pitzer and H. F. Schaefer, III, *J. Am. Chem. Soc.* **101,** 7176 (1979).
22. B. Åkermark, M. Alemark, J. Almlöf, J.-E. Bäckvall, B. Roos, and Å. Støgard, *J. Am. Chem. Soc.* **99,** 4617 (1977).
23. W. C. Swope and H. F. Schaefer, III, *J. Am. Chem. Soc.* **98,** 7962 (1976).
24. H. Basch, *J. Chem. Phys.* **56,** 441 (1972).
25. J. N. Murrell and C. E. Scollary, *J. C. S. Dalton Trans.* p. 1034 (1977).
26. W. C. Swope and H. F. Schaefer, III, *Mol. Phys.* **34,** 1037 (1977).
27. A. Haaland, *Acc. Chem. Res.* **12,** 415 (1979).
28. A. Veillard and J. Demuynck, *in* "Modern Theoretical Chemistry, Vol. 4, Applications of Electronic Structure Theory," (H. F. Schaefer III, ed.) pp. 203–206. Plenum, New York, 1977.
29. M.-W. Coutière, J. Demuynck, and A. Veillard, *Theor. Chim. Acta* **27,** 281 (1972).
30. M.-M. Rohmer, A. Veillard, and M. H. Wood, *Chem. Phys. Lett.* **29,** 466 (1974); see also Ref. 24, p. 213 (Table 11).
31. P. Bagus, U. I. Wahlgren, and J. Almlof, *J. Chem. Phys.* **64,** 2324 (1976).
32. S. Alexandratos, A. Streitweiser, Jr., and H. F. Schaefer, III, *J. Am. Chem. Soc.* **98,** 7959 (1976).

33. E. D. Jemmis, S. Alexandratos, P. von R. Schleyer, A. Streitweiser, Jr., and H. F. Schaefer, III, *J. Am. Chem. Soc.* **100,** 5695 (1978).
34. C. S. Ewig, R. Osman, and J. R. van Wazer, *J. Am. Chem. Soc.* **100,** 5017 (1978).
35. I. H. Hillier and V. R. Saunders, *Mol. Phys.* **23,** 449 (1972).
36. M. F. Guest, I. H. Hillier, B. R. Higginson, and D. R. Lloyd, *Mol. Phys.* **29,** 113 (1975).
37. M.-M. Rohmer, J. Demuynck, and A. Veillard, *Theor. Chim. Acta* **36,** 93 (1974); see also Ref. 24, pp. 199–202.
38. M.-M. Rohmer and A. Veillard, *J. C. S. Chem. Commun.* p. 250 (1973).
39. M. Bénard, *J. Am. Chem. Soc.* **100,** 7740 (1978).
40. I. H. Hillier and V. R. Saunders, *Mol. Phys.* **22,** 1025 (1971).
41. I. H. Hillier, M. F. Guest, B. R. Higginson, and D. R. Lloyd, *Mol. Phys.* **27,** 215 (1974).
42. J. Demuynck and A. Veillard, *Theor. Chim. Acta* **28,** 241 (1973).
43. G. R. Mitchesen and I. H. Hillier, *J. C. S. Faraday Trans. II* **75,** 929 (1979).
44. H. B. Jansen and P. Ros, *Theor. Chim. Acta* **34,** 85 (1974).
45. A. Veillard and J. Demuynck, in Ref. 28, pp. 196–199.
46. J. Demuynck, A. Strich, and A. Veillard, *Nouv. J. Chim.* **1,** 217 (1977).
47. R. S. Berry, *J. Chem. Phys.* **32,** 933 (1930).
48. J. A. Connor, L. M. R. Derrick, M. B. Hall, I. H. Hillier, M. F. Guest, B. R. Higginson, and D. R. Lloyd, *Mol. Phys.* **28,** 1193 (1974).
49. M. B. Hall, I. H. Hillier, J. A. Connor, M. F. Guest, and D. R. Lloyd, *Mol. Phys.* **30,** 839 (1975).
50. J. A. Connor, M. R. Derrick, I. H. Hillier, M. F. Guest, and D. R. Lloyd, *Mol. Phys.* **31,** 23 (1976).
51. P. J. Hay, *J. Am. Chem. Soc.* **100,** 2411 (1978).
52. J. Demuynck, E. Kochanski, and A. Veillard, *J. Am. Chem. Soc.* **101,** 3467 (1979).
53. M. F. Guest, M. B. Hall, and I. H. Hillier, *Mol. Phys.* **25,** 629 (1973).
54. J. Demuynck, A. Veillard, and G. Vinot, *Chem. Phys. Lett.* **10,** 522 (1979).
55. J. Demuynck, A. Veillard, and U. Wahlgren, *J. Am. Chem. Soc.* **95,** 5563 (1973); see also the review in Ref. 28, pp. 192–195.
56. J. Demuynck and A. Veillard, *Chem. Phys. Lett.* **6,** 204 (1970).
57. T. J. M. Smit, C. Haas, and W. C. Nieuwpoort, *Theor. Chim. Acta* **43,** 277 (1977).
58. A. Dedieu, M.-M. Rohmer, and A. Veillard, *J. Am. Chem. Soc.* **98,** 5789 (1976).
59. A. Dedieu, M.-M. Rohmer, M. Benard, and A. Veillard, *J. Am. Chem. Soc.* **98,** 3717 (1976).
60. L. Pauling, *Nature* (*London*) **203,** 182 (1964).
61. A. Dedieu, M.-M. Rohmer, and A. Veillard, *Proc. Jerusalem Symp. Quantum Chem. Biochem., 9th* p. 101 (1976).
62. A. Dedieu, M.-M. Rohmer, H. Veillard, and A. Veillard, *Bull. Soc. Chim. Belg.* **85,** 953 (1976).
63. B. D. Olafson and W. A. Goddard, III, *Proc. Natl. Acad. Sci. U.S.A.* **74,** 1315 (1977).
64. K. Morokuma, *Acc. Chem. Res.* **10,** 294 (1977).
65. P. A. Kollman, *in* "Modern Theoretical Chemistry, Vol. 4, Applications of Electronic Structure Theory," (H. F. Schafer III, ed.) Chap. 3. Plenum, New York, 1977.
66. H. Umeyama and K. Morokuma, *J. Am. Chem. Soc.* **99,** 1316 (1977).
67. P. Kollman, J. McKelvey, A. Johnson, and S. Rothenberg, *J. Am. Chem. Soc.* **97,** 955 (1975); see also W. C. Topp and L. C. Allen, *J. Am. Chem. Soc.* **96,** 5291 (1974).
68. R. C. Kerns and L. C. Allen, *J. Am. Chem. Soc.* **100,** 6587 (1978).
69. S.-I. Yamabe and K. Morokuma, *J. Am. Chem. Soc.* **97,** 4458 (1975).

70. See the bibliographies of W. G. Richards and collaborators, Oxford Univ. Press (Clarendon), London and New York, 1971 and Supplements 1970–1973, 1974–1977, and 1978–1980.
71. H. Popkie, H. Kistenmacher, and E. Clementi, *J. Chem. Phys.* **59,** 1325 (1973).
72. H. Kistenmacher, G. C. Lie, H. Popkie, and E. Clementi, *J. Chem. Phys.* **61,** 546 (1974).
73. H. Kistenmacher, H. Popkie, E. Clementi, and R. O. Watts, *J. Chem. Phys.* **60,** 4455 (1974).
74. O. Matsuoka, E. Clementi, and M. Yoshimine, *J. Chem. Phys.* **64,** 1351 (1976).
75. G. C. Lie, E. Clementi, and M. Yoshimine, *J. Chem. Phys.* **64,** 2314 (1976).
76. G. C. Lie and E. Clementi, *J. Chem. Phys.* **64,** 5308 (1976).
77. G. H. F. Diercksen, W. P. Kramer, and B. O. Roos, *Theor. Chim. Acta* **36,** 249 (1975).
78. G. H. F. Diercksen and W. P. Kraemer, *Chem. Phys. Lett.* **6,** 419 (1970).
79. D. R. Yarkony, S. V. O'Neill, H. F. Schaeffer, III, C. P. Baskin, and C. F. Bender, *J. Chem. Phys.* **60,** 855 (1974).
80. H. Lischka, *J. Am. Chem. Soc.* **96,** 4761 (1974).
81. G. H. F. Diercksen, W. P. Kraemer, and W. von Niessen, *Theor. Chim. Acta* **28,** 62 (1972).
82. K. Morokuma, *J. Chem. Phys.* **55,** 1236 (1971).
83. S. Swaminathan, R. J. Whitehead, E. Guth, and D. L. Beveridge, *J. Am. Chem. Soc.* **99,** 7817 (1977).
84. S. Iwata and K. Morokuma, *Chem. Phys. Lett.* **19,** 94 (1973); *J. Am. Chem. Soc.* **95,** 7563 (1973); see also J. E. Del Bene, where CH_3OH, NH_2OH, H_2O, and FOH dimers are also discussed, for n,π^* transitions (minimal basis).
85. S. Iwata and K. Morokuma, *J. Am. Chem. Soc.* **97,** 966 (1975).
86. S. R. Ungemach and H. F. Schaefer, III, *J. Am. Chem. Soc.* **96,** 7898 (1974).
87. S. W. Harrison, S. Swaminathan, D. L. Beveridge, and R. Ditchfield, *Int. J. Quantum Chem.* **14,** 319 (1978).
88. J. E. Del Bene, *Chem. Phys. Lett.* **24,** 203 (1974).
89. G. del Conde and O. A. Novaro, *Theor. Chim. Acta* **51,** 79 (1979).
90. G. Alagona, E. Scrocco, and J. Tomasi, *Theor. Chim. Acta* **47,** 133 (1978).
91. D. T. Clark and B. J. Cromarty, *Theor. Chim. Acta* **44,** 181 (1977).
92. E. Clementi, F. Cavallone, and R. Scordamaglia, *J. Am. Chem. Soc.* **99,** 5531 (1977).
93. G. Bolis and E. Clementi, *J. Am. Chem. Soc.* **99,** 5550 (1977).
94. R. Scordamaglia, F. Cavallone, and E. Clementi, *J. Am. Chem. Soc.* **99,** 5545 (1977).
95. G. Alagona, A. Pullman, E. Scrocco, and J. Tomasi, *Int. J. Pept. Protein Res.* **5,** 251 (1973); see also A. Johansson and P. A. Kollman, *J. Am. Chem. Soc.* **94,** 6196 (1972).
96. A. Pullman, G. Alagona, and J. Tomasi, *Theor. Chim. Acta* **33,** 87 (1974).
97. G. N. J. Port and A. Pullman, *Int. J. Quantum Chem., Quantum Biol. Symp.* **1,** 21 (1974).
98. T. Ottersen, H. H. Jensen, R. Johansen, and E. Wisløff-Nielsen, *J. Mol. Struct.* **30,** 379 (1976).
99. Y. Orita and A. Pullman, *Theor. Chim. Acta* **45,** 257 (1977).
100. D. Demoulin, A.-M. Armbruster, and B. Pullman, *Theor. Chim. Acta* **48,** 143 (1978).
101. D. Peters and J. Peters, *J. Mol. Struct.* **53,** 103 (1979).
102. E. Clementi, J. Mehl, and W. von Niessen, *J. Chem. Phys.* **54,** 508 (1971).
103. E. Ady and J. Brickmann, *Chem. Phys. Lett.* **11,** 302 (1971).
104. J. del Bene and W. L. Kochenour, *J. Am. Chem. Soc.* **98,** 2041 (1976).
105. S. Iwata and K. Morokuma, *Theor. Chim. Acta* **44,** 323 (1977).
106. S. Scheiner and C. W. Kern, *J. Am. Chem. Soc.* **101,** 4081 (1979).

107. F. Carnovale, M. K. Livett, and J. B. Peel, *J. Chem. Phys.* **71,** 255 (1979).
108. H. Hashimoto, S. Ikuta, and M. Imamura, *Chem. Phys. Lett.* **62,** 567 (1979).
109. M. Dreyfus, B. Maigret, and A. Pullman, *Theor. Chim. Acta* **17,** 109 (1970); see also Ref. 97; R. Janoschek, *Theor. Chim. Acta* **32,** 49 (1973).
110. M. Dreyfus and A. Pullman, *Theor. Chim. Acta* **19,** 20 (1970).
111. T. Ottersen and H. J. Jensen, *J. Mol. Struct.* **26,** 355 (1975).
112. A. D. Isaacson and K. Morokuma, *J. Am. Chem. Soc.* **97,** 4453 (1975).
113. J. Catalán, M. Yánez, and J. I. Fernández-Alonso, *J. Am. Chem. Soc.* **100,** 6917 (1978).
114. G. Karlström, H. Wennerström, B. Jönsson, S. Forsén, J. Almlöf, and B. Roos, *J. Am. Chem. Soc.* **97,** 4188 (1975).
115. G. Karlström, B. Jönsson, B. Roos, and H. Wennerström, *J. Am. Chem. Soc.* **98,** 6851 (1976).
116. E. Clementi and G. Corongiu, *Chem. Phys. Lett.* **60,** 175 (1979); *Int. J. Quantum Chem.* **16,** 897 (1979).
117. G. N. J. Port and A. Pullman, *FEBS Lett.* **31,** 70 (1973).
118. A. Pullman and D. Perahia, *Theor. Chim. Acta* **48,** 29 (1978).
119. B. Pullman, S. Miertus, and D. Perahia, *Theor. Chim. Acta* **50,** 317 (1979).
120. M. Tsuda, H. Touhara, K. Nakamishi, K. Kitaura, and K. Morokuma, *J. Am. Chem. Soc.* **100,** 7189 (1978).
121. J. E. Del Bene, *J. Am. Chem. Soc.* **97,** 5530 (1975).
122. J. E. del Bene, *J. Am. Chem. Soc.* **95,** 6517 (1972).
123. W. L. Jorgensen, *J. Chem. Phys.* **71,** 5034 (1979).
124. B. Åkermark, H. Johansen, B. Roos, and U. Wahlgren, *J. Am. Chem. Soc.* **101,** 5896 (1979).
125. A. Hinchcliffe, *Adv. Mol. Relaxation Processes* **13,** 309 (1978).
126. F. Carnovale, M. K. Livett, and J. B. Peel, *J. Am. Chem. Soc.* **102,** 569 (1980).
127. J. del Bene, *J. Am. Chem. Soc.* **101,** 7146 (1979).
128. E. Clementi and H. Popkie, *J. Chem. Phys.* **57,** 1077 (1972).
129. H. Kistenmacher, H. Popkie, and E. Clementi, *J. Chem. Phys.* **58,** 1689 (1973); **58,** 5627 (1973).
130. H. Kistenmacher, H. Popkie, and E. Clementi, *J. Chem. Phys.* **59,** 5842 (1973).
131. G. H. F. Diercksen and W. P. Kraemer, *Theor. Chim. Acta* **23,** 387 (1972).
132. G. H. Diercksen and W. P. Kraemer, *Chem. Phys. Lett.* **5,** 570 (1970).
133. G. H. F. Diercksen, W. P. Kraemer, and B. O. Roos, *Theor. Chim. Acta* **36,** 249 (1975).
134. C. W. David, *J. Mol. Struct.* **53,** 151 (1979).
135. R. L. Woodin, F. A. Houle, and W. A. Goddarrd, III, *Chem. Phys.* **14,** 461 (1976).
136. H. Berthod and A. Pullman, *Isr. J. Chem.* Spec. Issue (1979).
137. W. P. Kraemer and G. H. F. Diercksen, *Theor. Chim. Acta* **27,** 265 (1972).
138. W. P. Kraemer and G. H. F. Diercksen, *Theor. Chim. Acta* **23,** 393 (1972).
139. H. Kistenmacher, H. Popkie, and E. Clementi, *J. Chem. Phys.* **61,** 799 (1974).
140. R. O. Watts, E. Clementi, and J. Fromm, *J. Chem. Phys.* **61,** 2550 (1974).
141. J. Fromm, E. Clementi, and R. O. Watts, *J. Chem. Phys.* **62,** 1388 (1975).
142. J. W. Kress, E. Clementi, J. J. Kozak, and M. E. Schwartz, *J. Chem. Phys.* **63,** 3907 (1975).
143. E. R. Chan, J. W. Kress, and G. C. Lie, *J. Chem. Phys.* **62,** 2026 (1975).
144. E. Clementi and R. Barsotti, *Chem. Phys. Lett.* **59,** 21 (1975).
145. S. Ikuta, *Chem. Phys. Lett.* **68,** 179 (1979).

146. A. Pullman and A. M. Armbruster, *Int. J. Quantum Chem., Symp.* **5,** 169 (1974); *Chem. Phys. Lett.* **36,** 558 (1978); correction in *Int. J. Quantum Chem.* **11,** 701 (1977).
147. G. N. J. Port and A. Pullman, *Theor. Chim. Acta* **31,** 231 (1973).
148. G. Corongiu and E. Clementi, *J. Chem. Phys.* **69,** 4885 (1978).
149. J. O. Noell and K. Morokuma, *Chem. Phys. Lett.* **36,** 465 (1975); *J. Chem. Phys.* **80,** 2675 (1975).
150. B. M. Rode and G. J. Reibnegger, *J. C. S. Faraday Trans. II* p. 178 (1979).
151. J.-J. Delpuech, G. Serratrice, A. Strich, and A. Veillard, *Mol. Phys.* **29,** 849 (1975).
152. H. Veillard, J. Demuynck, and A. Veillard, *Chem. Phys. Lett.* **33,** 221 (1975).
153. B. Pullman, A. Pullman, H. Berthod, and N. Gresh, *Theor. Chim. Acta* **40,** 93 (1975).
154. B. Pullman, A. Pullman, and H. Berthod, *Int. J. Quantum Chem., Quantum Biol. Symp.* **5,** 79 (1978); *Theor. Chim. Acta* **47,** 175 (1978); H. Berthod and A. Pullman, *Chem. Phys. Lett.* **46,** 249 (1977).
155. P. Kollman, *Chem. Phys. Lett.* **55,** 555 (1978).
156. J. E. del Bene, *Chem. Phys.* **40,** 329 (1979).
157. A. Pullman, C. Giessner-Prettre, and V. V. Kruglyak, *Chem. Phys. Lett.* **35,** 156 (1975).
158. B. M. Rode and H. G. Kraft, *Chem. Phys. Lett.* **61,** 410 (1979).
159. D. Demoulin, A. Pullman, and B. Sarkar, *J. Am. Chem. Soc.* **99,** 8498 (1977); D. Demoulin and A. Pullman, *Theor. Chim. Acta* **49,** 161 (1978); A. Pullman and D. Demoulin, *Int. J. Quantum Chem.* **16,** 641 (1979).
160. A. Pullman, T. Ebbesen, and M. Rholam, *Theor. Chim. Acta* **51,** 247 (1979).
161. V. Kothekar, A. Pullman, and D. Demoulin, *Int. J. Quantum Chem.* **14,** 779 (1978).
162. D. Perahia, A. Pullman, and B. Pullman, *Theor. Chim. Acta* **42,** 23 (1976).
163. D. Perahia, A. Pullman, and B. Pullman, *Theor. Chim. Acta* **43,** 207 (1977).
164. A. Pullman, "Molecular and Quantum Pharmacology," p. 401. Reidel Publ., Dordrecht, Netherlands, 1974; *Int. J. Quantum Chem., Biol. Symp.* **1,** 33 (1974).
165. D. S. Marynick and H. F. Schaefer, III, *Proc. Natl. Acad. Sci. U.S.A.* **72,** 3794 (1975).
166. N. Gresh and B. Pullman, *Theor. Chim. Acta* **52,** 67 (1979).
167. A. Pullman and A. M. Armbruster, *Theor. Chim. Acta* **50,** 359 (1979).
168. W. P. Kraemer and G. H. F. Diercksen, *Chem. Phys. Lett.* **5,** 463 (1970).
169. M. D. Newton and S. Ehrenson, *J. Am. Chem. Soc.* **93,** 4971 (1971).
170. W. Meyer, W. Jakubetz, and P. Schuster, *Chem. Phys. Lett.* **21,** 97 (1973).
171. R. Janoscek, E. G. Weidemann, H. Pfeiffer, and G. Zundek, *J. Am. Chem. Soc.* **94,** 2387 (1972).
172. P. A. Kollman and L. C. Allen, *J. Am. Chem. Soc.* **92,** 6101 (1970).
173. J. Emsley, O. P. A. Hoyte, and R. E. Overill, *J. C.S. Chem. Commun.* p. 225 (1977).
174. W. J. Bouma and L. Radom, *Chem. Phys. Lett.* **64,** 216 (1979).
175. J. Emsley, O. P. A. Hoyte, and R. E. Overill, *J. Am. Chem. Soc.* **100,** 3303 (1978).
176. W. P. Kraemer and G. H. F. Diercksen, *Theor. Chim. Acta* **23,** 398 (1972).
177. B. O. Roos, W. P. Kraemer, and G. H. F. Diercksen, *Theor. Chim. Acta* **42,** 77 (1976).
178. I. Lukovits, A. Karpfen, H. Lischka, and P. Schuster, *Chem. Phys. Lett.* **63,** 151 (1979).
179. P. A. Kollman and S. Rothenberg, *J. Am. Chem. Soc.* **99,** 1333 (1977).
180. H. Umeyama and K. Morokuma, *J. Am. Chem. Soc.* **98,** 4400 (1976).

INDEX

B

C

I

K

L

M

N